Encyclopedic
Dictionary
of
Electronic
Terms

John E. Traister
Robert J. Traister

Prentice-Hall, Inc., Englewood Cliffs, New Jersey 07632

Library of Congress Cataloging in Publication Data

Traister, Robert J. (date)
  Encyclopedic dictionary of electronic terms.

  1. Electronics--Dictionaries. I. Traister, John E.
II. Title.
TK7804.T7 1984   621.381'03'21   84-2107
  ISBN 0-13-276998-0

Editorial / production supervision:
    Barbara H. Palumbo
Cover design:
    Lundgren Graphics
Manufacturing buyer:
    Anthony Caruso

Printed in the United States of America

10 9 8 7 6 5 4 3 2 1

ISBN 0-13-276998-0

PRENTICE-HALL INTERNATIONAL, INC., *London*
PRENTICE-HALL OF AUSTRALIA PTY. LIMITED, *Sydney*
EDITORA PRENTICE-HALL DO BRASIL, LTDA., *Rio de Janeiro*
PRENTICE-HALL CANADA INC., *Toronto*
PRENTICE-HALL OF INDIA PRIVATE LIMITED, *New Delhi*
PRENTICE-HALL OF JAPAN, INC., *Tokyo*
PRENTICE-HALL OF SOUTHEAST ASIA PTE. LTD., *Singapore*
WHITEHALL BOOKS LIMITED, *Wellington, New Zealand*

Preface

The term "Age of Computers" is often heard to describe the current trend in technology. To the authors, this reference is not completely accurate. The "Computer Era" is a natural outgrowth of what can be accurately described as the "Age of Electronics."

While certainly involving the near ultimate in complex technology, the study of electronics reveals that any electronic device, regardless of its complexity, is made up of a number of highly simple, basic circuits. It is through the understanding of basic principles, circuits, and applications that a fuller understanding of the complexities is accomplished.

The *Encyclopedic Dictionary of Electronic Terms* is offered as a quick reference source for the serious comprehension of the basics involved in electronics. Each topic has been chosen and its explanation written to aid the beginner as well as the experienced technician. Most major studies in the field of electronics are included in a format that facilitates quick referencing. This text was written to help the reader over the difficulties encountered in mastering a certain aspect of the science of electronics. It is hoped that the definitions and discussions will not only aid further progress in studying this science but will also allow the reader to obtain a fuller comprehension of those areas with which he or she is already somewhat familiar.

The *Encyclopedic Dictionary of Electronic Terms* may be used successfully as a tutorial, but long after this study has been committed to practical application, this same text will be used time and again as a quick reference source, workbench guide, and even as a troubleshooting guide on the system level.

It has been said that during the past century mankind has progressed more in the field of technology than it did in all of the centuries (combined) before. Fortunately, our means of communicating the nature of these advances has made similar strides. It is hoped that this text will prove a valuable asset in the area of such communication.

John E. Traister
Robert J. Traister

AAA battery: A power source about one-third the size of a double-A battery. See *AA battery*. A relatively new introduction to the electronics market, AAA cells are intended for microminiature flash units used for photography purposes. These cells are often used in battery banks composed of two units in a series connection. Each battery delivers approximately 1.5-V dc and will withstand only limited current drains. In flash units, they are especially attractive and offer long life for their size. Flash units draw larger amounts of current, but only for an instant. Thus, the overall drain is relatively low.

AA battery: A small 1.5-V power source with numerous applications. The most popular use for this cell has been in small electronic equipment, such as in AM/FM receivers, often called "pocket radios." The double-A cell is intended for relatively low current applications and is often used in small battery banks where two or more cells are combined in series to deliver a higher amount of voltage. Double-A cells also find application in camera flash units and for the powering of small electric motors used for hobby purposes. Many electronic toys employ these batteries, although the larger C and D cells are used when space permits. The latter provide much longer life due to their ability to store larger amounts of power.

Double-A batteries often are of dry cell design, although alkaline cells in this size are very popular due to their longer lifetime of operation. Alkaline cells cost more than dry cells and are used mostly where higher current is drawn over long periods of time. Rechargeable double-A batteries have come on the market in recent years. Their voltage potential is about 1.3- to 1.35-V dc. These units may be recharged many times by an ac-derived power supply. Even though they produce a slightly lower voltage, they are usable in most circuits that were originally designed to operate from a 1.5-V dry cell.

A battery: A power source used to supply power to tube filaments for heaters in battery-operated tube circuits. Connections from the tube elements often are designated as the A-plus or A-minus supply contacts. The voltage potential of the A battery will depend upon tube design requirements and can be anywhere from a fraction of a volt to 12 V or more. Battery-operated tube circuits are almost completely obsolete today. However, they were the rule during the early and middle stages of electron tube circuit design. A-battery supplies were used in transmitters during both World Wars and in commercial broadcast receivers up until about two decades ago. War surplus communications equipment still available today often specified the use of A-plus supplies. However, when this equipment is converted for modern applications, the A batteries are often replaced with ac-derived electronic power supply circuits.

Acceleration switch: A switch that operates automatically when the acceleration of a body to which it is attached exceeds a predetermined rate in a given direction.

Accelerator: A machine or device in which charged particles, such as neutrons, are given high velocity for use on atomic disintegration and kindred processes. It is also a name given to a substance that speeds up a chemical reaction. For example, the catalyst used with certain resins for encapsulating electronic equipment is called an *accelerator*.

Accelerometer: A transducer whose output voltage is proportional to the acceleration of the moving body to which it is attached.

Acceptor impurities: Particles introduced in preparing germanium or silicon crystal materials for solid-state devices that cause a deficiency or an abundance of electrons, thus increasing conductivity. Material that conducts due to a lack of electrons is called "P type"; material with an abundance of electrons is called "N type." These two types of materials are combined in transistors and all solid-state devices to arrive at desired operating characteristics.

A schematic representation of a bipolar transistor is shown in Fig. 1. At *A,* an NPN transistor is shown, with *B* representing the PNP variety. These designations indicate the polarity of the device as well as the crystal-type layering. A PNP transistor is composed of an N-type crystal between two layers of P-type material. The reverse is true of an NPN transistor, where the P-type material is positioned at the center.

Access arm: A mechanical device that serves as an interface between two other mechanical elements. A good example of an access arm is found in some computer storage units. Here, the arm is used to position the reading and writing mechanism across the surface of a circular drum used to retain data by

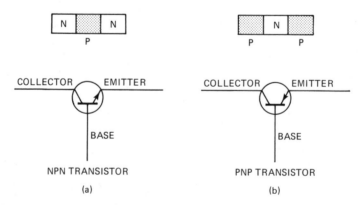

FIGURE 1   The P and N types of transistors utilizing acceptor impurities in their construction.

magnetic means. Another example is found on mechanical chart recorders that provide inked graphs based upon electronic input data. The electronic information is processed and converted to mechanical motion through a servo-unit. The access arm is attached to the servo and the write head, to the opposite end of the arm. The inked writing head is applied to the chart paper and moved back and forth by the servo and access arm. Access arms may serve to convert circular motion into linear movements, depending upon the coupling methods used.

Access time: **1.** The time it takes a computer to locate data or an instruction word in its storage section and transfer it to its arithmetic unit, where the required computations are performed. **2.** The time it takes to transfer information that has been operated on, from the arithmetic unit to the location in storage where the information is to be stored.

Accounting machine: A machine that reads information from one medium (cards, paper tape, magnetic tape, etc.) and produces lists, tables, and totals on separate forms or continuous paper.

Accuracy: The degree of exactness of an approximation or measurement. High accuracy implies low error. *Accuracy* normally denotes absolute quality of computer results; *precision* usually refers to the amount of detail used in representing those results. Thus, four-place results are less precise than six-place results; nevertheless, a four-place table might be more accurate than an erroneous computer six-place table.

Acetate: A tough thermoplastic material that is an acetic acid ester of cellulose. It is used as a dielectric and in the manufacture of phonograph records and photographic films. Cellulose acetate film once served as the base for the magnetic oxide coating in early recording tapes, but has been replaced—for the most part—by those with a polyester base.

Achromatic: A term applied to a black-and-white television signal. One of the basic signals of TV transmitters is called the *luminance signal* and contains only the brightness variations of the picture; that is, from black through gray to white. Even color television transmitters produce this luminance signal to permit a conventional black-and-white receiver to receive the signal and reproduce a color telecast without color. A more modern term for achromatic is *monochromatic*.

Acidity: The quality produced by an acid. A compound containing hydrogen replaceable by a metal, and whose water solution turns blue litmus paper red, or whose water solution contains positive hydrogen ions, is known as an acid. The acidity of pH of a solution is very important, and its close control is often a necessity in electrochemical processes. When pH is too low, quite often hydrogen is deposited at the cathode instead of metal; when too high, basic salts may be precipitated in the solution, or with the deposited metal, causing unsatisfactory plates. It has an effect, too, upon the quality of the deposit and many other factors governing the success of electrolytic reactions.

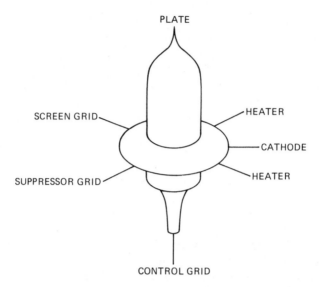

PLATE

SCREEN GRID

HEATER

CATHODE

SUPPRESSOR GRID

HEATER

CONTROL GRID

FIGURE 2 Acorn tube.

Water solutions of acids, bases, and salts conduct the electric current and are called *electrolytes*. These substances are decomposed by the electric current, and the products are liberated at the electrodes. This process is called *electrolysis*. Pure water, sugar solutions, alcohol, and glycerine do not conduct the electric current. Such liquids are called nonelectrolytes. Therefore, an electrolyte is a liquid or solution that conducts the electric current and is decomposed by it.

Acorn tube: A tiny glass-enclosed electron tube having terminal pins that extend from its ends and sides. See Fig. 2. Direct access to internal elements, such as the plate, grids, cathode, and heater, allows the acorn tube to be used efficiently for VHF and UHF communications due to the decreased inductive and capacitive effects of the leads. These are available as triodes, pentodes, and tetrodes and usually require direct solder connections between their external terminals and the rest of the circuit. Acorn tubes derive their name from their size and general bell-shaped physical structure, which closely resembles that of an acorn.

Acoustic feedback: A regeneration of sound waves experienced in public address systems and other audio circuits that utilize a microphone and speakers. The condition results in sound waves from the speaker reaching the microphone and being amplified over and over again. Feedback can also result from sound waves setting some part of an amplifier circuit into vibration, thus modulating the current flowing in the circuit.

Acoustic feedback usually starts at a low decibel level and increases in a matter of seconds to a loud howl or whistle. This unwanted condition can be avoided or controlled by decreasing the gain of the amplifier circuit or by placement of the microphone in such a position that it does not receive a great deal of acoustical transmission from the speaker system. Occasionally, acoustic feedback may be used in a controlled manner to produce a series of sound effects. These effects are controlled by adjusting the amplifier gain and/or circuit tone controls.

Acoustic transmission: The direct transmission of audio energy without the intermediary of an electric current. Generally, this means a sound produced mechanically. A common snare drum is a good example of an acoustic transmitter that produces sound when its diaphragm is struck. The force of the strike determines the degree of vibration created within the diaphragm, while the hollow body of the drum serves as an acoustical resonator, allowing the mechanical vibration to be transmitted or broadcast. The human voice is another example of acoustic transmission that utilizes a set of vibrating cords (vocal cords) that act in much the same manner as the diaphragm of a drum when air is passed over them. The human mouth serves as a resonator and controller of the acoustic transmission output. Most nonelectronic musical instruments (guitars, brass instruments, wind instruments, etc.) can be thought of as acoustic transmitters.

Acquisition time: The period between accessing the data from transducers, computers, and other devices and receiving the completed information. For example, if the problem of 1248/4.888 is punched into a calculator and the answer is printed out in 4 s, then the acquisition time is 4 s. In this case, acquisition time would be the period required to receive the correct answer after the (=) button is depressed.

Active communications satellite: A satellite containing a microwave receiver and transmitter designed to receive signals from the earth and to retransmit them back to another point on the earth. An active satellite is a type of repeater, in that it receives signals from one transmitter and then performs a retransmit procedure. Figure 3 shows the difference between an active com-

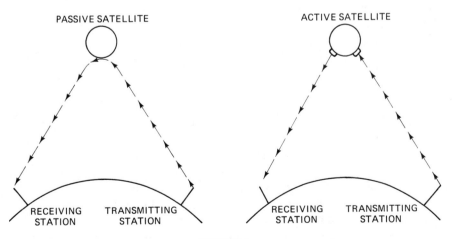

FIGURE 3

munications satellite and a passive one. The passive satellite simply reflects the transmitted signal back to earth. The same signal, then, that was originally transmitted is the same one that arrives back on the earth. The active satellite detects the information from the original signal and then feeds this to the input of a discrete transmitter. Active communications satellites provide stronger signals at the earth receiving site than do passive satellites. This is due to the space losses that are doubled in the latter system.

Activity coefficient: The ratio of the space occupied by an insulated conductor wound in a coil to the volume of solid copper that would occupy the same coil space. A thin insulation of high dielectric strength is desirable in order to achieve a high space factor. The relative space factors of magnet wires are expressed by the numbers of turns per square inch. A coil having a high activity coefficient contains copper of greater cross section for a given number of turns than a similar coil having a lower space factor. Therefore, the higher the activity coefficient (or space factor) of a coil, the greater the coil conductance, resulting in an increased number of ampere-turns per volt or per watt of power input.

Acyclic generator: A generator using a homopolar operation; that is, a single conductor continuously moved at a constant speed through a constant magnetic field to produce a constant dc voltage with no commutator required. Such generators are used for making longitudinal welds in the manufacture of large steel pipe from rounded plates.

ac/dc: The name sometimes given to a motor or an appliance that operates either on single-phase alternating current or direct current. The motor usually is a small series-wound type used to operate drill motors, grinders, vacuum cleaners, and similar small appliances. Certain hot plates, coffee makers, and other resistance-heat appliances are designed to operate on either 120-V dc or 120-V ac; thus, the name *ac/dc appliance*. The name *universal* is more common when the characteristic applies to a motor.

ac relay: A magnetically operated switch for alternating current that can be used to:

1. Control circuits distant from the operating point.
2. Control a relatively high-voltage or high-wattage circuit by means of a low-voltage circuit.
3. Obtain a variety of control operations not possible with ordinary switches.

The circuits controlled will be either normally closed or normally opened (when the relay coil is energized), depending upon the arrangement and connection of the relay contacts. When current flows through the relay coil, it magnetizes the iron core with a polarity that depends upon the connection of the coil to the source. This pole induces, in the iron section of the movable assembly, a pole of opposite sign, and the attraction between these operates

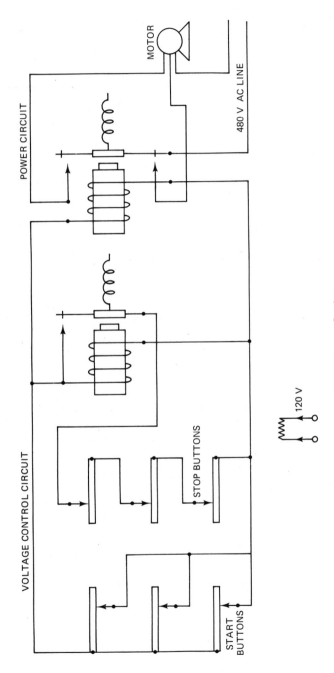

FIGURE 4 AC relays.

the relay switch. If the current through the coil is reversed, both poles are reversed; therefore, attraction always occurs. Relays can be designed to operate on either direct or alternating current, and while they may vary widely in mechanical construction, they all operate on the same principle.

Figure 4 shows how a motor may be operated from several different locations. In this arrangement, two relays are used to isolate completely the lower voltage (120-V) from the higher voltage (480-V) motor circuit.

ac ripple: A very small alternating current voltage or current occurring in a direct current component as an unavoidable by-product of rectification or commutation.

ac-to-dc converter: A rotating machine similar in appearance to a motor or generator that converts alternating current to direct current. The synchronous converter has a field and an armature. One side of the armature is connected to an ac system through slip rings; the other side of the armature has a commutator connected to a dc system. The alternating current is converted to direct current by the commutator.

Converters are used less frequently than stationary rectifiers because their efficiency is much lower than that of a rectifier. Maintenance expenses and installation costs of the rectifier are low, making it more desirable than the more costly converters. For these reasons, together with the fact that they operate without noise, rectifiers can be used where others would not be considered.

Adaptors: Fittings used to change either the connection or the size of a conduit, plug, jack, or socket to that of another. An adaptor is also required to provide a transition from one type or style of conduit or conductor to another: for example, polyvinyl chloride, PVC (plastic), conduit to rigid metal conduit, waveguide to coaxial line, etc. The term may also be used to describe an auxiliary system or unit used to extend the operation of another system, such as a citizens band radio adaptor for broadcast reception.

Adcock antenna: A directional transmitting and/or receiving system that behaves much like a loop antenna. It consists of two evenly spaced vertical antennas, as shown in Fig. 5. The two major elements are connected so the

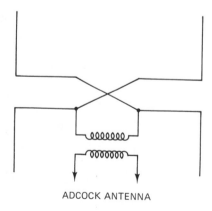

ADCOCK ANTENNA

FIGURE 5 Adcock antenna.

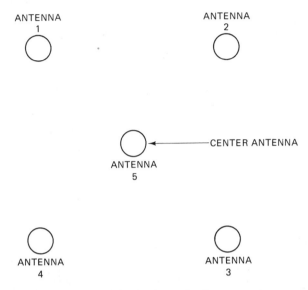

FIGURE 6   Adcock antenna Radio range system.

entire system discriminates against transmission waves that are horizontally polarized, with respect to the earth. This effectively cancels out most horizontally polarized waves. The output of the antenna is proportional to the vector difference of the signal voltages induced in the two vertical arms by the transmission lines and feed system.

The adcock antenna is sometimes used in radio direction-finding systems. Additionally, some radio range systems are based upon the application of four of these antenna systems. Each is situated at the corner of a square with a fifth antenna situated at the center point. (See Fig. 6.)

Adjustable speed drive: An electronic drive used in electric motors and consisting of a direct current motor with electronic rectifiers to supply power for the armature and field. Speed control is otained by adjusting either the armature voltage or the field current of the motor. This is done by phasing the firing angle of the electronic tubes to control the average output voltage. The horsepower and torque characteristics of this drive are the same as for the conventional dc adjustable voltage drives. The electronic drive is particularly well suited for feed drives, since a wide speed range with constant-torque characteristics is available.

Adjustable voltage: A means of controlling motors to obtain a wide range of speeds. Controlling a direct current motor by adjusting the voltage is a very efficient means of obtaining varying speeds. Varying the voltage of alternating current motors is also a practical means of controlling the speed, especially in fractional horsepower motors.

For example, the illustration (Fig. 7) shows a tapped autotransformer connected to the motor through a multiposition selector switch by means of

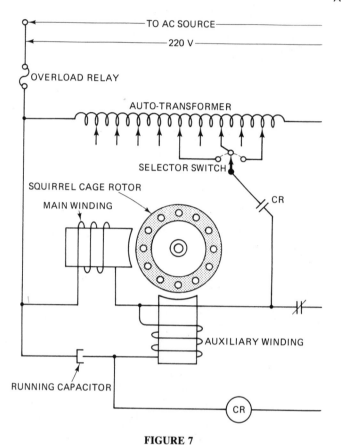

FIGURE 7

which any suitable voltage may be provided. The transformer may have a number of taps, only two or three of which are brought down to the switch. This provides two or three speeds, any or all of which may be changed to suit the needs of the application, by the proper choice of taps to connect to the switch. Additional contacts on the switch make it a complete starter and speed regulator.

Admittance: The ease with which alternating current flows in an electrical circuit. Admittance is measured in mhos and is the reciprocal of impedance. See *Impedance*.

Aerodynamics: The branch of physics that deals with the forces exerted by air or other gases in motion, especially upon bodies (such as aircraft) moving through these gases.

Aeronautical Broadcasting Service: The Aeronautical Broadcasting Service encompasses a complex network of specialized radio stations that broadcast information specifically relating to air navigation and air travel in general. Some of the aspects of this service include beacons that transmit signals that are received by specialized instruments within the aircraft and allow the pilot to determine a proper course or to change a flight heading. The Aeronautical Broadcasting Service transmits constantly updated meteorological data on an around-the-clock basis to all classes of aviators. This information is gathered from local weather stations, as well as from the National Weather Service.

Aerophysics: The physical science of the air or earth's atmosphere. This science is heavily applied to the electronics field as it relates to the effect of the atmosphere on the transmission and reception of radio waves. At certain frequencies, the physical properties of the atmosphere will cause radio waves to be reflected back to the earth; at other frequencies, atmospheric conditions allow these signals to pass out into space relatively unaffected.

Aerospace: The region above the earth that takes in the atmosphere and all extraterrestrial space. Aerospace, then, involves the entire universe, with the exception of the area occupied by the earth itself. Aerospace research has facilitated the production of communications equipment, satellites, and space ships, and is a means to understand better the relationship of the earth to the rest of the universe.

AFC: Abbreviation of *automatic frequency control.*

AGC: Abbreviation of *automatic gain control.*

Aging: With regard to properties of structural materials, a spontaneous change in properties of a metal after a heat-treatment or a cold-working operation. Aging tends to restore the material to an equilibrium condition and to remove the unstable condition induced by the prior operation and usually results in increased strength of the metal with corresponding loss of ductility. The fundamental action involved is generally one of precipitation of hardening elements from the solid solution, and the process can usually be hastened by a slight increase in temperature.

Agitation: A motion process used in electroplating. Agitation is often helpful in permitting a higher plating speed. However, some of the drawbacks include stirring up sediment and slimes as well as increasing the absorption of atmospheric carbon dioxide. In automatic plating machines, agitation is effected by the movement of the cathode through the plating solution. In still-tank plating, agitation may be accomplished by the mechanical reciprocating motion of the cathode bar. Agitation may also be combined with continuous filtration by circulating the plating solution from the tank, through a filter, and back into the original tank.

Air conditioning: A broad field including numerous processes, among which are refrigeration, heating and ventilation, humidification, electronic air filtering, and the like. It also includes electric resistance duct heaters used in electric furnaces and as auxiliary heat in heat pumps.

Air cooling: A process of ventilating in which air is circulated by a fan on the rotor. Centrifugal fans are sometimes used, but axial-flow fans are probably more common in modern designs. Most large turbogenerators have a recirculating ventilating system with air coolers. The air coolers are usually located in the pit beneath the generator, with ducts provided to guide the air to and away from the coolers.

Air-core reactors: A type of transformer with only one winding and no iron used to limit the current in a system or to supply reactive power to the line. Air-core reactors are built in two types, air insulated and oil insulated. An air insulated reactor is shown in Fig. 8 and consists of insulated cable (1) wound in a helical form and supported on a concrete structure (2) with columns of insulation concrete (3) to support the turns at frequent internals. The two terminals are shown at (4) and (5). The whole structure is insulated above ground on porcelain insulators, like the one shown at (6).

The winding, or coil, of an oil-immersed current-limiting reactor is built like the windings on an oil-immersed transformer of equivalent size. The coil is supported in an oil-filled steel tank but without magnetic material for a core. Since an air-core reactor sets up a large magnetic field, means must be provided for keeping the magnetic flux out of the steel tank, or excessive heating due to eddy currents will result. Two methods are used to keep the magnetic flux out of the tank. One is to line the tank with laminations of transformer steel to allow the flux to bypass the tank by providing a low-reluctance path for it. The

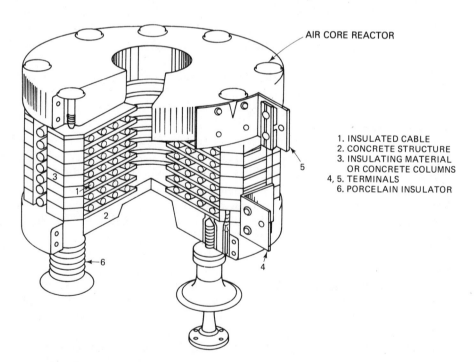

AIR CORE REACTOR

1. INSULATED CABLE
2. CONCRETE STRUCTURE
3. INSULATING MATERIAL
   OR CONCRETE COLUMNS
4, 5. TERMINALS
6. PORCELAIN INSULATOR

FIGURE 8   Distribution and power transformers: (a) insulated cable; (b) concrete structure; (c) insulating material or concrete columns; (d, e) terminals; and (f) porcelain insulator.

other is to provide a series of short-circuited secondaries of a transformer. These rings prevent most of the magnetic flux from getting into the tank wall because the current flowing in them sets up opposing ampere-turns.

Aircraft electric subsystems: Electric power devices used in airplanes to perform most of these functions: flight control; lighting; communication and navigation; actuators for flaps, landing gear, bomb doors, etc.; control systems engine, propeller warning, etc.; heating; pumps and blowers; instruments; offensive and defensive armament (radar, computers, etc.).

Electric power subsystems used on aircraft range in sizes from 28-V dc to 400-Hz three-phase 200 Y-connected 115 V. The power needed starts at only a few hundred watts in small aircraft to over 270 kW on large bombers. Table 1 lists the voltage ratings for various aircraft apparatus.

Aircraft generators: Power sources for electrical systems in aircraft. In aircraft, a ready means of power generation has always been available via the engine—either internal combustion prop engines or jets. Such means of power generation are not as readily available on missiles and spacecraft. The latter crafts usually must rely on batteries for electric power.

Alternating current generators found on modern-day aircraft are almost exclusively of the synchronous type. Characteristics of typical aircraft three-phase ac generators with integral excitators, as supplied by the General Electric Company, are shown in Table 2.

Air dielectric: A device that uses empty space (or air-filled space) to serve as an insulator between two or more elements at a certain potential and/or frequency. The term *air dielectric* may be applied to capacitors, coaxial cable, and other such electronic components. While the term indicates that air is an insulator, this is not true in all cases. High-voltage potentials can easily be conducted over short air gaps. In this instance, the air becomes a conductor. This is true of any dielectric; therefore, all insulators are said to be nonconductors of electricity only at specified potential maximums and within a stated frequency range.

Air gap: A small space characteristic of the induction motor. The size of the air gap has an important bearing on the power factor of the motor. Doing anything to affect the gap, such as grinding the rotor laminations or filing the stator teeth, results in increased magnetizing current, with resultant lower power factor.

Good maintenance procedure calls for periodically checking the air gap with a feeler gauge to ensure against a worn sleeve bearing that might permit the rotor to rub the laminations. These measurements should be made on the motor's shaft end. On large machines, it is desirable to keep a record of these checks. Four measurements should be taken approximately 90.028 apart, one of these points being the load side; i.e., the point on the rotor periphery that corresponds with the load side of the bearing. A comparison of the new measurements with those previously recorded will permit the early detection of bearing wear. A slight rub will generate heat sufficient to destroy the coil insulation.

TABLE 1 Voltage Ratings for Various Aircraft Apparatus

	Direct current†			Alternating current		
Nominal system designation (volts)	14	28	120	115	115/200	230/400
Generators:						
Rated voltages	15	30	125	120	120/208	240/416
Voltage adjustment range, %	+0 to −15	+0 to −15	+0 to −15	115 ± 5	115/200 ± 5	230/400 ± 5
Continuous-duty devices:						
Rated voltages	13	27	115	115	115/200	230/400
Voltage range, %	± 10	± 10	± 10	± 5	± 5	± 5
Intermittent-duty devices:						
Rated voltages	12.5	26	115†	115	115/200	230/400
Voltage range, %	± 10	± 10	± 10	+5 to −10†	+5 to −10	+5 to −10
Battery-operated devices (devices that must operate while generators are not in use):						
Rated voltages	11.5	23				
Voltage range, %	± 25	25				
Emergency-duty devices:						
Rated voltages				115	115/200	230/400
Voltage range, %				+5 to −15	+5 to −15	+5 to −15
Dielectric tests (rms volts for 1 min at 60 c or 120% of value for 5 s):						
Factory-test volts	1500	1500	1500	1500	1500	1800
Field test or retest before and after use (clean and dry only):						
(75% of factory-test volts)	1125	1125	1125	1125	1125	1260

TABLE 2

	10	20*	20	30	30	30	40	60
Continuous rating, kva	10	20*	20	30	30	30	40	60
Rated power factor	0.75	0.75	0.75	0.9	0.75	0.75	0.75	0.75
Min. rated speed, rpm	5700	7600	5700	4000	4800	5700	5700	5700
Max. rated speed, rpm	6300	8400	6300	8000	7200	6300	6300	6300
Frequency	380/420	380/420	380/420	400/800	320/480	380/420	380/420	380/420
Cooling air, S.L. temp., °C	80	120	80	40	40	80	80	80
Cooling air, rated cfm	140	220		300	330	330	330	330
Frame diam., in.	7	8¼	11	11	11	11	11	11
Length with axial blast cap, in.	16 15/16	14 5/16	15⅝	17⅝	16⅞	16⅞	17⅝	19 1/32
Weight, lb	45	38*	62	99	87	75	87	115
Lb per kva	4.5	1.9*	3.0	3.5	3.1	2.7	2.3	2.0
Max. harmonic and % of fundamental	5-1.90	5-1.45	5-1.45	7-1.40	5-0.75	5-0.75	5-1.4	5-0.44
Phase voltage unbalance with ⅓ current unbalance, max. %	3.7		3.6	3.9	3.8	3.0	2.9	2.9
3-phase S.C. current at base speed, p.u.	5.0		3.6	3.3	3.5	3.0	3.3	3.5
Reactances at 400 cycles, p.u.:†								
Synchronous, direct axis (x_d)	1.20	2.02	2.13	1.12	2.51	2.14	1.99	1.61
Synchronous, quad. axis (x_q)	0.68	1.16	1.25	0.65	1.35	1.20	1.12	0.79
Transient, direct axis (x'_d)	0.20	0.38	0.42	0.22	0.37	0.37	0.34	0.23
Subtransient, direct axis (x''_d)	0.14	0.28	0.24	0.14	0.21	0.20	0.19	0.12
Negative sequence (x_2)	0.18	0.36	0.34	0.16	0.23	0.21	0.20	0.12
Zero sequence (x_0)	0.28	0.34	0.012			0.17	0.011	0.15
Negative sequence resistance (r_2)	0.035		0.057			0.042	0.039	0.025
Zero sequence resistance (r_0)	0.026	0.32	0.039			0.026	0.024	0.020
Efficiency, full load, 6000 rpm	77	85	77	80	78	83	83	

*This machine is designed for static excitation and has no integral exciter.

†Calculated values.

Air/ground radio control: The operation of an aircraft or other similar device by radio control from a ground station. The operation of gas-powered model airplanes by radio is a good example of this type of operation. A small transmitter on the ground beams coded signals to a miniature receiver mounted within the model aircraft. This receiver, in turn, controls servounits connected to the air surface controls and to the throttle. While the plane is aloft, the ground operator can control its direction and flight characteristics as long as the transmitter and receiver units are within range.

An air/ground control radio station is part of the Aeronautical Broadcasting Service and is defined as a station for aeronatical telecommunications relating to the operation and control of local aircraft. Air/ground radio control is always utilized at controlled airport facilities and in most local airports where arrival and departure instructions as well as runway conditions are communicated between the fixed base operation and the pilot. See *Aeronautical Broadcasting Service.*

Air-spaced coax: Coaxial cable in which air serves as the basic dielectric or insulator.

Air terminals: Single-pointed heavy rods used to attract lightning. Rods should be placed on upward projections like chimneys, towers, etc.; on flat roofs 50 ft off center (o/c); and on the edges of flat roofs and ridges of pitched roofs 25 ft o/c. Rods should be from 10 to 60 in. above flat roofs and ridges; from 10 to 14 in. above upward projections.

Alarm functions: The varied uses of systems designed to warn of an intrusion, fire, or the like. Magnetic contacts are used on doors and windows in closed-protective circuits, in direct-wire systems, and in open-circuit applications. Movable elements within the switch unit of the magnetic contacts usually consist of a single flexible contact arm that provides a solid metal circuit path from the terminal screw to the contact-point end. The circuit continuity should not depend upon conduction across a hinge joint or through a coil spring.

When magnetic contacts are mounted on either non-co-planar or ferromagnetic surfaces, magnet and/or switch units should be held away from their respective mounting surfaces as necessary to:

1. Bring switch and magnet into close proximity when the door, window, etc., is closed.
2. Reduce the shunting effect of ferromagnetic materials so that positive switch pull-in occurs when the magnet approaches to within 18 in. of the switch.

Mechanical contacts are used as emergency, panic, or fire-test switches. Ball contacts and plunger contacts are used in both closed- and open-circuit applications. See Fig. 9. Mercury contacts are sometimes used in the low-energy alarm of signal systems to detect tilting of any horizontally hinged window, door, cover, access panel, etc. Due to the different items to be

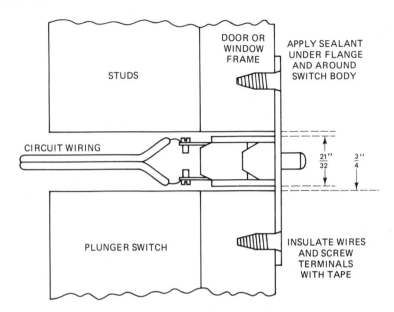

FIGURE 9 Alarm with mechanical contacts.

protected, it is best to install mercury contacts that can be adjusted to sensitivity after installation. For combined detection of either opening or breakthrough, cord-mounted contacts with foil connected to takeoff terminals should be used. Wiring diagrams of mercury contacts are shown in Fig. 10. Holdup switches are usually installed under counters or desks in banks or stores, where an employee observing a holdup may be able to signal for help.

In banks and similar places where large amounts of money are exchanged, a money-clip alarm device is sometimes used. This device automatically triggers an alarm when all bills are removed from a cash drawer. A bill inserted in the clip holds its switch in the normal position. Additional bills on top of the clip keep it concealed. Bills may be added or removed as required for normal business operations as long as one remains in the clip. However, the removal of all bills trips the clip switch to signal an alarm.

Window foil is used extensively in commercial applications. For fixed windows, the connections to the building alarm system are usually made through foil blocks. For movable windows and doors, a retractable door cord must be used. Ultrasonic motion detectors for commercial applications are usually extended in range above those used for residential applications and a typical coverage of one type is shown in Fig. 11. Commercial telephone dialers are available to dial emergency numbers and deliver voice messages. Most distinguish between burglar and fire alarm channels. A typical wiring diagram is shown in Fig. 12.

Digital alarm transmitters are becoming more popular for both commer-

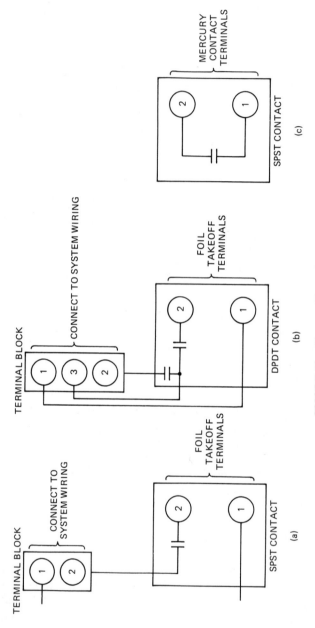

FIGURE 10 Alarms with mercury contacts.

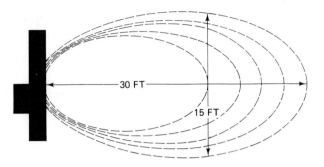

FIGURE 11 Alarm coverage for commercial use.

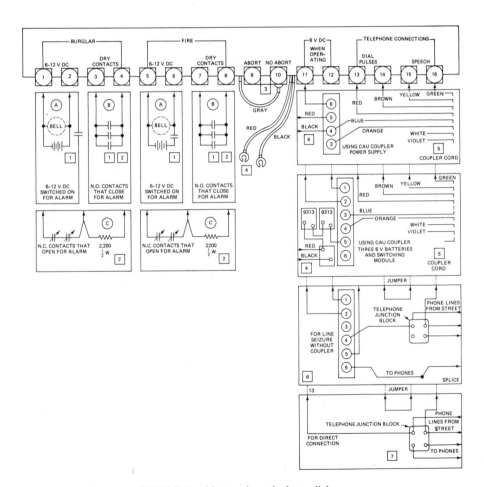

FIGURE 12 Alarm using telephone dialer.

cial and industrial applications. They can be programmed on memory chips to meet the exact requirement of any business. Sample printing formats for a digital alarm system are shown in Fig. 13.

Vibration detectors are often used on industrial buildings to detect vibrations caused by forced entry. Such detectors have been used on a variety of construction materials such as hollow tile, plaster and lath, brick, concrete, and wood. Once mounted in place, they may be adjusted with a set screw for the desired sensitivity.

Fence guard detectors use a vertical-motion detector that is sensitive to movement created by climbing or cutting the fence. Normal side motions such as wind or accidental bumping do not affect the detector or cause false alarms.

```
DEC   2    2 02 A M          JAN   3    2 02 P M
LOCATION  ||||               LOCATION  ||||
FIRE                         FIRE
HOLD-UP                      HOLD-UP
BURGLARY  |                  BURGLARY
BURGLARY  2                  AUXILIARY  |
AUXILIARY                    AUXILIARY  2
OPEN                         AUXILIARY  3
CLOSED                       AUXILIARY  4
LOW BATTERY                  LOW BATTERY
RESTORED                     RESTORED
```

```
APR   4    5 05 P M          DEC   3    4 04 P M
LOCATION  ||||               LOCATION  ||||
FIRE                         POLICE
HOLD-UP                      FIRE
BURGLARY  |                  SECURITY  |
BURGLARY  2                  SECURITY  2
SUPERVISORY                  CONTROL
OPEN                         ARMED ROBBERY
CLOSED                       BURGLARY
LOW BATTERY                  LOW BATTERY
RESTORED                     RESTORED
```

```
DEC   2    2 05 P M          APR   7    6 26 P M
LOCATION  ||||               LOCATION  ||||
BURGLARY                     FIRE
FIRE                         HOLD-UP
PANIC                        NITE  |
FAILURE                      NITE  2
SYSTEM OFF                   SUPER
SYSTEM ON                    DAY NORMAL
AUXILIARY                    NITE NORMAL
LOW BATTERY                  LOW BATTERY
RESTORED                     RESTORED
```

FIGURE 13

They are normally mounted about midway up the fence, and every 10 ft of fence length. Most of these devices set off the alarm if they are tampered with or if the wire is cut. They may be connected to a control panel, and the alarm will "sound" in the form of a bell or horn, or it will silently dial the local law enforcement agency.

The outdoor microwave detector is used for protecting large outdoor areas such as car lots, construction sites, and factory perimeters. In operation, a solid circular beam of microwave energy extends from a transmitter to the receiver over a range of up to 1500 ft. Any movement inside the beam will activate the alarm.

Other protective devices include thermistor sensors and ultraviolet radiation fire detectors.

Alarm systems: Systems designed to warn of an intrusion, a fire, or other undesired occurrence. A signal circuit used for an alarm system may be classified as open circuit or closed circuit. The former is one in which current flows only when a signal is being sent, while the latter has current flowing continuously, except when the circuit is opened to allow a signal to be sent.

Regardless of the type, all alarm systems have three functions in common: detection, control, and annunciation signaling. Many systems incorporate switches or relays that operate when entry, movement, pressure, infrared-beam interruption, and other intrusions or happenings occur. The control senses the operation of the detector with a relay and produces an output that may operate a siren, bell, silent alarm such as telephone dialers to law enforcement agencies, or other signals. The control frequently contains ON/OFF switches, test meters, time delays, power supplies, standby batteries, and terminals for connecting the system together. The control output usually provides power on alarm to operate signaling devices or switch contacts for silent alarms. See Fig. 14.

An example of a basic closed-circuit alarm system is shown in Fig. 15. The detection, or input, subdivision in this diagram shows exit/entry door or window contacts, detectors, switch mats, ultrasonic detectors, etc. The control subdivision for the system in Fig. 15 consists of switches, relays, a power supply, a reset button, and related wiring. The power supply shown is a 6-V

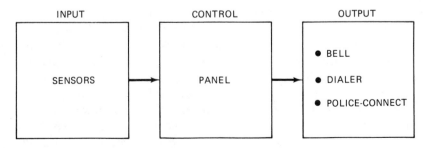

FIGURE 14 Alarm systems.

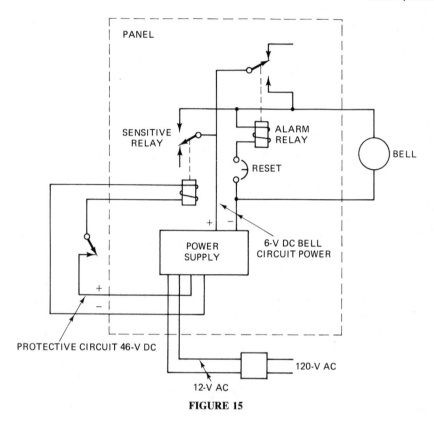

FIGURE 15

nickel-cadmium battery that is kept charged by a plug-in transformer unit. Terminals are provided on the battery housing to accept 12-V ac charging power from the plug-in transformer which provides 4- to 6-V power for the detection circuit and power to operate the alarm or output division.

A closed-loop protective circuit consists of an annunciator connected to a special design contact on each door and window and a relay so connected that, when any window or door is opened, it will cause current to pass through the relay. The relay, in turn, will operate to close a circuit on a bell, horn, or other type of annunciator that will continue to sound until it is shut off.

Wire sizes for the majority of low-voltage alarm systems range from No. 22 to No. 18 AWG. However, when larger-than-normal currents are required or when the distance between the outlets is long, it may be necessary to use wire sizes larger than specified to prevent excessive voltage drop.

The wiring of any alarm system is installed like any other type of low-voltage system; that is, locating the outlets, furnishing a power supply, and finally interconnecting the components with the proper size and type of wire. Most closed systems use two-wire cables and are color-coded to identify them. A No. 18 pair normally is adequate for connecting bells or sirens to controls if the run is 40 ft or less. Many installers, however, prefer to use No. 16 or even No. 14 wire.

A summary of the various components for a typical alarm system is depicted in Fig. 16. These components are:

1. *Control station:* This is the heart of any security or alarm system, since it is the circuitry in these control panels that senses a broken contact and then either sounds a local bell or horn or omits the bell for a silent alarm. Most modern control panels use relay-type controls to sense the protective circuits and regulate the output for alarm-sounding devices. They also contain contacts to actuate other deterrent or reporting devices and a silent holdup alarm with dialer or police-connected reporting mechanism.

2. *Power supplies:* Power supplies vary for different systems, but in general, they consist of rechargeable 6-V ac input from a plug-in or otherwise connected transformer to a 120-V circuit. The better power supplies have the capability of operating an armed system for 48 h or more without being charged and still have the capacity to ring an alarm bell for 30 min or longer. Power supplies are obviously used in conjunction with a charging source and supply power for operation of the alarm system through the control panel.

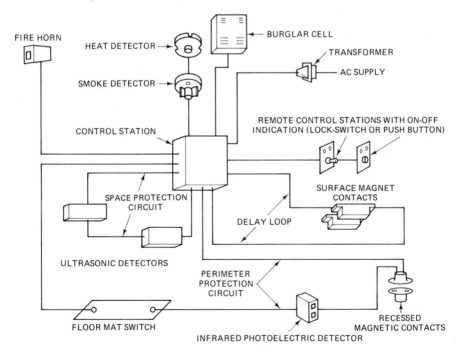

FIGURE 16 Alarm system.

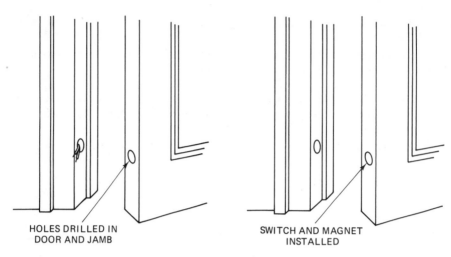

HOLES DRILLED IN
DOOR AND JAMB

SWITCH AND MAGNET
INSTALLED

FIGURE 17

3. *Recessed magnetic contacts:* Holes are drilled in doors and windows and also their casings, one directly across from the other, and a pair of wires from the positive side of the protective circuit is run out through the switch holes. The switch and magnet are then installed with no more than a 1/8 in. gap between them. See Fig. 17.

4. *Surface-mounted magnetic contacts:* A switch is mounted on the door or window casing with a magnet on the casing and one on the door or window. As long as the switch and magnet are parallel and in close proximity when the opening is shut, they may be oriented side-to-side, top-to-side, or top-to-top. See Fig. 18.

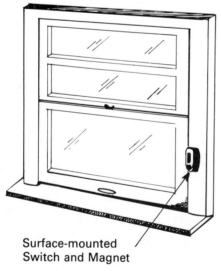

Surface-mounted
Switch and Magnet

FIGURE 18

5. *Conductive foil:* A self-adhesive foil block (terminator) on the door or window is connected to a similar unit on the door or window frame by a short length of flexible cord to allow for movement. The foil is connected in the positive conductor of the protective circuit and is adhered to the glass parallel to and about 3 in. from the edge of the glass, using recommended varnish. Breaking the glass breaks the foil and opens the circuit. To provide more coverage, a double circuit of foil may be taken from the foil block. Coiled, retractable cords are available for use between foil blocks to allow for sliding window and door travel.

ALC: An abbreviation for *Automatic level control.*

Algorithmic: Pertaining to a constructive calculating process usually assumed to lead to the solution of a problem in a finite number of steps.

Aligned grid: An arrangement in a vacuum tube in which at least two grids are supported in such a manner that one is concentric to the other. This construction is used in electron tube design and is especially applicable to beam power tubes that use an aligned grid arrangement to produce high input sensitivity. Aligned grid construction allows the wires of one grid to shade those of the other. This produces excellent control of electron flow by application of relatively small amounts of outside power. Beam power tubes are often known as aligned-grid tubes due to their general use of this type of construction. See *Beam power tube.*

Alkaline battery: An energy storage device composed of alkaline cells as opposed to the zinc and carbon construction of most standard dry cells. Alkaline batteries are manufactured to the same physical sizes as most of the popular types of dry cells and can be used interchangeably in most applications. Each individual cell in an alkaline battery consists of the negative electrode, which is granular zinc mixed with potassium hydroxide (alkaline) electrolyte. The positive electrode of the alkaline cell is in direct electrical contact with the outer metal can portion of the battery proper. The zinc and alkaline electrodes are divided by a porous separator. Most alkaline cells deliver a voltage potential of 1.5-V dc but have higher capacities than a dry cell of the same physical size.

Alkaline batteries exhibit a relatively flat discharge curve under a constant load, and thus are more desirable for higher current applications than would be appropriate with carbon/zinc cells. Alkaline cells often are advertised as long-life or extended-life batteries, which is appropriate when comparing them to standard dry cell components. This type of battery is sometimes referred to as an Edison cell after Thomas Alva Edison. Specifically, an Edison cell uses a negative plate material composed of powdered iron oxide mixed with cadmium and a potassium hydroxide (alkaline) electrolyte. Typical voltage potential for the true Edison cell is about 1.25 V at full charge.

Some persons confuse alkaline cells with rechargeable batteries, which are constructed in a completely different manner. Alkaline cells, as offered today, are not rechargeable and can be damaged when connected in a circuit that incorporates a constantly active recharger. Alkaline cells are more expensive than conventional dry cells, but the added cost may be more than compensated for by the extended operational life in circuits that are best designed to take advantage of the characteristic flat discharge rate of the long-life battery.

All-pass filter: An electronic network that is ideally designed to pass all frequencies induced at its input.

This type of filter is said to have zero attenuation at any frequency. In practice, this is not possible, but most all-pass filters have relatively little attenuation effect over a broad frequency spectrum. All-pass filters are used in applications where a phase shift is desired. They are seen often in various types of time delay applications.

Alloy-difused components: Electronic systems elements that consist of a number of solid-state devices such as transistors to establish contact points through a diffusion process combining gas into a semiconductor material. Another junction in the same component is formed by alloying a suitable material such as indium with the semiconductor. An alloy-diffused device, then, uses two different processes, alloying and diffusion, to establish its junction.

Alpha: Current amplification factor (emitter to collector) evidenced by a change in collector current for a change in emitter current. This refers to the current amplification between emitter input and collector output, and is a property of the grounded-base circuit. This amplification is designated by α, the Greek letter alpha.

Alpha, transistor: A rating defining the current gain of a specific common-based connector bipolar transistor. Alpha is the ratio of the differential in collector current to the differential in emitter current. For most junction transistors, alpha is always less than unity, or one, but very close to it. Alpha is the transfer ratio in a common-base forward current transistor circuit, whereas beta is the same ratio in a common emitter circuit. In the common-base circuit, the emitter serves as the input electrode, while the collector serves as the output. The alpha or beta of a bipolar transistor is more often referred to as the amplification factor. See *Amplification factor*.

Alphabetic-numeric: The characters that include letters of the alphabet, numerals, and other symbols such as punctuation or mathematical symbols.

Alternate energy: The name given to forms of energy that differ from the conventional types such as electric, oil, and gas. See Figs. 19–21. Falling under the category of alternate energy are solar, wind, and water power. Wood burning for space heating and cooking is also considered a form of alternate energy, although it was the principal means of space heating and cooking before the early part of this century.

In many areas in the United States and other parts of the world, home power units have been constructed and used for a long time. Windmills have

FIGURE 19 The Amernault wind generator.

been used for centuries for grinding grain and pumping water. In the 1930s, they were used extensively in rural areas for pumping water and to charge 32-V electric battery systems for lights and small electrical loads around the farm.

Water power—utilizing the well-known water wheel—has also been used quite a lot in the past for grinding grain, running saw mills, and for generating electricity.

Methane generators also are not new. In India, with over 3000 methane generators in use on farms and in villages, the feasibility of such power units has been demonstrated over and over again. England, Germany, and Africa have similar methane-generating units that supply valuable gas, compost and waste disposal systems necessary for survival.

Actually, "alternate" energy is nothing new. In many cases, what is considered alternate energy today was commonplace only a few decades ago;

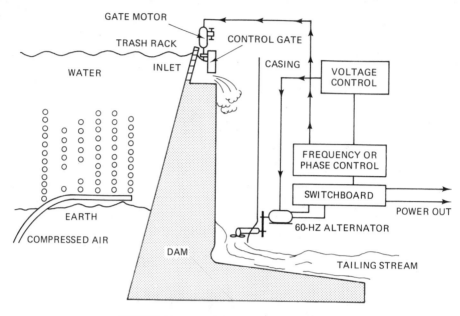

FIGURE 20 A small hydroelectric installation.

but now, with the world facing an energy shortage, almost all concerned people are once again looking at alternate energy sources as a means to save money, and more important, as a means to survive. In order that future generations will have life support systems, other sources of energy—like the sun, wind, water—must be found and put to practical use.

Alternate routing: The ability of a switching system to recognize that a route or routes are blocked and choose an alternative. In telephone, telegraph, and data systems, automatic switching can be accomplished by means of a rotary dial at the station in a switch that will follow pulses generated by it. An improvement in service can be achieved by alternate routing.

The rotary dial, which generates pulses of direct current, has been the general means for a telephone customer to transmit the desired number to his serving central office. In recent years, however, a push-button system generates combinations of audible frequencies and is being applied to all types of switching systems and will eventually replace the rotary dial. These audible frequencies can be transmitted over a completed telephone connection whereas rotary-dial pulses cannot, since the dial pulses are interruptions of direct current.

Alternating current: An electrical current, from alternators, that produces a voltage to change the polarity of the generator terminals at regular, very short intervals. If a terminal is positive at one instant, it will become negative in the next instant, and the other terminal will be positive. While the polarity of the terminals alternates, the direction of the current in the circuit

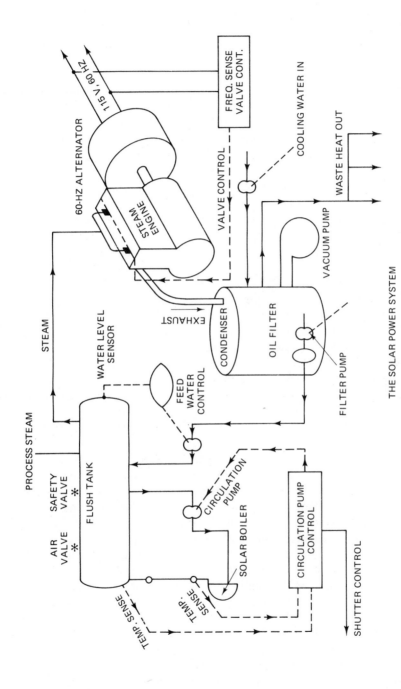

FIGURE 21 The solar power system—alternate energy.

THE SOLAR POWER SYSTEM

29

connected to the terminal changes every time the terminal polarity changes. Such a current is an alternating current, and the voltage is an ac (alternating current) voltage. An alternating current changes in both strength and direction in a given period of time. Such a current starts from zero, gradually increases to its highest value, and gradually decreases to zero; then it starts to flow in the opposite direction, when it again increases to its highest value and returns to zero. A rise and fall in one direction followed by a rise and fall in the opposite direction is called a cycle. The number of cycles that occur per second is called the frequency of an alternating current. The frequency of the alternating current supplied for lighting and power purposes in the United States and Canada is usually 60 cps (cycles per second). See Fig. 22.

Alternation: A half-cycle in alternating current. There are two alternations in a complete cycle—one in the positive direction and one in the negative direction. See Fig. 23. As applied to the flow of electricity, an alternating current may be defined as a current that reverses its direction in a periodic manner, rising from zero to maximum strength, returning to zero, and then going through similar variations in strength in the opposite direction. These changes comprise the cycle that is repeated with great rapidity.

The properties of alternating currents are more complex than those of continuous currents, and their behavior is more difficult to predict. This arises from the fact that the magnetic effects are more important than those of steady currents. With the latter, the magnetic effect is constant and has no reactive influence on the current once the direct current is established. The lines of force, however, produced by alternating currents are changing as rapidly as the current itself, and they thus induce electric pressures in neighboring circuits, and even in adjacent parts of the same circuit. This inductive influence in alternating currents renders their action very different from that of continuous current.

Alternator: A rotary device that, when powered, produces alternating current. The main parts of an alternator are the field and the armature. The field is composed of several electromagnets that are so arranged that the north and south poles are alternately distributed in the field. The electromagnets are coils on iron cores that obtain direct current in such a direction that the desired polarity always appears at a certain end of the coil. The poles are so designed in strength and number that they provide the necessary number of lines of force that must link the armature conductors. In most modern alternators, the field, or the poles, are placed on the rotating part of the machine. It is more practical to use such an arrangement than to let the conductors move through the magnetic field. The result is the same: Relative movement of magnetic lines and conductors induces a voltage in the conductors. Since the group of poles is rotating, the alternator part that carries the poles is called the rotor. An alternator rotor that carries 36 poles is shown in Fig. 24. The electromagnets, or poles, one of which is indicated at 1, receive the current through the leads 2, shown on one of the spokes of the rotor spider 3, through the collector rings 4.

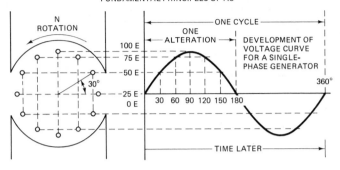

FUNDAMENTAL PRINCIPLES OF AC

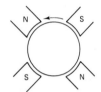

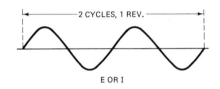

INDUCTION MOTORS

NO. OF POLES	CYCLES PER REV.	REV. PER SEC FOR 60	RPM FOR 60	RPM OF MAGNETIC FIELD	RPM OF ROTOR
2	1	60	3600	3600	3450
4	2	30	1800	1800	1740
6	3	20	1200	1200	1160
8	4	15	900	900	860
10	5	12	720	720	690
12	6	10	600	600	580

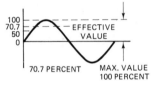

$$\text{POLES} = \frac{120 \times \text{FREQUENCY}}{\text{RPM}}$$

$$\text{RPM} = \frac{120 \times \text{FREQUENCY}}{\text{POLES}}$$

$$\text{FREQUENCY} = \frac{\text{POLES} \times \text{RPM}}{120}$$

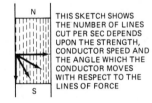

THIS SKETCH SHOWS THE NUMBER OF LINES CUT PER SEC DEPENDS UPON THE STRENGTH, CONDUCTOR SPEED AND THE ANGLE WHICH THE CONDUCTOR MOVES WITH RESPECT TO THE LINES OF FORCE

DEVELOPMENT OF VOLTAGE CURVES FOR A 3-PHASE GENERATOR

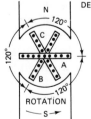

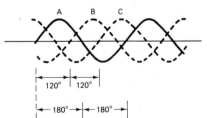

FIGURE 22 Alternation.

31

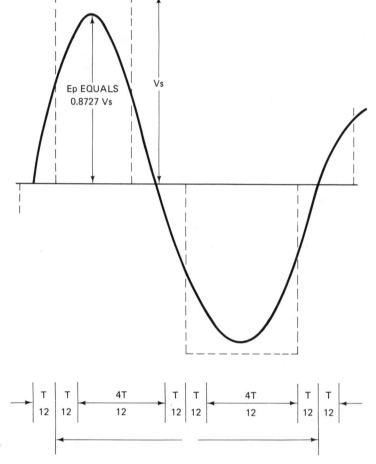

FIGURE 23 Fundamental principles of AC.

In a large alternator, there are many coils of conductors in which the voltage is generated. All these conductor coils are supported by an iron frame, and they build the part called the armature of the alternator. The armature is the stationary part of the alternator, and for that reason it is also called the *stator*. There are machines that have the field placed on the stator and the conductors on the rotor, but here, when we refer to a rotor, we mean the field poles.

A stator is shown in Fig. 25. The conductors are placed in the slots of the iron frame and so connected that they build closed circuits. An alternator with four poles is shown in Fig. 26. The rotor 1 carries the poles N, S, N_1 and S_1. The exciting direct current for the poles is obtained from the exciter 2, which is a small dc generator. The current flows from the terminals of the exciter 2 to the

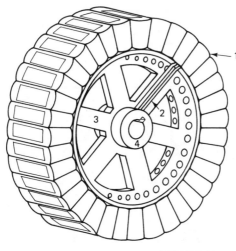

1. ELECTROMAGNET
2. LEADS
3. ROTOR SPIDER
4. COLLECTOR RINGS

FIGURE 24

collector rings 3, through the brushes 4. The arrows on the windings around the poles indicate the direction of the current that causes the pole cores to be magnetized and to act as north or south poles. The rotor 1 rotates on a shaft (not shown) within the stator 5. The stator has an iron core with slots in which the conductors are placed. The alternator shown in Fig. 26 is a three-phase alternator, which means that its conductors are connected in three circuits. There are three lines leading from the alternator to the delta-connected load 6.

The alternator rotor is revolved by an outside source of mechanical

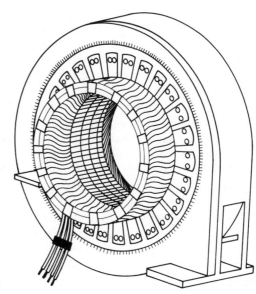

FIGURE 25

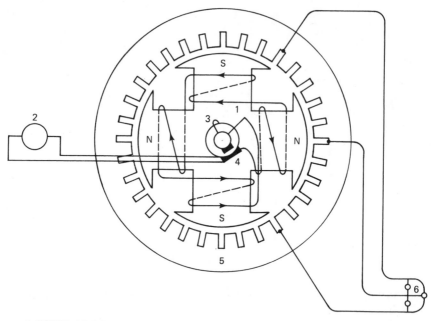

1. ROTOR, OR FIELD
2. EXCITER
3. COLLECTOR RINGS
4. BRUSHES
5. STATOR, OR ARMATURE
6. LOAD

N, N_1 = NORTH POLES
S, S_1 = SOUTH POLES

FIGURE 26 (a) Rotor, or field; (b) exciter; (c) collector rings; (d) brushes; (e) stator, or armature; (f) load. N, N_1 = north poles; S, S_1 = south poles.

power. The shaft must be turned at a constant speed by a machine that is called a *prime mover.* The prime mover can be of various types, such as a steam engine or a turbine. The steam for a steam engine or turbine is produced in a boiler that obtains heat from coal, gas, or nuclear reaction. A generating plant that uses heat to move the prime mover is called a *thermal plant.* A hydroplant uses the power of falling water to turn the turbine or water wheel that serves as the prime mover. The engines that are used as prime movers are diesel, gas, or steam engines, and they range in speed from 180 to 600 rpm. Alternators that run at higher speeds usually have hydraulic or steam turbines as prime movers.

Hydraulic turbines are of three principal types: Francis, propeller, and impulse. Most Francis and propeller turbines are made with vertical shafts; most impulse turbines are made with horizontal shafts. An alternator driven by a Francis or a propeller turbine must, therefore, be constructed with a vertical shaft. Several hundred pounds per square inch lie on the thrust bearings of a vertical-shaft alternator, and the bearings must, therefore, be designed with great precision. Under the most unfavorable conditions, a Francis or an impulse turbine may attain a speed that is about 170 to 200 percent of normal; a propeller turbine may reach 300 percent of normal speed. An alternator driven by one of these prime movers (Fig. 27) must be so designed mechanically that it can stand sudden increases in speed.

Since an alternator requires a constant field that acts continuously in the same direction, its field windings must be connected to an outside source of direct current. If one is available in the electric station, or power plant, it will be utilized. Otherwise, it will be necessary to have an auxiliary dc generator driven by a separate prime mover. Such an auxiliary generator is called an *exciter*. Except in large plants, the most common practice is to couple the exciter directly to the alternator shaft. This construction means that each alternator will be equipped with its own exciter. The usual exciter is rated at 125 V, but on a large machine, the exciter voltage may be 250, 375, or 500 V.

For proper operation of an alternator, it is important to provide constant direct current to the field. A precise regulation of the excitation is obtained by an exciter with multiple field windings. Excitation may also be obtained by an electronic system that supplies high-speed response to any change in the voltage.

The stationary part of the alternator consists essentially of a laminated iron core, its supporting structure or yoke, and stator coils that are placed in slots provided on the inner, cylindrical surface of the laminated core. End shields or guards that extend over the stator-coil ends and protect them from mechanical injury are attached to the yoke. The yoke is usually fabricated from a steel plate to provide the necessary stiffness and mechanical strength to the machine. The stator core is constructed of laminations punched from thin steel sheets. The core is made of steel to provide a low-impedance path for the magnetic flux; the thin laminations are used to minimize the losses that accompany the passage of a changing magnetic flux in steel. The magnetic loss is called *core loss,* and consists of eddy-current loss and hysteresis loss. The laminations are commonly 0.014 or 0.018 in. thick and have a thin coat of insulating varnish. To prevent hysteresis loss, which does not depend on the thickness of laminations, a small amount of silicon is added to the steel alloy used for the core.

The stator coils most often used in alternators are double-layer lap windings that employ formed coils, which are accurately formed during manufacture to fit in the slots of the stator core. The winding is so arranged that two coil sides are placed in each slot; hence, the name double-layer winding.

The coils are made of solid or stranded copper wire with sufficient current-carrying capacity. Each turn must be insulated from others, and the entire coil must be insulated from the stator core. The amount and kind of insulation depends on the voltage that the insulation must withstand.

If the alternator is constructed to generate single-phase voltage, all stator coils are connected in series to form one closed circuit. However, most alternators are wound for three-phase operation. Therefore, the stator coils are connected to form three distinct and independent windings that are displaced 120° from each other. The total voltage generated in the coils of one such winding, or phase, is called the phase voltage of the alternator. The three windings are connected to a common point to build a star-connected alternator or end-to-end to form a delta-connector alternator. As a rule, three terminal leads are brought out from interconnected windings. In a star-connected alternator, the neutral point can also be brought out to be grounded for protection. According to the ac circuit theory, the delta-connected alternator has the line voltage equal to the phase voltage; in the star-connected alternator, the line voltage is 1.732 × phase voltage.

The rotor commonly carries the poles, or the field. The number of poles and the form of the rotor depend on the speed at which the rotor will revolve. The two forms of rotors are the salient pole and the round, or cylindrical pole. A salient-pole rotor carries several laminated pole cores that are bolted or dovetailed to a laminated rotor body. All rotor coils are wound in the same direction. In order that adjacent poles will be of opposite polarity, current flows from top to bottom in one coil and from bottom to top in the adjacent coil. The leads that connect the coils go from the bottom of one coil to the bottom of the next and from the top of that coil to the top of the coil that follows.

In some alternators, there is a pole-face winding in addition to the field winding. Pole-face windings—called damper, or amortisseur, windings—consist of copper or bronze bars inserted in holes provided for them in the punching near the air-gap surface of the pole. The bars in any one pole are connected together at their ends by brazing them to the pole end plates or to a short-circulating bar. Occasionally, they are connected from pole to pole. The bars are not insulated from the poles. Damper windings serve two purposes: to dampen the fluctuations in speed and to prevent excessive unbalanced voltages

FIGURE 27

during short circuits. The damper winding is important when the prime mover is a diesel or other reciprocating type of engine.

Alternators that use steam turbines as prime movers run at high speeds and, instead of salient-pole rotors, use cylindrical, or round, rotors. A round rotor is made comparatively narrow and long in order to withstand the mechanical stresses that are due to the high speed. It also has the field windings placed in radial slots in the rotor body, instead of being wound around salient-pole bodies. Several coils connected in series are used to provide the necessary magnetic force.

Hard copper strap bent on edge is used for coils, although some designs call for aluminum alloy. The conductors are bare, and turns are insulated from each other by strips of mica or a similar insulating material.

The ends of the field winding are connected to slip rings, or collector rings, mounted on and insulated from the shaft. Stationary carbon brushes held against the surface of the rings conduct the direct field current from the exciter through the rotating rings to the field coils.

Brushes are made of a suitable material, usually a relatively soft electrolytic graphite. They are supported in brush holders that keep them at the proper angle to the collector rings. The brushes are free to move radially in the holders, while spring pressure keeps them firmly on the rings. Proper brush pressure is necessary for satisfactory brush performance. A brush spring must be capable of accurate adjustment of brush pressure, and it must maintain constant pressure over the full wearing length of the brush. The brush holder and the spring assembly are usually supported on an insulated stud from the housing. The number of brushes depends on the amount of exciting current the brushes are to carry. Attached to each brush is a flexible braided-wire pigtail, or shunt, which is connected to the brush holder or to a copper bus bar to provide a low-resistance path for current from the brush to the external circuit.

Alternator transmitter: A radio transmitter that generates power with a radio frequency (RF) alternator. In general, radio transmitters generate RF energy at a definite frequency and convey this energy to the transmitting antenna for radiation. To obtain a useful radiated signal, information must be superimposed on the radio waves. In continuous-wave (CW) transmitters used for radio telegraphy, the desired information is added by interrupting the radio-frequency oscillations in accordance with a telegraphic code. In radio-telephone (modulated) transmitters, the information is added by modulating either the amplitude or the frequency of the radio-frequency carrier wave with the speech or music to be transmitted.

Altimeter: An instrument used for measuring elevation. An altimeter does not necessarily indicate elevation above ground level but rather, the height above sea level. Altimeters depend upon barometric pressure for proper operation when constructed from aneroid capsules. An aneroid is a disk-shaped metallic device used as a pressure-sensing element. Aneroid barometers are designed for aircraft use because they offer compact size coupled with excellent sensitivity to changing barometric pressures. The aneroid elements

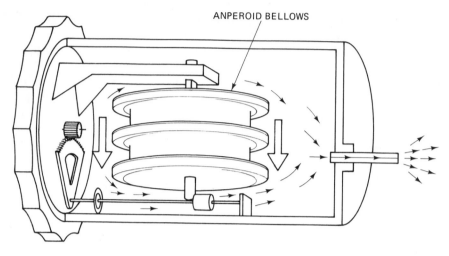

ANPEROID BELLOWS

FIGURE 28 Altimeter.

serve as a bellows, which is coupled to an indicating meter by an access arm. See Fig. 28. As the atmosphere pressure inside the altimeter decreases, the aneroid bellows expands. This results in the pointer's moving higher on the scale. At lower altitudes, air moves into the altimeter case due to an increase in outside pressure, causing the pointer to move down the scale.

Electronic altimeters often incorporate thermistor elements, replacing the mechanical bellows. An absolute altimeter does not depend upon barometric pressure and therefore is capable of indicating absolute altitude, which is the distance above the surface of the earth as opposed to height above sea level. This latter type of altimeter transmits a radio signal from the aircraft to the ground and measures the time interval between this transmission and the reception of the reflected signal on a receiver that is incorporated in the altimeter circuitry. This time element can be quite accurately used to determine actual height above terrain on a direct readout meter.

The major uses of altimeters involve aircraft of all types. Privately-owned planes normally use the simpler mechanical units, whereas military and commercial aircraft are beginning to make more use of the highly complex and expensive absolute altimeters.

Aluminization: The act of depositing a thin layer of aluminum on the back of a picture-tube screen to brighten the picture image and reduce ion-spot formation.

Aluminum: A chemical element, Al, atomic number 13, and atomic weight 26.9815. Aluminum is a low-density metal that exhibits high reflectance. Its ability to form a protective oxide surface coating that resists corrosion increases the metal's usefulness. Electrical conductor (EC) grade aluminum, having a purity of approximately 99.5% and a minimum conductivity of 61.0% IACS, is used for conductor purposes. Specified physical properties are obtained by closely controlling the kind and amount of certain impurities.

Aluminum-clad steel: Wires with a relatively heavy layer of aluminum surrounding and bonded to the high-strength steel core. The aluminum layer can be formed by compacting and sintering a layer of aluminum powder over a steel rod or by electroplating a dense coating of aluminum on a steel rod, and then cold drawing to wire.

Amauroscope: A highly complex biomedical network that interfaces electronic circuits to the human brain. Used as an electronic aid to blind persons, the amauroscope incorporates photocells in a pair of goggles that are fitted over the eyes. The photocells convert light images into electric pulses, which are impressed upon the visual receptors of the brain. These receptors are accessed by placing electrodes in the nerves above each eye. Using this system, blind persons are able to distinguish between varying light levels and to receive crude images of properly illuminated subjects.

American Electric Power Company System: An interconnecting group of electric power-generating companies. Electric power companies in the United States and Canada have found it advantageous, for reasons of reliability and economy, to interconnect and to operate in parallel. Each interconnecting system is made up of one or more control areas, each of which is defined as that portion of an interconnected system to which a common generation control scheme is applied. The American Electric Power Company System has four operating companies. The ties between the companies and their inter-ties with other control areas are shown in Fig. 29.

American or International Morse code: A telegraph code developed by Samuel Morse consisting of "dit-dah" signals. The principle of the Morse code is that every letter is made up of combinations of short and long signals. Each letter can be made by short or long sounds, short or long light flashes, or short or long electrical impulses. The Morse code is:

Letters		
A di-dah	F di-di-dah-dit	K dah-di-dah
B dah-di-di-dit	G dah-dah-dit	L di-dah-di-dit
C dah-di-dah-dit	H di-di-di-dit	M dah-dah
D dah-di-dit	I di-dit	N dah-dit
E dit	J di-dah-dah-dah	O dah-dah-dah
P di-dah-dah-dit	T dah	X dah-di-di-dah
Q dah-dah-di-dah	U di-di-dah	Y dah-di-dah-dah
R di-dah-dit	V di-di-di-dah	Z dah-dah-di-dit
S di-di-dit	W di-dah-dah	

Numerals	
1 di-dah-dah-dah-dah	6 dah-di-di-di-dit
2 di-di-dah-dah-dah	7 dah-dah-di-di-dit
3 di-di-di-dah-dah	8 dah-dah-dah-di-dit
4 di-di-di-di-dah	9 dah-dah-dah-dah-dit
5 di-di-di-di-dit	0 dah-dah-dah-dah-dah

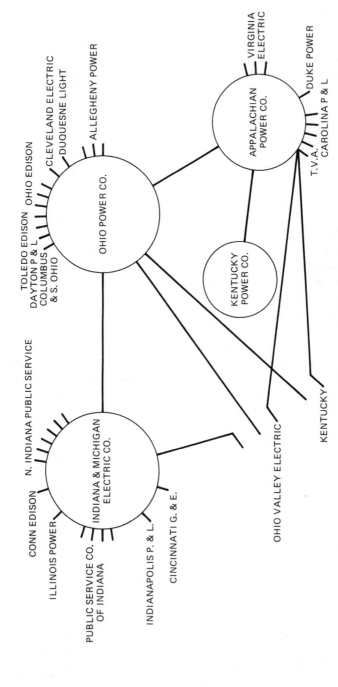

FIGURE 29 Utilities.

American Standard Association: See USA Standard Code.

American wire gauge: (AWG) Also called Brown and Sharpe or B&S gauge, the standard method of designating various wire sizes. Wire is listed according to gauge numbers from 0000 (4/0) to 40.

Ammeter: A device used to measure current in a circuit. The name is a contraction of *ampere meter* because the device measures the current in units called amperes. Very low current values are measured in milliamperes and microamperes, as follows:

$$1 \text{ A (ampere)} = 1000 \text{ mA (milliamperes)}$$
$$= 1{,}000{,}000 \text{ } \mu\text{A (microamperes)}$$
$$1 \text{ mA} = 1000 \text{ } \mu\text{A}$$
$$= 0.001 \text{ A}$$
$$1 \text{ } \mu\text{A} = 0.001 \text{ mA}$$
$$= 0.000001 \text{ A}$$

A milliammeter is an instrument that measures the current in milliamperes, and a microammeter measures the current in microamperes. An ammeter is always connected in a circuit in series with other devices. The current is the same in each device connected in series, and the ammeter will receive the same current as any other device and indicate the circuit current.

Ampacity: Current-carrying capacity of electric conductors expressed in amperes.

Ampere: The rate at which electricity flows through a conductor is represented by a unit. It may be compared to the rate of flow of water through a pipe in gallons per second. For all practical purposes, the unit strength of an ampere is represented when an electrical current passing through a specific solution of nitrate of silver in water deposits silver at the rate of 0.001118 g/s. If twice that amount of silver is deposited during 1 s, the current is 2 A, and so on.

Ampere-hour meter: A graphic or recording instrument used to produce a continuous and/or permanent record of the current in a given electrical circuit. Such an instrument has a meter element similar to the conventional indicating ammeters, but in addition, it is equipped with a pen or other marking device so that a curve is drawn as current changes occur. The marking device on the recording instrument replaces the pointer on scale-indicating meters.

The charts are either circular or in strip form, depending on the type of instrument. Recorder instruments are available as portable instruments or as permanent rack- or panel-mounted recording meters. Circular charts are preferred for approximate work where convenience in handling and filing is important. The strip recorder is capable of greater accuracy.

Ampere's law: A law of electromagnetism that expresses the contribution of a current element of length, *dl* to the magnetic induction (flux density), *B* at a point near the current.

Ampere's law, sometimes called Laplace's law, was derived by A. M.

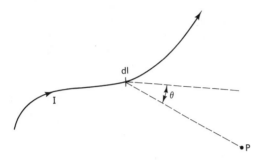

FIGURE 30   Representation of
Ampere's law.

Ampere after a series of experiments during 1820–1825. Whenever an electric charge is in motion, there is a magnetic field associated with that motion. The flow of charges through a conductor sets up a magnetic field in the surrounding region. Any current may be considered to be broken up into infinitesimal elements of length, *dl*. Each element contributes to the magnetic induction at every point in the neighborhood. The contribution, *dB*, of the element is found to depend upon the current, *I;* the length, *dl*, of the element; the distance, *r*, of the point, *P*, from the current element; and the angle, 0, between the current element and the line joining the element to the point, *P*. See Fig. 30.

Ampere's law expresses the manner of the dependence by this equation

$$dB = k\frac{I\,dl\,\sin 0}{r^2}$$

Ampere turns: The amount of turns of wire in a coil to produce lines of flux. Rowland's law is normally used for the calculation. To illustrate, assume that we wish to find the ampere-turns *(IN)* necessary to produce 20,000 lines of flux in a cast-steel ring having a cross-sectional area of 4 cm^2 and an average length of 20 cm.

Flux density is expressed as

$$B = \frac{Q}{A} = \frac{200{,}000}{4} = 5000 \text{ lines/cm}^2$$

The corresponding value of *H* for cast steel is 3.9. The formula for *H* is

$$H = \frac{1.257\,IN}{l}$$

$$IN = \frac{Hl}{1.257}$$

Substituting 3.9 for *H* and 20 for *l* in the preceding equation

$$IN = \frac{3.9 \times 20}{1.258} = \text{ampere-turns}$$

Amperian whirl: The stream of electrons produced when voltage is applied to a single-turned wire loop. The wire loop acts as an elementary electromagnet—provided that any amount of current flows through the wire.

Amplidyne generator: A two-stage electrical power amplifier. If a direct current generator designed as shown in Fig. 31a and operated with a very weak field is driven at constant speed, the main brushes may be short-circuited as indicated. This action results in relatively heavy currents in the armature that

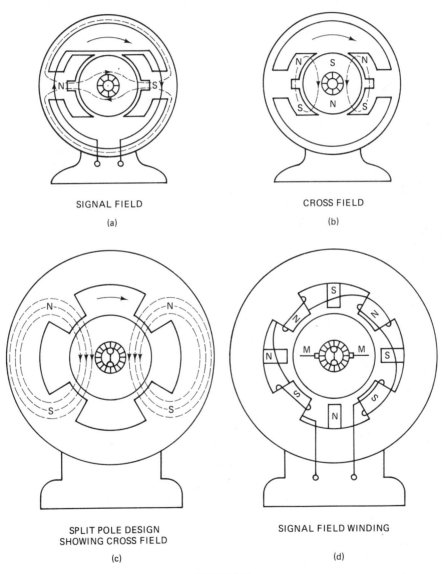

SIGNAL FIELD

(a)

CROSS FIELD

(b)

SPLIT POLE DESIGN
SHOWING CROSS FIELD

(c)

SIGNAL FIELD WINDING

(d)

FIGURE 31

in turn produce an intense armature cross field with the polarities shown. If the poles are especially designed to provide a magnetic circuit of low reluctance to this cross field in Fig. 31b, a strong magnetic field will be developed in the air gap. The armature, rotating in this field, produces a relatively high voltage at right angles to the normal brush axis, and if brushes are placed as shown, almost equivalent to the normal rating of the machine may be obtained.

As the operating point for the field magnetism is set on the steep part of the magnetization curve, a small variation in the magnetizing force produced by the field coils will produce a relatively great change in the short-circuit current produced by the armature, and this in turn will greatly increase the generated output voltage. Therefore, if special control coils are placed on the poles and if these coils are fed from a low-voltage or low-power source, the variations that these coils produce may be caused to reappear in the output circuit in a greatly amplified form. This is the principle of operation of the amplidyne generator.

The amplidyne generator's use is concerned with control situations in which small controlling impulses are employed to handle equipment that demands a large amount of power to operate it. The small control power is fed to the field coils where it effects a relatively high variation in field magnetism; this variation is amplified in the cross field and again in the output circuit. Amplifications of 20,000 to 1 are common and 1,000,000 to 1 are possible. Thus, a variation of 1 W in the input control circuit may produce a change in generator output of 20 kW, a range impractical for any electronic amplifier. The range may be extended by the use of a preamplifier using ordinary radio tubes.

The arrangement indicated in Fig. 31c shows the constructional features of a modern amplidyne unit. Although four poles are shown, adjacent groups are wound with the same polarity, and the machine is therefore a two-pole unit. Figure 31d shows the construction of an amplidyne unit using interpoles. Although several field windings are employed in an actual machine, only the signal winding is shown. The brushes, M, are the output brushes from which the amplified energy is obtained.

Amplification factor: The ratio of the change in plate voltage to the change in grid voltage in order to effect equal changes in plate current of a vacuum tube. The amplification factor of a typical electron tube could range from 2.5 to more than 100, depending upon the number of elements and type of construction. The amplification factor is usually designed by the Greek letter, μ (mu). Amplification is often referred to as the mu, or mu factor. A high-μ electron tube is designated as one with an amplification factor of 30 or more. Medium-μ tubes have amplification factors in the 8 to 30 range, with low-μ designs falling below a factor of 8. For solid-state devices, such as bipolar transistors, amplification may be expressed as alpha or beta.

Amplifier, paraphase: Also called a two-stage inverter, a circuit that outputs two signals of equal amplitude 180° out of phase with each other. The

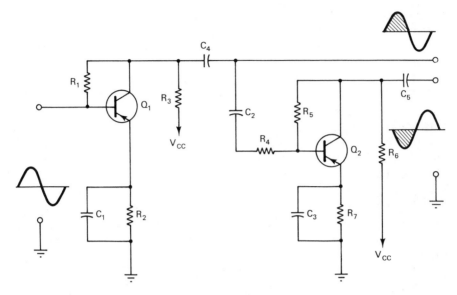

FIGURE 32

transistor version of this circuit is shown in Fig. 32. The circuit consists of two identical CE configurations.

R_1 develops forward bias for Q_1; C_1 and R_2 aid in the selection of the operating point. Q_1 is the first-stage amplifying device, and R_3 is its collector load resistor. C_2 is an interstage coupling capacitor, and R_4 is an attenuator resistor that reduces signal amplitude to Q_2. R_5 establishes forward bias for Q_2. C_3-R_7 is the self bias network for Q_2. C_4 and C_5 are the coupling capacitors for the two output signals.

Assume that each transistor configuration produces a voltage gain of 20 and that R_4 produces a 20 to 1 voltage reduction. Assume that a negative half-cycle is applied to the input at a 0.1-V amplitude. Because of the 180° phase reversal in the CE configuration, the signal at the collector of Q_1 will be positive going. The signal will have been amplified and will appear in the collector circuit as a 2-V signal. This 2-V positive going signal is coupled through C_4 as one of the required outputs. It is also coupled through C_2 to R_4 and Q_2. The action of R_4 will be such that it will drop 1.9 V and allow only a positive going 0.1-V signal to be felt on the base of Q_2. Q_2 will amplify the signal and a negative going 2-V signal will appear on its collector. This signal is coupled out through C_5 as the second of the two required signals.

Since two identical CE configurations are used, the source impedances are equal for two input circuits of the push-pull output stage. In addition, the amount of power than can be delivered by the two-stage paraphase amplifier is much greater than that of the split-load phase inverter. Also, rather than having a voltage gain of less than unity, the paraphase amplifier is capable of producing voltage amplification at the same time that the phase-splitting action is taking place. In the case where the gains of Q_1 and Q_2 are not quite equal, R_4 may be made adjustable so that it can provide additional or less attenuation as required.

Figure 33 shows the electron tube version of the two-tube paraphase amplifier. Again, circuit purposes and operational characteristics are essentially the same as those of the transistor version. C_1-R_1 is the input coupling network for V_1, R_2-C_2 is the self bias network for V_1, whereas R_3-C_3 serves the same function for V_2. R_4 and R_5 are the plate load resistors for the two stages. C_4 and C_5 are interstage coupling capacitors. R_6-R_7 forms a voltage divider network that develops the correct input voltage for V_2. R_3 is a simple output voltage developing resistor.

For a numerical example of the operation of this circuit, assume that V_1 and V_2 each have a gain of 20, R_8 has a resistance of 40,000 Ω and the resistance of the combination of R_6 plus R_7 is also 40,000 Ω. The output voltage of V_1 would be 20 V if there was an input voltage of 1 V. In order to produce equal output voltages, the input voltages to the two grids must be identical. The input to V_1 is assumed to be 1 V, so it is therefore necessary to have a 1-V input for V_2. As there is a 20-V output of V_1, R_6-R_7 must form a voltage divider network that will provide 1 V to the grid of V_2 with a 20-V signal applied to the voltage divider. Therefore, R_7 must be $\frac{1}{20}$ of the total resistance of R_6 plus R_7. As the total resistance is 40,000 Ω, the value of R_7 must be 2,000 Ω. Consequently, R_6 must be 38,000 Ω. It is possible to make either R_6 and R_7 variable (usually by replacing R_7 with a potentiometer) in order to adjust for differences in tube gains. Triodes are normally preferred to pentodes in this circuit. A pentode has

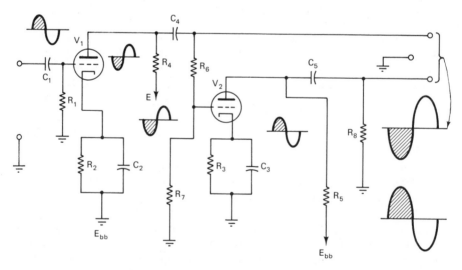

FIGURE 33   Two-tube paraphase amplifier.

a higher gain than a triode, but a higher gain would make the value of R_7 too critical to be practical.

The output voltage of V_2 has more amplitude distortion than the output of V_1. This is due to V_1 and V_2 being connected in cascade. Any distortion in the output of V_1 is fed to V_2 and amplified. In addition, V_2 introduces additional distortion of its own.

Amplifier noise: Signals that can be measured in the output of an amplifier when no input signal has been applied. This may also be known as ambient circuit noise. In an electronic circuit, noise is thought of as those signals that are not induced into the input of the circuit but that are incurred from the power supply and other extraneous sources. Commercial radio stations are required to perform noise measurements on their equipment yearly as part of the proof-of-performance test. Here, amplifier noise is an overall product of audio and RF sections in the control board, the transmitter, and all interfacing circuits.

Amplifier noise is measured in decibels and more often in negative decibels below a preset point that is established by inducing an input tone, establishing a specific level, and then removing the tone. Any noise that is indicated is logged as being so many dB below the set point. For example, 52 dB below unity would be expressed as -52 dB.

Amplitude: The extent to which an alternating or pulsating current or voltage swings from zero or from another set of values. Figure 34 shows a graph of a varying sine wave indicating various amplitude points in relation to others on the same chart.

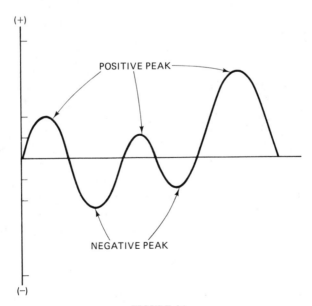

FIGURE 34

Amplitude separator: A term used to describe a circuit that is common to all television receivers. Along with the video signal, commercial television transmissions contain control pulses that are demodulated and acted upon by the receiver circuitry. These pulses, like the video signal, are amplitude modulators and travel on the same carrier wave. The amplitude separator circuit divides the incoming video signal, separating the control pulse information.

Analog computer: A computational system in which data to be measured, computed, transmitted, or controlled is represented in continually variable quantities. Physical conditions such as voltage, resistance, flow, temperature, pressure, and position are translated into the proper mechanical or electrical measurements. Several examples of analog systems are speedometers, clocks, volume controls on audio equipment, electric mixers, detectors, and slide rules. Analog devices represent quantities only approximately and are subject to degrees of variance. This is in contrast to digital computers, which deal with precise units and are generally more accurate and versatile.

Analog technique: The measuring or recording of electrical or mechanical events in an analog fashion.

Analog-to-digital converter: Any signal or device that performs analog-to-digital conversion.

Analysis: The examination of an activity, procedure, method, technique, or a business to determine what must be accomplished and how the necessary operations may best be accomplished.

Analyzer: A computer routine whose purpose is to analyze a program written for the same or for a different computer. This analysis may consist of summarizing instruction references to storage and tracing sequences of jumps.

AND gate: A signal circuit with two or more input wires in which the output wire gives a signal, if and only if all input wires receive coincident signals.

Anderson bridge: An alternating current bridge circuit that has six impedances. Shown in Fig. 35, this circuit permits the value of an unknown inductance to be determined in terms of a standard capacitance. In short, the Anderson bridge compares the unknown inductance with a known capacitance. Nonreactive resistors are used in order to make the circuit independent of input frequency.

Angle of beam: The angle that encloses the major lobe of the transmitted energy from a directional antenna. The angle originates at the center of the antenna and extends outward, enclosing the radio wave. Fig. 36 shows the radiation pattern of a highly directional antenna and its angle of beam.

Angle of conduction: A condition measurement used to describe current flow in an amplifier circuit. Also called angle of flow, it describes the number of degrees of an excitation-signal cycle applied to an amplifier during

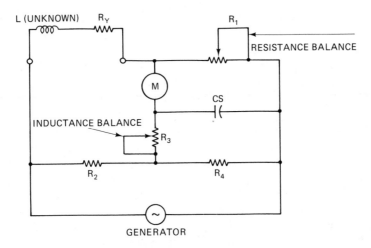

FIGURE 35 Anderson bridge.

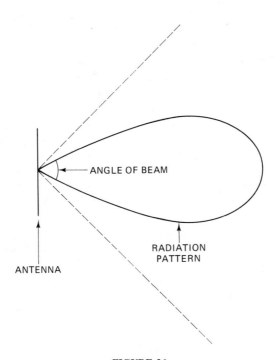

FIGURE 36

which output current flows. Angle of conduction is also the number of degrees of any sine wave at which conduction of a device begins.

Angle of convergence: A graphical representation, the angle formed by any two lines or plots that come together at a point. This can also be applied to oscilloscope representation. In light-optical systems, the angle formed by the light paths of two photocells focused upon the same object is also referred to as angle of convergence.

Angle of declination: The angle formed between a horizontal line and a descending line. It is measured in degrees and can be closely compared with an angle of elevation made with a horizontal line that is intersected by an ascending line.

Angle of deflection: The angle between the electron beam in a cathode-ray tube when it is at rest and when in a new position resulting from deflection. See Fig. 37.

Angle of departure: In transmitting antenna applications, the angle made by the line of propagation of a transmitted radio wave from the horizontal surface of the earth measured at the transmitter site. Figure 38 shows a representation of angle of departure. This measurement partially determines whether the antenna transmits at a low or high angle of radiation. See *angle of radiation*.

Angle of divergence: In cathode-ray tubes, the angle formed by the spreading of an undeflected electron beam as it extends from the gun to the phosphor dots at the screen. The angle of divergence can be used to describe the fine focus properties of the particular cathode-ray tube circuit. For example, circuits with electrostatic deflection of the beam will often have larger angles of divergence than those circuits incorporating magnetic deflection.

Angle of incidence: The angle formed by an arc between a radio-frequency wave striking a reflecting surface and a line that is drawn perpendicular to that surface. See Fig. 39. In the illustrated example, the electromagnetic energy strikes the reflecting surface at an angle of 45° to the surface. This is also 45° from a line drawn perpendicular to the surface. In this case, the angle of arrival and the angle of incidence are the same. However, if the angle of arrival were larger, the angle of incidence would be correspondingly smaller.

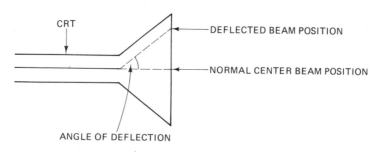

FIGURE 37

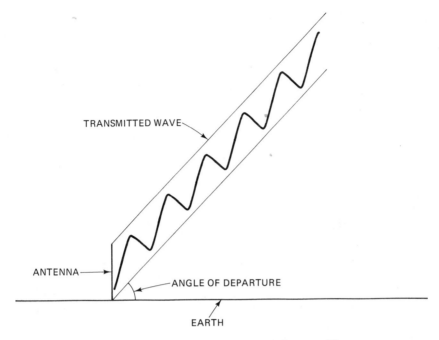

TRANSMITTED WAVE

ANTENNA

ANGLE OF DEPARTURE

EARTH

FIGURE 38   Two-stage transistor paraphase amplifier.

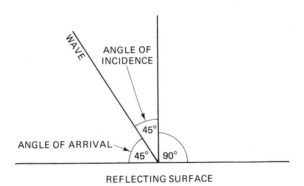

WAVE

ANGLE OF
INCIDENCE

45°

ANGLE OF ARRIVAL

45°   90°

REFLECTING SURFACE

FIGURE 39   Two-stage electron tube paraphase amplifier.

Angle of lag: The angle of lag, indicated by the Greek letter φ (phi), is the angle whose tangent is equal to the quotient of the inductance expressed in ohms or *spurious resistance* divided by the ohmic resistance; that is

$$\text{Tan } \phi = \frac{\text{Reactance}}{\text{Resistance}} = \frac{2\,fL}{R}$$

Angle of lead: The angle of phase differences where one component precedes another in time measured in degrees or radians. This assumes that both components are at the same operating frequency. This is closely aligned with phase angle.

Angle of radiation: In antenna terminology, the angle at which the principal lobe of a wave leaves a transmitting antenna. This angle is measured with respect to the horizontal surface of the earth. Angle of radiation may also be thought of as the angle of a transmitting or receiving antenna's optimum sensitivity. Angle of radiation is a determinant in how signals transmitted from an antenna will behave during the total length of the transmitting path. Vertical antennas exhibit high angles of radiation, whereas horizontal designs usually offer low angles. High angles of radiation are usually associated with long hop transmissions of several thousand miles or more at high frequencies. Low-angle radiators generally provide shorter single-hop distances at the same frequencies.

Angle of refraction: The angle made by a ray or the line of propagation of a wave and perpendicular to the surface of a medium of propagation as the ray passes out of the medium.

Angle structure towers: Structures often used for high-voltage power transmission and designed to provide clearance from transmission conductors to the structure and to the guys under all conditions, and at the same time, attaching guys as close to the load as possible to keep bending stresses down to a conservative value.

Steel tower construction is now used to a great extent for high-voltage transmission lines over long distances and over mountainous terrain. The primary purpose of steel tower construction is for the support of conductors and not the mounting of equipment on the towers, since the transformers and switching equipment are located at substations or the switchyards of generating plants.

Angular acceleration: The rate of change of angular velocity. It is usually expressed as the number in radians per second of a rotating vector or radius. In other words, angular acceleration is the second derivative of angular velocity.

Angular displacement: The separation, in degrees, between two waves in an alternating current circuit. The lag or lead between the instant that one alternating quantity reaches its maximum value and the instant when another alternating quantity reaches its maximum value is known as the phase angle. The instants and their difference (lag or lead) are measured in degrees or radians along the horizontal axis of the time-versus-magnitude graph of the ac quantity. The separation between the two waves is the angular displacement.

Angular frequency: Frequency expressed in radians per second equivalent to frequency in hertz multiplied by 2π.

Angular velocity: The rate at which an angle changes expressed in radians per second. The angular velocity of a periodic quantity is the product of the frequency multiplied by 2π.

Annealing: The gradual heating and gradual cooling of glass, metals, or other materials to reduce brittleness and increase flexibility. Copper wire, for example, is commercially supplied hard, partially annealed, and fully annealed. A stress-relief anneal is usually required to remove the stress introduced by fabrication and to restore the magnetic properties of fully processed electrical steel. When semiprocessed electrical steel is to be used, it must first be annealed by either continuous annealing, box annealing, or open-coil annealing.

Anode: The positive electrode of a discrete device or circuit. This is the electrode toward which electrons move during current flow. The anode of an electron tube, then, is the plate. The anode connection of a dry cell battery is the positive terminal. The opposite of an anode is a cathode, which generally is the negative electrode of a device or circuit.

Antenna: A conductor or system of conductors that serves to radiate or intercept energy in the form of electromagnetic waves. In its elementary form, an antenna, or aerial, may be simply a length of elevated wire like the common receiving antenna for an ordinary broadcast receiver. However, for communication and radar work, other factors make the design of an antenna system a more complex problem. For instance, the height of the radiator above ground, the conductivity of the earth below it, and the shape and dimensions of an antenna all affect the radiated-field pattern in space. Also, the antenna radiation often must be directed between certain angles in either the horizontal or the vertical plane, or both.

An antenna may be constructed to resemble a resonant two-wire line with the wires so arranged that the fields produced by the currents in the wires add in some directions instead of canceling completely. One way to prevent cancellation of the fields is by making the earth one conductor. This permits considerable separation of the conductors. In this manner, the fields resulting from the current expand considerably farther into space than if the other conductor were nearby, and therefore can be detached from the radiating conductor by rapid reversals more easily.

Another way to accomplish the radiation is to spread the ends of the two wire lines so these ends are 180° apart. The currents that cancel the fields of each other then aid in producing a field in space. However, if the antenna extends vertically above the earth, it is possible to elevate the effective radiation field a greater distance in all directions by operating the antenna at some harmonic, such as the third, fifth, or seventh harmonic of the fundamental frequency. The result is that the intensity of the radiated field at various points in space is considerably changed when compared with the field of the simple dipole.

Nonresonant lines also can be expanded to antennas, but they are not efficient radiators (except in a given direction between fixed stations). Resonant conductors are more efficient omnidirectional radiators because they have large standing waves of voltage and current, and hence, they produce intense fields with a minimum of generator current and voltage.

An antenna that is cut to an electrical half-wavelength also radiates other frequencies, but its effectiveness as a radiator diminishes as the standing waves of current and voltage decrease.

If an antenna is made of very small wire and is isolated perfectly in space, its electrical length corresponds closely to its physical length. Thus, in free space, a one-wavelength antenna for 10 m would be 10 m in length, and a half-wavelength antenna for the same signal would be 5 m in length. In actual practice, however, the antenna is never isolated completely from surrounding objects. For example, the antenna will be supported by insulators with a dielectric constant greater than 1. Therefore, the velocity of the wave, along with the conductor, is always slightly less than the velocity in space, and the physical length of the antenna is correspondingly less (by about 5%) than the corresponding wavelength in space. The physical length, L, in feet, of a half-wavelength antenna for a given frequency is

$$L = \frac{300 \times 3.28 \times 0.95}{2f} = \frac{468}{f}$$

where f is the frequency in megacycles, 3.28 ft equals 1 m, and 0.95 represents the velocity of the wave in the antenna compared to that in free space. This formula does not apply to antennas longer than one half-wavelength.

Antenna ground system: An antenna system consisting of a low-resistance connection to the earth's surface as well as the antenna itself. This contrasts with a straight antenna system, which has no connection to the earth's surface.

Antenna pattern: A graphical polar plot of antenna performance that indicates field strength as opposed to angle of azimuth. Figure 40 shows the antenna pattern of a perfect vertical antenna. Note that field strength is equal at all points around the antenna, which lies at the center. Figure 41 shows the plot of a directional, horizontally polarized beam antenna. The major lobe of radiation is toward the north, with several minor lobes occurring at the south and on either side.

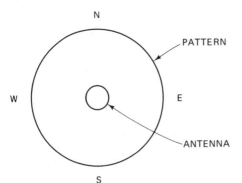

FIGURE 40

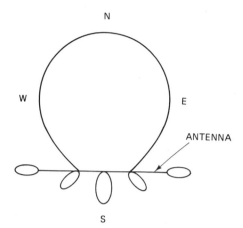

FIGURE 41

Antenna patterns are plotted by using field strength metering. Measurements are taken from all compass points around the antenna and in relatively close proximity to it. The field strength readings are then used to plot points on a graph. By interconnecting these points, the pattern is established.

Antenna tuning: The adjustment of the length of an element in a single-element antenna system or adjustment of the length and/or spacing in a multielement system in order to arrive at resonance at a particular frequency or range of frequencies. This often involves the physical pruning of a single antenna element in order for its physical size to match the formula required for resonance. Lumped capacitance may be inserted between the transmission lines and the antenna element to facilitate matching to the transmission line after resonance has been achieved. This is often called antenna tuning, but could more accurately be described as transmission line matching.

Special antenna construction techniques may be used in designing systems that will exhibit resonance over a wide range of frequencies. This usually involves multielement designs and sophisticated LC networks placed between the system and the transmission line.

An antenna impedance bridge may be used to facilitiate tuning, although the most popular method involves the use of an SWR bridge between the transmitter and transmission line input to the system. This is shown in Fig. 42. This latter method has led to many false impressions about antenna tuning, as the SWR meter simply matches the impedance of the system to that of the transmitter. The SWR meter is designed to operate into a specific impedance, which may not be the same impedance of the antenna feed point at element resonance.

A more accurate method of antenna tuning is to use a field strength meter at a point a short distance away from the antenna element or elements. The antenna is adjusted for maximum field strength. This means that the output from the element is at its maximum. An SWR meter may indicate ratios of 1.5

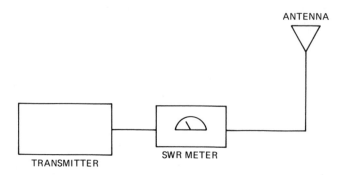

FIGURE 42   SWR Bridge connection.

to 1 or even 2 to 1 when the antenna is properly tuned. Most persons attempt to achieve a perfect 1 to 1 ratio, but to do so, the antenna element is actually being detuned in order to arrive at an impedance that perfectly matches the transmission line and SWR meter.

Apparent power: The power, in watts, obtained by using Ohm's law ($P = EI$) or multiplying together the simultaneous voltmeter and ammeter readings. True power is the wattage actually measured with a wattmeter. The power factor of a circuit is the amount of wattage indicated by a wattmeter, divided by the apparent watts, and may be expressed as being equal to

$$\frac{\text{True power}}{\text{Apparent power}} = \frac{\text{True watts}}{\text{Apparent watts}} = \frac{\text{True watts}}{\text{Volts} \times \text{amperes}}$$

Application: The system or problem to which a computer is applied. Reference is often made to an application as being either of the computational type, wherein arithmetic computations predominate, or of the data processing type, wherein data handling operations predominate.

Aquadag: The conductive graphite coating inside the bell of a cathode-ray tube. This coating provides shielding from stray fields that might interfere with the electron beam, prevents light from striking the back of the screen, and most importantly gathers the secondary electrons emitted when the phosphor is bombarded by the electron stream, returning them to the cathode through the accelerating anode power supply.

Arc chutes: Barriers used in electrical switches to confine, cool, and deionize the arc. Switches used in the earliest low-voltage electric circuits were generally of the hand-operated, knife-blade type, but as currents and voltages increased, the arc burn that occurred while opening the switch damaged or destroyed the contacts. Switches were then developed that opened and closed rapidly (by spring or gravity action), thus reducing the time or arcing and the amount of burning. Arc chutes were added later.

Arc-controlled devices: Devices used for oil circuit breakers. All make use of oil pressure generated by gas created by the arc to force fresh oil through the arc path in such quantity as to provide the necessary insulation at current zero to prevent a restrike of the arc and thereby interrupt the circuit. The gases that are generated, after being effectively cooled, pass out through the tank vent pipe to the open air.

Arc converter: A radio-frequency oscillator consisting of a dc electric arc and a tuned LC circuit. The arc oscillates by virtue of its negative resistance. Prior to the existence of high-powered tube-type transmitters, the arc converter was the most widely used device in early long-distance, high-powered radiotelegraph and radiotelephone transmitters, often delivering up to 100 kW at frequencies of 50 to 1000 kHz.

Arcing horn: Small part attached to clamps to prevent damage to insulators in alternating current power transmission systems. Several devices have been developed in attempting to ensure that an arc will hold free of the insulator spring. One of the first protective measures consisted of attaching small horns to the clamp. The horns, however, had to be very large to be effective. Especially under lighting impulse, the arc tends to cascade the string, and tests have shown that the gap between horns should be considerably less than the length of the insulator string. Protection by arcing horns thus results in either a reduced flashover voltage or an increase in the number of units and length of the string.

Arc-lamp carbons: Elements burned to produce light in direct current and alternating current lamps. The carbons of a direct current arc burn to shape as shown in Fig. 43a—the positive burning away twice as fast as the negative. Most of the light comes from the "crater" in the positive carbon. This gives a natural light distribution.

In an alternating current arc (see Fig. 43b), each of the carbons is alternately positive and negative, and both carbons have practically the same

ARC LAMP CARBONS

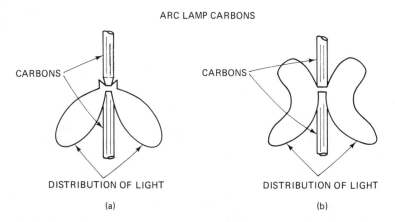

FIGURE 43   (a) Direct-current open carbon arc, showing shape of carbons and natural distribution of light carbons; (b) Alternating-current open carbon arc, showing shape of carbons and natural distribution of light.

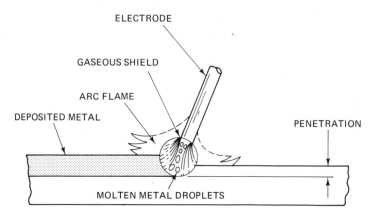

FIGURE 44   Arc welding.

temperature and are therefore equally luminous. In addition, the natural distribution is the same above and below the arc.

Arc transmission: See *Arc converter*.

Arc welding: Usually today, welding that uses a consumable metallic rod. This rod is used up as it fuses with the parts being welded, becoming a filler material. The arc is a spark formed between the work and the electrode. The metal work must be attached to a ground clamp from one pole of the electric current, whereas the other pole of the circuit is connected to the electrode. When the electrode comes into contact with the work, which is attached to the electrode, an arc forms at the tip of the electrode, where the temperature is about 7000°F. See Fig. 44.

Area: The amount or extent of surface such as described in the following table:

	Square meter	Square inch	Square foot
1 m²	1	1.550.0	10.764
1 in² =	6.3516×10^{-4}	1	6.9444×10^{-3}
1 ft² =	0.092×903	144	1
1 circular mil = 5.067	1×10^{-4} mm²		

Arithmetic unit: The portion of the hardware of a computer in which arithmetic and logical operations are performed. The arithmetic unit generally consists of an accumulator, some special registers for the storage of operands, and results supplemented by shifting and sequencing circuitry for implementing multiplication, division, and other desired operations. Synonymous with *ALU*.

Armature: The rotating member of a motor and also in some types of generators. The armature of a motor is made of iron and is always of laminated construction, or built up of thin iron sheets pressed tightly together. The laminated construction is used to prevent the flow of induced eddy currents in the armature core. The core has a number of slots around its entire outer surface in which the armature coils are placed. The iron armature core provides a magnetic path for the flux of the field poles, and also carries the coils that are rotated at high speed through the field flux.

In a generator, it is the cutting of these coils through the flux that produces the voltage. In a motor, it is the reaction between the field flux and the flux around the armature conductors that causes the torque or turning effort.

Small armatures are often constructed of laminations in the form of complete disks that merely have a hole through their center for the shaft, and possibly bolt holes for clamping them. This makes a core which is solid clear to the shaft. In the larger machines, it is not necessary to have the entire core solid, so the laminations are assembled like the rim of a wheel on the outside of short spokes, as shown in Fig. 45. This wheel or center framework is called *the spider,* and the sections of core laminations are dovetailed into the spider, as shown in the illustration. Heavy clamping rings at each end of the group, drawn tight by bolts, hold the entire core in a solid, rigid unit.

The illustration shows a sectional view through such a spider and core. Note the spaces or air duct that are left between the laminations for ventilation and cooling of the core and windings.

Array: Any of a number of elements such as antennas, diodes, memory cells, resistors, etc., which are combined in such a manner as to produce a single functioning module. For example, an antenna array, shown in Fig. 46, consists of a number of properly dimensioned and spaced elements. Acting as a single unit, they produce, transmit, and receive properties that are generally more efficient and directional than a one-element design could achieve.

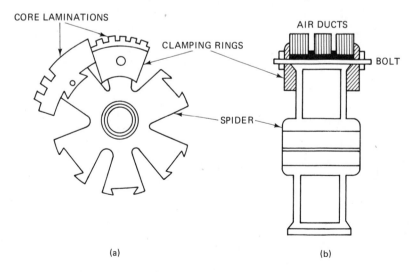

FIGURE 45 Armatures.

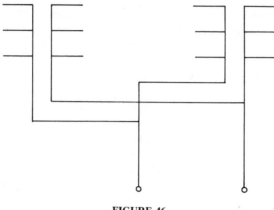

FIGURE 46

Diode arrays, such as the one shown in Fig. 47, are often used in sophisticated computer memory systems.

Arresters, lightning: Essentially, a discharge gap with one side connected to a line at the primary side of a transformer and the other side connected to the ground. If lightning strikes a power line near a transformer, the easiest path may be through the transformer windings or over its case to the grounded secondary neutral. If it strikes the line at some distance from a transformer, or even if it strikes near the line, a high-voltage surge, with less energy, may run along the line and this may also be sufficient to damage the

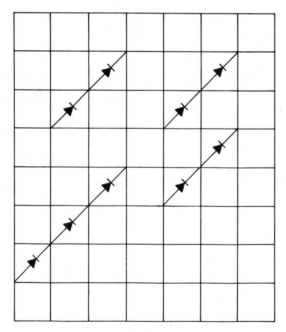

FIGURE 47

transformer. A lightning arrester is set so that it will not break down under normal voltage on the line, but will break down and discharge when a lightning surge is impressed on it. This reduces the surge voltage so that the transformer is better able to withstand it.

In order that the normal voltage on the line will not continue the discharge to ground after the lightning surge has passed, a device with a low impedance at high current and a high impedance at low current is placed in series with the gap in the discharge path to ground. Such a connection seals off the path as the normal voltage passes through the zero value in its cycle, and it thereby stops the flow of current from the system to the ground.

The ideal lightning arrester will hold the voltage practically constant after the gap breaks down, regardless of the amount of current discharged, and at a value below the voltage for which the strength of the transformer insulation has been designed. Commercial arresters make use of various materials and constructions for this purpose. They approach the ideal, even though the voltage does actually rise with the current to some extent. See Fig. 48.

Artificial daylight: An approximation of sunlight produced by artificial means. Daylight varies in color, depending upon the time of day and the atmospheric conditions. Color temperature of direct sunlight varies from

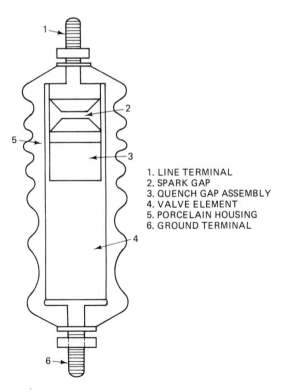

1. LINE TERMINAL
2. SPARK GAP
3. QUENCH GAP ASSEMBLY
4. VALVE ELEMENT
5. PORCELAIN HOUSING
6. GROUND TERMINAL

FIGURE 48

about 1900°K with the sun near the horizon to about 5300°K with the sun near the zenith. Clear blue sky ranges from about 11,000 to 50,000°K. Overcast sky ranges from 6000 to 7000°K. The color temperature of a perfectly white horizontal surface illuminated by overcast sky or sun and clear blue sky is of the order of 6500°K.

Incandescent light, in general, is more yellow and less blue than daylight. The light from tungsten-filament vacuum lamp and gas-filled tungsten lamps is 2350 to 3100°K. To correct the light from tungsten-filament lamps to color temperature of, say, 6500°K involves the use of absorption filters that transmit only about one-tenth of the light. Where accurate equivalence of daylight is desired, this subtractive method of correcting tungsten-filament incandescent light is well worthwhile in spite of its physical inefficiency.

Fluorescent lamps that most nearly produce daylight quality illumination are an acceptable substitute where critical matching of colors is not involved. The presence of the mercury lines in the light of fluorescent lamps operates to make them a less exact substitute for daylight than is desirable for critical work.

Artificial grounding: The method of providing an operating ground using the radials or disk of a ground-plane antenna, as opposed to actual ground, which is the earth itself.

Asbestos: A fibrous crystalline material whose composition formula is $3MgO, 2SiO_2, 2H_2O$. The composition, however, varies from about 37 to 44% silicon oxide, 39 to 44% magnesium oxide, 12 to 15% H_2O, and up to 6% iron oxide. It is a fibrous mineral having a specific gravity of 2.2 to 2.6, a hardness of 2.5 to 4.0 on a Hohs scale. Individual fibers are very strong, with tensile strengths up to 400,000 lb in^2. The mineral loses its water of crystallization at approximately 400°C and becomes weaker.

Asbestos is used as an insulation, but it is not a good insulator unless it is totally dry. It absorbs moisture too readily for many electrical applications. It is dried and impregnated with varnishes or inorganic binders such as phosphate solutions. Refining processes have been developed to remove water-soluble electrolytes and iron oxide. The fibers are then dispersed and formed into thin sheets, as in papermaking. These products are known as *Quinterra* and *Novabestos*. The materials are then generally used in a treated manner, either by impregnation or in laminates.

Untreated asbestos is often used for low-voltage insulation or for barriers to provide separation where heat-resisting or arc-resisting properties are required. Asbestos-insulated cords are frequently used for heating appliances such as irons.

Askarel: A generic term for a group of nonflammable synthetic chlorinated hydrocarbons used as electrical insulating media. Askarels of various compositional types are used. Under arcing conditions, the gases produced, although they consist predominantly of noncombustible hydrogen chloride, can include varying amounts of combustible gases, depending upon the askarel type.

Asperities: The tiny points on the surface of an electrode where the electric field is intensified. These points are also the area from which discharges are likely to occur.

Astable multivibrator: A device that employs two vacuum tubes or transistors whose inputs and outputs have been cross-coupled to cause conduction to alternately switch between the two. The term *astable* refers to *free-running*.

Asymmetrical conductivity: The action of a device whereby conduction occurs in only one direction. A common example of this type of operation can be found in a diode.

Asynchronous: A term applied to induction motors because the armature does not turn in synchronism with the rotating field, or, in the case of a single-phase induction motor, with the reciprocating field (considering the latter in the place of a rotating field).

Atom: The smallest particle or an element capable of existing alone. In the study of chemistry, it soon becomes apparent that the molecule is far from being the ultimate particle into which matter may be subdivided. The salt molecule may be decomposed into radically different substances—sodium and chlorine. The particles that make up molecules can be isolated and studied separately. More than 100 elements have been identified. They can be arranged into a table of increasing weight, and can be grouped into families of materials having similar properties. This arrangement is called the Periodic Table of the Elements.

The idea that all matter is composed of atoms dates back more than 2000 years to the Greeks. Many centuries passed before the study of matter proved that the basic idea of atomic structure was correct. Physicists have explored the interior of the atom and discovered many subdivisions in it. The core of the atom is called the nucleus. Most of the mass of the atom is concentrated in the nucleus. It is comparable to the sun in the solar system, around which the planets revolve. The nucleus contains protons, which are positively charged particles, and neutrons, which are electrically neutral.

Most of the weight of the atom is in the protons and neutrons of the nucleus. Whirling around the nucleus are one or more smaller particles of negative electric charge. These are the electrons. Normally, there is one proton for each electron in the entire atom so that the net positive charge of the nucleus is balanced by the net negative charge of the electrons whirling around the nucleus. Thus, the atom is electrically neutral.

The electrons do not fall into the nucleus even though they are attracted strongly to it. Their motion prevents it, as the planets are prevented from falling into the sun because of their centrifugal force of revolution.

The number of protons, which is usually the same as the number of electrons, determines the kind of element in question. Figure 49 shows a simplified picture of several atoms of different materials based on the conception of planetary electrons describing orbits about the nucleus. For example,

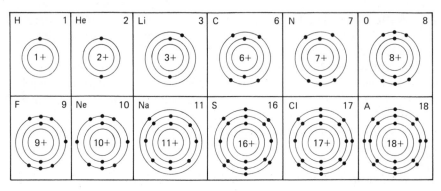

FIGURE 49

hydrogen has a nucleus consisting of 1 proton, around which rotates 1 electron. The helium atom has a nucleus containing 2 protons and 2 neutrons with 2 electrons encircling the nucleus. Near the other extreme of the list of elements is curium (not shown in the illustration), an element discovered in the 1940s, which has 96 protons and 96 electrons in each atom.

The periodic table of the elements is an orderly arrangement of the elements in ascending atomic number (number of protons and neutrons in the nucleus). The various kinds of atoms have distinct masses or weights with respect to each other. The element most closely approaching unity (meaning 1) is hydrogen, whose atomic weight is 1.008, as compared with oxygen, whose atomic weight is 16. Helium has an atomic weight of approximately 4; lithium, 7; fluorine, 19; and neon, 20; as shown in Fig. 50 (see p. 66).

In Fig. 51 (see p. 68), visible matter, at the left, is broken down first to one of its basic molecules, then to one of the molecule's atoms. The atom is then further reduced to its subatomic particles—the protons, neutrons, and electrons. Subatomic particles are electric in nature; that is, they are the particles of matter most affected by an electric force. Whereas the whole molecule or a whole atom is electrically neutral, most subatomic particles are not neutral (with the exception of the neutron). Protons are inherently positive and electrons are inherently negative. It is these inherent characteristics that make subatomic particles sensitive to electric force.

When an electric force is applied to a conducting medium, such as copper wire, electrons in the outer orbits of the copper atoms are forced out of orbit and impelled along the wire. The direction of electron movement is determined by the direction of the impelling force. The protons do not move, mainly because they are extremely heavy. The proton of the lightest element, hydrogen, is approximately 1850 times as heavy as an electron. Thus, it is the relatively light electron that is most readily moved by electricity.

When an orbital electron is removed from an atom, it is called a free electron. Some of the electrons of certain metallic atoms are so loosely bound to the nucleus that they are comparatively free to move from atom to atom.

Thus, a relatively small force or amount of energy will cause such electrons to be removed from the atom and become free electrons. It is these free electrons that constitute the flow of an electric current in electrical conductors.

Attack: In electricity and electronics, the length of time required for a signal to reach maximum value from an off state. This term may also be used to describe the time required for an automatic gain control circuit to suppress a sudden gain surge. Another term that is closely aligned is *attack time,* which is the interval required for a signal that suddenly increases in amplitude to attain a value that is equivalent to 63 percent of its maximum or ultimate peak value.

Attenuation: The reduction of signal amplitude. Signal attenuation is desirable in some applications and undesirable in others. Attenuation is created by an obstruction in the path of a signal flow. Mismatched impedances between circuits can create this condition. Attenuation is measured in decibels (dB) and is a comparison of the signal input level to a system with the signal output. In some instances, it is desirable to attenuate signals to make them more usable at the inputs of various types of circuits. Sophisticated multielement networks are used to accomplish this task without introducing distortion to the signal content.

Unwanted harmonics at a radio transmitter output can create interference problems, so most output networks in the final stages of signal generation are designed to pass a certain band of frequencies while suppressing or attenuating those that lie outside of this range. A low-pass filter may be added to the transmitter output and between it and the transmission line to provide further attenuation. All signals below a certain frequency are passed without attenuation, while all signals above the present frequency are greatly attenuated. The purpose of any filter is to pass a desired range of frequencies while attenuating those which lie outside of its passband.

Attenuation characteristics: A measurement of the reduction in the amplitude of a signal as compared to the frequency. This is most often expressed in decibels per octave. This type of measurement can be applied to an amplifier, an individual component, or even a network of devices. In performing the same measurement on a transmission line, the decrease of amplitude in the signal is taken per unit length. In this case, the attenuation characteristic is measured either in decibels per mile or decibels per 100 ft, whichever is more applicable.

Attenuation distortion: The occurrence of some variation in attenuation when compared with frequency in a given range, such as in a circuit or system.

Attenuation network: A combination of components such as resistors, capacitors, or inductors that provide a constant signal attenuation with negligible phase shift over a specific frequency band. The selection of component values will determine the degree of attenuation and the output impedances. Figure 52 shows a simple attenuation network called a *pi* because of its physical appearance, which resembles the Greek symbol for pi (π). Depending

Name	Symbol	Atomic number	Atomic weight	Name	Symbol	Atomic number	Atomic weight
Actinium	Ac	89	—	Mercury	Hg	80	200.59
Aluminum	Al	13	26.9815	Molybdenum	Mo	42	95.94
Americium	Am	95	—	Neodymium	Nd	60	144.24
Antimony	Sb	51	121.75	Neon	Ne	10	20.183
Argon	Ar	18	39.948	Neptunium	Np	93	—
Arsenic	As	33	74.9216	Nickel	Ni	28	58.71
Astatine	At	85	—	Niobium	Nb	41	92.906
Barium	Ba	56	137.34	Nitrogen	N	7	14.0067
Berkelium	Bk	97	—	Nobelium	No	102	—
Beryllium	Be	4	9.0122	Osmium	Os	76	190.2
Bismuth	Bi	83	208.980	Oxygen	O	8	15.994†
Boron	B	5	10.811†	Palladium	Pd	46	106.4
Bromine	Br	35	79.904‡	Phosphorous	P	15	30.9738
Cadmium	Cd	48	112.40	Platinum	Pt	78	195.09
Calcium	Ca	20	40.08	Plutonium	Pu	94	—
Californium	Cf	98	—	Polonium	Po	84	—
Carbon	C	6	12.0115†	Potassium	K	19	39.102
Cerium	Ce	58	140.12	Praseodymium	Pr	59	140.907
Cesium	Cs	55	132.905	Promethium	Pm	61	—
Chlorine	Cl	17	35.453$_b$	Protactinium	Pa	91	—
Chromium	Cr	24	51.996$_b$	Radium	Ra	88	—
Cobalt	Co	27	58.9332	Radon	Rn	86	—
Copper	Cu	29	63.546$_b$	Rhenium	Re	75	186.2
Curium	Cm	96	—	Rhodium	Rh	45	102.905
Dysprosium	Dy	66	162.50	Rubidium	Rb	37	85.47
Einsteinium	Es	99	—	Ruthenium	Ru	44	101.07
Erbium	Er	68	167.26	Samarium	Sm	62	150.35

Element	Symbol	Number	Weight	Element	Symbol	Number	Weight
Europium	Eu	63	151.96	Scandium	Sc	21	44.956
Fermium	Fm	100	—	Selenium	Se	34	78.96
Fluorine	F	9	18.9984	Silicon	Si	14	28.086†
Francium	Fr	87	—	Silver	Ag	47	107.868b
Gadolinium	Gd	64	157.25	Sodium	Na	11	22.9898
Gallium	Ga	31	69.72	Strontium	Sr	38	87.62
Germanium	Ge	32	72.59	Sulfur	S	16	32.064†
Gold	Au	79	196.967	Tantalum	Ta	73	180.948
Hafnium	Hf	72	178.49	Technetium	Tc	43	—
Helium	He	2	4.0026	Tellurium	Te	52	127.60
Holmium	Ho	67	164.930	Terbium	Tb	65	158.924
Hydrogen	H	1	1.00797†	Thallium	Tl	81	204.37
Indium	In	49	114.82	Thorium	Th	90	232.038
Iodine	I	53	126.9044	Thulium	Tm	69	168.934
Iridium	Ir	77	192.2	Tin	Sn	50	118.69
Iron	Fe	26	55.84‡	Titanium	Ti	22	47.90
Krypton	Kr	36	83.80	Tungsten	W	74	183.85
Lanthanum	La	57	138.91	Uranium	U	92	238.03
Lead	Pb	82	207.19	Vanadium	V	23	50.942
Lithium	Li	3	6.939	Xenon	Xe	54	131.30
Lutetium	Lu	71	174.97	Ytterbium	Yb	70	173.04
Magnesium	Mg	12	24.312	Yttrium	Y	39	88.905
Manganese	Mn	25	54.9380	Zinc	Zn	30	65.37
Mendelevium	Md	101	—	Zirconium	Zr	40	91.22

FIGURE 50

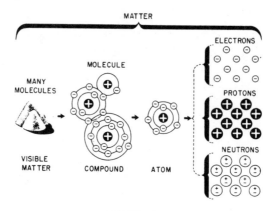

FIGURE 51

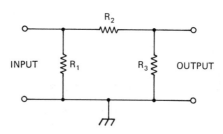

FIGURE 52   Attenuation network.

upon the value of the resistors, total attenuation may range from a very low to a very high level.

Audio image: An undesired interfering signal that is produced at a frequency to one side of the desired signal. This terminology is also applied to CW reception in superheterodyne receivers. It is the product of the audio frequency created by the internal oscillator's output beating with the frequency of the received signal.

Audio mixer: An amplifier circuit with one output and several inputs. It is used for combining two or more audio-frequency signals, such as those delivered by microphones or other low-level input devices. Figure 53 shows a typical audio-mixer circuit.

Audio-signal generator: Devices that produce stable frequency signals from about 20 to 200,000 Hz. They are used primarily for testing audio sections of equipment. The major components of an audio signal generator are an oscillator (or oscillators), one or more amplifiers, an output control, and a power supply. In addition, voltage regulator circuits are necessary to ensure stability of the oscillator, because the alternating current line voltage sources may vary.

A representative audio-frequency signal generator is shown in Fig. 54. This unit is intended primarily for bench testing of electronic equipment. It

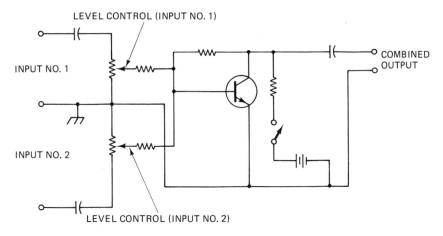

FIGURE 53

operates from line voltage (115-V ac) and produces the output frequencies over a continuous range in conjunction with a four-position multiplier.

Any frequency from 20 to 200,000 Hz may be selected by setting the main tuning dial and the range switch so that the two readings, when multiplied together, equal the desired frequency. For example, to select an output frequency of 52,000 Hz, set the main tuning dial to 52 and the range switch to X1000. Voltages from 0 to 10 V may be selected by using the output level control in conjunction with the attenuator switch. The attenuator is calibrated in seven decade steps so that with the output meter set to 10, output voltages of 10 V to 10 μV can be obtained by simply switching the attenuator. For intermediate value of output voltage, the output level control is varied so that the output meter reads the desired voltage. The attenuator switch is then set so that its value, multiplied by the output meter reading, gives the desired output voltage level. For example, to obtain an output voltage of 0.04 V, set the meter by means of the output level control to read 4 and set the attenuator switch to the 0.01 position. The output voltage will then be the meter reading multiplied by the attenuator setting, or 0.04 V.

Oscillator frequency calibration may be checked, during operation, on the first two bands, by means of the built-in frequency meter. To check the operation of the oscillator at 60 Hz, set the main tuning dial to 60 and the range switch to X1. This sets the frequency of the oscillator at 60 Hz. Turn the frequency meter switch on and move the main dial back and forth slightly until the reed vibrates with maximum amplitude. This point is rather critical, and care must be used to see that the point of maximum response is located. The main tuning dial should indicate 60 within one division, if the frequency calibration is correct. Similarly, the output frequency may be checked at 400 Hz by means of the 400-Hz reed in the frequency meter.

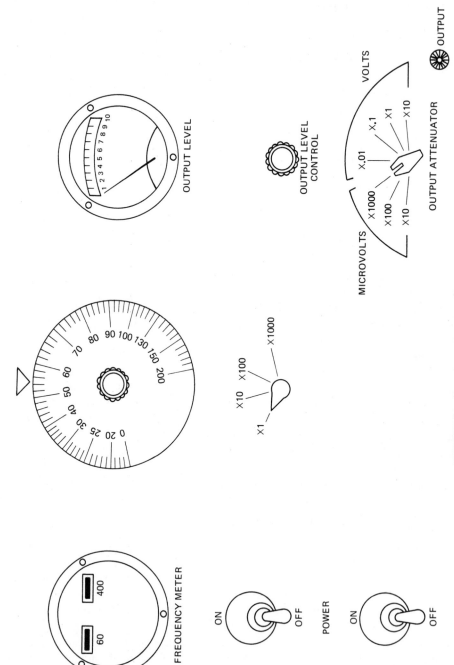

FIGURE 54 Representative audio signal generator.

Audio taper: A semilogarithmic variation in resistance, as opposed to angular rotation in volume and tone controls used in audio circuits. Figure 55 shows a chart with percentage of resistance logged on its vertical scale and percentage of rotation on the horizontal scale. It can be seen that the percentage of rotation is nonlinear in relationship to the percentage of resistance. Audio-taper controls are available specifically for use as volume and tone controls. Linear taper controls`produce a generally linear rotation/resistance curve and are not suitable for audio purposes.

Auditory inhibition: The inherent ability of a device to cancel or partially reject sound waves. The rejection is dependent upon the wave intensity, direction of impact, or the intervals between transmission and arrival. Directional microphones utilize auditory inhibition to receive and pass audio signals along to the associated amplifier circuitry. These devices normally use direction of impact as a means of producing their limited response.

Aurora: A patterned scattering of light produced by particle reflection. Often, a prismatic effect is produced which divides light into its principle colored bands. Certain types of auroras have the ability to absorb radio waves. This is called *auroral absorption* and is a phenomena about which scientists are still learning more.

Automatic code: A code that allows a machine to translate or convert a symbolic language into a machine language for automatic machine or computer operations.

Automatic frequency control: A method used in circuits containing a phase detector that compares the difference in frequency or phase between the horizontal deflection voltage and the horizontal sync pulses. The output of the phase detector is a direct current control voltage that is proportional to this frequency. When applied to the grid of the horizontal blocking oscillator, it

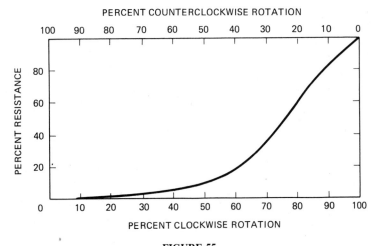

FIGURE 55

changes its bias by the amount necessary to bring the oscillator back into synchronism with the average frequency of the horizontal sync pulses. The horizontal sweep frequency is then amplified by the horizontal deflection amplifier and is finally applied to the horizontal deflection coils in the yoke. This produces the width dimension of the scanning raster in accordance with the amplitude of the horizontal deflection current.

Automatic gain control: A system that holds the gain and, accordingly, the output of a receiver or amplifier substantially constant, in spite of input-signal amplitude fluctuations. A rectifier samples the alternating current signal output and delivers a direct current voltage proportional to the output. The dc voltage is then applied in correct polarity as bias to the early stage in the system to reduce the gain when the output swings beyond a predetermined level, and vice versa.

Automatic level control: A circuit that adjusts the input gain of a magnetic-tape recording device compensatorially.

Automatic modulation control: A circuit used in a radio transmitter for automatically varying the gain of the speech/modulator channel to compensate for fluctuating input audio. The action prevents overmodulation.

Automatic noise limiter: Often abbreviated *anl,* a circuit used for clipping noise peaks, especially in a receiver system. A maximum amplitude for received signals is predetermined within the circuit, and all signals that exceed this value are automatically shorted to ground. This is done without further manual control by the operator.

Automatic repeater: A radio station that receives signals and then automatically and simultaneously retransmits them at a different frequency. Automatic repeater stations are very popular in business and amateur radio communications in the VHF and UHF frequency range. Figure 56 shows a typical repeater operation whereby the radio station is placed atop a mountain. When a mobile unit transmits, its frequency is detected by the repeater receiver and simultaneously retransmitted from the mountain top location at a different frequency and normally at a higher power level. The activation of the mobile transmitter breaks the squelch at the repeater receiver. This automatically activates the transmitter, and the detected audio is channeled directly to the transmitter input section. The mobile operator's signals are then transmitted over a wider area than would be possible using the mobile equipment alone. When the mobile transmission is ended, the repeater transmitter is automatically cut out, and its receiver returns to the squelched position, awaiting another mobile transmission.

Automatic scanning: The nonmanual, repetitive tuning and adjustment of a circuit throughout a given frequency range. Automatic scanning receivers are very popular today for receiving police, fire, and other emergency radio channels without the necessity of having to manually switch from frequency to frequency. The internal circuitry of the receiver automatically

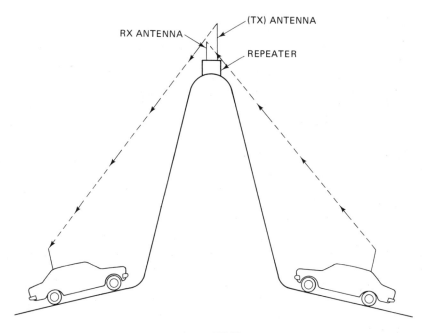

FIGURE 56

switches from frequency to frequency until the squelch is broken. This occurs when an active channel is accessed during the scan. When the squelch is broken, the receiver locks onto the active channel, demodulates it, and passes the audio information on to the speaker.

Automatic speed regulation: The alteration of a wound rotor motor to accelerate, decelerate, or maintain certain speeds called for by installed pilot devices. An example is given in Fig. 57. Assuming that the wound rotor motor is coupled to a fluid pump in a liquid controlled system, operation involves the following.

For automatic operation in maintaining liquid level, the selector switch is placed in the automatic position. As the liquid rises, the master float switch (MFS) operates to close the circuit to the control switch. As the fluid continues to rise, the float switch (FS_1) operates to energize the control relay (CR_1). CR_1 closes the main starter contacts (M), starting the motor in slow speed and energizing the timing relay (T_1). If this motor speed is too slow to effect proper delivery, the liquid level in the tank will eventually close the third float switch (FS_2). This energizes CR_2 through the now closed contacts of T_1 to operate the first accelerating contactor (1A), shunting out the first bank of resistance and also energizing the second time delay relay (T_2). This process continues until a motor speed is reached that will cause the liquid level in the tank to remain constant. If the control selector switch is placed on "manual," the motor must

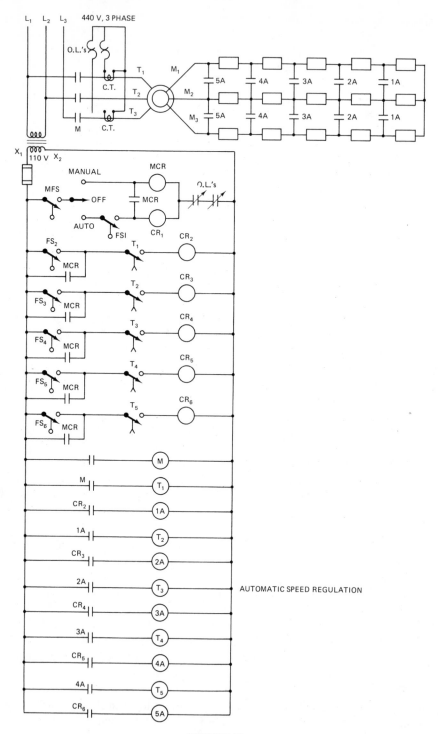

FIGURE 57

start with all resistance in the secondary circuit and follow the described timing sequence until all resistance has been shunted out for maximum operation of the pump.

Automation: The technique and application of implementing a process in such a manner that human intervention is minimized. The electronic computer can be a useful tool in the automation of the manufacture and inspection of almost any product. It can be programmed to operate milling and drilling machines, turret latches, and other machine tools with more speed and accuracy than is possible with human operators.

An act or process can be inspected using suitable transducers; data from the transducers can be used by the computer to control and adjust the machine tool or valve. An example of automation on a liquid pipeline illustrates the concepts of data collection, analysis, and control. Four valves are used for controlling the flow rate within the pipeline, and pressure transducers are used to monitor the liquid pressure on the pipeline walls. Data from the pressure transducers are used by the computer to control the inlet valve openings.

Automotive electrical system: A power system for vehicles using a lead-acid energizer or storage battery. The battery is an electrochemical device for converting chemical energy into electrical energy. It is not a storage tank for electricity as it is often believed, but instead, stores electrical energy in chemical form.

Active materials within the battery react chemically to produce a flow of direct current whenever lights, radio, cranking motor, or other current-consuming devices are connected to the battery terminal posts. This current is produced by chemical reaction between the active materials of the plates and the sulfuric acid of the electrolyte.

The battery performs at least three functions in automotive applications:

1. It supplies electrical energy for the cranking motor and for the ignition system as the engine is started.

2. It supplies current for the lights, radio, heater, and other accessories when the electrical demands of these devices exceed the output of the generator.

3. The battery acts as a voltage stabilizer in the electrical system. Satisfactory operation of the vehicle is impossible unless the battery performs each of these functions.

See *Batteries, lead-acid.*

A diagram of a typical automotive integral charging system is shown in Fig. 58. The system is made up of two basic components: a generator with a built-in solid-state voltage regulator and an energizer (battery). The basic operating principles follow.

When the switch is closed, current from the battery flows to the generator No. 1 terminal, through resistor R_1, diode D_1, and the base-emitter of transistor TR_1 to ground, and then back to the battery. This turns on transistor TR_1, and

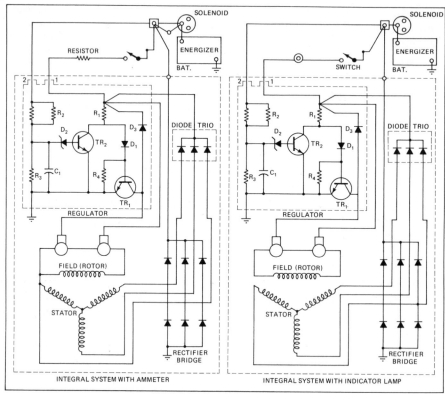

INTEGRAL CHARGING SYSTEM CIRCUITRY

FIGURE 58   Automotive electrical system.

current flows through the generator coil and TR_1 back to the battery. The indicator lamp then turns on.

With the generator operating, alternating current (ac) voltage is generated in the stator windings, and the stator supplies direct current (dc) field current through the diode trio, the field, TR_1, and then through the grounded diodes in the rectifier bridge back to the stator. Also, the six diodes in the rectifier bridge change the stator ac voltages to a dc voltage, which appears between ground and the generator "BAT" terminal. As generator speed increases, current is provided for charging the battery and operating electrical accessories. Also, with the generator operating, the same voltage appears at the "BAT" and No. 1 terminals, and the indicator lamp goes out to indicate that the generator is producing voltage.

The No. 2 terminal on the generator is always connected to the battery, but the discharge current is limited to a negligible value by the high resistance of R_2 and R_3. As the generator speed and voltage increase, the voltage between R_2 and R_3 increases to the point where zener diode D_2 conducts. Transistor TR_2 then turns on and TR_1 turns off. With TR_1 off, the field current and system voltage decrease, and D_2 then blocks current flow, causing TR_1 to turn back on. The field current and system voltage increase, and this cycle then repeats many times per second to limit the generator voltage to a preset value.

Capacitor C_1 smooths out the voltage across R_3, resistor R_4 prevents excessive current through TR_1 at high temperatures, and diode D_3 prevents high-induced voltages in the field windings when T_1 turns off.

Ignition systems used in automotive electrical systems are generally of two types: the standard breaker-point type consisting of a coil, condenser, distributor, switch, wiring, spark plugs, and a source of electrical energy; and electronic ignitions. The function of either type is to produce high-voltage surges and to direct them to the spark plugs in the engine cylinders. The sparks must be timed to appear at the plugs at the correct instant near the end of the compression stroke with relation to piston position. The spark ignites the fuel-air mixture under compression so that the power stroke follows in the engine.

Electronic ignitions do not require the conventional points, condenser, or distributor cam. These systems use a magnetic-sensing unit and a transistorized control module to interrupt the primary ignition circuit, thereby creating a spark at the spark plugs. Each of the systems is somewhat different in design, and any work performed should first be checked against the recommended procedures in the shop manual for the particular system that is going to be serviced.

Autopatch: A frequently used term for a remotely controllable system that patches a radio-communications system into a telephone land-line network. All mobile telephone systems use autopatch to enable persons traveling in automobiles to make standard telephone calls to homes, businesses, etc. An automatic repeater station (see *Automatic repeater*) is almost always used in this application. The mobile unit transmits to a repeater receiver, but instead of retransmitting the input through another radio system, the receive output is switched directly to a standard telephone line equipped with the proper interface. The answering party's voice is received at the repeater site by land line and is channeled to the input section of the repeater transmitter, which broadcasts this signal back to the mobile unit where it is heard by the person initiating the call via the mobile receiver.

Autopilot: A self-correcting control and guidance device used within an aircraft to relieve the pilot of some manual control responsibilities. By true definition, an autopilot does not receive information from the ground after the flight is initiated. Autopilots are used for automatic management of many types of corporate and commercial aircraft and in guided missiles. These devices normally encompass a sophisticated electronic sampling system whose outputs are used to drive motors, solenoids, and other electromechanical devices attached directly to aircraft control circuitry.

Autotransformer: A type of transformer used usually as a compensator for motor starting boxes, for balancing three-wire systems, and for tying two power systems of different voltage. In this class of transformer, there is only one winding that serves as both primary and secondary. Autotransformers

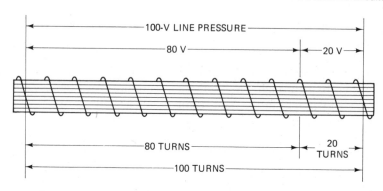

FIGURE 59 Autotransformer.

are used where the ratio of transformation is small, as a considerable savings in copper and iron can be effected and the whole transformer can be reduced in size as compared with one having separate windings. Figure 59 shows the electrical connections and the relations between the volts and number of turns.

By using the end wire and tapping in on turn No. 20, a current at 20-V pressure is readily obtained and may be used for starting motors requiring a large starting current and yet not draw heavily on the line. Since the primary is connected directly to the secondary, it would be dangerous to use an autotransformer for lighting service.

Avalanche breakdown: The sudden, marked increase of reverse current at the bias voltage is a reverse-biased semiconductor at which avalanche begins. This occurrence closely resembles a breakdown, but when the current is limited by external means, the action is nondestructive in nature.

A schematic representation of a typical avalanche breakdown is shown in Fig. 60.

Avalanche conduction: An enhancement of conduction properties through a series of semiconductor materials due to avalanche effect. See *Avalanche breakdown.*

Avalanche impedance: The reduced impedance of a diode during periods of avalanche operation. Measured in ohms, avalanche impedance is present in the breakdown region of the device. Avalanche impedance is also known as *breakdown impedance.*

Average noise figure: The ratio of total noise generated at the output of a circuit to the noise that is generated by thermal effect at the circuit input. The total noise is summed over frequencies from dc to a very high point. The noise temperature of the input termination is usually figured as 290°K a standard value.

Axial leads: The centrally located leads protruding from the ends of cylindrical electronic components. Figure 61 shows several examples of axial leads in capacitors and resistors. The leads serve as the attachment points for the remainder of the circuit in which each device is connected.

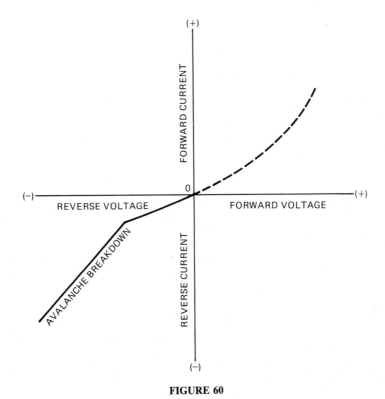

FIGURE 60

Axis: A coordinate in a graphical representation. Figure 62 shows a chart that includes a horizontal and a vertical axis. The *x*-axis on an oscilloscope describes the representation produced by the vertical input channel, while the y-axis is derived from the horizontal input. A *z*-axis is sometimes included in oscilloscope terminology. This refers to the intensity of the trace line.

Ayrton-Mather galvanometer shunt: A step-adjustable universal

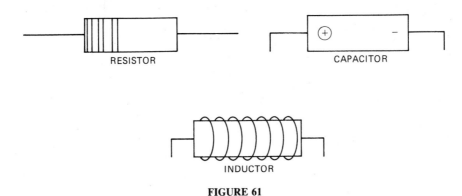

FIGURE 61

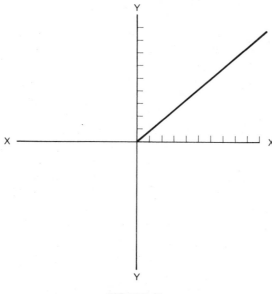

FIGURE 62

shunt resistor that serves to keep the galvanometer critically damped. Shown in Fig. 63, this shunt is used in microammeters, ammeters, and in multirange milliammeters.

Ayrton-Perry winding: A noninductive winding of two conductors connected in parallel. See Fig. 64. Thus, current is conducted in opposite directions, which causes the inductance of each to cancel each other out, or neutralize the magnetic field.

Ayrton shunt: See *Ayrton-Mather galvanometer shunt.*

Azimuth alignment: In the adjustment of tape-recording units, the positioning of the playback and record tape head gaps in such a manner that

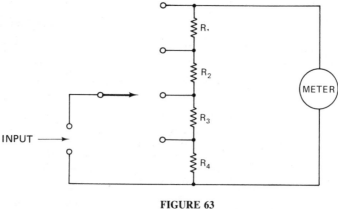

FIGURE 63

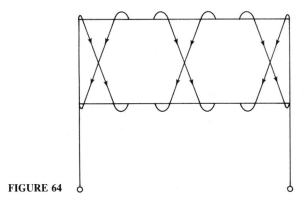

FIGURE 64

their center lines are in parallel. When these gaps are misaligned, the audio output from the recorder will be void of many of the higher frequency signals. These are greatly suppressed or lost entirely by the misalignment. Azimuth alignment is a standard preventive maintenance procedure conducted by technicians who maintain tape-recording equipment, especially in commercial applications.

Azusa: A continuous-wave phase comparison electronic tracking system that operates at microwave frequencies. One station is used in this complex system, providing slant range and two-direction cosines for location and tracking purposes.

B + : Also referred to as B plus or B positive, the term designates the positive terminal of a B battery. It may also refer to the positive dc voltage that is required for certain electrodes of tubes, transistors, etc. In regard to the positive polarity of a device, the B-plus terminal is the point at which the positive side of the anode voltage source is connected.

Babble signal: Basically, a jamming signal used in electronics primarily as a deceptive method of confusing the transmission so that it appears unintelligible to anyone who may be attempting to receive information that is confidential or private and designated for reception at one receiving station only.

Back-wave oscillator: A microwave oscillator that closely resembles a traveling wave oscillator in that it contains a helical transmission line.

Shown in Fig. 65, electron bunching in the electron beam is the result of interaction between the beam and the RF field, causing reflection to occur at

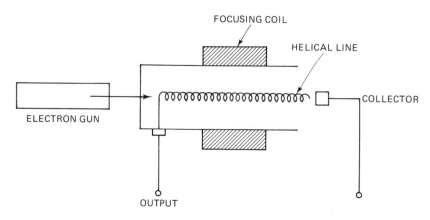

FIGURE 65   Back wave oscillator.

the collector. This results in the wave's moving backward from collector to cathode; thus, oscillation is sustained, since the backward wave is in phase with the input. Output is taken from the cathode end of the helix. By varying the helix dc voltage, a 2 to 1 bandwidth in the 1 to 40 GHz range can be obtained.

Baffle: A single opaque or translucent element to shield a light source from direct view at certain angles, or to absorb unwanted light. A series of baffles that may be arranged in a geometric pattern is a louver or louver grid.

Balance coil: A device used in an ac circuit to compensate for unequal loads. It consists of an autotransformer that allows a three-wire ac circuit to be supplied from a two-wire line. This is accomplished by means of a series of taps around the center of the winding of the autotransformer.

Ballast: A coil of insulated copper wire wound on a frame, or core, made up of thin layers of iron stampings. Ballast is designed to limit the current flow through a lamp to the value for which the lamp is designed. Each fluorescent lamp, for example, is rated for a definite current, no matter what the line voltage is. The ballast must be connected in series with each lamp to cause a drop in the line voltage and provide the desired lamp voltage, which in turn determines the rated current in the lamp. The ballast is built in and concealed in the fixture designed to hold fluorescent lamps.

The voltage drop across the ballast is about as much as the drop across the lamp. Therefore, a circuit with a fluorescent lamp and its ballast will operate with a low power factor. To eliminate this disadvantage, lamps are connected in pairs that use one ballast. Such a ballast is called a two-lamp ballast. One of the lamps is ballasted by inductive reactance and the other by capacitive reactance, in series. This means that one will operate with a lagging and the other with a neutralizing leading power factor, and the overall power factor of the circuit will be quite high. Also, the flicker, or fluctuation, in light output will be reduced. Each ballast is rated for specific wattage, frequency, and voltage.

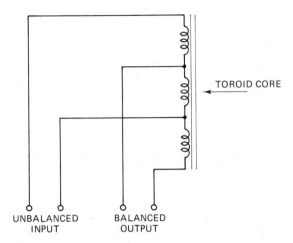

FIGURE 66 Toroidal balun.

Balun: A device for matching an unbalanced coaxial transmission line to a balanced, two-wire antenna system. A dipole antenna is, by nature, a balanced system. However, a coaxial transmission line that often serves as the connection between a single-ended transmitter output tank circuit and the dipole antenna is unbalanced. This means that only one antenna element section will receive the majority of output power. The other is at ground potential with the transmitter chassis.

A balun is actually a transformer with an unbalanced input winding and a balanced output. Its construction is shown in Fig. 66. When connected between the transmission line and the center feed of the dipole, the rf distribution is equalized in both element portions. Baluns are often constructed around ferrite cores in order to decrease size and to allow for broadband operation. Air-wound baluns may be used for single-frequency applications.

Bandpass filter: A device designed to pass currents of frequencies within a continuous band, limited by an upper and lower cutoff frequency, and substantially to reduce, or attenuate, all frequencies above and below that band. A simple bandpass filter is shown in Fig. 67. The curves of current versus frequency are given in Fig. 68. The high-Q circuit gives a steeper current curve; the low-Q circuit gives a much flatter current curve.

The series and parallel-tuned circuits are tuned to the center frequency of the band to be passed by the filter. The parallel-tuned circuit offers a high impedance to the frequencies within this band, whereas the series-tuned circuit offers very little impedance. Thus, the desired frequencies within the band will travel onto the load without being affected, but the currents of unwanted frequencies—that is, frequencies outside the desired band—will meet with a high series impedance and a low shunt impedance so that they are in a greatly attenuated form at the load.

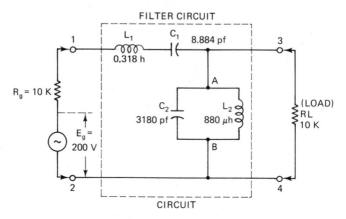

FIGURE 67   Bandpass filter.

   Barkhausen effect: A series of jumps or changes in the magnetization of a ferromagnetic material. These changes, which are abrupt in nature, occur when the magnetizing force is either increased or decreased in variation over a continuous range.
   Barrier capacitance: The interelement capacitance that exists within a transistor between the emitter and the base and shunts any portion of the tank

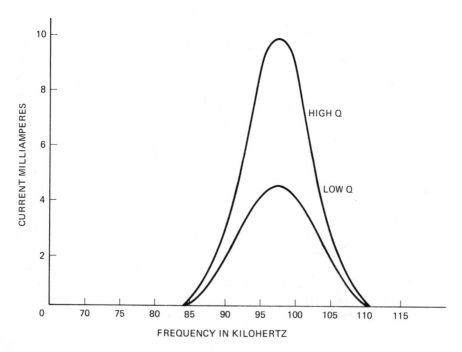

FIGURE 68

circuit or feedback network connected between these elements. It is sometimes referred to as *input capacitance.*

The value of this capacitance is dependent upon temperature, emitter current, and frequency. The additional phase shift in a transistor is primarily caused by the barrier capacitance. Any change in the operating point, temperature, current, etc., will alter the value of the barrier capacitance. Since the barrier capacitance is in shunt with the feedback circuit, or part of the tank, variation of this capacitance will affect the frequency of oscillation.

The frequency stability of a transistor oscillator can be improved by minimizing the effect of barrier capacitance variation. This is accomplished by the use of a regulated power supply and by the stabilization of emitter current and bias voltages.

In oscillators operated at high frequencies (near alpha cutoff), the effect of diffusion capacitance variations may be minimized by inserting a relatively large swamping capacitor across the emitter-base elements. The total capacitance of the two results in a circuit that is less sensitive to variations. The added capacitor may be part of a tuned circuit, as in the Colpitts oscillator.

The use of common bias source for both collector and emitter electrodes maintains a relatively constant ratio of the two voltages. In effect, a change in one voltage is somewhat counteracted by the change in the other, since an increased collector voltage causes an increase in the oscillating frequency and an increased emitter voltage causes a decrease in the oscillating frequency. However, complete compensation is not obtained, since the effects on the circuit parameters of each bias voltage differ.

Barrier voltage: The amount of electromotive force (EMF) required in a circuit or other device to initiate conduction of electricity through the junction of two dissimilar materials. A good example of this type of device would be a P-N junction diode.

Bass-reflex enclosure: A loudspeaker cabinet that has a critically dimensioned duct or port. The opening serves to permit back waves to be radiated in phase with front waves thus serving to eliminate any unwanted signal cancellation.

Battery: An electrochemical device for converting chemical energy into electrical energy. It is not a storage container for electricity, but instead stores electrical energy in chemical form. Active materials within the battery react chemically to produce a flow of direct current whenever a resistance is connected to the battery terminal posts. This current is produced by chemical reaction between the active materials of the plates and the sulfuric acid of the electrolyte.

The simplest unit of a *lead-acid storage battery* is made up of two unlike materials, a positive plate and a negative plate, and are kept apart by a porous separator. This assembly is called an *element.* When this simple element is put in a container filled with a sulfuric acid and water solution (called *electrolyte),* a 2-V cell is formed. Electricity will flow when the plates are connected to an electrical load.

An element made by grouping several positive plates together and several negative plates together with separators between them also generates 2 V but can produce more total electrical energy than a simple cell.

When six cells are connected in series, a *battery* of cells is formed, which produces six times as much electrical pressure as a simple cell, or a total of 12 V—the voltage of most automotive batteries. If a battery continuously supplies current, it becomes discharged and must be recharged by some means to restore the chemical energy to the battery. This is accomplished by sending current through the battery in a direction opposite to that during discharge. This reversing current reverses the chemical actions in the battery and restores it to a charged condition.

As mentioned previously, the electrolyte is a dilute solution of sulfuric acid and water and can be supplied in almost any desired specific gravity. A hydrometer can be used to measure the specific gravity of the electrolyte in each cell. The hydrometer measures the percentage of sulfuric acid in the battery electrolyte in terms of specific gravity. As the battery drops from a charged to a discharged condition, the acid leaves the solution and enters the plates, causing a decrease in specific gravity of the electrolyte. The specific gravity of the electrolyte varies not only with the percentage of acid in the liquid but also with temperature. As temperature increases, the electrolyte expands so that the specific gravity is reduced. As temperature drops, the electrolyte contracts so that the specific gravity increases. Unless these variations in specific gravity are taken into account, the specific gravity obtained by the hydrometer may not give a true indication of the concentration of acid in the electrolyte.

In general, a fully charged battery will have a specific gravity reading of approximately 1.270 at an electrolyte temperature of 80°F. If the electrolyte temperature is above or below 80°F, additions or subtractions must be made in order to obtain a hydrometer reading corrected to the 80°F standard. For every 10° above 80°F, add 4 specific gravity points (0.004) to the hydrometer reading. For example, a hydrometer reading of 1.260 at 110°F would be 1.272 corrected to 80°F—indicating a fully charged battery. For every 10° below 80°F, subtract 4 points (0.004) from the reading. If specific gravity readings show a difference between the highest and lowest cell (of a multicell battery) of 0.050 (50 points) or more, the battery is defective and should be replaced.

The electrolyte of the *nickel-cadmium-alkaline battery* is a strong caustic solution and if spilled on the skin, it should be washed away immediately with a copious amount of water. Acid should never be added to the cells of this type of battery because doing so may ruin the cells. Furthermore, any utensils, such as hydrometer syringes or thermometers, that have been used with acid cells, should never be used with cells containing alkaline.

The standard nickel-cadmium cell can be charged by either the constant-current method, the constant-voltage method, or operated by the floating method. The voltage, under the constant-current method, rises to 1.4 to 1.6 V during the first 60 percent of the charging period, provided the battery is fully discharged at the start. It then rises rapidly until it reaches a final voltage of about 1.75 V; the value of the final voltage depends on the temperature of the electrolyte and on the charge rate.

When the floating method is used, the net input into the battery takes care of the internal losses and any steady load in the circuit. Float voltage is 1.40 to 1.45 V per cell, depending somewhat on the temperature. The constant-voltage charge is accomplished by impressing across the battery terminals a constant voltage appreciably higher than the floating voltage. The resulting high initial charge current requires a generator or rectifier of excessively high capacity. The modified constant-voltage method is the most satisfactory one for charging two or more batteries in parallel from a single dc source. The charging current is automatically regulated to an average of normal rate, and a device is normally employed to open the charging circuit automatically when the battery reaches a charged condition.

A battery that is not kept in regular service or is used infrequently may become temporarily sluggish and deliver less than the capacity of which it is capable. This fault may be corrected by cycling the battery as follows:

1. Discharge the battery through variable resistances to keep the rate at normal until the potential of the battery falls to the equivalent of 1 V per cell.
2. Charge the battery for 15 h at normal rate.
3. Discharge the battery at the normal rate of 1 V per cell.
4. Short circuit each tray and let it stand until the electrolyte cools to not more than 5°F above room temperature.
5. Charge the battery at the normal rate for 15 h after adding water as necessary to bring the electrolyte level up to the recommended level.
6. Discharge the battery at the normal rate, keeping a record of the time until the voltage falls to the equivalent of 1 V per cell.
7. If all cells fail to deliver 1 V at the end of 5 h of discharge, repeat steps 3 and 4 until no further improvement is noted.

The electrolyte temperature should be kept below 115°F except during step 4. Modern electronic equipment, using solid-state technology, uses far less current than did the vacuum-tube circuits of a few years ago. Filament power for the tube circuits used up large amounts of current from storage batteries even when the electronic equipment was in an idle or standby situation. Inefficiencies also existed in converting the available battery voltage into voltage that would fit the operating parameters required by the tube components. Most storage batteries deliver multiples of 6-V outputs (6, 12, 24, etc.), whereas most vacuum-tube circuits require voltage on the order of 150-V dc and possibly more.

Devices that converted this basic storage battery voltage to higher values included the dynamotor, which was a battery-driven dc generator; the vibrator-type dc to dc converter; and in more modern times, the dc to dc converter of solid-state design.

Of these devices, the dynamotor is the least efficient, offering an average conversion efficiency of only about 40 percent. If such a device requires 100-W output, the power output at the new voltage will be only about 40 W. Converters of the vibrator design offer a better conversion factor efficiency of about 65 percent. However, the solid-state converters are the most efficient with an efficiency of 85 to 80 percent for most designs.

In spite of the gains in conversion efficiency obtained by different devices over the years, one fact still remains in front when considering the use of batteries for electronic purposes: to convert from one voltage (from the storage battery) to another without a loss incurring. The battery will have to deliver more current to the voltage conversion device than will be delivered solely to the electronic load. Even at 90 percent efficiency, if 10 A of direct current is delivered to the voltage converter, only 9 A will be delivered to the load—1 A has been lost in the conversion process. Obviously, if the voltage conversion factor can be eliminated, better power efficiency will be obtained, and batteries will be able to power electronic equipment for longer periods of times.

Transistors and other solid-state components are basically low-voltage devices and operate efficiently from voltage levels presently available from most storage batteries and also dry-battery combinations. No conversion is necessary, no filament voltage is required, and battery operation on a portable basis has finally come of age.

In addition to the automobile storage battery or lead-acid battery already discussed, solid-state electronic equipment may be powered from simple *dry cells,* which have been used for years to power flashlights, small toys, and small motor-driven devices. Dry cells of the "D" size, when wired in series or series-parallel configurations, offer very adequate battery packs for low-powered solid-state equipment.

Rechargeable *Nicad batteries* are now available on the electronics market for competitive prices and may be recharged many times before replacement is necessary. Rechargers (battery chargers) are also available for prices that are low enough so that no real economic advantage is gained by constructing a similar unit from the various components.

Nicad batteries are available in the various dry-cell sizes of normal use—AA, C, and D. Their shelf life (the period of time they may be stored unused) is good, and their supply of operating current is excellent. Generally, Nicad batteries are of slightly lower voltage for their size than are the standard dry cells, but this minor difference usually is insignificant for use on most electronic circuits.

It should be remembered that many transistorized circuits, such as small oscillators and control units, draw tiny amounts of current. Devices of this type will allow the battery being used to last almost its expected shelf life. Other transistor and solid-state device circuits may draw large amounts of current while the unit is operating.

Audio devices, for example, may draw many amperes of current during periods of audio peaks, while drawing only a few milliamperes in a standby position. Radio transmitters may call for medium to large amounts of current for only a minute or so and then be in a standby position for an equal or longer length of time. The life of the battery charge will be wholly dependent upon the type of service or equipment that is powered. By using only the minimum amount of power required to operate the electronic device, extended hours of operation will usually be provided.

Digital watches use very small batteries that can supply a steady amount of current of very low value for over a year. Here again, the life of the battery is wholly dependent upon the service. A very small amount of current is drawn by the watch to keep the correct date and time, but a larger amount is required to display the date or time if the watch is an LED type. The less the display is called on, the longer the battery will last. This is normally referred to as the battery duty cycle.

Lantern-type batteries are still used to power some of the medium-power solid-state circuits. These are available in various sizes and current and voltage ratings. Some types will supply almost 100 V, but most are rated to deliver current at 6 and 12 V. Rechargeable batteries of this type are available but are usually very expensive and have a useful life of only about 20 full recharges. These types of batteries are very popular for powering portable television sets and low-powered transmitting and receiving equipment. From 12 to 17 h of charging is usually adequate to bring a "dead" battery back up to a full charge.

In the search for a battery that will supply more current per ounce than the conventional types available, some very unusual configurations and chemical combinations have been tried. Many of these test types have provided tremendous current delivery advantages over the lead-acid battery, but they have also presented some new and interesting disadvantages.

A sodium-sulfur battery, as its name implies, uses sodium and sulfur as its chemical agents. Liquid sodium and liquid sulfur are stored in two separate containers, one inside the other. Other chemicals have also been used successfully in battery designs: silver-zinc, lithium-metal sulfide, and others. Each delivers much more current per ounce than does the lead-acid battery, but each must also operate at extreme temperatures over 500°F. If the temperature of the chemicals drops below this point, no electricity is produced. Then too, many of the chemicals used are explosive if they contact each other. Therefore, extremely good containment and insulation is necessary to prevent a dangerous situation. Also, the expense of most of these newer types of experimental batteries makes them impractical.

Battery charging: A process involving an electrochemical action within the internal plates of cells. Charging is accomplished in a lead-acid

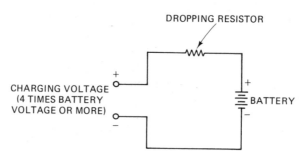

FIGURE 69   Battery charging.

battery, for example, by forcing a current backward through the cell. This reverses the electrochemical action of discharging. There are two basic types of battery charges, constant current and constant voltage. A constant current battery charger is shown schematically in Fig. 69. The charging rate is determined by the value of the voltage-dropping resistor, which will vary depending upon the total charge to be had.

Figure 70 shows a constant-voltage charger, which is identical to the constant-current charger, with the exception of the voltage-dropping resistor's being deleted and the source's having a value more in line with the voltage rating of the battery to be charged. As the battery charges, its voltage value rises and approaches that of the source voltage. This results in a tapering off of the charging current as internal resistance increases.

Most modern battery chargers are electronic in nature and use a step-down transformer that is driven from the ac line. The transformer secondary usually contains several taps so the different charging voltages may be selected. Most modern chargers offer 6- and 12-V outputs for the two most common types of lead-acid batteries.

Many modern chargers offer low, moderate, and high charge rates that determine the amount of current that is pushed through the internal cells. A slow, constant charge is much better for the battery and provides a longer service life. Fast and moderate chargers are used in situations where it is necessary to bring a battery up to full charge in a short period of time. Fast charge should never be used on a completely discharged battery nor for long

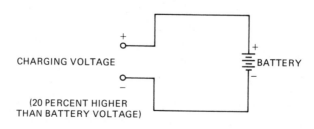

FIGURE 70

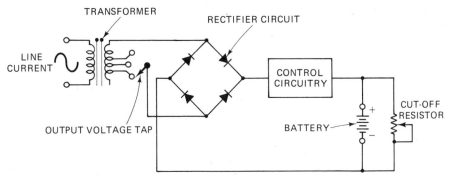

FIGURE 71

periods of time. Fast charging liberates the gas on the surface of the plate, which bubbles up and dislodges sulfur crystals that can create a sludge buildup in the battery case.

Fully charged batteries that must be stored in the charged state are often attached to trickle chargers. These latter devices are standard chargers in most instances with a very low charge position that applies a few milliamperes to the plate at all times to maintain the full charge. Some will automatically go to a higher charge rate when more current is required to maintain the full charge value. Figure 71 shows a typical electronic battery charging circuit.

Baudot code: One of the earliest machine codes in general use and which was used in most of the earlier models of the teletypewriter machines. Baudot has been largely replaced by another more modern code known as *ASCII,* although many Baudot machines are still in daily use. Amateur radio computer buffs on the lookout for a reliable, low-cost, hard-copy printer have kept the prices of these older machines high, even though this is primarily a surplus market.

Most teletypewriters will have a keyboard with a shift key, much the same as a regular typewriter. In most cases, the capital letters are on the shifted keyboard, and the lower case letters are on the unshifted keyboard. However, some teletypewriters are arranged in just the opposite manner.

The Baudot code uses five bits, so it is capable of representing only 2^5, or 32, different characters or control signals. A shift control, much like that found on regular typewriters, allows another 32 characters, for a total of 64. Note that not all 64 characters are used on all Baudot-encoded teletypewriters.

B battery: A battery used in a battery-operated circuit that utilizes vacuum tubes to supply the required dc voltage to the tube plates and screens.

Beam power pentode: An electron tube with five electrodes so designed that the electron emission from the cathode is focused toward the plate area. See *Beam power tube.*

Beam power tube: A vacuum tube requiring less driving power than triodes for the same power output and thus having a greater power sensitivity.

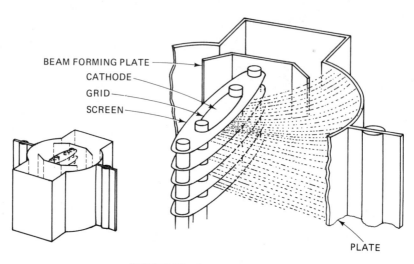

BEAM FORMING PLATE
CATHODE
GRID
SCREEN

PLATE

FIGURE 72 Beam power tube.

Power sensitivity is a term that is used when dealing with power amplifiers. It is the ratio of the power output in watts to the grid signal voltage causing it (when no grid current flows). If grid current flows, the term usually means the ratio of plate power output to grid power input.

Power amplifier circuits normally use vacuum tubes that are specifically designed for purposes of power amplification. One such tube is the beam power tube. Its special design gives it the ability to handle very high values of current. The plate characteristics of the beam power tube are similar to the characteristics of a pentode tube. The primary difference between these two tubes is that in the beam power tube, the electrons are concentrated into sheets as they are attracted to the plate. The sheets, or beams of electrons, are formed by a set of beam-forming plates located inside the tube.

The location and configuration of all the elements of a beam power tube are shown in Fig. 72. The cathode is large and flat on two sides to provide a large emitting surface. The plate is usually corrugated to increase the effective plate area, thereby increasing its power dissipation capability. Another basic difference in the construction of the beam power tube is the way in which the grids are wound. In the beam power tube, the screen grid is wound directly in line with the control grid, which reduces the likelihood of electrons striking the screen grid on their way to the plate. Figure 73 shows how the control grid and screen grid wires in an ordinary tetrode determine the electron paths. The wires are out of alignment so that many of the electrons that pass through the control grid wires are deflected from their paths, striking the screen grid. This produces a screen current that reduces the value of the plate current. The need for a very large plate current in power tubes makes this characteristic undesirable.

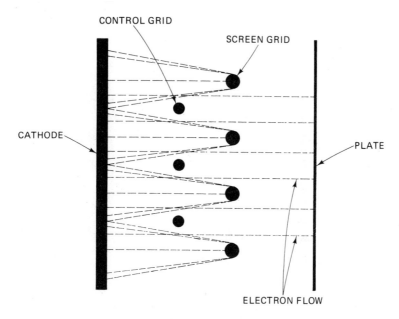

FIGURE 73   Tetrode.

In Fig. 74, the screen grid wires are wound directly in line with the control grid wires so that the screen grid is shaded from the electron stream. As a result, the screen grid intercepts fewer electrons. This result of a lower value of screen grid current in the "shadow" of the control grid is called *shading*.

The overall result of the addition of the beam-forming plates, the shading of the grids, and the use of a corrugated plate is a tube that can handle a substantial amount of electrical power without a great deal of distortion. The plate and control grid of the beam power tube are electrically isolated, the plate current is high, and the plate resistance is relatively low.

Another important function of the beam-forming plates is the suppression of secondary electrons from the plate. Figure 75 shows how the electrons pass through the control grid and screen grids, past the ends of the beam-forming plates, and finally, to the plate. Since the beam-forming plates are connected to the cathode, they are at the same potential as the cathode; that is, highly negative with respect to the plate. Because of this, the beam-forming plates produce an effect equivalent to a space charge in the area between the screen and the plate. This area of effect, shown in Fig. 76 as dashed lines joining the ends of the beam-forming plates, is identified as the virtual cathode. Its effect is to repel secondary electrons emitted from the plate, preventing them from reaching the screen grid.

Figure 77 symbolizes the beam power tube with the beam-forming plates, whereas Fig. 78 shows the version in which a grid replaces the beam-forming plates. As can be seen, there is no difference between the schematic represen-

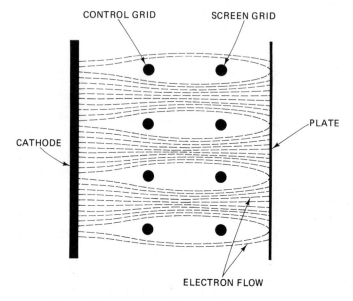

FIGURE 74 Beam power.

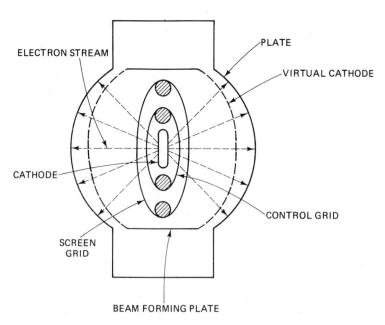

FIGURE 75 Top view of beam power tube.

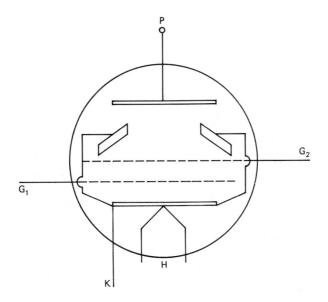

FIGURE 76

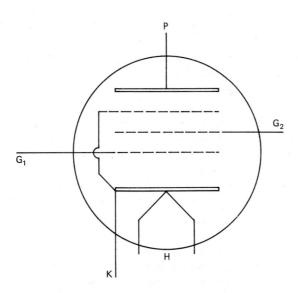

FIGURE 77 Power cathode.

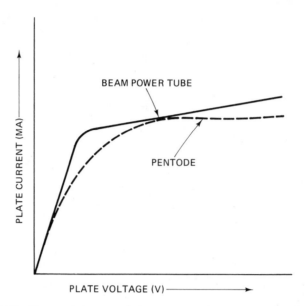

FIGURE 78   Characteristic curve of a beam power tube as compared to the curve for an ordinary pentode.

tation of the power pentode and the schematic symbol used to represent the ordinary pentode.

The plate characteristic curve for the beam power tube differs from that of the ordinary pentode, as shown in Fig. 78. Notice the rapid rise in plate current for the beam power tube, as indicated by the solid line. The more gradual rise in plate current for the normal pentode shown by the dashed line is an important detail relative to power-handling ability with minimum distortion. The solid-line curve shows that the zone, in which the plate current is primarily a function of the plate voltage, is much more limited. The plate current becomes substantially independent of plate voltage at a much lower value of plate potential. This characteristic enables the beam power tube to handle greater amounts of electrical power at lower values of plate voltage than the ordinary pentode. In addition, the beam power tube produces less distortion than the ordinary pentode while accommodating a larger grid swing and plate current change.

Beverage antenna: A horizontal long-wire antenna one or more wavelengths long and terminated at one end in its characteristic impedance (approximately 500 Ω). Generally by using poles, this type of antenna is suspended about 10 to 20 ft from the ground and is terminated to ground in its characteristic impedance at the end nearest the transmitter (when the antenna is used for receiving). The antenna may be considered a one-wire transmission line with a ground return.

The beverage antenna is illustrated in Fig. 79. Because of its termination, the antenna is nonresonant and may be used for the reception of several

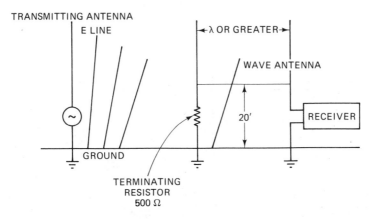

FIGURE 79 Beverage antenna.

different frequencies without too much variation in directivity. Because the ground losses are high, this type of antenna is not very efficient for transmitting, but is often used for receiving.

The vertical transmitting antenna radiates a vertically polarized wavefront. As the wavefront travels toward the receiving antenna, it is tilted forward because of the absorption of the earth and the resultant decrease in propagation velocity. When the wavefront arrives at the receiving antenna, the E lines are no longer vertical. Instead, they have both a vertical and a horizontal component. The horizontal component is effective in inducing a voltage in the receiving antenna. Therefore, the more the wavefront is tilted, the greater the horizontal component and the greater the voltage induced in the antenna.

The amount of tilt depends on the wavelength and the type of earth over which the wave passes. At standard broadcast frequencies, the tilt may be from 6 to 12°; at higher frequencies the tilt may be 20° or more. At the lower frequencies and over good ground (salt marshes or salt water), the tilt is scarcely perceptible. At the higher frequencies and over poor ground, the tilt is maximum.

Binary counter: A series of bistable devices (flip-flops, magnetic cores, etc.) interconnected in such a fashion that the counter will increment and/or decrement by one in binary fashion each time a pulse is applied on its input line. Figure 80 illustrates one way in which a four-stage binary counter may be configured. Since this circuit has four stages and counts in binary fashion, it has the capability of registering and/or indicating all binary numbers from 0000 through 1111, giving it a modulus of $10,000_2$ (16_{10}).

In operation, the count will be incremented by one each time a pulse of the proper amplitude and polarity—a high—appears at the pulse input. However, if the circuit is to function properly—that is, provide the proper count—two conditions must be met: The counter must be enabled and the counter be reset to zero at the start of the counting period.

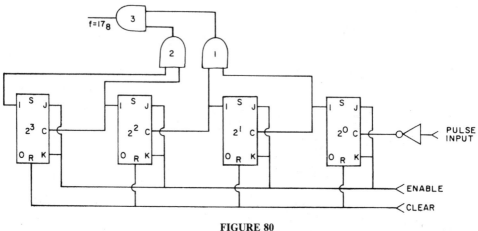

FIGURE 80

The counter is enabled by applying a voltage of the proper amplitude and polarity on the enable line. This may be for the entire counting period or only during the actual time the counter is to be incremented. The counter may be reset by enabling the counter and applying a momentary high to the clear input, which will cause all flip-flops in the counter to be set to their clear states.

Assuming these conditions are met, each time a high—a count pulse—appears at the pulse input, it will be inverted to a low by the inverter/amplifier and applied to the C input of the LSD flip-flop, where in conjunction with the enable signals applied at the J and K inputs of all flip-flops, it will cause the count to be incremented.

Bit: An abbreviation of binary digit used to designate a single character in a binary number, a single pulse in a group of pulses, or a unit of information capacity of a storage device. The capacity in bits is the logarithm to the base two of the number of possible states of the device.

Bit rate: In a computer system, the number of individual data bits that can be processed in a given period of time. The period of time most frequently used as a measurement of the bit rate of a system is 1 s.

Bit slice: An approach in structuring microprocessors in such a manner that the resulting microcomputers are put together using a building block technique. A typical processor using bit-slice chips might use four 4-bit microprocessor chips. Some other bit-slice microprocessors are only two bits wide. The bit-slice approach lets the user configure the microprocessor and requires the development of a specialized instruction set during the initial design phase. A 16-bit processor can be built up with two dozen or less chips that will handle programs designed for popular minicomputers at similar speeds. The processor consumes only about 10 W of power with an instruction time of about 1 ns and a cycle time of 300 ns.

Block diagram: In electronics, a simplified diagram of an electronic system in which stages are shown as functional two-dimensional boxes with the detail circuitry and wiring not included. In computer applications, a block diagram is a graphical representation of any operational circuit or system in which each functional element is presented as a box or block, with the relationship of each element to the other elements depicted by connected lines in relationship to hierarchy.

Blondel's theorem: A statement of power measurement in polyphase circuits based on a theorem stated by Blondel in 1893. The theorem states that the total power delivered to a load system by means of N conductors is given by the algebraic sum of the indication of N wattmeter elements so connected that each of the N wires contains one wattmeter current coil, its potential coil being connected between that wire and some point of the system in common with all the potential loads is on one of the N wires, the total power is obtainable from the indications of $N-1$ elements.

For three-wire systems, therefore, the power can be obtained by reading from two elements, and for four-wire systems, from three elements. The theorem presupposes the general case and holds good for every possible condition of unbalance. In fact, Blondel's theorem holds good even if the electrical loads in the respective branches of the circuit are of different frequencies, or if the load components are partly direct current and partly alternating current. The theorem does not exclude the possibility that, under certain conditions, the power in a circuit could be obtained by even fewer elements than stated in the theorem.

Bode plot: In computer applications, a term used to define a method of plotting control element transfer functions. This method utilizes logarithms of gain or phase angles versus the logarithm of the frequency of the plotted function. For closed-loop system control, it is desirable to obtain the maximum gain while retaining control loop stability. The higher the gain of the system, the faster the system response and thus, the better the control of the controlled variable.

Analysis of a closed-loop control system usually begins with an assessment of the frequency response of each component in the system. To obtain this information, if it is not known, it is necessary to conduct frequency-response tests. With this information regarding the characteristics of the frequency response, it will be possible to calculate the maximum gain for loop stability through the use of a bode plot.

In the bode plot, the gain and phase of the control component are plotted as a function of the log frequency. The frequency at which the phase is $-180°$ is referred to as the critical frequency, and the gain at this frequency must be less than one for loop stability.

Bode plots have a number of advantages with regard to control system analysis. First, since logarithms are used, the expressions for gain are additive. Also, for many electronic elements, the shape allows representation of the exact plot by straight line asymptotes. Yet another advantage of bode plots is that, since the plot is quite easy to construct, it provides a convenient starting point for methods of control analysis, which are more complex.

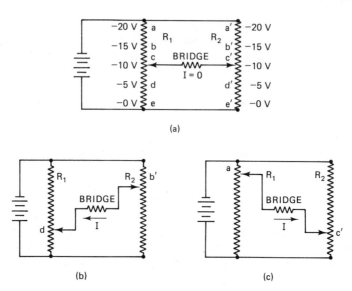

FIGURE 81 Bridge circuit resistance.

Bridge circuit, resistance: A network in which a cross connection or "bridge" is placed between two resistors. A resistance bridge circuit in its simplest form is shown in Fig. 81. Two identical resistors, R_1 and R_2, are connected in parallel across a 20-V source. Voltage across both resistors is dropped at the same rate, because the resistors are identical. Therefore, points a-a', b-b', c-c', and d-d' are at equal potentials. If the bridge is connected between points of equal potential, as in Fig. 81a, no current will flow through the bridge. However, if the bridge is connected between points of unequal potential, current will flow from the more negative to the less negative end, as shown in part b of that illustration. In Fig. 81b, current flows right to left from b' to d. In Fig. 81c, current flows left to right from a to c'.

Thus, it can be seen that the direction of bridge current is controlled by the difference in potential between the two ends of the bridge resistor. When the bridge resistor is across points of equal potential, no current flows through the bridge resistor and the bridge is said to be balanced. When it is across unequal potentials, current flows through the bridge resistor and the bridge is said to be unbalanced. The bridge may be unbalanced in either or both of two ways: by connecting the bridge to unequal potentials or by using resistors of unequal value.

Brightness: The characteristic relating specifically to the flux issuing from a light source or from some surface that is illuminated by a light source. (See Fig. 82, p. 102.) There are two general methods of expressing the brightness of a surface: intrinsic brightness and surface brightness. The term *intrinsic brightness* applies to surfaces that actually generate light, and *surface brightness* applies to surfaces that are visible only because of the light that they reflect.

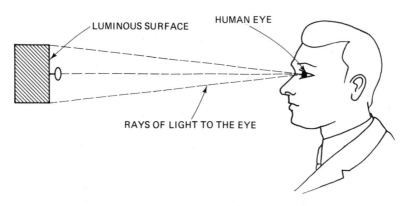

FIGURE 82   Brightness.

Brightness control: In any cathode-ray tube (CRT) device such as a television receiver or oscilloscope, a potentiometer that varies the negative bias voltage on the control grid of the CRT. Since this is the control voltage connection, the brightness of the image is inversely proportional to the voltage value. As the voltage drops, the intensity of the image created by the electron beam is increased. As the control grid voltage increases, the image dims.

British thermal unit: A unit of heat more commonly expressed as *Btu*. It is defined as the amount of heat required to raise the temperature of 1 lb of water 1°F, as shown in Fig. 83. In this illustration, 1 lb of water at 70°F is placed in a position to be heated. Note the thermometer reading. The burner is lighted, and the thermometer reading is watched. At 71°F, the burner is turned off. One Btu of heat has been added to the water.

Broadside array: An antenna arrangement so designed that maximum-signal radiation is in the direction broadside to the array. The desired beam widths are provided for some VHF radars by a broadside array (Fig. 84, p. 103), which consists of two or more half-wave dipole elements and a flat reflector. The elements are placed one-half wavelength apart and parallel to each other. Because they are excited in phase, most of the radiation is broadside to the plane of the elements. The flat reflector is located approximately one-eighth wavelength behind the dipole elements and makes possible the unidirectional characteristics of the antenna system.

Bubble memory: In computer applications, a new concept in nonvolatile storage device modules. A bubble memory is approximately the same size as an integrated circuit (IC) and is operated in a magnetic field, using tiny magnetized areas called *bubbles in a crystalline material*. It requires less power and space than core memory but currently is much slower than semiconductor and core memory. However, there are signs that bubble memories may replace certain auxiliary storage devices.

Buffer: An internal portion of a data processing system serving as intermediary storage between two storage or data handling systems with different access times or formats; usually to connect an input or output device

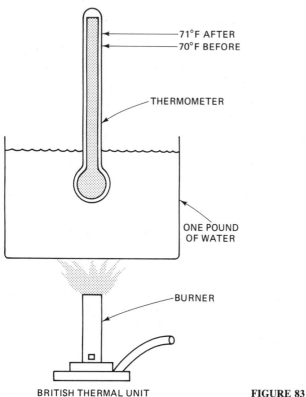

71°F AFTER
70°F BEFORE

THERMOMETER

ONE POUND
OF WATER

BURNER

BRITISH THERMAL UNIT   **FIGURE 83**

with the main or internal high speed storage. Also, a logical OR circuit, an isolating component designed to eliminate the reaction of a driven circuit on the circuits driving it, or a diode.

Buffer stage: Each of a series of buffer circuits used to isolate stages in some electronic circuits. A buffer is an amplifier used for isolation purposes between two portions of a complex circuit. Often, the amplifier gain is unity, so no signal increase is effected. Buffers are typically used between an oscillator and driver circuit in radio frequency transmitters to prevent the latter from loading the frequency-determining portion of the circuit, causing a frequency shift.

Bus: A circuit over which data or power is transmitted, often one that acts as a common connection among a number of locations. The term is synonymous with *trunk*. A bus is also a communications path between two switching points. In computer systems, a bus is a representation of a group of individual lines that have a common function in the functional diagram, such as address, and it transfers information around the system. The bus may contain many individual lines or few. Some signals flow both to and from the block.

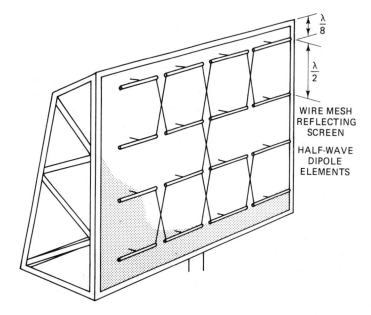

FIGURE 84   Broadside antenna array.

These are designated by double arrows pointing in opposite directions, but most of the blocks handle data in one direction only. There is always a source and a destination for the data. There may be more than one destination, but never more than one source.

All devices that can drive the bus must be tristate. That is, there are three states that the output of the device can take: high, low, and off. The off state is the high-impedance state, and basically disconnects the chip from the bus. This prevents loading of the bus and prevents loading of the unselected circuit outputs.

Bus-bar: An electric conductor that serves as a common connection between load circuits and the source of electric power of one polarity in direct-current systems or of one phase in alternating-current systems. They must be designed to carry the continuous current without overheating. The highest continuous current can be in the order of hundreds of thousands of amperes. Bus-bars are designed to withstand the mechanical forces caused by short-circuit currents.

Bushing, insulating: Devices used in high-voltage circuits to support component parts that must be insulated from ground or where leads must pass through the internal shielding of the apparatus frame. They are constructed of ceramic material with a highly glazed surface. An insulator is no better than its surface, so deposits of foreign substances on the surface will materially reduce the insulation value of the bushing.

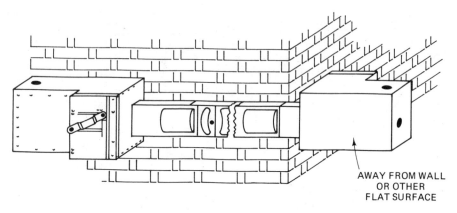

AWAY FROM WALL
OR OTHER
FLAT SURFACE

FIGURE 85   Typical busway system.

Insulating bushings are used as supports for tube sockets and for high-voltage leads; also, for high-voltage terminals, transformers, and capacitors. They are also used as mountings for resistors and high-voltage circuits and as supports for panels that mount other parts. The condition of insulating bushings that are used solely as panel supports is not too critical, but the condition of bushings used as high-voltage insulators is extremely important.

Busways: Various types of bus duct systems with self-contained conductors of copper strap buses used as substitutes for conduit and wire or cable systems for specialized feeder and power installations, especially in industrial applications. Such systems are installed exposed with the proper hangers after the building structure is completed. Busway systems are useful in applications where there is a need for surface wiring that provides ready access to conductors in order to obtain power for electrical equipment that is subject to change at frequent intervals. A busway system also provides a mechanical protection around the conductors, as shown in Fig. 85.

Busway systems may be installed only for exposed work. Busways should also not be installed where they will be subject to severe physical damage or corrosive vapors, in hoistways, in any hazardous location, nor outdoors or in wet or damp locations unless specially approved for the purpose. The system should be securely supported at intervals not exceeding 5 ft. Where the busway is installed in a vertical position, the supports for the bus-bars must be designed for such an installation. Branches from busways may be made with either conduit or other busways. When approved fittings are used, even metal-clad cable and suitable cords may be used.

Butt-welding: A type of joint in which the weld is between the surface planes of both fused parts. The basic butt joint is the square or plain. When joints are prepared and welded from both sides, they are classed as double joints. See Fig. 86.

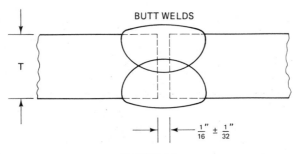

BUTT WELDS

T

$\frac{1}{16}'' \pm \frac{1}{32}''$

FIGURE 86

Candela: The unit of intensity (I) that may be compared to pressure in a hydraulic system. The term is sometimes called *candlepower* and describes the amount of light in a unit of solid angle, assuming a point source of light. It can be seen in Fig. 87 that while the light travels away from the source, the solid angle covers a larger and larger area. The angle itself remains the same, as does the amount of light it contains. Therefore, in a given direction, intensity is constant, regardless of distance.

$$I = \frac{\text{Light energy}}{\text{Solid angle}}$$

Candlepower: See *Candela*.

Capacitance: The property of an electrical device or circuit that tends to oppose a change in voltage. Capacitance is also a measure of the ability of two conducting surfaces, separated by some form of nonconductor, to store an

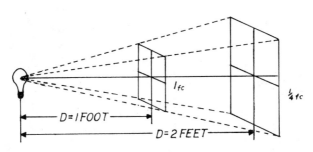

I_{fc}

$\frac{1}{4} fc$

D = I FOOT

D = 2 FEET

FIGURE 87

FIGURE 88

electric charge. The device used in electrical circuits to store a charge by virtue of an electrostatic field is called a *capacitor* and is shown schematically in Fig. 88. The larger the capacitor, the larger the charge that can be stored.

The simplest type of capacitor consists of two metal plates separated by air. A free electron inserted in an electrostatic field will move. The same is true, with qualifications, if the electron is in a bound state. The material between the two charged surfaces of Fig. 89—air in this case—is composed of atoms containing bound orbital electrons. Since the electrons are bound, they cannot travel to the positively charged surface. Therefore, the resultant effect will be a distorting of the electron orbits. The bound electrons will be attracted toward the positive surface and repelled from the negative surface. This effect is illustrated in Fig. 90. There is no difference in charge across the plates and the structure of the atom's orbits is undisturbed. If there is a difference in charge across the plates, the orbits will be elongated in the direction of the positive charge.

As energy is required to distort the orbits, energy is transferred from the electrostatic field to the electrons of each atom between the charged plates. Since energy cannot be destroyed, the energy required to distort the orbits can be recovered when the electron orbits are permitted to return to their normal positions. This effect is analogous to the storage of energy in a stretched spring. A capacitor can thus store electrical energy. See *Capacitor*.

Capacitive coupling: The transfer of ac energy between two circuits or devices by a capacitor or a capacitance effect. Capacitive coupling is achieved by selecting the coupling capacitors that present a high-capacitive reactance at the resonant frequency. In this manner, the signal may be routed from one location to another without excessively loading the tank circuit and lowering the *Q*. Slight adjustments may be necessary, as capacitive coupling will often have slight effects on the frequency of operation.

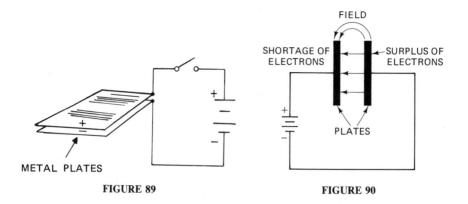

FIGURE 89   FIGURE 90

Capacitive reactance: A certain opposition to the flow of alternating current in a capacitor measured in ohms. The total charge accumulated over a period of time when a capacitor is connected to dc is $Q = it$, or equivalently, the current is the rate of flow of charge per unit time, $i = Q/t$.

When a capacitor is connected to ac, the rate of flow of charge, and hence the current, is constantly changing. We, therefore, must take the ratio of a very small change in charge over a short period of time to obtain the instantaneous current at any time. This is written mathematically, $i = \Delta Q/\Delta t$, where Δ stands for *a small change*. Since the charge is the product of the capacitance and the voltage impressed between the plates of the capacitor ($Q = CV$), we can substitute in the expression for the instantaneous current:

$$i = \frac{\Delta Q}{\Delta t} = \frac{\Delta CV}{\Delta t} = C\frac{\Delta V}{\Delta t}$$

where we have taken C out of the expression, since it is a constant. To obtain the instantaneous value of the current, multiply the capacitance by the rate of change of the applied voltage with time. Since the applied voltage is a sinewave ac voltage of the form

$$V = V_m \sin wt$$

its rate of change is computed with time by the methods of elementary calculus. It is shown there that the rate of change (or derivative) of the above expression for V

$$\frac{dV}{dt} = wV_m \cos wt$$

where dV/dt stands for the rate of change in calculus symbols. Equating this expression with the one previously obtained for the instantaneous current

$$i = C\frac{\Delta V}{\Delta t} = wCV_m \cos wt$$

This resulting expression tells us that the instantaneous current in a capacitive ac circuit varies as a cosine wave; that is, it has the same waveshape as the sine-wave voltage, but it leads the voltage by an angle of 90°, since a cosine wave leads a sine wave by 90°. Moreover, the maximum (peak) value of the current, I_m, is obtained when $\cos wt - 1$, so that we can write

$$I_m = wCV_m \text{ (substituting } \cos wt - 1\text{)}$$

or

$$\frac{V_m}{I_m} = \frac{1}{wC}$$

Finally, since the ratio of the maximum values of the voltage and current equals the ratio of the effective (rms) values, i.e.

$$\frac{V_m}{I_m} = \frac{1.414\ V}{1.414\ I} = \frac{V}{I}$$

we get the result

$$\frac{V}{I} = \frac{1}{wC}$$

Just as in the case of inductive reactance, the ratio of voltage to current represents the opposition to the current, this ratio in a capacitive circuit defines its opposition to current flow or the capacitive reactance, X_c. Thus, we obtain the final result

$$\frac{V}{I} = \frac{1}{wC} = X_c$$

or

$$X_c = \frac{1}{wC} = \frac{1}{2\pi\ fC} = \frac{1}{6.283\ fC}$$

where we have substituted for the angular velocity $w - 2\pi f$, as before. This expression shows that the capacitive reactance of a circuit decreases with increasing capacitance and increasing frequency of the supply voltage.

EXAMPLE 1:

What is the capacitive reactance of a 0.002-μF capacitor at a frequency of 2.5 megacycles (2,500,000 cps)?

Solution:

$$X_c = \frac{1}{6.283\ fC} = \frac{1}{6.283 \times 2.5 \times 10^6 \times 0.002 \times 10^{-6}} = 31.8\ \Omega$$

The example shows that when the capacitance is given in microfarads and the frequency in megacycles, the factors of 10^6 and 10^{-6} can be omitted, since they cancel out, and the result is obtained directly in ohms.

EXAMPLE 2:

What is the magnitude of the current when a 220-V, 60-cycle ac voltage is applied across a 25-UF capacitor?

Solution: The capacitive reactance

$$X_c = \frac{1}{6.28 \times 60 \times 25 \times 10^{-6}} = 106\ \Omega$$

Hence,

$$I = \frac{V}{X_c} = \frac{220}{106} = 2.075 \text{ (rms) A}$$

Capacitor: Essentially, a system in which two or more metal plates (conductors) are placed in close proximity to each other and are separated by an insulating material called the *dielectric*. When the plates of the capacitor are connected to a voltage source, there will be a surplus of electrons on the plate connected to the negative side and a shortage of electrons on the plate connected to the positive side of the voltage source. The surplus of electrons on the negative plate will repel the electrons on the other plate—driving them back toward the positive side of the voltage source. The positive plate will attract electrons to the negative plate from the voltage source. The electron flow will continue until the negative and positive charges on the capacitor plates are equal to the voltage source. When this condition exists, the capacitor is said to be *charged*.

When the voltage source is disconnected, the condition of unbalance that has been set up on the capacitor plates will remain, thus providing a means of storing electricity in the capacitor. The ratio between the magnitude of the charge on the plates and the voltage difference between the plates is called the capacitance, C, and is defined

$$C = \frac{Q}{E}$$

or

$$Q = CE$$

where C = capacitance in farads
Q = magnitude of charge in coulombs
E = potential difference in volts.

One farad (F) is a unit of capacitance that is capable of storing a charge of 1 C for every volt of applied difference of potential. A farad is an extremely large unit of capacitance and for practical applications, smaller capacitance units such as microfarad (mF) and picofarad (pF) are normally used.

$$1 \text{ F} = 1,000,000 \text{ }\mu\text{F}$$

$$1 \text{ F} = 1,000,000,000,000 \text{ pF}$$

$$1 \text{ }\mu\text{F} = 1,000,000 \text{ pF}$$

An illustration of a simple capacitor and its schematic symbol are shown in Fig. 91. The conductors that form the capacitor are called plates. The material between the plates is called the dielectric.

In part b, the two vertical lines represent the connecting leads. The two

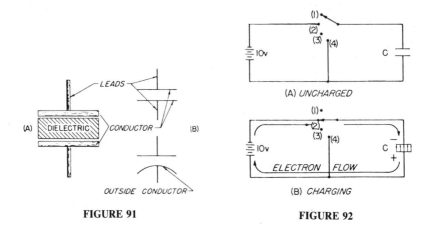

FIGURE 91   FIGURE 92

horizontal lines represent the capacitor plates. Notice that the schematic symbol (b) and the simple capacitor diagram (a) are similar in appearance. In a practical capacitor, the parallel plates may be constructed in various configurations (circular, rectangular, etc.); but the cross-sectional area of the capacitor plates is tremendously large in comparison to the cross-sectional area of the connecting conductor. This means that there is an abundance of free electrons available in each plate of the capacitor. If the cross-sectional area and plate material of the capacitor plates are the same, the number of free electrons in each plate must be approximately the same.

It should be noted that there is a possibility of the difference in charge becoming so large as to cause ionization of the insulating material to occur (causing bound electrons to be freed). This places a limit on the amount of charge that can be stored in the capacitor.

In order to understand the action of a capacitor in conjunction with other components better, the charge and discharge action of a purely capacitive circuit will be analyzed first. For ease of explanation, the capacitor and voltage source used in Fig. 92 will be assumed to be perfect (no internal resistance, etc.), although this is impossible in practice.

In Fig. 92a, an uncharged capacitor is shown connected to a four-position switch. With the switch in position 1, the circuit is open and no voltage is applied to the capacitor. Initially, each plate of the capacitor is a neutral body, and until a difference of potential is impressed across the capacitor, no electrostatic field can exist between the plates.

To charge the capacitor, the switch must be thrown to position 2, which places the capacitor across the terminals of the battery. Under the given conditions, the capacitor would reach full charge instantaneously. However, the charging action will be spread out over a period of time in the following discussion so that a step-by-step analysis can be made.

At the instant the switch is thrown to position 2 (Fig. 92b), a displacement of electrons will occur simultaneously in all parts of the circuit. This electron

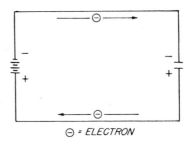

FIGURE 93 $\ominus$ = *ELECTRON*

displacement is directed away from the negative terminal and toward the positive terminal of the source. An ammeter connected in series with the source will indicate a brief surge of current as the capacitor charges.

If it were possible to analyze the motion of individual electrons in this surge of charging current, the action would be as shown in Fig. 93. At the instant the switch is closed, the positive terminal of the battery extracts an electron from the bottom conductor, and the negative terminal of the battery forces an electron into the top conductor. At this same instant, an electron is forced into the top plate of the capacitor and another is pulled from the bottom plate. Thus, in every part of the circuit, a clockwise displacement of electrons occurs in the manner of a chain reaction.

As electrons accumulate on the top plate of the capacitor and others depart from the bottom plate, a difference of potential develops across the capacitor. Each electron forced onto the top plate makes that plate more negative, while each electron removed from the bottom causes the bottom plate to become more positive. Notice that the polarity of the voltage that builds up across the capacitor is such as to oppose the source voltage. The source forces current around the circuit of Fig. 93 in a clockwise direction. The EMF developed across the capacitor, however, has a tendency to force the current in a counterclockwise direction, opposing the source. As the capacitor continues to charge, the voltage across the capacitor rises until it is equal in amount to the source voltage. Once the capacitor voltage equals the source voltage, the two voltages balance one another and current ceases to flow in the circuit.

In studying the charging process of a capacitor, it must be emphasized that no current flows through the capacitor. The material between the plates of the capacitor must be an insulator. To an observer stationed at the source or along one of the circuit conductors, the action has all the appearances of a true flow of current, even though the insulating material between the plates of the capacitor prevents having a complete path. The current that appears to flow in a capacitive circuit is called *displacement current.*

To provide a better understanding of charging action, a capacitor can be compared to the mechanical system in Fig. 94. Part a of the diagram shows a metal cylinder containing a flexible rubber membrane that blocks off the cylinder. The cylinder is then filled with round balls as shown. If an additional

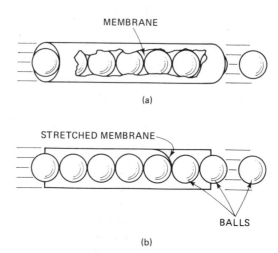

(a)

(b)

FIGURE 94

ball were then pushed into the left-hand side of the tube, the membrane would stretch and a ball would be forced out of the right-hand end of the tube. To an observer who could not see inside the tube, the ball would have the appearance of traveling all the way through the tube. For each ball inserted into the left-hand side, one ball would leave the right-hand side, although no balls actually pass all the way through the tube.

As more balls are forced into the tube, it becomes increasingly difficult to force in additional balls, due to the tendency of the membrane to spring back to its original position. If too many balls are forced into the tube, the membrane will rupture, and any number of balls can then be forced all the way through the tube.

A similar effect occurs in a capacitor when the voltage applied to the capacitor is too high. If an excessive amount of voltage is applied to a capacitor, the insulating material between the plates will break down and allow a current flow through the capacitor. In most cases, this will destroy the capacitor, necessitating its replacement.

When a capacitor is fully charged and the source voltage is equaled by the CEMF across the capacitor, the electrostatic field between the plates of the capacitor will be maximum. Since the electrostatic field is maximum, the energy stored in the dielectric will be maximum.

If the switch is now opened as shown in Fig. 95a, the electrons on the upper plate are isolated. Due to the intense repelling effect of these electrons, no electrons will return to the positive plate. Thus, with the switch in position 3, the capacitor will remain charged indefinitely. At this point, it should be noted that the insulating dielectric material in a practical capacitor is not perfect and a small leakage current will flow through the dielectric. This current will eventually dissipate the charge. A high-quality capacitor may hold its charge for a month or more, however.

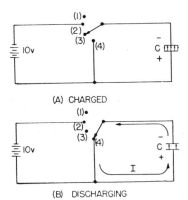

FIGURE 95   (B) DISCHARGING

To review briefly, when the capacitor is connected across a source, a surge of charging current will flow. This charging current develops a CEMF across the capacitor that opposes the applied voltage. When the capacitor is fully charged, the CEMF will be equal to the applied voltage, and charging current will cease. At full charge, the electrostatic field between the plates is at maximum intensity, and the energy stored in the dielectric is maximum. If the charged capacitor is disconnected from the source, the charge will be retained for some period of time. The length of time the charge is retained depends on the amount of leakage current present. Since electrical energy is stored in the capacitor, a charged capacitor can act as a source.

To discharge a capacitor, the charges on the two plates must be neutralized. This is accomplished by providing a conducting path between the two plates (see Fig. 95b). With the switch in position 4, the excess electrons on the negative plate can flow to the positive plate and neutralize its charge. When the capacitor is discharged, the distorted orbits of the electrons in the dielectric return to their normal positions, and the stored energy is returned to the circuit. It is important to note that a capacitor does not consume power. The energy the capacitor draws from the source is recovered when the capacitor is discharged.

Capacitor bank characteristics: The electrical properties of capacitors combined, in series or in parallel, to form a capacitor bank. The electrical characteristics of this bank will depend upon the type or types of connections made. A circuit consisting of a number of capacitors in series is similar in some respects to one containing several resistors in series. In a series capacitive circuit, the same displacement current flows through each part of the circuit and the applied voltage will divide across the individual capacitors. Figure 96 shows a circuit containing a source and two series capacitors. When the switch is closed, current will flow in the direction indicated by the arrows on the diagram. Since there is only one path for current, the amount of charge current in motion is the same in all parts of the circuit. This current is of brief duration and will flow only until the total voltage across the capacitors is equal to the source voltage.

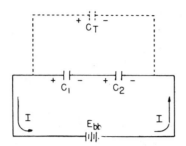

FIGURE 96

Since the charge (Q) is the same in all parts of the circuit

$$Q_t = Q_1 = Q_2 = \ldots Q_n \tag{1}$$

also

$$C = \frac{Q}{E} \tag{2}$$

transposing

$$E = \frac{Q}{C} \tag{3}$$

Since the sum of the capacitor voltages must equal the source voltage (Kirchhoff's law)

$$E_t = E_1 + E_2 + \ldots E_n \tag{4}$$

Substituting Eq. (3) into Eq. (4)

$$\frac{Q_t}{C_t} = \frac{Q_1}{C.} = \frac{Q_2}{C_2} + \ldots \frac{Q_n}{C_n} \tag{5}$$

Since by Eq. (1) all charges are the same, dividing each term of Eq. (5) by Q_t yields

$$\frac{1}{C_t} = \frac{1}{C_1} = \frac{1}{C_2} + \ldots \frac{1}{C_n} \tag{6}$$

Taking the reciprocal of both sides

$$C_t = \frac{1}{\dfrac{1}{C_1} + \dfrac{1}{C_2} + \ldots \dfrac{1}{C_n}} \tag{7}$$

where C_t, C_1, etc. are in farads.

Equation (7) is the general equation used to compute the total capacitance of capacitors connected in series. Note the similarity between this equation and the one used to find equivalent resistance of parallel resistors. If

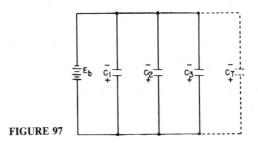

FIGURE 97

the circuit contains only two capacitors, the product over the sum formula can be used:

$$C_t = \frac{C_1 C_2}{C_1 C_2} \qquad (8)$$

where C_t, C, etc., are in farads.

As might be anticipated from the equations, the total capacitance of series connected capacitors will always be smaller than the smallest of the individual capacitors. When capacitors are connected in parallel, one plate of each capacitor is connected to the other terminal of the source. In Fig. 97, since all the negative plates of the capacitors are connected together and all the positive plates are connected together, C_t appears as a capacitor with a plate area equal to the sum of all the individual plate areas. Capacitance is a direct function of plate area. Connecting capacitors in parallel effectively increases plate area and thereby the capacitance.

For capacitors connected in parallel, the total charge is the sum of all the individual charges

$$Q_t = Q_1 + Q_2 + Q_3 + \ldots Q_n \qquad (9)$$

Transposing Eq. (2)

$$A = CE \qquad (10)$$

Substitute Eq. (10) into Eq. (9)

$$C_t E = CE + C_2 E + C_3 E \qquad (11)$$

Divide both sides by E

$$C_t = C_1 + C_2 + C_3 + \ldots C_n \qquad (12)$$

If capacitors are connected in a combination of series and parallel, the total capacitance is found by applying Eqs. (8) and (12) to the individual branches.

Capacitor, ceramic: A capacitor so named because of the use of ceramic dielectrics. One type of ceramic capacitor uses a hollow ceramic cylinder as both the form on which to construct the capacitor and the dielectric material. The plates consist of thin films of metal deposited on the ceramic cylinder.

A second type of ceramic capacitor is manufactured in the shape of a disk. After leads are attached to each side of the capacitor, the capacitor is completely covered with an insulating moisture-proof coating. Ceramic capacitors usually range in value between 1 pF and 0.01 μF and may be used with voltages as high as 30,000 V.

Capacitor, electrolytic: A capacitor that uses an aluminum foil as an anode and a current-carrying fluid or electrolyte as a cathode. A coating of aluminum oxide forms the dielectric, while another aluminum foil is used as an electrical contact with the electrolyte. Since the electrolyte is often a liquid, it is contained in a porous paper that is wound between layers of foil to prevent leaking. By using many sheets of foil as a common contact, several capacitors can be built into a single, compact container.

The aluminum oxide and the foil form a special type of semiconductor that enables current to pass through the oxide film toward the foil. This can be accomplished in one direction only; thus, an electrolytic capacitor is a polarized device. It cannot be used with alternating current and sees most of its applications as part of the filtering circuit in a dc power supply. A less common type of electrolytic capacitor is nonpolarized and can be used in ac circuits. This type has the oxide film on both aluminum foils with an electrolytic fluid forming a floating negative plate.

Electrolytic capacitors are available in many different capacitance and voltage values, as well as case designs and configurations. Values range from a fraction of a microfarad to several thousand and in voltage ratings from 3-V dc to approximately 600-V dc. Figure 98 shows the typical construction of a single unit electrolytic with the anode and cathode connections labeled accordingly.

While electrolytic capacitors are sometimes used in transistor circuit coupling and as audio-frequency bypass capacitors, their most recognized use

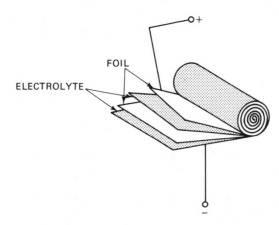

FIGURE 98

is in the filtering of the pulsating dc output of a rectifier circuit. Often, due to the low-voltage characteristics of the electrolytic capacitor, it is necessary to series-connect them, forming a capacitor bank. This is almost always necessary when these units are used in high-voltage power supplies. For example, 10 100 μF capacitors, each with a voltage rating of 450-V dc, may be connected in series to form a bank with an overall value of 10 μF and a voltage rating of 4500-V dc. In this arrangement, each capacitor drops a dc value of 450 V, thus remaining within its voltage tolerance level.

Capacitor microphone: Also called a condenser microphone, a microphone consisting of a tightly stretched metal diaphragm that forms one plate of an air-dielectric capacitor. A closely situated metal plug forms the other plate. A dc bias voltage is applied to this arrangement. Sound waves cause the diaphragm to vibrate, varying the capacitance between the two plates. This causes the output current to vary accordingly.

The simplest form of frequency modulation uses a capacitor microphone, which shunts the oscillator-tank circuit, LC, as shown in Fig. 99. The capacitor microphone is equivalent to an air-dielectric capacitor, one plate of which forms the diaphragm of the microphone. Sound waves striking the diaphragm compress and release it, thus causing the capacitance to vary in accordance with the spacing between the plates.

This type of transmitter is not practical, but is useful in explaining the principles of frequency modulation. The oscillator frequency depends on the inductance and the capacitance of the tank circuit, LC, and therefore varies in accordance with the changing capacitance of the capacitor microphone.

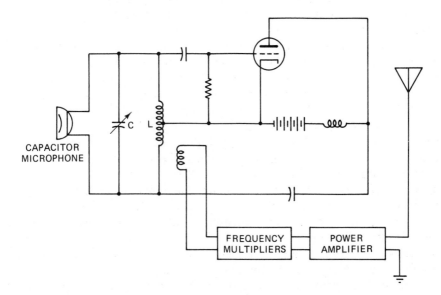

FIGURE 99 FM transmitter modulated by a capacitor microphone.

If the sound waves vibrate the microphone diaphragm at a low frequency, the oscillator frequency is changed only a few times per second. If the sound frequency is higher, the oscillator frequency is changed more times per second. When the sound waves have low amplitude, the extent of the oscillator frequency change from the no-signal or resting frequency is small. A loud AF signal changes the capacitance a greater amount and therefore deviates the oscillator frequency to a greater degree. Thus, the deviation frequency of the oscillator tank depends upon the amplitude of the modulating signal.

Capacitor-start motor: A split-phase-type motor in which the phase-splitting action is obtained by the insertion of a condenser in series with the starting winding. Such a motor starts and runs as a two-phase motor. The autotransformer connected across the condenser applies a comparatively high voltage to the condenser, thereby giving a higher capacity effect, and making possible the use of a smaller condenser than would otherwise be necessary. (Fig. 100)

During starting, the centrifugally operated switch is in the "start" position. This applies about 500 V to the condenser, giving a high-capacity effect and producing a comparatively high starting torque. When the motor has reached about 75 percent of normal full speed, the switch is thrown over to the "run" position, applying about 350 V to the condenser, thereby reducing its capacity effect to a value that will maintain a high-power factor during operation.

This motor will develop approximately four times normal full-load torque with seven times normal full-load current. Compared with the repulsion-start-induction motor, the capacitor motor has a lower starting torque and a much higher starting current, about the same full-load efficiency and a higher full-load power factor.

For equal rating, capacitor motors cannot stand as long a starting period as the repulsion type. Capacitor motors are widely used in household refrigeration and may be used where repulsion-start-induction motors are applicable, except where very high-starting torque and long-starting periods are involved, in which case the repulsion-start-induction motor is used.

The small diagrams a, b, and c in Fig. 100 are schematic diagrams of capacitor motors. Figure 98 is the circuit for the large diagram (capacitor start, capacitor run motor), whereas b and c represent two other types which do not use an autotransformer. Figure 98 uses a condenser on starting only, whereas c uses two condensers on starting, with only one remaining in the circuit when running.

The electrolytic type of condenser is used on condenser-start motors only. This type of condenser must not be left in the circuit for more than 3 or 4 s if breakdown is to be avoided. Condensers marked "X" may be electrolytic, but the others shown must be the metal foil and paper type.

Carrier terminal: The equipment for generating, modifying, or utilizing the carrier energy located at each end of a carrier-current line or cable. The

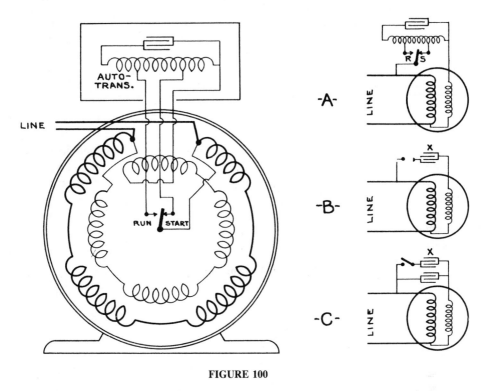

FIGURE 100

carrier terminal with its coaxial cable is matched to the impedance of a power line by an adjustable impedance-matching transformer. The coupling capacitor may be used for more than one frequency if more tuning elements are added.

Carrier wave: The basic frequency or pulse repetition rate of a signal, bearing no intrinsic intelligence until it is modulated by another signal that does bear intelligence. A carrier may be amplitude, phase, or frequency modulated; e.g., in a typical mercury delay line storage of a digital computer, the 8-megacycle/s sound wave carrier is amplitude- or pulse-modulated by a 1-megacycle/s pulse code signal, and the presence or absence of a pulse determines whether or not a one or a zero is present in the binary number being represented.

Carry, end-around: A signal or expression that arises in adding when the sum of two digits in the same digit place equals or exceeds the base of the number system in use. If a carry into a digit place will result in a carry out of the same digit place, and if the normal adding circuit is bypassed when generating this new carry, it is called a high-speed carry, or standing on nines carry. If the normal adding circuit is used in such a case, the carry is called a *cascaded carry*. If a carry resulting from the addition of carries is not allowed to propagate, e.g., when forming the partial product in one step of a multiplication process, the process is called a *partial carry*. If it is allowed to propagate, the process is called a *complete carry*. If a carry generated in the most significant place is sent directly to the least significant place, e.g., when adding two negative numbers using nine complements, that carry is called an *end-around carry*.

Cartridge fuse: An overcurrent device that destroys itself when it interrupts a circuit. Fuses are made of a low-melting-temperature metal and are so calibrated that they melt at a specific current rating. Since they are connected in series with the load, they open the circuit when they melt. All fuses have an inverse time characteristic. A fuse rated at 30 A should carry 30 A continuously, but with about 10 percent overload it would melt in a few minutes, and with 20 percent overload it would melt in less than a minute. A 100 percent overload would require only a fraction of a second to cause the element to melt and open (or clear) the circuit. Some fuses have a higher time lag so that overloads of short duration, say several seconds, will not cause the fuse to "blow." Cartridge fuses have two voltage classifications, 250 and 600 V, as shown in this table:

Table 3   Classifications of cartridge fuses

Rating (A)	Type of retainer	Dimensions (in.) 250 V	600 V
Up to 30	Ferrule	9/16 × 2	13/16 × 5
35–60	Ferrule	13/16 × 3	1 1/16 × 5 1/2
70–100	Knife blade	1 × 5 7/8	1 1/2 × 7 7/8
110–200	Knife blade	1 1/2 × 7 5/8	1 3/4 × 9 5/8
225–400	Knife blade	2 × 8 5/8	2 1/2 × 11 5/8
450–600	Knife blade	2 1/2 × 10 3/8	3 × 13 3/8

Cascade control: An automatic control system in which various control units are linked in sequence, each control unit regulating the operation of the next control unit in line.

Cathode, electron tube: The electrode in an electron tube that emits electrons. The cathode may be a filament directly heated by ac or dc, or through an indirectly heated sleeve. Only a few substances can be heated to the high temperatures required to produce satisfactory thermionic emission without melting.

Tungsten, thoriated tungsten, and oxide-coated emitters are the only types that are commonly used in electron tubes. Tungsten has a great durability as a cathode but requires a large amount of heating power and a high operating temperature for satisfactory emission. Tungsten cathodes are used primarily in high-power electron tubes like those in high-power radio-transmitting equipment.

A thoriated-tungsten cathode has a thin layer of thorium on the surface of the tungsten. The layer of thorium is monomolecular—that is, only one molecule thick. Thoriated-tungsten cathodes have greater electron emission at a lower operating temperature than a cathode of pure tungsten and are normally used in tubes that are operated at plate voltages of 500 to 5000 V. These tubes and others like them are used extensively in low-power radio transmitters.

Oxide-coated cathodes consist of metal, such as nickel coated with a mixture of barium and strontium oxides, over which is formed a monomolecular layer of metallic barium and strontium. This is the most efficient type of cathode. It operates at a lower temperature than tungsten or thoriated tungsten and therefore requires less heating power, resulting in a longer life at a higher emission efficiency. It is used in almost all types of receiving tubes.

The graphs in Fig. 101a show electron emission as a function of cathode temperature for the three types of cathode materials discussed. The temperature at which emission becomes appreciable is called the *normal operating temperature*. The emission efficiency of the three types of cathode materials is shown in Fig. 101b.

The electron-emitting cathodes of electron tubes are heated in two ways—directly and indirectly. A directly heated cathode (Fig. 102a) receives its heat by the passage of a current through the filament itself, which serves as the cathode. An indirectly heated cathode (Fig. 102b) comprises a metal sleeve that surrounds the filament but is electrically insulated from it. The sleeve serves as the cathode and receives its heat primarily by radiation.

Directly heated cathodes are generally employed in portable equipment that is supplied from batteries. The filaments of these tubes are so constructed that the drain on the filament battery is low. Indirectly heated cathodes would require too much power for heating purposes. Because the filament current is steady, the heating is uniform; and the filament cross section is relatively small compared with ac filaments. Directly heated ac filaments require relatively large cross sections to reduce the temperature variations that occur at twice the power frequency. When ac power is available, it is common practice in receiving equipment to employ indirectly heated cathodes. The cathode in this type of tube is isolated from the ac heater supply, and thus hum occurring at the power frequency (or at twice the power frequency) is largely eliminated.

Cathode follower: A vacuum tube circuit in which the output signal is taken at the cathode and referenced to ground. The circuit is characterized by zero-phase shift, a maximum voltage gain of one, a low-impedance output, and very low stage distortion. One of the principal advantages of a cathode follower is that it can be used to match a high impedance to a low impedance. Thus, it can take the voltage developed across a high-impedance and supply a low-impedance load with only a slightly less voltage, but with a correspondingly large increase in current. One or more of the circuit elements of a cathode follower may be varied to achieve a more precise impedance match if the match is critical. When tubes having a high mutual conductance are used, the low value of output impedance extends the amplification into the upper range of frequencies because the shunting effects of interelectrode and distributed

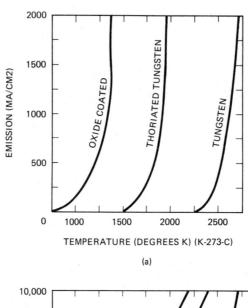

(a)

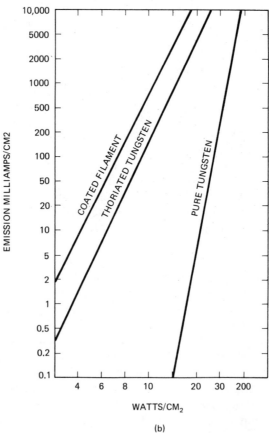

(b)

FIGURE 101 Emission vs. temperature curves for three types of cathodes.

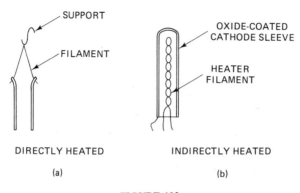

FIGURE 102

capacitances are proportionately smaller. The low-frequency response is improved by allowing the dc component of cathode current to flow in the load, thus avoiding the use of a series-blocking capacitor.

The degenerative effect caused by the unbypassed cathode resistor increases the input impedance. Thus, less shunting effect is offered to the previous stage, and a better overall frequency response is produced. The input and output voltages have the same instantaneous polarity. When pulses are used, it may be necessary to feed a positive- or a negative-going pulse to a load without polarity inversion and to afford an impedance match. Circuit stability is also improved, as in regular amplifiers, by degenerative feedback. Specifically, amplitude distortion occurring within the tube, the effect of plate-supply voltage variations, aging of tubes, production of harmonics, and other undesirable effects that occur within the stage are counteracted by this type of circuit. However, these advantages are achieved at the expense of an overall reduction in voltage gain. Normally, the voltage gain is slightly less than unity, but the circuit is capable of producing a gain in power.

Cathode-ray oscilloscope: See *Oscilloscope.*

Cathode-ray tube: A special type of electron tube, commonly abbreviated CRT, in which electrons emitted from the cathode are shaped into a narrow beam and accelerated to a high velocity before striking a phosphor-coated viewing screen. The screen fluoresces or glows at the point where the electron beam strikes and thus provides visual waveforms of current and voltage. The cathode-ray oscilloscope is a test instrument that uses the cathode-ray tube. Another common device that uses a CRT is the television receiver.

Cathodic protection: The most widely used method in corrosion prevention in which, by the application of an external dc voltage, the underground material to be protected (pipeline, cable, etc.) becomes lower in potential than the surrounding soil. Buried metal is thus made a cathode instead of an anode. It has been found that all usual forms of corrosion are prevented when the cathode protection makes the pipe or other metallic structure 0.25 to 0.30 negative to the soil or liquid surrounding the pipe.

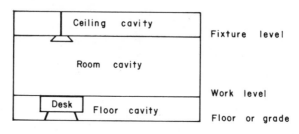

FIGURE 103

Cavity ratios: In the zonal-cavity method, the effects of room proportions, luminaire suspension length, and work-plane height upon the coefficient of utilization. They are respectively accounted for by the room cavity ratio, ceiling cavity ratio, and floor cavity ratio. The ratios are determined by dividing the room into three cavities, as shown in Fig. 103, and substituting dimensions (in feet or meters) in this formula:

$$\text{Cavity ratio} = \frac{5h(\text{Room length} + \text{Room width})}{(\text{Room length}) \times (\text{Room width})}$$

where $h = h_{rc}$ for the Room Cavity Ratio, RCR
$\qquad h_{cc}$ for the Ceiling Cavity Ratio, CCR
$\qquad h_{fc}$ for the Floor Cavity Ratio, FCR

Note that

$$CCR = RCR\frac{h_{cc}}{h_{rc}}$$

and

$$FCR = RCR\frac{h_{fc}}{h_{rc}}$$

Cellulose: The most widely used and one of the oldest of electrical insulating materials. It is used in the form of papers, fabrics, and pressboards. Common insulating papers are made chiefly from coniferous woods, but also from cotton and linen rags, rope, and other materials.

Centrifugal switch: A commonly used switch that has two separate parts—a stationary part attached to a bracket and a part attached to a rotor. In some recent designs, the stationary part has been combined with the terminal board and attached to the stator so that the bracket, or end shield, can be removed without disturbing the connections between the stator and the switch.

Channel capacity: The maximum number of binary digits or elementary digits to other bases that can be handled in a particular channel per unit time. It is the maximum possible information transmission rate through a channel at a specified error rate. The channel capacity may be measured in bits per second or bauds.

Character: One symbol of a set of elementary symbols such as those corresponding to the keys on a typewriter. The symbols usually include the decimal digits 0 through 9, the letters A through Z, punctuation marks, operation symbols, and any other single symbols that a computer may read, store, or write. It is also the electrical, magnetic, or mechanical profile used to represent a character in a computer and its various storage and peripherals devices. A character may be represented by a group of other elementary marks, such as bits or pulses.

Characteristic impedance: The surge impedance of a transmission line having infinite length measured in ohms at the operating frequency and presented by the line to the source feeding the line. This impedance across the input of a theoretically infinite line has a very valuable use. If a load equal to this impedance is connected to the output end of the line, regardless of the length of the line, the impedance presented to the source by the input terminals of the line is still equal to the characteristic impedance of the transmission line. Only one value of impedance for any particular type and size of line acts in this way.

A section of two-wire transmission line of unit length has a certain amount of resistance—no material is a perfect conductor—that varies directly with the length and inversely with the cross-sectional area of the conductor.

The same section of line has the property of distributed inductance. This property exists because of magnetic flux linkages that are established within the section when current flows. For example, an open line composed of two No. 12 conductors spaced 6 in. apart has an inductance of approximately 0.6 μH/ft. This section of line also has the property of capacitance because the two wires, separated by a dielectric, act as the two plates of a capacitor. The capicitance of the two-wire line in the previous example is approximately 1.7 pF/ft.

Finally, the transmission line of unit length has leakage resistance in the path through the insulating material that separates the two conductors—no substance is a perfect insulator. For convenience in working out problems dealing with longer lines, this property usually is expressed as the reciprocal of the leakage resistance, which is conductance. The conductance is of the order of a few picomhos per foot.

Character reader: A specialized device that can convert data represented in one of the type fonts or scripts read by human beings directly into machine language. Such a reader may operate optically or, if the characters are printed in magnetic ink, the device may operate magnetically or optically.

Charging: In a nickel-cadmium storage battery, the process of changing the active material of the negative plate (cadmium-oxide) to metallic cadmium (CdO to Cd). The active material of the positive plate (nickel-oxide) is changed to a higher state of oxidation (NiO to Ni_2O_3). As long as the charging current continues, this action occurs until both materials are completely converted. Toward the end of the charging process and during overcharge, the cell will emit gas due to the electrolysis of the water in the electrolyte. This action liberates four atoms of hydrogen gas ($2H_2$) at the negative plate for every two atoms of oxygen gas (O_2) liberated at the positive plate. The amount of gas liberated depends on the charging rate.

The electrolyte does not enter into any chemical reaction with the positive or negative plates. It acts simply as a conductor of current between the plates, and its specific gravity does not vary appreciably with the amount of charge. The effect of the reactions during charge is a transfer of oxygen from the negative to the positive plates.

In this respect, the nickel-cadmium storage battery is similar to the lead-acid battery. The specific gravity of the electrolyte remains constant, except at the end of a charge or on overcharge, when it increases because of the electrolysis of the water. The proper specific gravity can be obtained by adding distilled water.

Nickel-cadmium batteries can be charged by the constant voltage, constant current, and stepped constant-current methods. The most efficient performance is obtained when the charging rate is such that 140% of the rated ampere-hour capacity of the cell is delivered to the cell within a 3-h interval.

Choke: An iron-core device used in some power supply circuits. Ratings for these devices are expressed in henrys, millihenrys, or microhenrys. A choke-input filter will tend to act as a capacitive-input filter, unless the critical value of inductance is maintained. This value may be determined by the formula $L - E/I$, where L is the critical value desired, E is the output voltage of the power supply, and I is the current being drawn through the filter. Using this formula, the critical value of inductance will be expressed in henrys when current is expressed in milliamperes.

When the power supply is under no load, no current will be drawn through the filter. A critical value of inductance can only be maintained under loading conditions, as is stated by the formula. Thus, when using a choke filter, some current must be drawn at all times. The formula for minimum value of current is $I - E/L$, where L is the critical value of inductance in henrys, E is the output voltage from the power supply, and I is the unknown minimum current expressed in milliamperes.

This minimum current value is obtained when no load is being drawn from the power supply by the use of a bleeder resistor, which places a light load on the filter at all times. Adjustable resistors are often used to set this minimum load value.

Choke, longitudinal: A device used to reduce the longitudinal currents acting on series unbalances, by using a retard coil with two equal windings, one winding being connected in series with each line wire. The poling is such that the coil is noninductive to the metallic circuit and offers a high impedance to the longitudinal circuit.

Chronograph: A speed-recording instrument that provides a graphic record. In the usual forms, the record paper is placed on the surface of a drum that is driven at a certain definite and exact speed by clockwork or weights, combined with a speed-control device so that a specific distance on the paper represents a definite time. The pens that make the record are attached to the armatures of electromagnets. With the pens in contact with the paper and making a straight line, an impulse of current causes the pen to make a slight lateral motion and, therefore, a sharp indication in the record. This impulse can be sent automatically by a suitable contact mechanism on the shaft of the machine or by a key operated by hand. The time per revolution is then determined directly from the distance between marks.

Circuit: A system of conductors and related electrical elements through which electrical current flows. Also, a communications link between two or more points.

Circuit breaker: A device that protects a circuit against short circuits and overloading. The device is made by connecting the winding of an electromagnet in series with the load circuit to be protected and with the switch contact points.

The principle of operation is shown in Fig. 104. Excessive current through the magnet winding causes the switch to be tripped. The circuit to both the breaker and load is opened by a spring. When the circuit fault has been cleared, the circuit is closed again by manually resetting the circuit breaker.

Circular mil: The standard unit of wire cross-sectional area used in American and English wire tables. Because the diameters of round conductors, or wires, used to conduct electricity may be only a small fraction of an inch, it is convenient to express these diameters in mils to avoid the use of decimals. For example, the diameter of a wire is expressed as 25 mil instead of 0.025 in. A circular mil is the area of a circle having a diameter of 1 mil, as shown in Fig. 105. The area in circular mils of a round conductor is obtained by squaring the diameter measured in mils. Thus, a wire having a diameter of 25 mil has an area of 25^2 or 625 circular mil. By way of comparison, the basic formula for the area

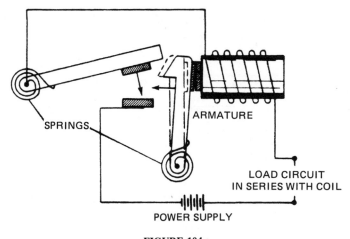

SPRINGS

ARMATURE

LOAD CIRCUIT
IN SERIES WITH COIL

POWER SUPPLY

FIGURE 104

of a circle is $A = \pi R^2$ and in this example, the area in square inches is

$$A = \pi R^2 = 3.14(0.0125)^2 = 0.00049 \text{ in.}^2$$

If D is the diameter of a wire in mils, the area in square mils is

$$A = \pi \left(\frac{D}{2}\right)^2 = \frac{3.1416}{4} D^2 = 0.7854 D^2 \text{ mils}^2$$

Therefore, a wire 1 mil in diameter has an area of

$$A = 0.7854 \times 1^2 = 0.7854 \text{ mil}^2$$

which is equivalent to 1 circular mil. The cross-sectional area of a wire in circular mils is therefore determined as

$$A = \frac{0.7854 D^2}{0.7854} D^2 \text{ circular mil}$$

where D is the diameter in mils. Thus, the constant $\pi/4$ is eliminated from the calculation.

In comparing square and round conductors, it should be noted that the circular mil is a smaller unit of area than the square mil. Therefore, there are more circular mils then square mils in any given area. The comparison is shown in Fig. 105. The area of a circular mil is equal to 0.7854 of a square mil. Therefore, to determine the circular-mil area when the square-mil area is given, divide the area in square mils by 0.7854. Conversely, to determine the square-mil area when the circular-mil area is given, multiply the area in circular mils by 0.7854.

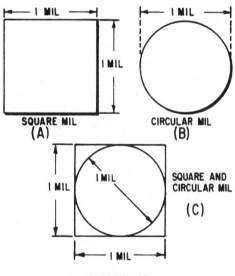

FIGURE 105

A wire in its usual form is a slender rod or filament of drawn metal. In large sizes, wire becomes difficult to handle, and its flexibility is increased by stranding. The strands are usually single wires twisted together in sufficient numbers to make up the necessary cross-sectional area of the cable. The total area in circular mils is determined by multiplying the area of one strand in circular mils by the number of strands in the cable.

Clamping, diode: The simplest type of clamping circuit that uses a diode in conjunction with an RC coupling circuit. Figure 106 shows a positive clamping in which the capacitor voltage is maintained at approximately the minimum applied voltage. If the cathode of a diode is made negative with

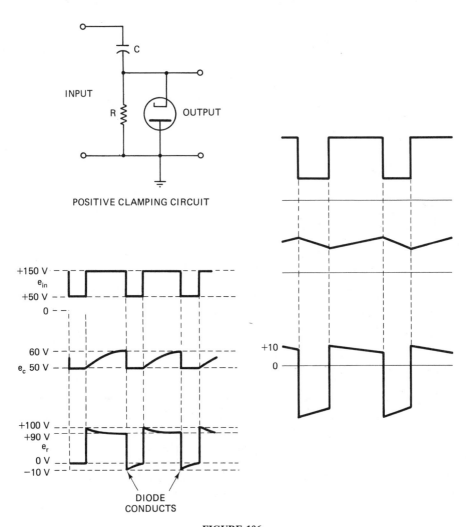

FIGURE 106

respect to the plate (or the plate positive with respect to the cathode), electrons flow from cathode to plate and the tube becomes a low resistance—in effect, a short circuit. If the cathode is made positive with respect to the plate, no current flows and the tube may be considered a high resistance—in effect, an open circuit.

The plate-voltage variations of a circuit producing a square-wave voltage is typical of the kind of input (e_{in}) applied to the clamping circuit of Fig. 107. In this clamping circuit, capacitor C charges gradually through the high resistance, R. After a period of time, depending on the RC time constant, the charge on the capacitor reaches 50 V, the base of the input waveform. The problem is to maintain the charge at this value in spite of the tendency of the capacitor to

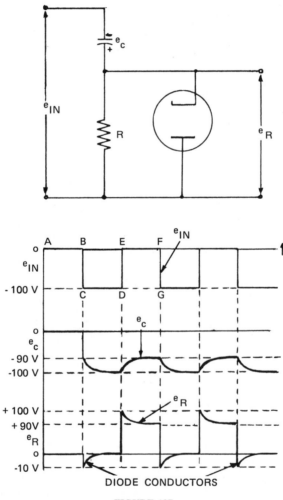

FIGURE 107

charge to a higher level when the applied voltage goes to $+150$ V. The waveforms shown at the right will exist if no diode clamper is used.

Assuming that a steady voltage equal in magnitude to that at time A has been applied for some time, the capacitor may then be considered to be charged to 50 V. During the time interval between A and B, the charge on the capacitor is equal to the applied voltage, and no current flows through R. Then, at point B, the applied potential suddenly increases to $+150$ V. Because it is impossible for the charge on the capacitor to change instantaneously, the difference between the $+150$ V applied and the 50 V across the capacitor must appear across R. This difference of 100 V becomes the output voltage e_r.

The fact that a voltage appears across R indicates that current flows through it. This current adds to the charge on C. Generally, the RC time constant is very long and the charge added to C is small. For simplicity, assume that the 150-V potential is applied for a time equal to $1/10$ RC— that is, the interval from point C to point D. Because the cathode of the diode is positive with respect to the anode (which is at ground potential), the tube is in effect an open circuit.

During a time equal to $1/10$ RC, the charge on the capacitor increases exponentially by 10 percent of 100 V, or 10 V, making the total charge on the capacitor 60 V. During the same time, the drop across the resistor decreases exponentially by 10 V to a value of 90 V, leaving the sum of e_r and e_c still equal to the applied potential of 150 V.

At point D, the applied voltage suddenly drops back to 50 V. The capacitor, however, is charged to 60 V. This would leave an output voltage (across R) of 10 V negative with respect to ground—a condition that must be avoided. In order for the output to return to zero very quickly, the capacitor must discharge the extra 10 V through a path having a very short RC time constant.

In Fig. 106, the cathode of the diode is connected to the high side of R, and the plate is grounded. Any output voltage that is negative with respect to ground makes the cathode negative with respect to the plate. Under this condition, the diode conducts and becomes, in effect, a very low-resistance discharge path for the capacitor until the charge is again equal to the applied voltage and the output voltage returns to zero. At this time, the diode becomes nonconducting. To illustrate the operation of the positive clamping circuit further, assume that the negative-going waveform shown in Fig. 107 is applied to the input of the clamping circuit. Because at point A the input voltage is zero, the output voltage is zero and remains so until point B is reached. At this time, the input voltage drops suddenly toward -100 V at point C. Because the capacitor cannot change its charge instantaneously, the output voltage across R also drops suddenly toward -100 V. When the cathode of the diode is sufficiently negative with respect to the plate, the tube conducts, charging the capacitor very rapidly through the short RC time constant of the conducting diode capacitor and reducing the voltage circuit across R to that across the conducting diode (almost zero). When the capacitor voltage becomes equal to the applied voltage, the diode becomes nonconducting. As long as the input remains at -100 V from point C to D, the output voltage remains at zero potential.

At point D, the input voltage changes back to zero, a rise of 100 V in the positive direction (-11 to 0). This rise produces a rise of 100 V (0 to $+100$) across R, because the capacitor again cannot change its charge instantaneously. The capacitor must now discharge very slowly because the diode is nonconducting, and the high-resistance path through R must be utilized.

Assuming again that the discharge time from points E to F is $1/10$ RC, the voltage across the capacitor at point F and thus the output voltage decrease to 90 V because the input is zero. At point F the input signal again drops to -100 V (point G). Instantaneously, the output across R goes to -10 V (input minus e_c). The diode conducts quickly, returning the charge on the capacitor to 100 V and the output to zero. The output voltage waveform is shown in the lower part of the figure. Note that no portion of the waveform is lost after the first cycle. The function of the clamping circuit is merely to shift the waveform from below to above the zero voltage reference level (ground).

A negative clamping circuit and its associated waveforms are illustrated in Fig. 108. This diode clamping circuit is capable of causing the output voltage to vary between some negative value (-100 V in this figure) and the zero reference voltage. The only difference between this circuit and the one illustrated in Fig. 106 is in the manner in which the diode is connected. In Fig. 108, the plate is grounded and the tube conducts if the cathode is made negative with respect to the plate. In Fig. 106, the cathode is grounded, and the tube conducts whenever the plate voltage rises above ground.

Clapp oscillator: See *Oscillator.*

Clock: A master timing device used to provide the basic sequencing pulses for the operation of a synchronous computer. A register that automatically records the progress of real time (or perhaps some approximation of it) and records the number of operations performed, and whose contents are available to a computer program.

Clock frequency: The master frequency of periodic pulses that schedules the operation of the computer.

Clock system: A master time and program system such as those used in schools and which has a primary control called the *master clock.* The master clock is normally a wall-mounted panel assembly installed in the office of a responsible school official. It is actually a master controller, performing two major functions:

1. It is wired to a central source of unswitched power and operates all other clocks in the system. Isolated power interruptions in a building will not affect remote secondary clocks. At fixed schedule periods, which might be hourly or every 12 h, the master controller will transmit synchronizing

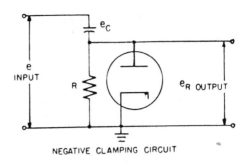

NEGATIVE CLAMPING CIRCUIT

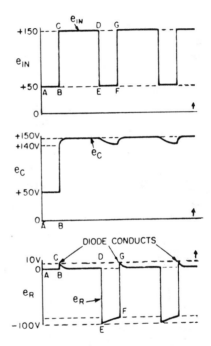

FIGURE 108

signals to all secondary clocks. This ensures that the time reading of all system clocks remain identical to that of the master.

2. The mechanism includes a unique multicircuit control feature called a *programmer*. This permits the master controller to transmit additional signal impulses to external circuits that connect to devices such as bells, horns, chimes, and buzzers. These are referred to as *program signals*. Program circuits are also utilized for on-off control of building utilities.

Program signal control is established and maintained by use of continuous-run, prepunched memory tapes. Each circuit is programmed independently and uses the same principle as those used in tape control of communication transmission networks and similar to that of punch-card data processing.

Clutch: A mechanical device normally used to disengage gears and/or drive belts during a change in speed, gearing, and the like. Clutches of various kinds have been used between synchronous motors and their loads, the motor being started light with the clutch open and the clutch engaged after the motor is in synchronism and fully excited. This results in little disturbance to the system, especially if other than across-the-line starting is used. Sometimes, the clutch is made an integral part of the motor. Combining the synchronous motor with clutches permitting speed variations opens the field of variable-speed drive to synchronous motor applications.

Clutch types include mechanically operated friction clutches, electrically operated friction clutches, resolving stator with brake, centrifugally operated friction clutches, squirrel-cage or eddy-current electrical clutches, planetary clutches, and hydraulic clutches.

COBOL: An acronym for *Common Business-Oriented Language,* a specific language by which business data processing procedures may be precisely described in a standard form. The language is intended not only as a means for directly presenting any business program in any suitable computer for which a compiler exists, but also as a means of communicating such procedures among individuals.

Code: A set of symbols sometimes used for communications. The Morse code of radiotelegraphy and wire telegraphy in which dots and dashes correspond to letters, numbers, and marks of punctuation is one example.

In a computer program, symbolically represented instructions are normally used. Electronic conductors are frequently color-coded for identification purposes. For example, conductors with white insulation usually indicate a neutral or common conductor.

Coding: The ordered list in computer code or pseudocode of the successive computer instructions representing successive computer operations for solving a specific problem.

Coefficient of coupling: A correction factor used in calculating the mutual inductance between two coils. In magnetic leakage between two adjacent coils, some of the magnetic lines through one coil do not link all the turns of the other coil. In practical cases, then, this flux leakage requires a correction factor to calculate the mutual inductance between two coils. The coefficient of coupling is the ratio of the actual value of mutual inductance to the ideal value that would apply in the absence of flux leakage. Thus, K always has a value less than 1. To include the effect of leakage, use the following equation

$$M = K\sqrt{L_1 L_2}$$

The relation between the mutual inductance (M) between two coils of self-inductance L_1 and L_2, respectively, and the coefficient of coupling, $K,$ is given by the above equation.

The coefficient of coupling, *K,* represents the relative amount of flux interlinkage between the coils equivalently or the absence of flux leakage. If all the flux produced by one coil links all the turns of the other coil, the flux leakage is zero and $K = 1$. This is the tightest possible coupling. If none of the flux of one coil links the other, $K = 0$, and there is no mutual inductance. (This may be achieved by placing the coils far away from each other and by placing their axes mutually perpendicular.)

Coincidence circuit: An electrical circuit that produces an output only when each of its selected inputs is of sufficient amplitude, proper polarity, and concurrent in time with its desired mate or mates. Coincidence circuits are also known as *NAND gates* when used in digital logic applications. Figure 109 depicts a three-input coincidence circuit and its associated waveforms. The polarities shown are correct for NPN transistors and electron tubes. For PNP transistors, all polarities would be reversed. All polarities are referenced to ground unless otherwise noted.

All the devices are connected in series and are biased into cutoff. There can be no current flow until all the devices are caused to conduct. During the period of time all inputs are sufficiently positive, the devices conduct heavily. This conduction produces a negative pulse at the output. If any one input is absent, no amplifier current flow is possible, and the output will be simply some constant value of positive voltage.

Figure 110a illustrates a pentode electron tube coincidence circuit. The control grid and suppressor grid are biased sufficiently negative with respect to the cathode to allow either grid to hold the tube cutoff. As shown in Fig. 110b, positive voltages must exist simultaneously on both the control grid and suppressor grid in order for the tube to conduct.

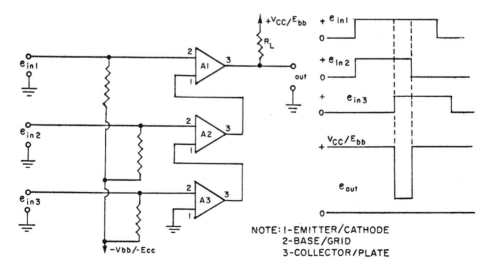

FIGURE 109   Basic coincidence circuit and waveforms.

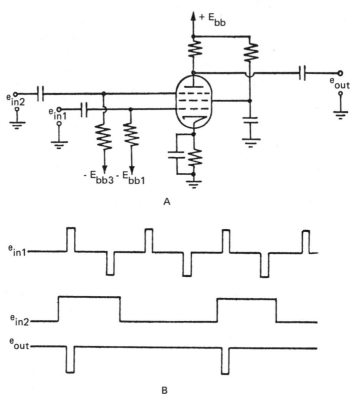

FIGURE 110

Cold-cathode voltage regulator tube: A diode filled with gas and containing no filament. Electrons are pulled from a cold cathode by the high anode voltage. This electron emission is obtained without cathode heating. Certain cold-cathode gas-filled diodes will maintain within limits a constant voltage drop across their elements, though the current through them varies.

Such a tube is illustrated in the basic voltage regulator circuit of Fig. 111. This circuit is similar to a zener regulator. Cold-cathode voltage regulator tubes, also called VR tubes, come in various operating voltages. Typical voltage ratings are 75-, 90-, 105-, and 150-V dc. Typical maximum current ratings are 30, 40, and 50 mA.

There are two voltages to be considered when discussing the operation of the basic cold-cathode regulator circuit. These are the firing voltage and the operating voltage. The firing voltage is that amount of voltage applied across the tube that will cause ionization of the gas and thus, conduction. The operating voltage is that voltage dropped across the tube when it is conducting. The firing voltage will be 30 to 40 percent greater than the operating voltage, and this difference in the two voltages is dropped across the series resistance, R_s, (Fig. 111) once the tube ionizes.

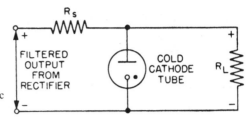

FIGURE 111   Cold cathode tube in basic voltage regulator circuit.

The ionization of the gas within the tube changes depending on the applied voltage. The greater the electron flow, the more ionization and the greater the current through the tube. When the applied voltage is decreased, the electron flow decreases and there is less ionization and less current. The gas tube regulator must be ionized at all times for proper regulation to occur. The amount of ionization might be changed due to a change in the unregulated input voltage, or a change in the load current, but in either case, the regulated output voltage will remain relatively constant at all times.

An increase in the input voltage causes an increase in the regulator tube current (decrease in tube resistance). This increase in tube current, passing through R_s, increases the voltage drop across R_s, leaving the voltage across the tube and the load constant. A decrease in the input voltage results in a decrease in tube current (increase in tube resistance), causing less total current to pass through R_s. This lowers the voltage drop across R_s and the output voltage remains constant.

For load changes, the results are the same—an increase in load current is an increase in total current, causing an increased voltage drop across R_s. This increased voltage drop across R_s leaves less voltage applied to the regulator tube and load. The decrease in voltage across the regulator tube causes the gas in the tube to deionize slightly, causing less current to pass through the tube. The current through series resistor R_s decreases by the same amount as the decrease through the tube. This results in a decrease in the voltage drop across R_s and thus, an increase in the voltage across the load to its original value.

If the load current should decrease, the gas within the tube would further ionize, causing an increase in current through the tube. Conditions opposite to those already discussed would occur to maintain the voltage across the load constant.

The limits within which the VR tube will operate are maximum and minimum tube current. If the maximum current is exceeded, the tube will be destroyed. If the tube current decreases below the minimum value, the tube will deionize, and the voltage across the tube must again be raised to the firing potential in order to reionize the tube.

Color: See *Light.*

Color burst: A signal that color television sets recognize and convert to the colored dots seen on the screen of a receiving set. Without the color burst signal, all pictures would be black and white.

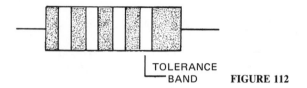

TOLERANCE
BAND   **FIGURE 112**

Color codes: A method of identifying carbon resistors. There are three or four colored bands on a carbon resistor. The resistors with only three color bands are 20 percent tolerance resistors, whereas resistors with the four color bands will be either 5 or 10 percent resistors, depending upon the color of the tolerance band. If the fourth band is a gold band, the resistor has a tolerance of 5 percent; if it is a silver band, the resistor has a tolerance of 10 percent.

The color bands are placed on the body of the resistor nearer to the one end than the other, as shown in Fig. 112. To read the color band, hold the resistor as shown in Fig. 112. The fourth band (if there is a fourth band) will be gold or silver, will be on the right, and indicates the resistor's tolerance. To read the value of the resistor, start at the left end. The first color band gives the first significant figure of the resistance value, the second gives the second figure, and the third tells how many zeros to add to calculate the resistance.

Values assigned the various colors are shown in Fig. 113. For example, a resistor with (left to right) red, red, black, and gold bands would have a resistance of 22 Ω. The first and second bands each indicate 2; the black band indicates no zeros. The gold band indicates 5 percent tolerance. If the resistor were colored orange, orange, red, the first and second bands would each indicate 3, and the red band two zeros, so the value would be 3300 Ω or 3.3 K. If the fourth band is silver, the tolerance is 10 percent; if gold, 5 percent. If there is no fourth band, the tolerance is 20 percent.

Color	1st Figure	2nd Figure	No. of Zeros
Silver			.01
Gold			.1
Black	0	0	none
Brown	1	1	0
Red	2	2	00
Orange	3	3	000
Yellow	4	4	0000
Green	5	5	00000
Blue	6	6	000000
Purple	7	7	
Gray	8	8	
White	9	9	

FIGURE 113

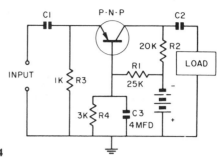

FIGURE 114

If the third color band on a resistor is gold, multiply the first two numbers by 0.1 to get the resistor value. Thus, a resistor coded red, red, gold, gold, is $22 \times 0.1 = 2.2\ \Omega$, 5 percent. If the third color band is silver, you multiply by 0.01. A resistor coded red, red, silver, gold is $22 \times 0.01 = 0.22\ \Omega$, 5 percent.

Common-base amplifier: A circuit that increases voltage without a phase shift. The common-base circuit is particularly useful in television receivers where some voltage amplification is required without a phase shift, as opposed to a common-emitter type of amplifier circuit which has a 180° phase shift. The circuit shown in Fig. 114 is a common-base circuit modified for use with a single battery. R_1 and R_4 make up a voltage divider that is connected across the battery. The forward bias, which is required for the emitter-base junction, is developed across R_4.

The emitter resistor, R_3, serves two purposes in the common-base amplifier. First, it is the impedance across which the input signal is developed. Second, it acts as a bias-stabilizing resistor. Resistors R_1 and R_4 are chosen such that the battery current through R_1 and R_4 is much greater than the base current that flows only through R_1. Thus, any variations in base current will have little effect on the base voltage that appears across R_4. Capacitor C_3 bypasses R_4 and assures that the base is grounded for ac signals.

Commutator: The portion of a motor or generator that switches a single input sequentially to a series of output terminals. The various parts of a commutator are shown in Fig. 115. They include a number of commutator bars, an equal number of mica segments, and an iron core consisting of two end rings and a connecting shell on which the bars and mica segments are placed. The commutator bars are made of high-grade copper. They are wedge-shaped, with the larger width on top. Toward the bottom, the bars are partly cut out on both sides in the shape of a V. Rings fit these V cuts to hold the commutator together.

Mica segments are used between bars to prevent adjacent bars from touching. Segments are cut from sheet mica of the proper thickness and placed between the bars. When these are replaced, the segments must be the same thickness as the original mica. Otherwise, the commutator will be either too loose or too tight. The end rings are made of iron insulated with mica and are called *V rings*. The rings fit into the V cuts on the commutator and hold all the

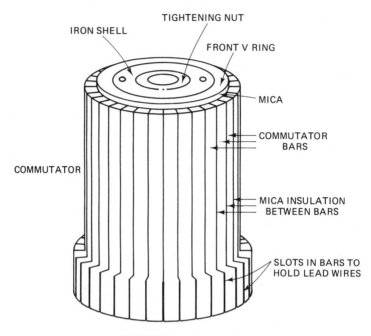

FIGURE 115 Commutator.

bars together. On one type of commutator, the V rings are tightened against the bars by means of a large nut that screws on the shell. The nut may be on either end of the commutator. Some commutators are tightened by means of large screws that extend from one ring to another. Still other types of commutators are riveted together and cannot be reinsulated.

Comparator: A device for comparing two different transcriptions of the same information to verify the accuracy of transcription, storage, arithmetic operation, or other processes, in which a signal is given dependent upon some relation between two items; i.e., one item is larger than, smaller than, or equal to the other. It is a form of verifier.

Compass test: A method to test reverses resulting from wrong connections between poles. In using the compass method, the stator is placed in a horizontal position, and a low dc voltage is applied to the winding. The compass is then held inside the stator and moved slowly from one pole to another. The compass needle will reverse itself at each pole, as shown in Fig. 116, if the winding is correctly connected. If the same end of the needle is attracted to two adjacent poles, a reverse pole is indicated.

Compensating winding: An extra winding in some repulsion motors used to raise the power factor and provide better speed regulation. The compensating winding is much smaller than the main winding and is usually wound in the inner slots of each main pole and connected in series with the armature. Figure 117 shows the compensating winding and its connections to the brushes. Four brushes are necessary. Two of these are connected together

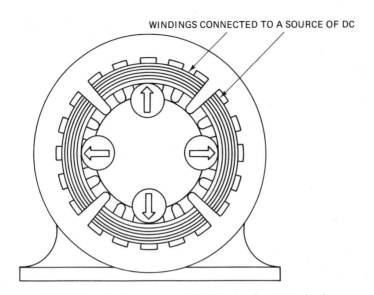

WINDINGS CONNECTED TO A SOURCE OF DC

FIGURE 116 The compass method of testing for reversed poles.

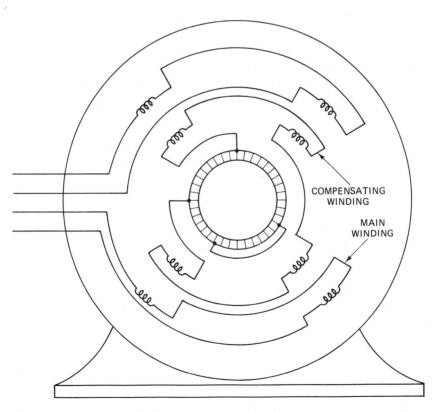

COMPENSATING
WINDING

MAIN
WINDING

FIGURE 117 A compensated repulsion motor.

and the other two are connected in series with the compensating winding. The motor illustrated may be connected for dual-voltage operation. To reverse this motor, it is necessary to reverse the compensating leads as well as shift the brush holder. A typical data layout diagram for a 36-V, six-pole motor of this type is shown in Fig. 118.

Compiler: A computer program more powerful than an assembler. In addition to its translating function, which is generally the same process as that used in an assembler, it is able to replace certain items of input with series of instructions, usually called *subroutines*. Thus, where an assembler translates item for item and produces as output the same number of instructions or constants that were put into it, a compiler will do more than this. The program that results from compiling is a translated and expanded version of the original.

Complement: A quantity expressed to the base N, which is derived from a given quantity by a particular rule; frequently used to represent the negative of the given quantity. A complement on N, obtained by subtracting each digit of the given quantity from $N-1$, adding unity to the least significant digit, and performing all resultant carrys; e.g., the twos complement of binary 11010 is 00110; the tens complement of decimal 456 is 544. Also, a complement on $N-1$, obtained by subtracting each digit of the given quantity from $N-1$; e.g., the ones complement of binary 11010 is 00101; the nines complement of decimal 456 is 543.

Complement arithmetic: In computers, the mathematical computation using subtraction or addition. An arithmetical complement is defined as the difference between a number and the power of the base next in series. Thus, in base 10

$$2 \text{ is the complement of } 8; (10 - 8 = 2)$$

$$26 \text{ is the complement of } 74; (100 - 74 = 26)$$

$$744 \text{ is the complement of } 256; (1000 - 256 = 744)$$

Referring to the definition, it is seen that complement arithmetic is not limited to base 10. Thus, in base 8

$$2 \text{ is the complement of } 6; (10 - 6 = 2)$$

$$4 \text{ is the complement of } 74; (100 - 74 = 4)$$

$$522 \text{ is the complement of } 256; (1000 - 256 = 522)$$

The relationship between any number and its complement in any base is then redefined by the following equation

$$C = B^D - n \tag{1}$$

where n = any number
   D = the number of digits in the number
   C = the complement
   B = the base of the system being used.

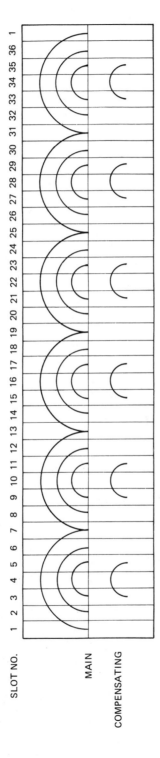

FIGURE 118 A layout of a six-pole compensated repulsion motor. Note the location of the compensating winding in relation to the main winding. The compensating winding is generally wound into the slots first.

Observe that 2_{10} is the complement of 8_{10}, and at the same time, 2_{10} is a number that has 8_{10} as a complement. Thus, developing a method of differentiating between 2 as a number and 2 as a complement of a number must also be accomplished before arithmetic operations can be performed using the complement method.

By modifying Eq. (1) as follows

$$C = B^{D+1} - n \tag{2}$$

an interesting situation develops.

Considering the previously developed complements and using Eq. (2) yields, in base 10

92 is the complement of 8; $(100 - 8 = 92)$

926 is the complement of 74; $(1000 - 74 = 926)$

9744 is the complement of 256; $(10000 - 256 = 9744)$

And in base 8

72 is the complement of 6; $(100 - 6 = 72)$

704 is the complement of 74; $(1000 - 74 = 704)$

7522 is the complement of 256; $(10000 - 256 = 7522)$

Observe here that the complement is preceded in each case by the highest digit in the base used. If all numbers that are not complements are preceded by 0, the identification of complemented numbers is solved. This removes the possible ambiguity that might develop in interpreting a positive 92_{10} and complement of 8_{10}.

Compound: A chemical combination in which atoms of an element combine to form molecules. The atom is the smallest particle of an element that can take part in chemical reactions. While scientifically the atom is composed of even finer particles—electrical charges—called neutrons, protons, and electrons, the atom is considered indivisible and unchangeable. When the atoms of the same or of different elements combine, they form a molecule that may have properties far different from any of the atoms in the individual elements. It is the energy of these molecules that makes up the heat energy of a material.

A small quantity of water contains many billions of molecules. Every molecule of water, ice, or steam is made up of two atoms of hydrogen and one atom of oxygen. Both elements (hydrogen and oxygen) are gases. Hydrogen is a very light and highly flammable gas, whereas oxygen, as a gas, supports combustion (burning). Water, which is a combination of these two gases, is a liquid having properties altogether different from either of the two elements.

Compressor: A machine that compresses air or gas from an initial intake pressure to a higher discharge pressure. Compressors are of the reciprocating, centrifugal, and rotary types. In a reciprocating compressor, the compressing elements have a reciprocating motion. This compressor is used primarily to handle relatively small volumes at high-pressure ranges. Reciprocating compressors can be designed in various sizes. They can be made very small for use in automobile garages, or very large for use with refrigerating machines in breweries and ice plants.

The centrifugal-type compressors apply the principle of using centrifugal force to compress some air or other gas. They are used for large volumes at comparatively low pressures. Centrifugal compressors run at high speeds and can therefore be smaller in size than reciprocating compressors handling the same volume of gas.

Rotary-type compressors are similar to rotary blowers, since the action of rotating elements compresses the air or gas. These compressors are suitable for handling moderate volumes or moderate pressure ranges.

Computer: A device capable of accepting information, applying prescribed processes to the information, and supplying the results of these processes. It usually consists of input and output devices, storage, arithmetic and logical units, and control units.

Condenser: A term commonly used to refer to a capacitor. The name *capacitor* is more widely used, since it is more descriptive of the operation of the device. The capacitor acts essentially as a storage unit; that is, it has the capacitor to store electricity. A condenser is also a device that transfers heat from the refrigeration system to a medium that can absorb and move it to a final disposal point. The condenser is the door through which the unwanted heat flows out of the refrigeration system. It is in the condenser that super-heated, high-pressure refrigerant vapor is cooled to its boiling (condensing) point by rejecting sensible heat. The additional rejection of latent heat causes the vapor to condense into the liquid state.

The three types of condensers are: air cooled, which uses air as the condensing medium; water cooled, using water; and evaporative, using both air and water.

Conductance: The ability of a material to pass electrons, or the ease with which a conductor permits a current to pass through it. The factors that affect the magnitude of resistance are opposite those for conductance. Therefore, conductance is directly proportional to area and inversely proportional to the length and specific resistance of the material. Assuming a constant temperature, the conductance of a material can be calculated if its resistance is known. The formula for conductance is

$$G = \frac{A}{pL}$$

where G = conductance measured in mhos
 A = cross-sectional area in cir mils
 L = length measured in feet
 p = specific resistance

The unit of conductance is the mho, which is *ohm* spelled backward, or more recently, the *siemens*. The relationship that exists between resistance and conductance is a reciprocal one

$$G = \frac{I}{R}$$ S (siemens) $= \Omega^{-1}$ (ohms)

or

$$R = \frac{I}{G}$$ Ω (ohms) $= S^{-1}$ (siemens)

If the resistance of a material is known, dividing its value into 1 will give its conductance. If the conductance is known, dividing its value into 1 will give its resistance.

Conduction band: The area of allowed energies in semiconductors. In the band theory of semiconductors, the individual atoms are considered to have discrete energy levels. These sharp energy levels are broadened into bands of allowed energy for electrons, separated by forbidden energy regions in which no electrons of these intermediate energies are allowed. The broadband semiconductors, such as germanium and indium antimonides, have a narrow band for forbidden energies and broad bands of allowed energies. The narrow-band semiconductors, of which the refractory oxides are good examples, have a broad band for forbidden energies and narrow bands of allowed energies.

Conductivity: The quality of permitting a flow of electrons by elements and compounds. Atoms are composed mainly of protons, electrons, and neutrons. Protons and neutrons form the nucleus, or core, of the atom, and electrons revolve around them. The distances through which the electrons move are very large compared with the size of the electrons, which is very small. Each of the elements, such as gold, copper, or hydrogen, has its own particular kind of atom. The important difference between the atoms of one element and another lies in the number of electrons associated with the nucleus.

Each proton has a small positive charge of electricity, and each electon has the same amount of negative charge. Neutrons are electrically neutral. The attraction between the positive and negative charges holds the atom together, and the positive charges neutralize the negative charges. Materials such as metals give up electrons easily and are therefore good conductors of electricity. Electrons farthest from the nucleus are attracted by neighboring atoms and are thereby separated from their own atom and become free electrons.

When a voltage is applied to the solid material, all of the free electrons, being negative, are strongly attracted toward the positive side of the circuit. As a result, the flow of electrons is in the direction from the negative to the positive side of the circuit. The resulting electron flow constitutes an electric current.

Conductor: Any element that will conduct current, but the term usually refers to those elements or devices that do so efficiently. Conductors are often

Table 4   Dimensions, Weight and Resistance of Pure Copper Wire

Gauge No. A.W.G.	Diameter (in.)	Area circular mils (d²) (1 mil = .001 in.)	Lb. per 1000 ft bare wire	Length ft/lb	Resistance at 77°F (Ω/1000 ft)
Stranded	1.151	1000000.	3090.	.3235	.0108
	1.029	800000.	2470.	.4024	.0135
	.963	700000.	2160.	.4628	.0154
	.891	600000.	1850.	.5400	.0180
	.814	500000.	1540.	.6488	.0216
	.726	400000.	1240.	.8060	.0270
	.574	250000.	772.	1.30	.0431
0000	.4600	211600.	640.5	1.55	.0500
000	.4096	167800.	507.9	1.97	.0630
00	.3648	133100.	402.8	2.48	.0795
0	.3248	105500.	319.5	3.13	.1002
1	.2893	83690.	253.3	3.95	.1264
2	.2576	66370.	200.9	4.98	.1593
3	.2294	52640.	159.3	6.28	.2009
4	.2043	41740.	126.4	7.91	.2533
6	.1620	26250.	79.46	12.58	.4028
8	.1284	16510.	49.98	20.01	.6405
10	.1018	10380.	31.43	31.82	1.018
12	.0808	6530.	19.77	50.59	1.619
14	.0640	4107.	12.43	80.44	2.575
16	.0508	2583.	7.82	127.90	4.094
Solid 18	.0403	1624.	4.92	203.40	6.510
20	.0319	1022.	3.09	323.4	10.35
22	.0254	642.	1.95	514.2	16.46
24	.0201	404.	1.22	817.7	26.17
26	.0159	254.	.77	1300.	41.62
28	.0126	159.8	.48	2067.	66.17
30	.0100	100.5	.30	3287.	105.2
32	.0080	63.2	.19	5227.	167.3
34	.0063	39.7	.12	8310.	266.0
36	.0050	25.0	.076	13210.	423.0
38	.0040	15.7	.047	21010.	672.6
40	.0031	9.89	.030	33410.	1069.0
42	.0025	6.22	.019	52800.	1701.
44	.0020	3.91	.012	82500.	2703.
46	.0016	2.46	.008	238800.	4299.
48	.0012	1.55	.004	229600.	6836.
50	.0010	0.97	.003	330000.	10870.

Table 5 Characteristics of Copper Conductors: Hard-drawn 97.3 Percent Conductivity

Size of Conductor		Outside diameter (in.)	Weight (lb/mi)	Approximate* current carrying capacity (A)	x'_a Shunt capacitive reactance at 1 ft (MΩ/mi)	r_a Resistance (Ω/cond./mi at 50°C, 60 cycles)	x_a Reactance at 1-ft spacing; 60 cycles (Ω/cond./mi)
Circular Mils	A.W.G. or B.&S.						
1,000,000	....	1.151	16,300	1,300	0.0901	0.0685	0.400
900,000	....	1.092	14,670	2,220	.0916	.0752	.406
800,000	....	1.029	13,040	1,130	.0934	.0837	.413
750,000	....	.997	12,230	1,090	.0943	.0888	.417
700,000	....	.963	11,410	1,040	.0954	.0947	.422
600,000	....	.891	9,781	940	.0977	.109	.432
500,000	....	.814	8,151	840	.1004	.130	.443
450,000	....	.770	7,336	780	.1020	.144	.451
400,000	....	.726	6,521	730	.1038	.162	.458
350,000	....	.679	5,706	670	.1058	.184	.466
300,000	....	.629	4,891	610	.1080	.215	.476
250,000	....	.574	4,076	540	.1108	.257	.487
211,600	4/0	.522	3,450	480	.1136	.303	.503
167,800	3/0	.464	2,736	420	.1171	.382	.518
133,100	2/0	.414	2,170	360	.1205	.481	.532
105,500	1/0	.368	1,720	310	.1240	.607	.546
83,690	1	.328	1,364	270	.1274	.765	.560
66,370	2	.320	1,071	240	.1281	.955	.571
52,630	3	.285	850	200	.1315	1.20	.585
41,740	4	.254	674	180	.1349	1.52	.599
33,100	5	.226	534	150	.1384	1.91	.613
26,250	6	.162	420	120	.1483	2.39	.637
20,800	7	.144	333	110	.1517	3.01	.651
16,510	8	.129	264	90	.1552	3.80	.665

*For: conductor at 75°C, air at 25°C, wind 1.4 mi/h (2 ft/s), 60 cycles.

Table 6 Characteristics of Copperweld and Copperweld-Copper Conductors

Conductor designation	Copper equivalent (cir. mils or A.W.G.)	Outside diameter (in.)	Weight (lb/mi)	Approximate [1] current carrying capacity (A)	x'_a Shunt capacitive reactance at 1 ft (MΩ/mi)	r_a Resistance [2] (Ω/cond./mi at 60 cycles)	x_a Reactance at 1-ft spacing; 60 cycles (Ω/cond./mi)
Copperweld—copper							
350 E	350,000	0.788	7,409	660	0.1012	0.204	0.456
250 E	250,000	.666	5,292	540	.1064	.278	.476
4/0 E	4/0	.613	4,479	490	.1088	.326	.486
3/0 E	3/0	.545	3,552	420	.1123	.406	.501
350 EK	350,000	.735	6,536	680	.1034	.188	.452
250 EK	250,000	.621	4,669	540	.1084	.261	.472
4/0 EK	4/0	.571	3,951	490	.1109	.308	.483
4/0 S	4/0	.633	4,210	490	.1079	.330	.477
2/0 S	2/0	.502	2,658	360	.1148	.513	.506
250 V	250,000	.637	4,699	530	.1077	.278	.480
4/0 V	4/0	.586	3,977	480	.1101	.325	.490
2/0 V	2/0	.465	2,502	360	.1170	.505	.518
4/0 F	4/0	.550	3,750	470	.1120	.320	.505
1/0 F	1/0	.388	1,870	310	.1224	.627	.547
1 F	1	.346	1,483	270	.1258	.785	.561
2 F	2	.308	1,176	230	.1293	.985	.575
2 A	2	.366	1,356	240	.1241	.978	.591
4 A	4	.290	853	180	.1310	1.544	.620
6 A	6	.230	536	140	.1379	2.44	.648
8 A	8	.199	392	100	.1422	3.87	.666
8 C	8	.179	320	100	.1460	3.87	.678
9½ D	9½	.174	298	85	.1462	5.43	.709
Copperweld—30% conductivity							
19 No. 9	76,000	0.572	3,696	370	0.1109	0.792	0.637
7 No. 4	89,300	.613	4,324	410	.1088	.672	.629
7 No. 5	70,800	.546	3,429	350	.1122	.848	.643
7 No. 6	56,100	.486	2,719	300	.1157	1.069	.657
7 No. 7	44,500	.433	2,157	260	.1191	1,348	.671
7 No. 8	35,300	.385	1,710	230	.1226	1.699	.685
7 No. 9	28,000	.343	1,356	200	.1260	2.14	.699
7 No. 10	22,200	.306	1,076	170	.1294	2.70	.713
3 No. 5	30,350	.392	1.467	220	.1221	1.963	.683
3 No. 6	24,100	.349	1,163	190	.1255	2.47	.697
3 No. 7	19,100	.311	922	160	.1289	3.12	.711
3 No. 8	15,150	.277	732	140	.1324	3.93	.725
3 No. 9	12,010	.247	580	120	.1358	4.96	.739
3 No. 10	9,528	.220	460	100	.1392	6.26	.753

(1) For: copperweld at 125°C, copperweld—copper at 75°C, air at 25°C, wind 1.4 mi/h (2 ft/s), 60 cycles.

(2) Copperweld conductors at 25°C, copperweld-copper at 50°C.

(3) 40% conductivity.

made of copper because of its relatively high conductivity of electric current. However, there is an increasing use of aluminum, particularly in the larger sizes. Because of its lower conductivity, a larger size aluminum conductor must be used in feeders and services and has not as yet proven practical for the usual branch circuit systems.

Conductors permit the free motion of a large number of electrons. It is the degree of difficulty in dislodging the planetary electrons from the outermost shell of an atom that determines whether the element is a conductor, an insulator, or a semiconductor.

The copper atom has 29 protons in the nucleus and 29 planetary electrons revolving in orbits within 4 shells around the nucleus. The first shell contains 2 electrons; the second, 8; the third, 18; and the fourth or outermost shell, 1 electron. The maximum number permitted in the fourth shell is 2×4^2, or 32. Thus, the single electron in the outermost shell of the copper atom is not very closely bound to the nucleus; it can be moved easily. In a copper conductor containing billions of atoms, it is easy to produce an electron flow of billions of electrons with little resistance.

An atom of an insulator contains a nucleus and 2 or more shells, with each shell completely filled with its quota of electrons. Thus, if the nucleus contains a net positive charge of 10 units, the first shell will contain 2 electrons and the second, 8 electrons. Since it is very difficult to move one of these electrons out of an atom, this material is called an *insulator* or *nonconductor*.

The important difference between conductors and insulators is that in a conductor, there are 1 or 2 electrons in the outer shell that are not tightly bound to the nucleus, whereas in the insulator, the outer shell is filled or almost filled and the electrons are tightly bound to the nucleus.

Most copper wire will be covered with a rubber coating or a plastic coating—an insulator. Its purpose is to keep the current flowing through the wire to the proper destination and prevent its traveling through another circuit. The tables give resistance and reactance constants of conductors, at 60 cycles. The compactness of the tables is secured by arranging the constants for the different conductors for 1-ft spacing and using additional tables of spacing factors to take care of other spacings. The formulae relating to these constants follow.

Three-phase circuit—impedance to neutral

$$A = r_a + j(x_a + x_d) \ \Omega/\text{mi}$$

Note: For unsymmetrical spacings, use an effective spacing equal to the cube root of the product of the three spacings.

EXAMPLE:

Determine the impedance to neutral of a 60-cycle line with 795,000 circular mil ACSR conductor, 54 aluminum strands, with conductor separation = 26 ft. From table of

ACRS:

$$r_a = 0.138 \text{ and } x_a = 0.401 \text{ } \Omega/\text{mi}$$

From reactance spacing factor tables

$$x_d = 0.395 \text{ } \Omega/\text{mi}$$

$$Z = r_a + j(x_s + x_d) = -0.138 + j\,0.796 \text{ } \Omega/\text{mi}$$

Shunt capacitive reactance

$$X' = Xa' + Xd'$$

Illustration: Determine shunt capacitive reactance of the above transmission line. From tables

$$Xa' = 0.0917 \text{ M}\Omega/\text{mi}$$

$$Xd' = 0.0967$$

$$X' = Xa' + Xd' = 0.1884 \text{ M}\Omega/\text{mi}$$

Single-phase circuit—without earth return
Total impedance of circuit $= 2r_a + j(x_a + x_d)$
With line and neutral wires, the latter grounded.

$$\text{Total impedance} = Z_A - \frac{M^2}{Z_N} \quad \text{when } Z_A = \text{line wire } Z \text{ and}$$
$$Z_N = \text{neutral wire } Z$$

where $Z = r_a + \dfrac{r_e}{3} + j\dfrac{x_e}{3} + x_a$ and

$$M = \frac{r_e}{3} + j\frac{x_e}{3} - x_d$$

Contact charging: The process by which a charged body such as a rubber rod can transmit some of its charge to other bodies by direct contact. If a rubber rod (negatively charged) is brought in contact with a neutral body, as shown in Fig. 119, some electrons will flow to the neutral body. The neutral body, having taken on a greater number of electrons, now becomes negatively charged.

The neutral body has an equal number of positive and negative charges. In contrast, the charged body has an excess of electrons. Since like charges repel one another, each excess electron on the charged body is repelled by every other electron it contains. When the charged body is brought in contact with the neutral body, some of the excess electrons are repelled or forced from the charged body to the neutral body.

When the negatively charged body is removed (Fig. 119 inset), the body that had been neutral now has acquired an excess of electrons and is negatively charged. In all cases of charging by contact, the body to be charged always takes on the same kind of charge as the body giving it the charge.

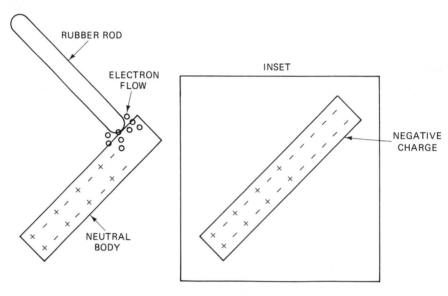

FIGURE 119

Contact EMF: The electromotive force (EMF) created when two dissimilar conductive materials are brought into physical contact with each other through movement of free electrons from the more dense material to the less dense material. This action is illustrated in Fig. 120.

An electrical potential difference is established between the two metals due to the displacement of electrons and their relative positions. This potential difference is known as contact potential difference. The addition of heat to the metals agitates their lattice structures and causes an increase in the transfer of free electrons. At a fixed temperature, however, the transfer of electrons soon reaches a static condition and ceases further movement.

In 1821, Thomas Johann Seebeck discovered that an EMF could be produced by purely thermal means in a circuit composed of two dissimilar

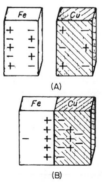

FIGURE 120

conductors when their junctions are kept at different temperatures. This form of producing an electrical potential is called the *thermocouple effect.* The device that works on this principle is called a *thermocouple,* and the EMF in the circuit is known as a thermal EMF.

A thermocouple circuit, such as shown in Fig. 121, provides a means for continuous replenishment of electrons from the material that gains the abundance of electrons to the material that becomes deficient in electrons. The direction of electron movement and the magnitude of the potential difference between the metals depends upon the type of metals used and the difference in temperature between their junctions.

A thermocouple of chromel-alumel will develop approximately 50 mV of electrical potential difference between the metals when one junction is at 0°C and the other at 1200°C. Obviously, this method of generating an EMF is not practical for producing source voltages for circuits required to perform an appreciable amount of work. Due to its sensitivity to temperature changes, however, the thermocouple is widely used for temperature measurement and temperature control devices.

Contacts: Switches used to either connect or disconnect branch circuits. They are magnetic in nature and are operated electromagnetically. Contacts, which are commonly used to switch loads such as lighting, heating, and controlling ac motors in which overload protection is dealt with separately, differ from a motor starter in a number of ways. The major difference is that a contact does not contain overload relays, whereas a motor starter does. Electromagnetic operation of switches presents a number of advantages over manual control. In situations where large contacts are necessary, such as in the case of remote control of high-current circuits, it would be quite costly to run power leads that would be required for manual control. Thus, electromagnetic operation used here would be less expensive.

Some of the devices that are commonly used to provide control for the operation of contacts are float switches, push buttons, limit switches, thermostats, pressure switches, and other pilot devices.

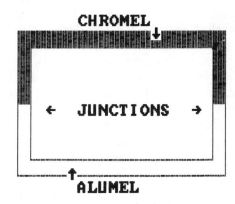

FIGURE 121

Continuity tester: A bell-and-battery device to test circuits after wiring is completed. In making such tests, connect a pair of dry cells with a bell in series, located at the center of the distribution panel board, to the main of the circuits to be tested. Then test each circuit separately as follows: Bare the ends of the wires at each outlet to be tested and, in addition, at the switch outlets. Then close the circuit by touching the proper wires together. If the circuit is complete, the bell will ring. If it is a large wiring installation, which makes it impossible to hear the bell when it is placed at the panel board, bare the ends of the branch-circuit wires, connect the wires together at the center of distribution, and connect the bell and battery in series to the branch-circuit wires at the outlet boxes. The bell will ring if the circuit is complete.

Each circuit should be opened immediately after it has been tested to avoid any possibility of later energizing the wiring system with the short circuits remaining on some branch circuits.

Control: 1. The part of a digital computer or processor that determines the execution and interpretation of instructions in proper sequence, including the decoding of each instruction and the application of the proper signals to the arithmetic unit and other registers in accordance with the decoded information. **2.** Frequently, one or more of the components in any mechanism responsible for interpreting and carrying out manually initiated directions. Sometimes called manual control. **3.** In some business applications, a mathematical check.

Controller: A device or group of devices that serves to govern, in some predetermined manner, the electric power delivered to the apparatus to which it is connected.

Convection: The transference of heat by the circulation of heated portions of a fluid. As a gas or a liquid gets heat from a hot surface, the heat expands the fluid, causing it to become less dense. The cooler and denser surrounding fluids settle and push the less dense portion upward. This method of transmitting heat by upward currents—convection currents—caused by heat is called *convection*.

One of the best examples of convection is the cooling of a refrigerator. The heat is transferred upward by convection, explaining why the warmest air is usually found near the top. This warm air near the top comes in contact with the cold evaporator. As it cools, it becomes denser and starts to settle.

Convection currents are thus set in motion. As the cool air settles, it picks up heat from the warm food and the walls of the refrigerator through which the heat has leaked. This heat is carried to the evaporator by the convection currents.

Conversion: The process of changing information from one form of representation to another, such as from the language of one type of machine to that of another or from magnetic tape to the printed page. It is also the process of changing from one data processing method to another, or from one type of equipment to another, e.g., conversion from punch card equipment to magnetic tape equipment.

Converter: A device that converts the representation of information, or that permits the changing of the method for data processing from one form to another, e.g., a unit that accepts information from punch cards and records the information on magnetic tape, and possibly includes editing facilities.

Copper loss: The power lost in heat in the windings due to the flow of current through copper coils. This loss varies directly with the armature resistance and as the square of the armature current. The armature resistance varies with the length of the armature conductors and inversely with their cross-sectional area. Armature conductor size is based on an allowance of from 300 to 1200 circular mil/A. For example, a two-pole armature that is required to supply 100 A may use a wire size based on 800 circular mil/A, or $100/2 \times 800 = 40,000$ circular mil. This value corresponds to a No. 4 wire. Very small armature windings may use only 300 circular mils per ampere with a resulting high current density. Large generators (5000 kW) require an allowance of 1200 circular mil/A with a resulting low current density in the windings. These variations are the result of the variable nature of the heat-radiating ability of the armature conductors.

Very small round conductors have a much higher ratio of surface-to-volume than do large round conductors. For example, a 0.1-in.-diameter round conductor of a given length has a surface-to-volume ratio of

$$\frac{4\pi D}{\pi D^2} \quad \text{or} \quad \frac{4}{0.1} = 40$$

A 1.0-in.-diameter conductor of the same length has a surface-to-volume ratio of 4/1, or 4. Since the heat-radiating ability of a round conductor varies as the ratio of its surface to volume, the 0.1-in.-diameter conductor has 40/4, or 10 times the heat-radiating ability of the 1-in.-diameter conductor, other factors being equal.

High-speed generators use a lower circular-mil-per-ampere allowance than low-speed generators because of better cooling. The temperature rise is limited by ventilating ducts, and in some cases by the use of forced ventilation, as in aircraft dc generators.

The hot resistance of an armature winding is higher than its cold resistance. A 2.5°C increase in the temperature of a copper conductor corresponds to an increase in resistance of approximately 1 percent. For example, if the no-load temperature of an armature winding is 20°C, and its full-load temperature is 70°C, the increase in resistance is $(70 - 20)/2.5$, or 20 percent. Thus, if the no-load resistance is 0.05 Ω between brushes, the hot resistance will be 1.2×0.05, or 0.06 Ω. If the full-load armature current is 100 A, the full-load armature copper loss will be $100^2 \times 0.06$, or 600 W. The armature copper loss varies more widely with the variation of electrical load on the generator than any other loss occurring in the machine. This is because most generators are constant-potential machines supplying a current output that varies with the electrical load across the brushes. The limiting factor in load on a generator is the allowable current rating of the generator armature.

The armature circuit resistance includes the resistance of the windings between brushes of opposite polarity, the brush contact resistance, and the brush resistance.

Coulomb: The unit quantity of electricity named for the French physicist, Charles Augustin de Coulomb. The coulomb may be defined as the quantity of electricity that passes any cross section of a circuit, in 1 s, when the current is maintained constant at 1 A strength. Hence, a current of electricity flowing through a circuit is measured in so many coulombs per second, just as the rate at which a quantity of fluid delivered by a pump is measured in so many gallons per second, or per minute.

The coulomb is the quantity of electricity, as the gallon or pound is the quantity of matter. In general practice, the current strength (meters) and not the quantity (coulombs) is dealt with; consequently, the coulomb is seldom used.

Coulomb's law: The electromagnetic principle stated by French physicist Charles Augustin de Coulomb as: "The force existing between two charged bodies is directly proportional to the product of the charges and inversely proportional to the square of the distance separating them."

When a charged body is brought into close proximity with another charged body, there is a force that causes the bodies to attract or repel one another. If the charged bodies possess the same sign of charge, a repelling force will exist between the two bodies. If they have unlike signs, there will be a force of attraction between them. The force of attraction or repulsion is caused by the electrostatic field that surrounds every charged body.

If a material is charged positively, it has a deficiency of electrons. If it is charged negatively, it has an excess of electrons. The direction of the electrostatic field is represented by lines of force drawn perpendicular to the charged surface and shown originating from the positive-charged material. Each line of force is drawn in the form of an arrow and is shown pointing from positive to negative.

If a test charge is inserted in an existing electrostatic field, it will move toward one or the other of the charged areas that is causing the field to exist. The direction of movement will depend on whether the test charge is positive or negative. A positive test charge placed in a field moves in the direction that the line of force points, from positive toward negative. In this case, the test charge will be an electron, and since the electron is negative, it will move in a direction opposite to that of the positive charge. In other words, an electron in an electrostatic field will move against the arrow from negative toward positive. This action is illustrated in Fig. 122.

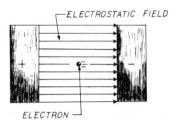

FIGURE 122   *ELECTRON*

If Coulomb's law is analyzed in connection with this figure, it can be seen that the greater the distance between the electron and the positive charge, the less the force of attraction.

Counter: A device, register, or location in computer storage for storing numbers or number representations in a manner that permits these numbers to be increased or decreased by the value of another number, or to be changed or reset to zero or to an arbitrary value.

Counter EMF: The opposing voltage that prevents excessive current flow after starting a large motor. Abbreviated *CEMF*. To start a large motor, it is necessary to place a resistance unit in series with the motor so that the starting current is reduced to a safe value. As the motor accelerates, this resistance can be gradually decreased. The resistance is not required after the motor has reached the desired speed because the motor is then generating a voltage in opposition to the impressed voltage, thereby preventing excessive current flow. This opposing voltage is called the *counter electromotive force (counter EMF)*, and its value will depend on the speed of the motor, which is greatest at full speed and zero at standstill.

For example, if the armature of a 240-V motor has a resistance of 2 Ω, the current flow at standstill will be, according to Ohm's law

$$I = \frac{E}{R} = \frac{240}{2} = 120 \text{ A}$$

If the motor is running and thus is generating a counter EMF of 100 V, the total voltage in the armature is 240 − 100, or 140 V. Therefore, the current is

$$I = \frac{E}{R} = \frac{140}{2} = 70 \text{ A}$$

The flow of current has been reduced considerably by the counter EMF. If the motor is running at full speed and is generating a counter EMF of 200 V, then the current is

$$I = \frac{E}{R} = \frac{230 - 200}{2} = 15 \text{ A}$$

Counterpoise: An artificial antenna ground composed of a long metal conductor usually stretched close to the surface of the earth and insulated from it. The length of the counterpoise should at least be equal to the above-ground height of the antenna with which it is associated. There are two types of counterpoise: radial and continuous. With either type, the conductor is buried a depth of 1 to 3 ft and may be connected solidly to each tower or through a small gap. The conductor material of the counterpoise is not important, except that it should be mechanically strong and resistant to corrosion.

It has been shown theoretically and by tests that the counterpoise has an initial transient surge impedance of 150 to 200 Ω, which decays to its final value of resistance, the transition time being the time required for the first reflection to return, traveling at one-third of the speed of light.

The continuous counterpoise may have some advantage in keeping lightning trouble localized, especially when areas of low resistance are more or less insulated from other areas by rock or similar nonconducting material.

Counter voltage: The counter electromotive force (EMF) or voltage induced in a motor by its armature conductor. As a motor rotates, its conductor will be cutting lines of force of the field. As the conductors of the motor in Fig. 123 are revolving in the same direction they did in the generator, the voltage induced in the coils will be in the opposite direction to the applied line voltage. This voltage is generated in the coils of any motor during operation.

The applied voltage is equal to the counter EMF plus the voltage drop in the armature, or $E = CEMF + IR$.

As the counter voltage opposes the applied line voltage, it regulates the amount of current the line will send through the armature. The resistance of the armature winding is very low, being only about 0.25 Ω in the ordinary 5-hp, 110-V motor. From this, it can be seen that if it were not for the counter voltage, an enormous current would flow through this armature.

Applying Ohm's law, or $E/R = I$, we find that $110/0.25 = 440$ A. Actually, a motor of this size would ordinarily draw only about 10 A when operating without mechanical load.

The counter voltage can be determined in the following manner: $I \times R = E$, so $10 \times 0.25 = 2.5$ V, or the voltage required to force 10 A through the armature resistance. Subtracting this from the applied voltage, $110 - 2.5 = 107.5$ V, counter EMF.

Counting circuit: A circuit that receives uniform pulses representing units to be counted and provides a voltage proportional to their frequency. Through slight modifications, the counting circuit is used in conjunction with a blocking oscillator to produce a trigger pulse that is a submultiple of the frequency of the pulses applied, thus acting as a frequency divider. The pulses applied to the counting circuit must be of the same time duration if accurate frequency division is to be made. Counting circuits are ordinarily preceded by shaping circuits and limiting circuits to ensure uniformity of amplitude and width. Under these conditions, the pulse repetition frequency constitutes the only variable, and frequency variations may be measured.

Coupling: A mechanical device that attaches a mechanical power

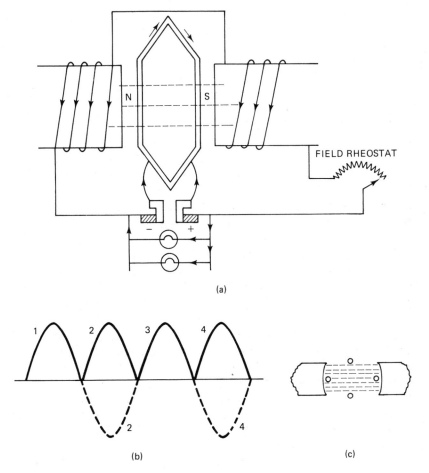

FIGURE 123 Counter voltage.

source, such as a motor, to the remainder of the power chain. There are two main types of couplings as related to electric motors: rigid and flexible.

The rigid type, in which the shafts of the motor and the machine are securely bolted or fastened together, is seldom used for motor applications, except on factory-assembled machines, such as motor-generator sets. Extremely accurate alignment of the shafts is necessary to prevent bearing difficulties.

Flexible couplings are made in a wide variety of types and designs to provide for slight misalignment between shafts, and also to absorb some of the shocks or vibrations. Most high-grade flexible couplings permit a reasonable amount of angular and parallel misalignment of the shafts and some axial movement.

Crossarms: Wooden or steel beams used to support insulators and conductors on some types of overhead power line construction. Crossarms vary in size and length depending upon the weight of the conductors or equipment, the number of conductors, and the spacing between conductors, depending on the circuit voltage and required climbing space. Crossarms are fastened to poles by means of machine bolts of the proper length and diameter, commonly called *through bolts,* with steel washers placed under the bolthead and nut. To provide adequate bearing surface against the pole, a flat notch or *gain* is cut in the face of the pole, or manufactured metal gains are used. The through bolt holes in the pole are usually bored on the job. Crossarm braces are used to maintain the crossarm in a horizontal position.

Crystal lattice filter: A type of filter commonly used in SSB equipment. This type of filter makes use of extremely high-Q quartz crystal resonators. Lattice filters often consist of six or more crystals. This type of device offers an economical method of selective filtering.

Figure 124 illustrates a single section crystal lattice filter and its equivalent electrical circuit. The series crystals, Y_1 and Y_2, are a matched pair, as are the shunt or lattice crystals, Y_3 and Y_4. The resonant frequency of Y_3 and Y_4 is higher than the resonant frequency of Y_1 and Y_2. The passband of the filter is primarily determined by the difference in frequency between the two sets of crystals.

The operation of this filter may be better visualized if the circuit is redrawn to form a bridge, as in Fig. 125. It may be observed that these circuits are electrically identical by referring to the lettered points in the two figures. The bridge will be balanced when the effective reactances of the crystals are equal in amplitude and of the same type (i.e., inductive or capacitive). Under this condition, there is zero output. The bridge is at its unbalanced limit when the effective reactances are equal in amplitude but opposite in type. Under such a condition, the output is maximum.

A graph of the effective reactances of the two sets of crystals versus frequency is shown in Fig. 126. Note that the parallel resonant point of the series crystals corresponds to the series resonant point of the lattice crystals.

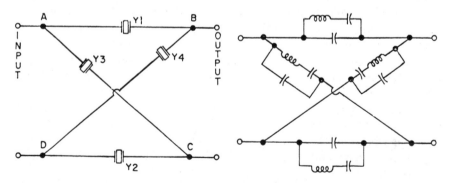

FIGURE 124   Crystal lattice filter.

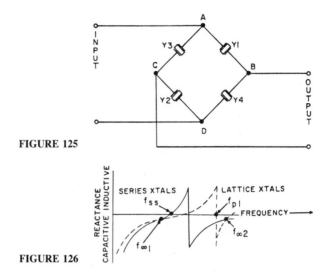

FIGURE 125

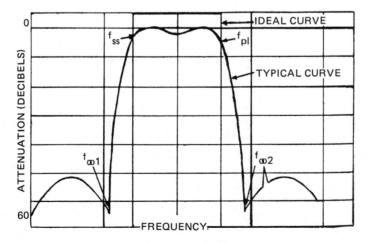

FIGURE 126

Figure 126 shows that the conditions previously given for an unbalanced bridge are satisfied between points f_{ss} and f_{pl}. This frequency range is the passband of the filter. At f_{oo1} and f_{oo2}, the conditions for a balanced bridge are satisfied. At these frequencies, output will be zero, and attenuation will be maximum. Below f_{oo1} and f_{oo2}, attenuation, though not maximum, will be very high and output will be very low.

A typical response curve for a crystal lattice filter is depicted in Fig. 127. It should be noted that the bridge balance and/or unbalance may not be absolute. This, of course, would result in an imperfect response curve. To improve the response, additional filtering sections are cascaded to form a multisection filter.

FIGURE 127   Crystal lattice filter response.

Current: The medium by which electricity carries out its mission produced in a material if its atoms have free charged particles (positive ions and negative ions) or electrons that actually can move from one atom to another. All metals have this characteristic. In the presence of an electric field, these free electrons stream from atom to atom through a conductor. The unit of current is the ampere (abbreviated A), named after the pioneer physicist, Andre M. Ampere. The symbol for current is I. Charge movement that causes a current flow in a conductor is often considered analogous to the fluid used in a hydraulic system as a medium to transmit power a moderate distance to perform some work function.

Current transformer: A device used to insulate the instruments and operator from the line voltage and to act as a multiplier for the instruments. In order to obtain maximum safety for men and apparatus, one secondary lead must be grounded; the metal case, if any, should be grounded; and connections must not be changed with voltage on. The primary of the transformer must be connected in the line and the secondary to the instruments; the secondary of the transformer must not be opened with the current flowing in the primary. The line voltage should not exceed the rated primary voltage of the transformer, and the line current should not exceed the rated current except for very short intervals.

In order to obtain satisfactory accuracy, the circuit frequency should not be less than the lowest rated frequency by more than a few cycles. The circuit frequency should never exceed 125 cycles. The line current should range from 10 to 125 percent of the rated current, and the secondary burden should not exceed 1 Ω (25 VA at 5 A).

The ammeters generally used with current transformers are rated at 5 A. Replacing such an ammeter by a lower-rated one for taking readings when the current is considerably below the transformer rating may often be as unsatisfactory as estimating the reading at the lower end of the 5-A ammeter. In such cases, a more suitable transformer should be obtained.

Most current transformers are designed for use at frequencies from 25 to 125 cycles. They are generally calibrated for ratio at 25 or 60 cycles, or both, with a secondary burden of one 5-A P3 ammeter, the current coil of one 5-A P3 wattmeter and 50 ft of leads (100 ft of No. 10 AWG wire), over a range of currents from 0.5 to 5.0 secondary A. Certificates giving the test results are furnished. Special tests are made when necessary in request from the person in charge of the test. The ratio will not, in general, differ from the certificate data by more than 1 percent at any frequency from 22 to 125 cycles. The addition of a 5-A ammeter or wattmeter will not raise the ratio by more than 0.25 percent on most transformers. It should be noted that the ratio generally differs at different secondary currents.

The formula for primary amperes, I, is

$$I - Ar \times CTr$$

where Ar is the ammeter reading corrected for scale error and CTr is the certified ratio of the current transformer.

It is preferable to use three current transformers and three ammeters to measure the three line currents on three-phase, three-wire circuits. If only two transformers are used, the current in the line without a transformer is represented by the reading of an ammeter placed in the common line short-circuiting the two secondaries, provided both transformers have the same rating and are connected as shown in Fig. 128.

The use of only two current transformers tends to unbalance the circuit when both the voltage and current are small, as when testing small induction motors. In case a current transformer should become accidentally magnetized, either by direct current flowing through either winding, or by alternating current flowing through either winding with the other winding open-circuited, it should be demagnetized by applying at least 50 percent of the rated primary current with 30 Ω or more in the secondary circuit. This resistance should then be gradually reduced to zero, in steps of 1 Ω or less. Magnetization affects the transformer accuracy.

Current, transistor cutoff: The current that flows in the collector circuit when the input current is zero. When the emitter lead is opened and no current flows through the emitter-base (input junction), there will still be a small amount of current flowing in the collector circuit due to the reverse current I_{CBO}.

I_{CBO} is the reverse current for a common-base configuration. If a common-emitter configuration is used, the reverse current will be termed I_{CEO}, or the current flowing in the collector circuit with the base lead open.

In any conductor, equal currents flowing in opposite directions will cancel and the resultant current in the conductor will be zero. Thus, in a common-emitter circuit, an input signal that exactly opposes the forward bias will cause a zero base current. The result is the same as if the base lead had been opened.

Even though the input current is zero, the output current will not necessarily be zero because of the reverse current. The area below the line marking the zero input current (either $I_E = 0$ or $I_B = 0$, depending on the

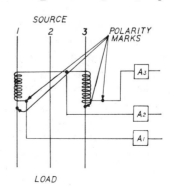

FIGURE 128   LOAD

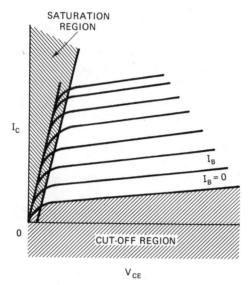

SATURATION
REGION

I_C

I_B

$I_B = 0$

0

CUT-OFF REGION

V_{CE}

FIGURE 129

configuration used) is called the *cutoff region*. The cutoff region is shown in Fig. 129.

Cybernetics: The field of technology involved in the comparative study of the control and intracommunication of information-handling machines and nervous systems of animals and man in order to understand and improve communication.

Damon effect: The change that the susceptibility of a ferrite undergoes under the influence of high RF power.

Damper, stockbridge: A device to reduce vibration on ac power transmission lines. Vibration is normally caused by wind eddies on the leeward side of the conductor that swing from the upper to the lower side at regular intervals—the rate depending on the diameter of the conductors and the velocity of the wind. Stockbridge dampers have been used on copper and steel current-carrying conductors, as well as ground wires.

Damper windings: Pole-face windings, also called *amortisseur windings,* on some alternators that consist of copper or bronze bars inserted in holes provided for them in the punching near the air-gap surface of the pole. The bars in any one pole are connected at their ends by brazing them to the pole end plates or to a short-circuiting bar. Occasionally, they are connected from pole to pole. The bars are not insulated from the poles.

Damper windings serve two purposes—to damper the fluctuations in speed that occur in each revolution when the prime mover is a diesel or other reciprocating-type engine, and to prevent excessive unbalance voltages during short circuits. The damping function is not important when the alternator is driven by a hydraulic or steam turbine. The voltage-balancing function is not usually considered important except on larger alternators, especially those connected to long transmission lines.

Damping: A characteristic built into electrical circuits and mechanical systems to prevent rapid or excessive corrections that may lead to instability or oscillatory conditions. An example is connecting a register on the terminals of a pulse transformer to remove natural oscillations or placing a moving element in oil or sluggish grease to prevent mechanical overshoot of the moving parts.

D'Arsonval meter: A stationary permanent magnet moving-coil meter that is the basic movement used in most measuring instruments for servicing electrical and especially dc equipment. It is so called because it was first employed by the Frenchman D'Arsonval in making electrical measurements.

The basic D'Arsonval movement consists of a stationary permanent magnet and a movable coil. When current flows through the coil, the resulting magnetic field reacts with the magnetic field of the permanent magnet and causes the coil to rotate. The greater the amount of current flow through the coil, the stronger the magnetic field produced. The stronger this field, the greater the rotation of the coil. In order to determine the amount of current flow, a means must be provided to indicate the amount of coil rotation. Either of two methods may be used—the pointer arrangement, or the light and mirror arrangement.

In the pointer arrangement, one end of the pointer is fastened to the rotating coil. As the coil turns, the pointer also turns. The other end of the pointer moves across a graduated scale and indicates the amount of current flow. A disadvantage of the pointer arrangement is that it introduces the problem of coil balance, especially if the pointer is long. An advantage of this arrangement is that it permits overall simplicity.

The use of a mirror and a beam of light simplifies the problem of coil balance. When this arrangement is used to measure the turning of the coil, a small mirror is mounted on the supporting ribbon (Fig. 130) and turns with the coil.

An internal light source is directed to the mirror and then reflected to the scale of the meter. As the moving coil turns, so does the mirror, causing the light reflection to move over the scale of the meter. The movement of the reflection is proportional to the movement of the coil. Thus, the amount of current being measured by the meter is indicated.

A simplified diagram of one type of stationary permanent magnet mov-

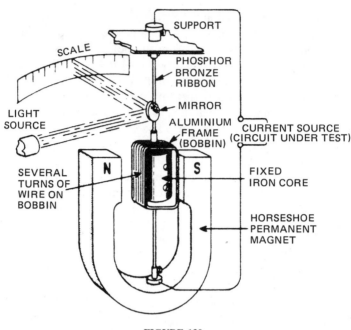

FIGURE 130

ing-coil instrument is shown in Fig. 130. Such an instrument is commonly called a *galvanometer*. The galvanometer indicates very small amounts (or the relative amounts) of current or voltage, and it is distinguished from other instruments used for the same purpose in that the movable coil is suspended by metal ribbons instead of shaft and jewel bearings.

The movable coil of the galvanometer in Fig. 130 is suspended between the poles of the magnet by means of thin, flat ribbons of phosphor bronze. These ribbons provide the conducting path for the current between the circuit under test and the movable coil. The restoring force, exerted by the twist in the ribbons, is the force against which the driving force of the coil's magnetic field is balanced in order to obtain a measurement of the current strength. The ribbons thus tend to oppose the motion of the coil and will twist through an angle that is proportional to the force applied to the coil by the action of the coil's magnetic field against the permanent field. The ribbons restrain or provide a counter force for the magnetic force acting on the coil. When the driving force of the coil current is removed, the restoring force returns the coil to its zero position.

If a beam of light and mirrors is used, the beam of light is swept to the right or left across a central-zero translucent screen (scale) having uniform divisions. If a pointer is used, the pointer is moved in a horizontal plane to the right or left across a central-zero scale having uniform divisions. The direction in which the beam of light or the pointer moves depends on the direction of current through the coil. This instrument is used to measure minute current, as in bridge circuits. In modified form, the basic D'Arsonval movement has the

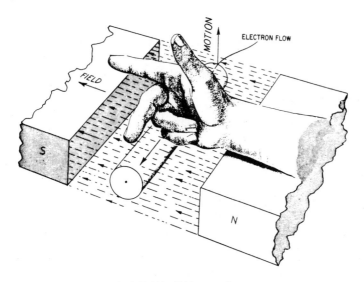

FIGURE 131 D'Arsonval meter.

highest sensitivity of any of the various types of meters in use today. To understand the operating principle of the D'Arsonval meter, it is first necessary to consider the force acting on a current-carrying conductor placed in a magnetic field. The magnitude of the force is proportional to the product of the magnitudes of the current and the field strength. The field is established between the poles of a U-shaped permanent magnet and is concentrated through the conductor by means of a soft-iron stationary member mounted between the poles to complete the magnetic circuit. The conductor is made movable by shaping it in the form of a closed loop and mounting it between fixed pivots so that it is free to swing about the fixed iron member between the poles of the magnets. A convenient method of determining the direction of motion for the conductor is by the use of the *right-hand motor rule for electron flow* (Fig. 131).

To find the direction of motion of a conductor, the thumb, first finger, and second finger of the right hand are extended at right angles to each other, as shown. The first finger is pointed in the direction of the flux (toward the south pole), and the second finger is pointed in the direction of electron flow in the conductor. The thumb then points in the direction of motion of the conductor with respect to the field. The conductor, the field, and the force are mutually perpendicular to each other. The force acting on a current-carrying conductor in a magnetic field is directly proportional to the field strength of the magnet, the active length of the conductor, and the intensity of the electron flow through it. Thus

$$F = \frac{8.85 \times BLI}{10^8}$$

where F is the force in pounds, B the flux density in lines per square inch, L the active length of the conductor in inches, and I the current in amperes.

In the D'Arsonval-type meter, the length of the conductor is fixed, and the strength of the field between the poles of the magnet is fixed. Therefore, any change in I causes a proportionate change in the force acting on the coil.

The principal of the D'Arsonval movement may be more clearly shown by the use of the simplified diagram (Fig. 132) of the D'Arsonval movement commonly used in dc instruments. In the diagram, only one turn of wire is shown. However, in an actual meter movement, many turns of fine wire would be used, each turn adding more effective length to the coil. The coil is wound on an aluminum frame or bobbin, to which the pointer is attached. Oppositely wound hairsprings (one of which is shown in Fig. 132) are also attached to the bobbin, one at either end. The circuit to the coil is completed through the hairsprings. In addition to serving as conductors, the hairsprings serve as the restoring force that returns the pointer to the zero position when no current flows.

As has been stated, the deflecting force is proportional to the current flowing in the coil. The deflecting force tends to rotate the coil against the restraining force of the hairspring. The angle of rotation is proportional to the force that the spring exerts against the moving coil (within the elastic limit of the spring). When the deflecting force and the restraining force are equal, the coil and the pointer cease to move. Because the restoring force is proportional to the angle of deflection, it follows that the driving force and the current in the coil are proportional to the angle of deflection. When current ceases to flow in the coil, the driving force ceases, and the restoring force of the springs returns the pointer to the zero position.

If the current through the single turn of wire is in the direction indicated (away from the observer on the right-hand side and toward the observer on the

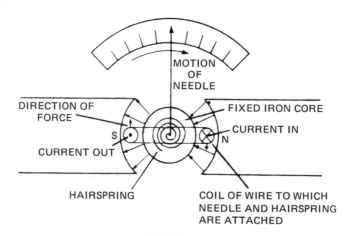

FIGURE 132

left-hand side), the direction of force, by the application of the *right-hand motor rule,* is upward on the left-hand side and downward on the right-hand side. The direction of motion of the coil and pointer is clockwise. If the current is reversed in the wire, the direction of motion of the coil and pointer is reversed. A detailed view of the basic D'Arsonval movement, as commonly employed in ammeters and voltmeters, is shown in Fig. 133. This instrument is essentially a microammeter because the current necessary to activate it is of the order of 1 μA.

The principle of operation is the same as that of the simplified versions discussed previously. The iron core is rigidly supported between the pole pieces and serves to concentrate the flux in the narrow space between the iron core and the pole piece (in other words, in the space through which the coil and the bobbin move).

Current flows into one hairspring, through the coil, and out of the other hairspring. The restoring force of the spiral springs returns the pointer to the normal, or zero, position when the current through the coil is interrupted.

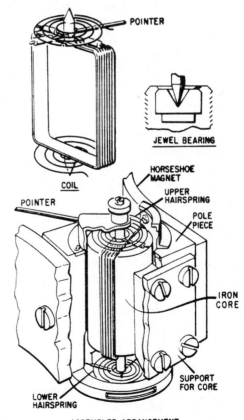

FIGURE 133   **ASSEMBLED ARRANGEMENT**

Conductors connect the hairsprings with the outside terminals of the meter. If the instrument is not damped, that is, if viscous friction or some other type of loss is not introduced to absorb the energy of the moving element, the pointer will oscillate for a long time about its final position before coming to rest. This action makes it nearly impossible to obtain a reading, and some form of damping is necessary to make the meter practicable. Damping is accomplished in many D'Arsonval movements by means of the motion of the aluminum bobbin upon which the coil is wound. As the bobbin oscillates in the magnetic field, an EMF is induced because it cuts through the lines of force. Therefore, according to Lenz's law, induced currents flow in the bobbin in such a direction as to oppose the motion, and the bobbin quickly comes to rest in the final position after going beyond it only once.

In addition to factors such as increasing the flux density in the air gap, the overall sensitivity of the meter can be increased by the use of a lightweight rotating assembly (bobbin, coil, and pointer) and by the use of jewel bearings as shown. While D'Arsonval-type galvanometers are useful in the laboratory for measurements of extremely small currents, they are not portable, compact, or rugged enough for use in the maintenance of military equipment. The Weston meter movement is used instead.

Data: A general term used to denote any or all facts, numbers, letters, and symbols, or facts that refer to or describe an object, idea, condition, situation, or other factors. The term connotes basic elements of information that can be processed or produced by a computer. Sometimes, data are considered to be expressible only in numerical form, but information is not so limited.

Data processing: The preparation of source media that contain data or basic elements of information and the handling of such data according to precise rules of procedure to accomplish such operations as classifying, sorting, calculating, summarizing, and recording. Also, the production of records and reports.

Data transmission equipment: The communications equipment used in direct support of data processing equipment.

Daylight: A term used for artificial lighting that approaches the target standard of the spectral quality of daylight at around 7500 K. Specifications for the best artificial daylighting for accurate work include: a large source of relatively low luminance, duplication of color of a moderately overcast north sky, and more light for inspecting dark-colored samples than light-colored samples.

db: Abbreviation of decibel. See *Decibel*.

dBa: Abbreviation of adjusted decibels, a noise level above a reference noise level measured at any point in a system with a noise meter that has previously been adjusted for zero according to specifications. All readings are in decibels. See *Decibel*.

dBm: Abbreviation of decibels referred to 1 mW, where the resistive load impedance is 600 Ω unless otherwise specified. To illustrate, 0 dBm is 1 mW across a 600-Ω resistor.

dB meter: A usually high-impedence ac voltmeter with a scale reading directly in decibels for measuring noise level.

dc generator: A direct current device to convert mechanical energy into electrical energy. Mechanical energy is used to rotate the armature coil through the magnetic flux between the field poles. Thus, an electrical pressure is induced in the coil, which will force current to flow if a load is connected to points marked *line*. If the load demands more current, more mechanical energy will be required to turn the armature coil. The device furnishing the mechanical energy (steam engine, diesel engine, electric motor, etc.) is known as the prime mover.

The windings on the iron field poles are connected to the brushes, as are the line wires. This places the field windings parallel with the armature and line. For this reason, it is called a *shunt-wound generator*. After the field poles have once magnetized, a small amount of magnetism will remain in the poles. This is residual magnetism.

A dc generator may be separately excited; that is, the field windings may receive their energy from an external source of supply. A simple conventional sketch of such a generator is shown in the upper left-hand corner of Fig. 134. This type is rarely used. More common is the self-excited type, which means that the energy to magnetize the field coils is obtained from the armature of the same machine. Loss of residual magnetism can be overcome by raising the brushes off the commutator and sending current through the field coils in the correct direction.

dc machines: Rotating machines that make use of the effects of electromagnetism. A generator converts mechanical power into electric power. A prime mover, which may be a water or steam turbine, a steam engine, or a motor, provides the power for moving the conductors in the generator. The voltage induced in the conductors by electromagnetic induction is used to supply electric current to electrical devices.

A motor converts electric power into mechanical power. The electric current applied to motor conductors in a magnetic field produces a twisting power, or torque, which can be applied to mechanical loads, such as blowers, pumps, cranes, elevators, and many other devices that were powered mechanically or manually before electricity was discovered.

Although for the most part alternating current is used for lighting systems, household appliances, and a majority of industrial motors, there are many instances in which direct current is used. Direct current motors have some characteristics, such as simpler speed control, which make them more suitable than ac (alternating current) motors for certain applications, even though an ac motor of a given rating costs less than a dc motor of the same rating.

dc motor: A motor that operates on the first law of magnetism, stating

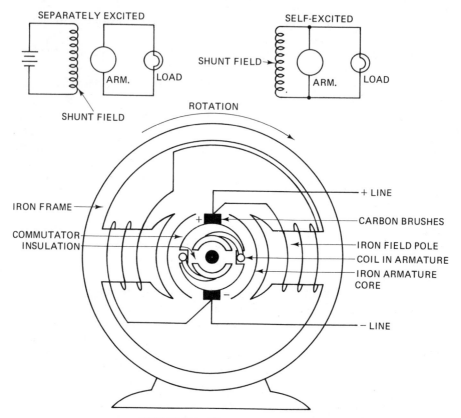

FIGURE 134 Direct current generator.

that like poles repel and unlike poles attract. Current flowing through the field coils produces the field poles, and current through the armature coils develops armature poles midway between the field poles. Attraction and repulsion between these two sets of poles produces rotation. The armature poles remain stationary in space. By reversing the direction of current flow through the fields or through the armature, the field poles of the armature poles will be reversed, and the direction of rotation changed.

Direct-current motors are built essentially the same as dc generators. They have a field, which carries the field poles energized by direct current, an armature that consists of an iron core with coils, a commutator mounted on a shaft, and brushes.

In motors as well as generators, the field coils are provided with direct current. In a generator, the generated voltage is utilized in an external circuit through the brushes; while in a dc motor, voltage is applied to the brushes, through them and through the commutator segments, to the armature coils. Since the armature is located in a magnetic field, the armature starts to revolve on the shaft when the current flows through the coils. The revolving motion of the shaft is mechanically transferred to perform mechanical work, such as moving a crane or a tool.

Dead band: A specific range of values in which the incoming signal can be altered without also changing the outgoing response. Synonymous with dead space, dead zone, and switching blank.

Dead coils: In armature windings of ac generators, coils left unconnected during manufacture to obtain operating characteristics not readily obtainable from adjustment of the number of coils and turns. Coils may sometimes be cut out of generators in order to permit operation after one coil or more has been damaged.

Decade: A group or assembly of 10 units; i.e., the readings given by a counter that counts to 10 in one column, or a resistor box that inserts resistance quantitites in multiples of powers of 10.

Decay time: The time in which a voltage or current pulse will decrease to one-tenth of its maximum value. Decay time is proportional to the time constant of the circuit.

Decibel: One-tenth of the international transmission unit, the bel, which is a unit of gain equivalent to a 10 to 1 ratio of power gain. Thus, the gain in bels is the number of times that 10 is taken as a factor to equal the ratio of the output power of an amplifier to the input power. For example, if output power is 100 times the input, the ratio is 100 to 1, or 10^2 to 1. The gain is therefore 2 bels; and the gain in decibels (dB) is 10 times 2, or 20 dB. Where the power ratio is 1000 to 1, it may also be written as 10^3 to 1, or 3 bels; and the gain in dB is 10×3, or 30 db. Where the power ratio has been increased 10,000 times, that gain is 10^4 to 1, or 4 bels; and the gain in decibels equals 40 dB.

The number of 10 factors contained in any ratio of the output power (P_2) to the input power (P_1) is the logarithm of the ratio to the base 10. The gain in decibels may therefore be expressed conveniently as

$$dB = 10 \log_{10} \frac{P_2}{P_1}$$

The human ear responds to ratio changes in intensity rather than to changes in absolute value. In other words, the ability of the human ear to detect changes in the intensity of sound is much greater at low levels of intensity than it is at high levels. A change in power level of 1 dB is barely perceptible to the human ear. For this reason, attenuators in audio systems are frequently calibrated in steps of 1 dB.

Because the ear responds logarithmically to variations in sound intensity levels, any practical system for measuring sound intensity levels must necessarily vary logarithmically. The decibel system of measuring power levels is based on this concept. Since the gains or losses in a system are expressed logarithmically, they are simply added or subtracted to determine the overall gain or loss. For example, transmission lines introduce a loss in power, amplifier stages produce a gain, and attenuators introduce a loss. The final result is the algebraic sum of the various gains and losses.

Primarily, the decibel is a unit of measure of power ratios. it can be used readily to compute current ratios as well, provided the resistances through which the current flow are taken into account. The dB gain or loss expressed in terms of the currents and resistances is determined as follows

$$dB = 20 \log_{10} \frac{I_2 \sqrt{R_2}}{I_1 \sqrt{R_1}}$$

where I_2 and I_1 are, respectively, the output and input voltages, and R_2 and R_1 are, respectively, the output and input resistances in ohms.

If the voltages and resistances are known, the dB gain or loss may be determined by direct substitution in the equation. If the resistances are equal, they may, of course, be cancelled out.

The same reasoning also applies to the voltage ratio, provided the resistances across which the voltages are applied are properly considered. The equation for dB gain or loss when voltages and resistances are employed directly is determined as follows

$$dB = 20 \log_{10} \frac{E_2 R_1}{E_1 R_2}$$

where E_2 and E_1 are, respectively, the output and input voltages, and R_2 and R_1 are, respectively, the output and input resistances in ohms. If the voltages and resistances are known, the dB gain or loss may be determined by direct substitution in the equation.

Considerable confusion has resulted from the use of various so-called zero-power reference levels. The term *zero reference level* is, in itself, somewhat confusing, because it does not mean that zero power is developed at that level. It means that the output level is referred to an arbitrary level designated as the reference, or zero, level. As such, it is perhaps one of the most convenient ways of expressing a power ratio. It is meaningless to say, for example, that a certain amplifier stage has an output of 30 dB, unless reference is made to some established power level.

It is common practice in naval and telephone work to consider 6 mW as the reference power level. Other values are also used in this and other fields, such as 1, 10, and 12.5 mW depending upon which unit is most convenient under the circumstances.

The voltage gain or loss of microphones, transmission lines, and voltage amplifiers is also generally expressed in decibels. In general, transmission lines introduce a loss and voltage amplifiers produce a gain. A reference voltage level and the resistance (if it differs from the one being compared) across which the signal appears must be given in order that the gain or loss may have meaning.

The voltage output of a microphone may be expressed in terms of decibels below 1 V/dyn/cm^2. In other words, 1 dyn acting on 1 cm^2 and producing an output of 1 V is taken as the zero-decibel output level.

When the voltage gain of an amplifier stage is given in decibels, the input and output impedances must be given, or they must be assumed to be equal. Thus, if they are assumed to be equal, the gain in decibels may be expressed as

$$dB = 20 \log_{10} \frac{E_2}{E_1}$$

where E_2 is the output voltage, and E_1 is the input voltage. When certain arbitrary power reference levels are used, the dB gain or loss is given a special designation. One of these designations is the dBm, or the power level in decibels referred to 1 mW, as in

$$dBm = 10 \log_{10} \frac{P}{1}$$

where P is the power output in milliwatts.

The volume level of an electrical signal made up of speech, music, or other complex tones is measured by a specially calibrated voltmeter called a *volume indicator*. The volume levels registered on this indicator are expressed in volume units (VU). The number of units is numerically equal to the number of decibels above or below the reference volume level. Zero VU represents a power of 1 mW dissipated in an arbitrary load resistance of 600 Ω (corresponding to a voltage of 0.7746 V). Thus, when the VU meter is connected to a 600-Ω load, VU readings in decibels can be used as a direct measure of power above or below a 1-mW reference level.

Decimal: A number, usually of more than one figure, representing a sum in which the quantity represented by each figure is based on the radix of 10. The figures used are 0, 1, 2, 3, 4, 5, 6, 7, 8, and 9.

Decoder: A device that determines the meaning of a set of signals and initiates a computer operation based upon such determination. Also, a matrix of switching elements that selects one or more output channels according to the combination of input signals present.

Decoding: Performing the internal operations by which a computer determines the meaning of the operation code of an instruction; also sometimes applied to addresses. In interpretive routines and some subroutines, it also may be an operation by which a computer determines the meaning of parameters in the routine. The term basically means translating a secretive language into a commonly understood one.

Definition: The resolution and sharpness of an image, or the extent to which an image is brought into sharp relief. Also, the degree with which a communication system reproduces sound images or messages. Also, the art of compiling logic in the form of general flowcharts and logic diagrams that clearly explain and present the problem to the programmer in such a way that all requirements involved in the run are presented.

Deflection sensitivity: In a cathode-ray tube, a constant that indicates how much the spot on the screen is deflected (in inches, centimeters, or millimeters) for each volt difference of potential that is applied to the deflection plates. For example, tube specifications may describe a certain tube as having a deflection sensitivity of 0.2 mV/V dc. This means that when the tube is operated according to the stipulated conditions, every volt of dc applied to the deflection plates causes the spot to move 0.2 mm from its undeflected position.

Deflection sensitivity is directly proportional to the length of the deflection plates and the distance between the deflection plates and the screen. It is inversely proportional to the separation between the deflection plate and the accelerating voltage. Deflection factor indicates the voltage required on the deflection plates to produce a unit deflection on the screen, and it is the reciprocal of deflection sensitivity. It is expressed in terms of a certain number of dc volts per centimeter (or per inch) of spot movement. For example, in a tube having a deflection sensitivity of 0.2 mm/V dc, the deflection factor is 50 V/cm. It is also common to express the deflection factor in terms of the second anode voltage. That is, the deflection factor is given as a certain amount for each kilovolt of second anode voltage that is used. For example, 60 V dc/in./kV of second anode voltage indicates that with 1 kV applied to the second anode, the factor is 60 V/in. With 20 kV applied to the second anode, the factor is 120 V/in.

Delay: The length of time after the close of a reporting period before information pertaining to that period becomes available. Delay may also cover the time to process data and prepare and distribute reports. It also refers to the retardation of the flow of information in a channel for a finite period of time.

Delta-matched antenna: A method of coupling a transmission line to the driven element of an antenna system to allow proper transfer of signals to take place.

Using this system, shown in Fig. 135, the impedance of the transmission line is gradually transformed into a higher value by the fanned-out feeder that attaches directly at two places to the antenna element.

Also called a *Y match,* the contact points on the element or tap points form a compromise between the impedance at the ends of the Y and the impedance of the transmission line proper.

The delta match is used with resonant antenna systems, and the contact points are quite critical in order to achieve the greatest transfer of power between the driven transmission line and the driven antenna element. This system is most often used with a half-wavelength radiator that is fed by a balanced transmission line with a characteristic impedance of 300, 450, or 600 Ω. For precise matching, the tap points on the antenna element are temporarily connected and then adjusted while observing an antenna tuning meter or SWR meter.

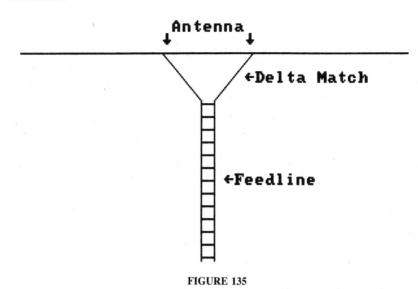

FIGURE 135

The delta match presents certain mechanical complications when wire elements are used, so this system is not used as much today as it was several decades ago before unbalanced coaxial line became readily available.

Demand factor: The ratio of the maximum demand of a system, or part of a system, to the total connected load of a system or the part of the system under consideration.

Demand meter: A device for measuring the maximum or peak rate of energy usage during a certain period, which can typically be either 15, 30, or 60 min. The demand charge based on this reading can be as much as 35 percent of a total electrical bill. From the utility's viewpoint, such charges can easily be justified, as a large investment is required to maintain reserve generators and power lines to handle peak loads which may occur for only a few hours each month.

Demand controllers are sometimes used to measure demand and turn off preselected or sheddable loads when the demand approaches some predetermined figure. The demand charges thereby avoided may be sufficient to pay for the controller.

Demodulation: The process of recovering the intelligence from a modulated wave. When a radio carrier wave is amplitude-modulated, the intelligence is imposed on the carrier in the form of amplitude variations of the carrier. The demodulator of an amplitude-modulated (AM) wave produces currents or voltages that vary with the amplitude of the wave. Likewise, the frequency-modulation (FM) detector and the phase-modulation detector change the frequency variations of an FM wave and the equivalent phase variations of a phase-modulated wave into currents or voltages that vary in amplitude with the frequency or phase changes of the carrier.

An instantaneous amplitude, e_o, of the carrier may be represented as

$$e_o = E_o \sin (2 f_o t + 0)$$

where E_o is the maximum amplitude of the original carrier, f_o is the frequency of the carrier, and 0 is the phase angle. (For AM signals, 0 may be considered as 0.)

One or more of the independent variables (those on the right-hand side of the equation) may be made to vary in accordance with the modulating signal to produce a variation in E_o. However, the general practice is to vary only one of the values (e_o for AM, f_o for FM, or 0 for phase modulation) and to prevent any variation in the others. The detector in the receiver must therefore be designed so that it will be sensitive only to the type of modulation used at the transmitter.

Demodulation data: The useful information retrieved from a transmitted waveform. The data are superimposed upon the radio carrier at the transmitter. Demodulation takes place at the receiver. Many different types of data can be retrieved and should be the equivalent of the modulation data applied at the transmitter.

In a typical voice communication system, the retrieved data will be the voice of the person speaking into the transmitter microphone. Demodulation data, then, can be the voice of a radio newscaster. This voice is converted to a varying current by the microphone at the broadcast studios. The signal is superimposed upon the carrier of the radio transmitter. The radio receiver acts as the demodulator that retrieves the useful modulation data from the radio carrier. This is then fed to the speaker as a varying signal. The speaker converts the signal back into an acoustical waveform, which is directly detected by the human ear.

Computer data may be sent by radio waves. This is often in the form of varying audio sounds that travel with the carrier in the same fashion as did the human voice. At the receiving end, this demodulation data is separated from the radio carrier and fed, as audio tones, to various pieces of processing equipment. Other examples of demodulation data include television audio and video, radioteletype driving pulses, and many different types of telephone communications.

Demodulator: A device that receives tones from a transmission circuit and converts them to electrical pulses or bits that may be accepted by another circuit.

Device: A unit of an electrical system that is intended to carry, but not utilize, electric energy.

Dielectric heating: The process of heating a nonconductor of electricity where the material is placed between two metallic plates. This action is similar to that of capacitors. When an alternating current is applied to two capacitor plates separated by insulating material, called the *dielectric*, the alternating current sets up strains in the dielectric. Heat is produced by the rapid reversal of these strains at high frequencies. When the frequency is increased to the region of 50,000,000 cps, a large amount of heat can be generated. This principle is applied in dielectric heating.

Dielectric heating is the only known method of developing heat uniformly throughout a material that does not conduct electricity. For this reason, these applications would almost always use dielectric heating, regardless of its cost:

1. Heating plastic preforms. In molding most plastics, the material used in the press is made into cylindrically shaped preforms. In order to reduce the time needed to heat the material in the mold, the material is preheated. Preheating was originally done in ovens. This type of preheating resulted in excessive drying and uneven heating. Now, however, by the use of dielectric heating, the preforms are heated uniformly, resulting in better molded products and fewer rejects.

2. Curing plywood. Plywood consists of several layers of thin wood glued together. The use of dielectric heating makes it possible to generate heat uniformly throughout the wood instead of having the heat applied to the outside surfaces and then conducted slowly through the material. The result is faster curing and a more superior product than could be obtained otherwise.

3. Grain sterilization to eliminate the eggs of weevils and grubs. Dielectric heating does this efficiently. Any other method involving transmission of heat by conduction would result in overheating the outside areas before the center areas could be brought to the proper temperature.

Dielectric hysteresis: A lag of change or discharge in a capacitor similar to magnetic hysteresis in an inductor. Dielectric hysteresis may be considered as if it were a kind of molecular friction, manifesting itself as a heating effect in the dielectric as the current flows in and out of the capacitor. Another name for dielectric hysteresis is *dielectric absorption.*

Dielectric material: The material used between capacitor plates that has a pronounced effect on the amount of capacitance. For example, suppose that a given pair of plates at a given separation constitute a capacitor of 15-pF capacitance when the plates are separated by air. If a sheet of mica as thick as the air space is inserted between the plates, the capacitance is in the range of 30 to 105 pF. That is, the mica-dielectric capacitor has a capacitance from two to seven times as great as that of an equivalent air-dielectric capacitor. The dielectric constant of mica is in the range of two to seven, depending upon the quality of the mica. The dielectric constant helps compare the effectiveness of various dielectric materials. The dielectric constant compares a given dielectric material with air (or with a vacuum); air is assigned an arbitrary dielectric constant of unity.

The dielectric constant of a material depends on how easily an electric field can displace the planetary-electron orbits from their normal positions.

Since the distorted planetary orbits can bind charges to the plates, the orbit displacement in a dielectric adds to the ability of the plates to acquire and hold a charge. Another important quality of the dielectric, although not directly related to capacitance, is the voltage required per unit thickness to break down the dielectric or spark across it. This quality is called the *dielectric strength* of the material.

Dielectric strength: See *Dielectric material.*

Dielectric test: A method used to determine whether or not insulation will withstand voltage stresses occurring during normal or assumed abnormal conditions of operation. The assurance provided by a dielectric test may warrant the risk involved in applying it as preventive maintenance prior to a critical period of operation. The damage resulting from an insulation failure at the time of a properly applied dielectric, or high-voltage, test is likely to be very little compared to the loss caused by a breakdown while apparatus is carrying an important load.

As an example, in the chemical industry, such a breakdown might ruin the product of many days of operation, and more days would then be required to bring up the process to the stage at the time of the failure, even if other electrical apparatuses were immediately available. Furthermore, the additional cost of repairs resulting from the failure under load must be given consideration.

The insulation of electrical apparatus that has been repaired or rewound should be given a dielectric test. Also, samples of insulating oil taken at regular intervals from transformers, induction regulators, and circuit breakers should be tested for breakdown value.

AIEE Standards for testing rotating machinery are now combined into ASA Standard C50.1, C50.2, C50.4, C50.5, and C50.6. (This supersedes AIEE Standard Nos. 5, 7, 8, 9, and 10.) Similarly, tests for transformers and regulators are now combined in ASA Standard C57.12 (which supersedes AIEE Standard No. 13). All specify a dielectric test (for the main winding insulation of new apparatus) of 1000 V plus twice the rated voltage of the apparatus.

The authorized dielectric test for field windings of synchronous machines is 10 times the exciter voltage. The standard test for Railway Motors (No. 11) shall be twice the rated voltage plus 1500 (grounded circuits) or twice the voltage plus 1000 (ungrounded). Each AIEE Standard lists exceptions. If special conditions exist, reference should be made to the standard that applies to the apparatus involved.

Dielectric tests are generally made with alternating voltage of commercial frequency (25–60 cycles) for a period of 60 s or equivalent dc potentials based on a 1.6 ratio.

Dielectric test voltage values have not been established for apparatus that has been operated in commercial service, nor is there authorization for dielectric tests for windings that have been repaired. It is the general practice to apply from 65 to 75 percent of the authorized test-voltage for new apparatus. As many of the conditions affecting such tests are extremely variable, the decision concerning procedure is generally made by the maintenance engineer.

Diffusion: The effect produced by light coming from all directions without preference. In interior lighting, diffusion prevents shadow of any kind and lights all surfaces in the room or area equally. Absence of shadow in the proper quantity helps destroy texture and form. Diffusion results, particularly at higher levels, in extreme direct glare that can be readily detected by looking at a bare incandescent lamp that is not inside-frosted.

Digital-analog multiplexing: A system in which two or more messages may be sent simultaneously in one or both directions over the same line. For infrequent transmission of coded data, digital telemetering may time-share a tone channel with a slow-changing signal of approximate numerical information, such as a generator output. The analog telemetering is interrupted periodically and the digital readings are transmitted over the tone channel for several seconds.

Digital circuitry: A circuit affording a dual-state switching operation such as ON or OFF, FAST or SLOW, HIGH or LOW, etc.

Digital computer: A computer that processes information represented by combinations of discrete or discontinuous data, as compared with an analog computer for continuous data. More specifically, it is a device for performing sequences of arithmetic and logical operations, not only on data, but its own program. A digital computer is capable of performing sequences of internally stored instructions, as opposed to calculators such as card-programmed calculators, on which the sequence is impressed manually.

Digital control: A technique for computer operations used to execute the complete assignment of area regulation and economic dispatch with a properly programmed digital computer without an intermediate analog control console. This approach is especially useful in large control areas.

Diode, semiconductor: A device resulting when P and N types of semiconductor materials are combined in manufacture with characteristics similar to the electron-tube diode. If properly biased, the semiconductor diode will conduct heavily in one direction and very little in the other. Semiconductor diodes, like electron tube diodes, are used for rectification and detection. In addition, they have special properties that make them useful in other applications, such as voltage regulators, switching, and multivibrator circuits. They vary in size from small devices with current ratings of less than 1 mA, to large 500-A types. Semiconductor diodes are classified according to construction and may be of the junction type or the point contact type.

A junction semiconductor diode is made by taking a single germanium or silicon crystal and adding a donor impurity to one region and an acceptor impurity to another region. This produces a single crystal with an N section and a P section. The point where the two sections meet is called the *junction*. Contacts are fastened to the two ends of the crystal.

The end (ohmic) contacts are large surfaces that make a good connection with the crystal. An ohmic contact is a contact between two materials possessing the property that the voltage drop across it is proportional to the current passing through it. If the connections were not good, there might be rectifying properties where they come in contact with the crystal.

There are two methods of combining the two types of material—the grown junction and the diffused junction. A grown junction is formed by growing a single crystal from a melt. At the start, the melt contains donor impurities and N-type material is formed. In the middle of this forming process, acceptor impurities are added to form the P-type portion of the crystal. The diffused (or alloy) junction diode is formed by placing a small amount of indium on a slab of N-type germanium. This combination is then heated to a specific temperature for an amount of time that allows the indium to fuse to the germanium. This fusion produces a P-type area of germanium in the slab immediately below the indium dot.

In the PN junction illustrated in Fig. 136a, the P material is shown at the left and the N material is shown at the right. The P material contains acceptor impurity atoms. These atoms are represented as negative charges enclosed in circles (Fig. 136b). As mentioned before, these atoms take on electrons from the pure crystal, leaving holes as the current carriers. The holes are represented in the P material as small circles scattered between the acceptor atoms. The N material contains donor atoms that give up electrons when they become a part of the crystal lattice. The donor atoms are represented as positive charges enclosed in circles. The free electrons in the N material are represented as dots scattered between the donor atoms.

When a PN junction is formed, free electrons in the N-type region diffuse across the junction and fill holes near the junction in the P region. Holes diffuse across the junction from the P region to the N region and capture free electrons near the junction in the N region. When an electron leaves the donor atom in the N region and moves over to the P region, the atom has fewer electrons than it needs to neutralize the positive charge on the nucleus and it becomes charged (ionized). It has one extra positive charge equal to the negative charge of the electron, which it lost. Similarly, when a hole leaves an acceptor atom in the P region, the atom takes on a negative charge because the hole has been filled by an electron and the atom has one more electron than it needs to neutralize the charge on its nucleus.

These charged atoms, or *ions,* are fixed in place in the crystal lattice structure and cannot move. Thus, they make up a layer of fixed charges on the two sides of the junction. On the N side of the junction, there is a layer of positively charged ions; on the P side of the junction, there is a layer of negatively charged ions.

Note in Fig. 136b that there is a barrier of negative ions on the P side of the junction. This negative barrier will repel electrons from the immediate vicinity of the junction and will prevent the diffusion of any more electrons

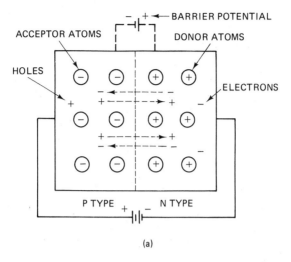

(a)

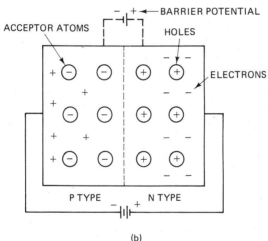

(b)

FIGURE 136 Biased junctions.

from the N side over into the P side of the crystal. Similarly, on the N side of the junction, there is a barrier of positive ions that will repel holes away from the immediate vicinity of the P side of the junction and prevent the diffusion of any additional holes across the junction from the P material into the N material.

The two layers of ionized atoms form a barrier to any further diffusion across the junction. Because the charges at the junction force the majority carriers away from the junction, the barrier is known as the *depletion layer.* It is also known as the *barrier layer,* or *barrier potential.*

The charge on the impurity atoms is distributed across the PN junction as shown in Fig. 137b, curve 1. In the P region, the ionized acceptors have a

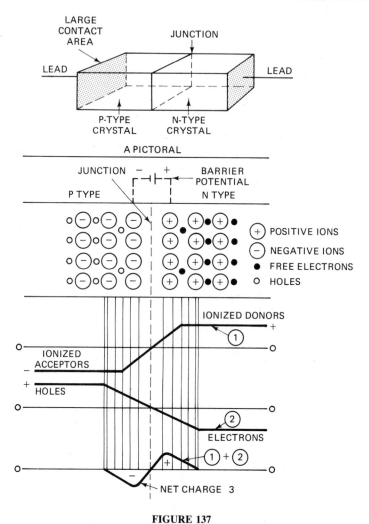

FIGURE 137

negative charge; and in the N region, the ionized donor atoms have a positive charge. At the junction, the charge is zero. However, in the P region, there are holes that have a positive charge. In the N region, there are free electrons that have a negative charge. This distribution of holes and free electrons is shown by curve 2 in Fig. 137b.

The potentials at the junction have driven the holes away from the junction in the P region and the electrons away from the junction in the N region, so the charges in the P region and N region are moved farther apart. Thus, the slope of curve 2 is more gradual than that of curve 1. The charge at the junction is zero, but the rise of either side is more gradual than that of curve 1. Moving further into the P region, the charge becomes positive due to the holes; moving into the N side of the crystal, the charge is negative due to the electrons.

The net charge on the crystal in the P region is equal to the difference between the charge on the ionized acceptor atoms and the charge on the holes. The net charge on the crystal in the N region is equal to the difference between the charge of the ionized donor atoms and the electrons. These charges cancel except in the immediate region of the junction. This arrangement is indicated by curve 3, which is the sum of curves 1 and 2 (Fig. 137b).

In the area near the junction, there is a negative charge in the P region and a positive charge in the N region. As stated earlier, they act as a barrier to prevent further diffusion of holes from the P region into the N region and the diffusion of electrons from the N region into the P region. This potential barrier is a potential difference (or voltage) across the junction and is of the order of a few tenths of a volt. It is sometimes referred to as the *potential hill battery* and may be represented as a dotted battery with the negative terminal connected to the P material and the positive terminal connected to the N material (top of Fig. 137b).

This barrier potential is like the plate-cathode voltage of an electron-tube diode. If the plate is made positive with respect to the heated cathode, the diode can be made to conduct current. If the plate is made negative with respect to the cathode, the diode will block the flow of current.

A limitation is imposed on the junction diode as the result of hole-electron pairs that are being formed at random within the crystal due to energy imparted to the crystal by heat, light, and electromagnetic radiation. Away from the depletion layer, these carrier pairs will recombine without materially affecting the carrier concentration in the crystal. In other words, the holes will remain the majority carriers in the P material, and electrons will remain the majority carriers in the N material. There are minority carriers in both regions (holes in the N material and electrons in the P material). The holes produced in the N material near the junction are attracted by the negative ions on the P side of the junction and pass across the junction. These holes will tend to neutralize the negative ions on the P side of the junction. Similarly, free electrons produced on the P side of the junction will pass across the junction and neutralize positive ions on the N side of the junction. This action is an example of *intrinsic conduction,* which is undesirable.

This flow of minority carriers weakens the potential barrier around the atoms that they neutralize. When this occurs, majority carriers are able to cross the junction at the location of the neutral atom. This means that holes from the P material will cross over to the N material, and electrons from the N material will cross over to the P material.

This action results in both holes and electrons crossing the junction in both directions. These motions cancel each other out, and the net movement contributes nothing toward the net charge or current flow through the junction. Because of intrinsic conduction, the junction is no longer a rectifier when an external voltage is applied across it. It is analogous to an electron tube in which not only the cathode emits electrons, but the plate is heated to the point where it also will emit enough electrons to break down the rectifying properties of the diode.

If a battery is connected across the PN junction, the battery potential will bias the junction. If the battery is connected so that its voltage opposes the barrier potential across the junction, it will aid current flow through the junction and the junction is said to be *biased* in the forward direction (low resistance). If the battery is connected across the junction so that its voltage aids the barrier potential across the junction, it will oppose current flow through the junction and the junction is said to be *reverse-biased,* or biased in the reverse direction. This is the direction of high resistance.

The forward bias connection is illustrated in Fig. 137. Here, the positive terminal of the bias battery is connected to the negative side of the barrier potential (P-type side of the junction), and the negative terminal of the battery is connected to the positive side of the barrier potential (N-type side of the junction).

The positive terminal of the battery connected to the end of the P-type germanium repels holes toward the junction and attracts electrons from the negative ions near it. The combination of holes moving toward the junction to neutralize charged negative ions on the P side of the junction and electrons taken from the negatively charged acceptor atoms tends to neutralize the negative charge of the barrier potential on the P side of the junction.

On the N side of the crystal, the negative terminal of the battery repels electrons toward the junction. These electrons tend to neutralize the positive charge on the donor/atoms at the N side of the junction. At the same time, the negative terminal of the battery attached to the N side of the crystal attracts holes away from the charged positive ions (donor atoms) on the N side of the junction. Both of these actions tend to neutralize the positive charge on the donor atoms at the junction, thereby reducing the barrier potential.

The effect of the battery bias voltage in the forward direction is to reduce the barrier potential across the junction and allow majority carriers to cross the junction. Thus, more electrons flow from the N-type material across the junction. At the same time, more holes travel from the P-type germanium across the junction where they combine with the electrons from the N-type material.

At the same time that the hole and electron movements are going on in the crystal, electrons are moving from the negative terminal of the bias battery in the external circuit to the N-type terminal, and electrons are moving from the P-type terminal in the external circuit to the positive terminal of the battery.

It is important to remember that in the forward-biased junction condition, conduction is by the majority carriers (holes in the P-type material and electrons in the N-type material). Increasing the battery voltage will increase the number of majority carriers arriving at the junction, and the current flow increases. The only limit to current flow is the resistance of the material on the two sides of the junction. If the battery voltage is increased to the point where the barrier potential across the junction is completely neutralized, heavy current will flow and the junction may be damaged from the resulting heat. Therefore, the voltage of the bias battery is limited to a relatively small voltage.

With reverse bias applied to the junction diode (Fig. 137), the negative terminal of the battery is connected to the P-type section and the positive terminal is connected to the N-type section. The negative terminal attracts holes away from the junction and depletes the holes in the P material. At the same time, the positive terminal of the battery attracts electrons away from the junction and increases the shortage of electrons on the N side of the junction. This action increases the barrier potential across the junction, because there are fewer holes in the P side of the junction to neutralize the negative ions, and fewer electrons on the N side to neutralize the positive ions formed on this side of the junction. The increase in barrier potential helps to prevent current flow across the junction by majority carriers.

The current flow across the barrier is not zero, however, because of minority carriers crossing the junction. Holes forming in the N side of the depletion layer are attracted by the negative potential applied to the end of the P-type section; electrons breaking loose from their outer shells in the atoms of the P material are attracted by the positive voltage applied to the end of the N-type section of the germanium.

This situation is described as intrinsic conduction or hole-electron pairs continually forming at random within the crystal (before any bias is applied) due to the energy of the crystal. With no bias applied to the crystal, minority carriers neutralize ions near the junction and allow majority carriers to cross the junction.

With bias applied, the minority carriers are attracted away from the junction by the potential applied across the crystal, so that all of the minority carriers do not remain near the junction to neutralize charged atoms. Hence, the minority carriers no longer allow the passage of an equal number of majority carriers in the opposite direction. The flow of minority carriers across the junction is not fully offset by a flow of majority carriers in the opposite direction. Therefore, there is a small current flow across the junction caused by the minority carriers crossing the junction. This current flow is small and nearly constant at normal operating voltages.

Reverse bias (also called *backward* or *back bias*) applied across a junction diode increases the barrier potential, making it more difficult for majority carriers to cross the junction. However, some minority carriers will still cross the junction, with the result being that there will be a small current.

This action is indicated in the static curve for a germanium crystal diode in Fig. 137. The forward portion of the curve indicates that the diode conducts easily when the potential across the junction is in the direction of forward bias (P side positive and N side negative).

The diode conducts poorly in the high-resistance direction (backward bias, P side negative and N side positive). For this condition, the holes and electrons are drawn away from the junction, causing an increase in the barrier potential. However, if the backward bias is increased beyond a critical value (about 70 V for some typical designs), the reverse current increases rapidly due to *avalanche breakdown.*

Avalanche breakdown occurs when the applied voltage is sufficiently large to cause the covalent bond structure to break down. At this point, a sharp rise in reverse current occurs. The acceleration of the few holes and electrons continues to such a point that they have violent collisions with the valence bond electrons of the germanium crystal atoms releasing more and more carriers. The maximum reverse voltage of the semiconductor diode corresponds to the peak inverse voltage of an electron-tube diode. Static characteristics for several semiconductor diodes are illustrated in Fig. 137. Semiconductor diodes operated in the reverse voltage breakdown region (lower left portion of static characteristics) are used as voltage regulators.

Diode, vacuum tube: A simple tube that contains a heated cathode and a cold plate. *Di* is a prefix signifying two. *Ode* is a suffix, as in electrode, cathode, and anode. The plate collects electrons when the cathode is heated in a vacuum and positive potential exists on the plate with respect to the cathode.

The original diode was constructed by Thomas Edison, inventor of the incandescent electric lamp, shortly after his invention of the lamp itself. He added a metal plate inside his evacuated lamp and provided an external terminal from it for use as an electrode. He then used a heated filament as another electrode.

One version of the diode is shown in Fig. 138a with its two elements indicated as plate and filament. Another version of a diode is shown in Fig. 138b. Its filament serves only as a heater. In an electronic circuit, the two electrodes of a diode act in the manner of a flow valve in a water pipe. (The British called it a valve instead of a tube.) The behavior of a diode is observed after connecting the plate and cathode elements in series with a battery and milliammeter, as shown in Fig. 139, carefully observing polarity changes of that battery when used in arrangements a and b, respectively. The cathode is brought up to normal temperature by applying rated voltage across the heater terminals. If the battery is connected so that the plate is positive with respect to the cathode (Fig. 139a), the meter will indicate a current flow. This phenomenon, the emission of electrons from hot bodies, was first observed by Edison in 1883 and is known as the *Edison effect.*

When the battery is connected so that the plate is negative with respect to the cathode (Fig. 139b), the meter will indicate no plate current flow. The total number of electrons emitted by the hot electrode at a given operating temperature is always the same, regardless of the plate voltage. This same condition exists regardless of the plate polarity, because the electrons fly into the space surrounding the cathode to produce a cluster or cloud, which is in turbulence (great agitation). This cloud constitutes a negative space charge that constantly

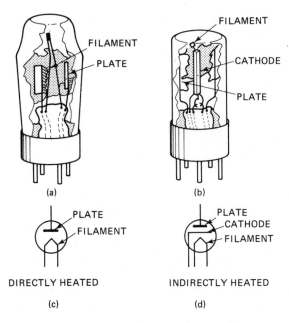

(a)   (b)

DIRECTLY HEATED   INDIRECTLY HEATED

(c)   (d)

FIGURE 138 Cutaway of two-element tubes.

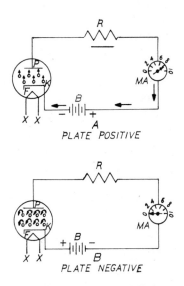

FIGURE 139 Action of diode.

tends to repel the electrons toward and into the cathode as fast as they are being emitted. The negative charge on the plate of Fig. 139b only repels the nearby electrons within the cloud, but the action is so effective that none of the electrons reaches the plate (regardless of the amount of voltage) as long as the plate remains negative.

With low values of positive plate voltage, only those electrons of the space-charge cloud that are nearest to the plate are attracted to it, and the plate current is low. As the plate voltage is increased (the cathode temperature remains constant), greater numbers of electrons are attracted to the plate and, correspondingly, fewer of those being emitted are repelled back into the cathode.

Eventually, a plate voltage value (saturation voltage) is reached at which all the electrons being emitted are in transit to the plate and none is repelled back into the cathode. The corresponding value of current is called the *saturation current.* Any further increase in plate voltage can cause no further increase in plate current flowing through the tube.

The relationship between the plate current and the plate voltage in a diode for different cathode temperatures for oxide-coated, tungsten, and thoriated-tungsten cathodes is shown in Fig. 140. At high plate voltages, the flow of plate current is practically independent of plate voltage, but is a function of the cathode temperature. However, at lower values of plate voltage, the plate current is controlled by the voltage between the plate and cathode and is substantially independent of the cathode temperature.

In other words, with a fixed plate voltage, electron emission and plate current will increase with cathode temperature until at some value of temperature, the plate current is limited by the space charge. Thus, more electrons are being emitted by the cathode than are being attracted by the plate. Continued increase of cathode temperature fails to produce any further increase in plate current. The temperature at which the plate current stops increasing is called the *saturation temperature.*

The dotted portion of the curves (Fig. 140) is representative of tungsten and thoriated-tungsten cathodes, and the solid curves are typical of oxide-coated cathodes. It is unlikely that the plate current in a tube employing an

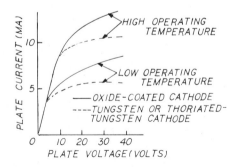

FIGURE 140

oxide-coated cathode will ever be entirely independent of the plate voltage. Before the plate voltage could be increased sufficiently to produce emission saturation, it is probable that the cathode would be seriously damaged.

Diodes that have been discussed thus far are of the high-vacuum type. There are other types of diodes that contain gas at a relatively low pressure. For example, hot-cathode mercury-vapor rectifier tubes are used to provide dc power for transmitters. The original use of the word diode was restricted to electron tubes. Scientific research has produced other devices that perform the same functions as earlier diodes, although they are not electron tubes. These devices are *semiconductor diodes.*

Since current can flow in only one direction through a diode, its basic use is as a rectifier. If a battery is replaced with an alternating voltage source, current will flow through the load resistor in the plate lead only on alternate half-cycles when the plate is positive with respect to the cathode. This unidirectional characteristic of the diode is also used when the tube is employed as a detector.

DIP: Acronym for *dual in-line package,* the most common container for an integrated circuit. DIPs have two parallel rows of pins spaced on 0.10-in. centers. DIPs usually come in 14-, 16-, 18-, 20-, 24-, and 40-pin configurations.

Direct coupling: A coupling that has two parts or halves, one mounted on a motor shaft and the other on the shaft of the driven mechanism. The parts may be bolted or rigidly fastened together so that there is no possibility of any relative movement between them. A coupling so constructed is called a *solid coupling.* The halves may also be fastened together by means of some other part in such a way that there may be a slight relative movement between them. Such a coupling is known as a *flexible coupling.* Many varieties of flexible coupling are made.

The motor should be properly aligned with the driven machine. When the alignment is correct, the center line of the motor or driving shaft and that of the driven shaft are in a straight line. A good alignment, whether of a flexible or solid coupling, avoids excessive bearing pressures that might cause damage. Since the bearings are not affected by the load, they do not limit the torque or speed in this type of drive. They should, however, be able to support the weight of the armature.

Direct current: A current having fixed polarity. In a direct current circuit, one pole is always positive, while the other is negative. The electron flow will originate from the negative pole and flow toward the positive. In alternating current circuits, the poles are constantly changing. Therefore, the direction of current flow reverses itself with each alternation. In a direct current circuit, current always flows in the same direction.

Disconnecting switch: A device by which a system can be completely disconnected from the source of power for emergency, repair, maintenance, or other work. As a safety precaution, the location of the switch must be known. Most electrical codes specify that each electric motor circuit shall be provided with a disconnect switch.

Disk storage: The storage of data on the surface of magnetic disks.

Diversity reception: Obtaining a steady signal by combining energy from each of a number of fluctuating signals. All tropospheric scatter systems use diversity reception. To obtain signals over different paths that fade and vary independently, some or all of these methods may be used:

1. Space diversity, which is comprised of receiving antennas separated by 50 wavelengths or more at the signal frequency (usually 10 to 200 ft is sufficient).
2. Frequency diversity, or transmission on different frequencies. Different frequencies fade independently even when transmitted and received through the same antennas.
3. Angle diversity that uses two feed horns producing two beams from the same reflector at slightly different angles. This results in two paths based on illuminating different scatter volumes in the troposphere.

Signals obtained over two or four independent paths by the above methods are combined in the receiver in such a way as to make use of the best signal at all times.

Doorknob capacitor: A device designed for high-voltage applications, often found in the flyback circuits of color television receivers and in blocking circuits of high-powered transmitter amplifier sections. Shown in Fig. 141, this device assumes the physical shape of a tiny doorknob. Doorknob capacitors are primarily used as bypass devices and for coupling from one stage to another. A typical capacitance value for a doorknob capacitor is 500 pF, which is a relatively high capacitance value for a device of this size and voltage rating. Figure 141 shows that the device is constructed of a molded plastic outer shell and is terminated in a variety of screw-in terminals.

This nonpolarized capacitor is used in an RF amplifier circuit between the plate and the output coupling network. In this mode, it is called a *blocking capacitor* and allows radio-frequency energy to travel from the plate circuit into the coupling circuit while blocking the flow of the dc plate potential.

Due to the noncritical nature of most applications for this type of capacitor, the stated capacitance values may differ by + or − 20 percent of the actual value. These devices are usually available in ratings of 5, 10, 20, and 30 kV and may be series-connected for increased voltage ratings or connected in parallel for an increase in exhibited capacitance.

FIGURE 141

Doping, semiconductor: The control of quality and quantity of impurities in semiconductor materials. In the pure form, semiconductor materials are of no use as semiconductor devices. When a certain amount of impurity is added, however, the material will have more (or less) free electrons than holes. Both forms of conduction will be present, but the majority carrier will be dominant. The holes are called *positive carriers,* and the electrons are called *negative carriers.* The one present in the greatest quantity is the *majority carrier;* the other is the *minority carrier.* Added impurities will create either an excess or a deficiency of electrons, depending on the kind of impurity added.

The impurities that are important in semiconductor materials are those that align themselves in the regular lattice structure, whether they have one valence electron too many or one valence electron too few. The first type loses its extra electron easily. In so doing, it increases the conductivity of the material by contributing a free electron. This type of impurity has five valence electrons and is called a *pentavalent impurity.* Arsenic, antimony, bismuth, and phosphorous are pentavalent impurities. Because these materials give up or donate one electron to the material, they are called *donor impurities.* The second type of impurity tends to compensate for its deficiency of one valence electron by acquiring an electron from its neighbor.

Doppler navigation system: A navigation radar that automatically and continuously computes and displays ground speed and drift angle of an aircraft in flight without the aid of ground stations, wind estimates, or true airspeed data. This is done by utilizing auxiliary inputs from an altitude rate sensor and from the aircraft's vertical reference system. Doppler navigation radars do not sense range and bearing (direction) as ordinary search radars do. They employ continuous-carrier or pulse-transmission energy and determine the forward and lateral velocity components of the aircraft by utilizing the apparent frequency change phenomenon (Doppler shift) in the RF range.

The radar radiates CW or pulsed energy (depending on the type of system being used) at one RF frequency, and the radiated energy is beamed fore-aft and left-right of the aircraft. These beams of energy strike the earth's surface and are reflected. Energy waves returning appear to be spaced differently than are the waves that were transmitted.

The receiver of the Doppler radar system detects this apparent change in frequency. This is accomplished by the use of two signal inputs—one from the transmitter and the other from the receiver antenna. The input signal from the transmitter is at all times the same frequency as the signal being radiated. The receiver antenna signal frequency will vary with the forward and lateral velocity of the aircraft. The two frequencies are composed in the receiver, and the difference frequency is representative of the ground speed and drift angle of the aircraft.

Doppler principle: The principle discovered by and named for Christian Doppler, which describes the effect of motion on a wave that has been generated and propagated. This effect is based on the fact that all waves, once they have been propagated into space or into a conducting medium, travel at a velocity independent of the wave source. The effect was first noted in connection with sound waves in 1842. It has subsequently been observed throughout the entire frequency spectrum. In all cases, a shift in the apparent frequency occurs when the wave source and the detector move toward or away from each other. If the distance separating the source and the detector remains constant, no frequency shift occurs. The present discussion is limited to the implications of this principle in electronic applications.

Assuming that a wave source is moving in the direction of its propagation, it partially overtakes one wave front before the next is emitted. This has the apparent effect of decreasing the distance between corresponding points on successive waves. (The speed of wave travel is independent of the motion of the source and need not be considered here.) The wavelength measured by a stationary detector located ahead of the source is shorter than the original frequency would indicate. The detector apparently receives a frequency higher than that propagated. If the source moves in a direction opposite to the direction of propagation, the frequency is decreased in a corresponding manner. The same effect is noted in the case of a stationary source and a moving detector. In this case, the detector is overtaking part of the wavefront, and the time and wavelengths are shortened and frequency is apparently increased.

Three aspects of a given radar beam may be considered. These are the generated signal wave, the incident wave, and the received wave.

An examination of 1 μs of the signal radiated from a CW radar transmitter shows that the signal is composed of many cycles of RF energy. In the time of 1 μs, a radar that operates at 13.3 GHz will radiate 13,300 cycles of RF energy, only a few of which are shown in Fig. 142a. Each cycle occupies a definite period of time and has a definite physical wavelength. At the frequency mentioned, each cycle occupies 75 ps and has a physical wavelength of 22.5 mm. An expanded view of one cycle of RF energy at 13.3 GHz is shown in Fig. 142b. This cycle will later be shown with one cycle of incident signal superimposed on it to demonstrate the expansion due to Doppler effect.

At this point, two approaches to explain the effect of signal expansion are used—the action of an energy cycle as it leaves the antenna, and the shape of each cycle as an arc of wave energy. First, consider a single cycle as it leaves the antenna.

The cycle shown in Fig. 142b occupies a certain discrete distance in space and is, in effect, the wavelength of the signal at the frequency shown. When transmitted from a stationary antenna, the leading edge will move out into space 22.5 mm by the time the trailing edge leaves the antenna. However, if that cycle is emitted from an antenna pointed aft from a moving aircraft, then the antenna will move a small distance during the time the cycle is being transmitted. The trailing edge, therefore, will be a greater distance away from the leading edge than it was when the cycle was transmitted from the stationary antenna. This is expansion.

Figure 143 illustrates this effect and shows by the dotted line the space

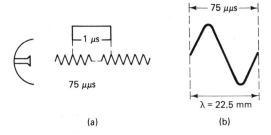

(a)   (b)

FIGURE 142

that a cycle would have occupied had the antenna been stationary. Since the cycle has been expanded, it occupies more distance in space, the wavelength has been increased, and the frequency has been proportionately decreased. The difference in frequency for this first Doppler shift is proportional to the sine of the radar beam angle multiplied by the ratio of the aircraft velocity to the generated wavelength. In symbols, this would be

$$D.S. = \frac{V}{\lambda} \times \sin \phi$$

When the signal is reflected from the ground and returns to the antenna, a similar expansion occurs. A slightly longer time is required for the cycle to reenter the moving antenna than would be required if the antenna were stationary. Thus, a double expansion has occurred. This double frequency expansion can be measured by the formula

$$D.S. = \frac{2V}{\lambda} \times \sin \phi$$

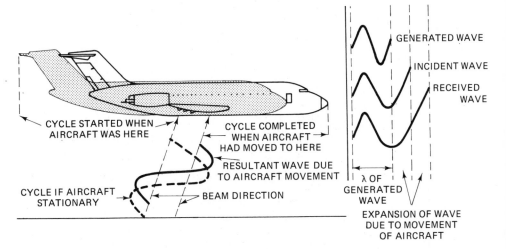

FIGURE 143 One-cycle expansion.

The change in frequency is slight, but measurable, and is a direct indication of the speed of the aircraft.

Compression of the signal will result if the antenna is pointed in the forward direction. The explanation is similar to that of expansion, except that the leading and trailing edges of the cycle will be closer together (cycle compressed). The effect will be to shorten the wavelength and increase the frequency.

Another approach to compression and expansion is to consider that a CW signal consists of waves of energy radiated from the antenna. Figure 144 illustrates the spacing between the waves of a signal from both stationary and moving antennas. The effect is the same as that described previously. The increased spacing (wavelength) between adjacent waves of energy is a direct indication of frequency change. The received signal frequency, when mixed with the generated signal frequency, produces a difference frequency that falls in the audio spectrum. This audio frequency is a direct indication of the aircraft's speed.

Double-conversion receiver: A receiving device that converts an incoming signal to the final intermediate frequency in two steps. The output of the RF strip, along with the frequency-multiplied signal from the first local oscillator, is fed to the first mixer (Fig. 145). The first local oscillator and frequency multiplier are made variable in order to provide tracking between the oscillator and the input signal.

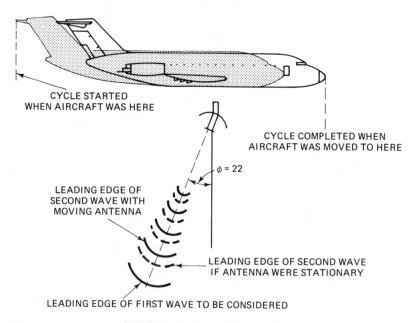

FIGURE 144   Energy wave spacing.

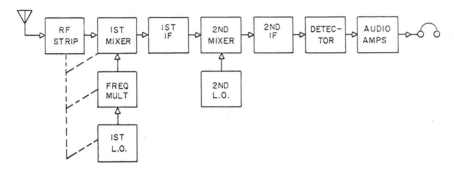

FIGURE 145 Block diagram of basic UHF receiver.

The output of the first mixer is a modulated IF, which is amplified in the first IF strip and used as the RF input to the second mixer. Heterodyning of this signal with that of the second local oscillator will produce a lower IF signal, which, in this case, is the final IF. If a lower final IF were desired, a third conversion system would be added by incorporating a third mixer, local oscillator, and IF strip. This would be called triple conversion. Note that only the first local oscillator is variable, since the output of the first mixer is a constant frequency.

The purpose of multiple conversion is maximum image rejection, increased selectivity, and improved gain. In a double-conversion receiver, the first intermediate frequency supplies the wide separation between the desired signal and its image. The second intermediate frequency supplies the increased gain and selectivity. The higher gain and better selectivity are more easily acquired in the second IF due to the lower frequency of these circuits.

Down-counter: A circuit used to count down from a preset number. In order to perform certain functions, digital equipment may be required to cycle through a predetermined number of operations and then stop. Such a circuit is illustrated in Fig. 146. This circuit is a modulus 16 down counter. It will count down from 1111 to 0000. The operation of this circuit may be understood by considering all the flip-flops to be set to the one state (1111) and referring to the waveforms depicted in Fig. 147.

Drier: A device designed to remove moisture from a refrigerant. Driers are placed in the liquid line unless some unusual condition indicates otherwise. The drying material is called the *desiccant*. A desiccant is a solid substance capable of removing moisture from a gas, liquid, or solid. The desiccant may be

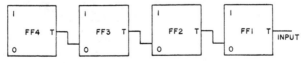

FIGURE 146 Down counter.

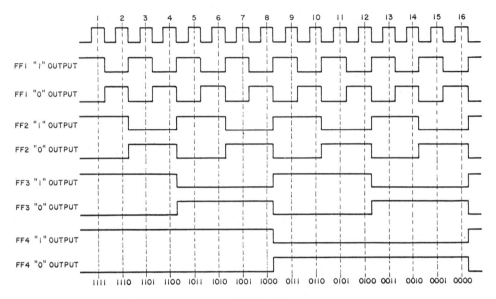

FIGURE 147

molded into a special form to fit a drier container or used as granules or in ball form.

A small scale unit drier and a cutaway view to show its internal construction is illustrated in Fig. 148. The desiccant is a molded block designed to eliminate the disadvantages of a loose desiccant. Another type is the angle unit, which permits the use of a replaceable desiccant core. The exploded view shows the assembly of this unit (Fig. 149). Presently, silicon gel, activated alumina, and *Drierite* are the most frequently used desiccants. Drierite is the tradename for anhydrous calcium sulfate.

Drum switch: A general-purpose controller designed for starting, speed regulating, and reversing series, shunt- and compound-wound dc motors. Typical applications include machine tools, mill motors, and crane

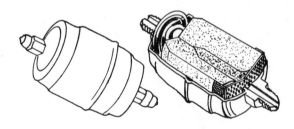

DRIER UNIT WITH CUTAWAY VIEW
TO SHOW CONSTRUCTION

FIGURE 148   Drier unit with cutaway view to show construction.

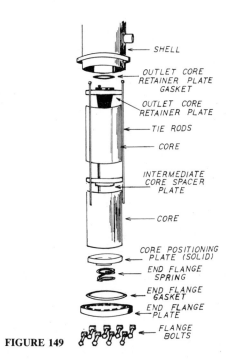

SHELL

OUTLET CORE
RETAINER PLATE
GASKET

OUTLET CORE
RETAINER PLATE

TIE RODS

CORE

INTERMEDIATE
CORE SPACER
PLATE

CORE

CORE POSITIONING
PLATE (SOLID)

END FLANGE
SPRING

END FLANGE
GASKET

END FLANGE
PLATE

FLANGE
BOLTS

FIGURE 149

motors for bridges and trolleys. They can also be applied to control hoist motors when there is no overhauling load, either through the use of worm gearing or with an automatic mechanical brake.

Forward and reverse armature points, off-position reset, and limit switch protection are standard on most controllers, whereas armature shunt and/or drift points are usually optional.

Dynamic-braking-lowering drum controllers are designed to control series or compound-wound crane hoist motors where power or dynamic lowering and light hook down-drive are required.

The ac nonreversing drum controllers are used with wound rotor induction motors on pumps, blowers, crushers, kilns, and similar nonreversing devices. They control the secondary circuit of the motor only, and a suitable primary control must be added. For primary control, linestarters are normally used. Pilot circuits in the drum permit use of the starter with or without a pushbutton station. Provision is also made to control the magnetic lockout on a circuit-breaker primary, when used. It is impossible to start the motor unless the drum is in the *resistance-all-in* position.

Duplexer: An electronic switch used to transfer an antenna connection from a receiver to a transmitter during the transmitted pulse and back to the receiver during the return (echo) pulse. Whenever a single antenna is used for both transmitting and receiving, as in a radar system, there arises the problem of ensuring that maximum use is made of the available energy. The high pulse repetition rate of radar systems makes the use of a mechanical switch impossible. Therefore, electronic switches, commonly called TR (transmit-receive) switches, are used. The device that performs the switching is called the *duplexer.*

When selecting a switch for the task of switching the antenna from transmitter to receiver, it must be remembered that protection of the receiver is as important as the power efficiency consideration. At frequencies where receiver RF amplifier tubes are used, such tubes are chosen to withstand relatively large input powers without damage. However, in microwave receivers, the crystal mixer at the input circuit is easily damaged by large signals and must be carefully protected.

In general, if the receiver input circuit is properly protected, the remaining receiver circuits can be prevented from blocking or overloading as the result of strong signals. However, a very strong main pulse signal will appear in the receiver's output unless additional precautions are taken to eliminate it. This can be done by a receiver gate signal that turns on the receiver during the desired time.

The requirements of a radar duplexing switch are:

1. During the period of transmission, the switch must connect the antenna to the transmitter and disconnect it from the receiver.

2. The receiver must be thoroughly isolated from the transmitter during the transmission of the high power pulse to avoid damage to the sensitive converter elements.

3. After transmission the switch must rapidly disconnect the transmitter and connect the receiver to the antenna. If targets close to the radar are to be seen, the action of the switch must be extremely rapid.

4. The switch should absorb an absolute minimum of power, both during transmission and reception.

Therefore, a radar duplexer is the microwave equivalent of a fast, low-noise, single-pole, double-throw switch. The devices that have been developed for this purpose are similar to spark gaps where high-current microwave discharges furnish low-impedance paths. A duplexer usually contains two switching tubes (spark gaps) connected in a microwave circuit with three terminal transmission lines—one each for the transmitter, the receiver, and the antenna (Fig. 150). These circuits may be connected in parallel (a) or series (b). One tube is called the transmit-receive tube or TR tube; the other is called the anti-transmit-receive tube or ATR tube. The TR tube has a primary function of disconnecting the receiver, and the ATR of disconnecting the transmitter.

Dynamometer test: A means of measuring the torque exerted by an induction motor at speeds intermediate between normal operation and standstill. The test is performed by directly coupling the motor to a direct-current

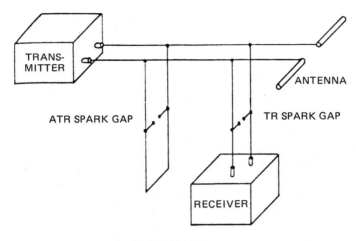

A. PARALLEL

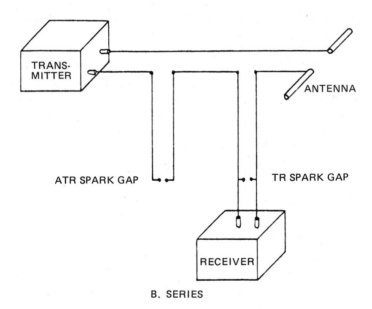

B. SERIES

FIGURE 150

generator, the stator of which is mounted on bearings and held from turning by the arm of a weighing scale. By proper adjustment of the field and armature voltages of the direct-current generator, the induction motor can be made to run at any desired speed, and its torque can be measured by means of the weighing scale.

Dynamometer tests should be taken at full voltage wherever possible, permitting the motor to run *light* between readings in order to cool. Where the capacity of the dynamometer is insufficient, a complete curve should be taken at reduced voltage and a curve at full voltage taken up to the capacity of the dynamometer. Normal frequency should be maintained throughout.

The usual object of taking a dynamometer test is to ascertain the shape of the speed-torque curve and to locate the minimum and maximum torque points. It is important, therefore, to search carefully for low-torque points in the region from zero to one-quarter speed and also to take a number of points in the region of maximum torque. Once these salient points have been located, the rest of the curve may be sketched in more freely. It is always desirable to take a few points at a low speed with backward rotation to locate the standstill-torque points better, because the dynamometer does not give an accurate reading at standstill due to the friction.

Dynamotor: A small, self-contained motor/generator. The motor and generator portions are enclosed in a common housing, giving the entire assembly the appearance of a simple motor. The dynamotor performs the dual functions of motor and generator, changing a relatively low voltage of a power supply into a much higher value for the plates and screens of electron tubes.

The dynamotor usually employs two windings on a single rotor. The two windings occupy the same set of slots and terminate in two or more separate commutators. The armature rotates in a single field frame with a conventional field winding to provide the excitation for both motor and generator. The motor armature winding is connected to the power source and develops driving torque, which rotates the rotor as a motor. The generator winding is connected to the plates and screens of the electron tubes of the associated equipment and generates the relatively high voltage for these loads.

Earphone: An electroacoustic transducer that like a loudspeaker, changes an electrical signal into sound waves. Also referred to as an audio receiver, the device may be a single unit or a pair of units usually attached by a band that passes over the head to hold the receiver(s) in place. Earphones are commonly used by telephone and switchboard operators, radio and ham operators, and recording personnel.

ECG: A photographic waveform record of a person's heartbeat indicating muscle contraction resulting from electrical stimulation. Acronym for *electrocardiogram.* As the heart beats, it discharges electrical charges that cause a voltage drop that can be measured on the surface of the body by a sensitive medical device called the *electrocardiograph,* which is actually a type of galvanometer.

Electrodes, which are placed on the patient's skin, are used to measure the various contractions of the heart that cause a changing EMF that rises and falls with heart action. The charges developed during the contractions are recorded on photographic paper as a series of zigzag lines called waves, with the various positive and negative peaks within one waveform cycle being labeled P, Q, R, S, and T. This notation aids in the diagnosis and analysis of possible heart disorders.

Echo: A reflected sound. Sound waves may strike an object, such as a building, canyon wall, or mine shaft; under certain conditions, they will bounce back to the source, where they are heard again as echo. Echoes prove to be a nuisance in such places as churches, theaters, and concert halls where clear sound transmission is quite important. As a result, costly and elaborate measures are sometimes taken to trap and absorb unwanted sound waves in order to reduce the occurrence of echoes. There are instances, however, when echoes are advantageous, such as in sea navigation, geological research, aviation tracking, and the ability of bats to miss obstacles in flight.

In the examples given, echoes are measured and recorded as a means of keeping track of distances of energy transmission and comparing reflected sound energy to directly transmitted waves.

Echo box: A test device that furnishes a standard reference signal for periodically checking radar systems. Acting as a microwave pulse generator, it is a high-Q resonant cavity with provisions for coupling the signal in and out. Cavity length is adjustable by a plunger, which, in turn, controls the frequency.

When the transmitter is pulsed, the cavity resonator is energized and shock excited into oscillation. As a result of the high Q, oscillation continues for some time after the trailing edge of the transmitter pulse and returns a signal to the receiver, appearing as an artificial target on the indicator screen. The time required for the oscillations to drop from maximum to a level below the sensitivity of the receiver is called *ring time*. Ring time depends on power output from the transmitter, as well as the sensitivity of the receiver. It is conceivable that losses could exist in the cavity resonator, but with the rugged tolerance requirements, the Q of the cavity is so high that any such losses can be completely ignored. In the motor-tuned echo box, the cavity is periodically tuned through resonance for a brief period of time. The resulting target appears as a series of radial spokes on a plan position indicator. If the echo box is at fixed resonance, the target appears as an intensified circle in the middle of the PPI screen. The spoke length or circle diameter is measured when the system is known to be in good operating condition.

With such a standard reference available, comparisons can be made with patterns produced at some later date. A decrease in spoke length is indicative of a decrease in performance of the system.

Echo splitting: That phenomenon that occurs in some radar equipment as the echo return is split, presenting itself on the screen of the radar indicator as a double indication. Special electronic circuits that are associated with the antenna lobe switching mechanism accomplish echo splitting. The target bearing is read from a calibrated scale when the two echo indications appear at equal height.

Edison effect: The thermionic emission of electrons (negative particles) from a hot filament that is sealed in an evacuated bulb. The electrons are attracted by a cold, positive-charged metal plate in the bulb. Thomas Edison discovered this action in 1888 when he was looking for the cause of his newly invented incandescent lamp's becoming black inside. This discovery formed the basis of the vacuum tube, upon which electronics rested until the coming of modern semiconductor devices in the 1950s.

Edison, Thomas Alva: One of the greatest and most admired of American inventors (1847–1931). Though not a brilliant scientist, Edison did much work in the field of electrical energy, basing his experiments and work upon the theories of other men. Using his cleverness and persistence, he was able to contribute much to society through his many inventions. Some of his first triumphs were a vote-recording machine, a carbon telephone transmitter, and the phonograph.

Edison's greatest achievement was the discovery of a practical light bulb, the incandescent lamp. He can also be commended for his work involving improvements in dynamos, motors, and power plants, all of which helped bring about the beginning of the electrical age. Other noted work included an alkaline (or Edison) storage battery, a magnetic iron-ore separator, and a synthetic carbolic acid.

Effective radiated power: The RF power output delivered by an antenna in terms of antenna gain. This is a product of the input power to the antenna multiplied by the antenna power gain. For example, if an antenna has a power gain of 3 and is fed with an input of 1000 W from the transmitter, the effective radiated power (ERP) will be 3000 W. This term especially applies to directional and beam antennas that radiate input power in one or two specific directions.

Einstein's equation: Energy equals mass times the square of the speed of light (c), or $e = mc^2$. Albert Einstein devised this equation after considerable work and experimentation in the area of physics, a field in which he contributed many ideas about the nature of space, time, and energy. Einstein's theory states that the speed of light in a vacuum never varies in spite of changes in the speed or motion of its source. Furthermore, the mass of an object increases with the object's velocity.

This idea was the first attempt at explaining mass in the form of energy, and it also showed that a very small change in mass could result in a very large change of energy. This work heralded the appearance of the nuclear reactor, the atomic bomb, and the field of nuclear physics.

Einthoven string galvanometer: A simple galvanometer in which a silvered glass filament carrying current is mounted in a magnetic field set up by either a permanent magnet or an electromagnet. The current causes the filament to be deflected through a distance proportional to the current strength, the deflection being observed through a microscope.

E layer: A level of density of the ionosphere. Actually, there is thought to be no sharp dividing line between layers, but for the purposes of discussion, such a demarcation is indicated. The E layer is shown in Fig. 151. This is the band of atmosphere at an altitude between 50 and 90 mi. It is a well-defined band with greatest density at an altitude of about 70 mi. This layer is strongest during daylight hours and is also present, but much weaker, at night. The maximum density of the E layer appears at about noon local time.

The ionization of the E layer at the middle of the day is sometimes sufficiently intense to refract frequencies of up to 20 mHz back to the earth. This action is of great importance for daylight transmissions for distances up to 1500 mi. In addition to the layers of ionized atmosphere that appear regularly, erratic patches of ionized atmosphere can occur at E-layer heights in the manner that clouds appear in the sky. These patches are referred to as sporadic-E ionizations. They are often present in sufficient number and intensity to enable good VHF transmissions over distances not normally possible. Sometimes, sporadic-E ionizations appear in considerable strength at varying altitudes and actually prove harmful to electronic transmissions.

Electret microphone: A microphone whose input sound waves cause a small dielectric disk or slab to vibrate in accordance with the audio frequency. The heart of the electret microphone is the dielectric disk, which is permanently polarized and possesses a permanent electric field. This is the electrical equivalent of a permanent magnet. When a vibration is established from sound waves, a change in the electric field is induced, and an audio-frequency output voltage is generated. Electret microphones are small in physical size and are often used in communications work and as clip-on devices for commercial broadcast interviews.

Electric charge: The force that exists between the electron and the nucleus of an atom. Electrons revolving around an atom are pulled to the nucleus center due to natural gravitation. However, due to their speed of rotation, they are forced to follow an orbital path around the nucleus. The electron is said to possess a negative charge, and the nucleus has a positive

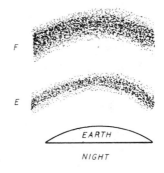

F

E

EARTH

FIGURE 151   *NIGHT*

charge, which is carried by protons. The number of electrons rotating about the nucleus equals the number of protons found within the nucleus of that atom.

The neutron is found in the nucleus of the atom, but it possesses no electrical charge. Under normal conditions, the atom remains stable as the forces in the atom are balanced. The positive charge of the nucleus (provided by the protons) attracts the electrons. However, the speed and energy of the electrons enable them to maintain their orbital paths.

Electrodynamometer: A meter in which two fixed coils and two movable coils produce the magnetic field. The two fixed coils are connected in series and positioned coaxially, with a space between them. The two movable coils are also positioned coaxially and are connected in series. The two pairs of coils are further connected in series with each other. The movable-coil unit is pivot-mounted between the fixed coils.

This meter arrangement is illustrated in Fig. 152. The central shaft on which the movable coils are mounted is restrained by spiral springs that hold the pointer at zero when no current is flowing through the coil. These springs also serve as conductors for delivering current to the movable coils. Since these conducting springs are very small, the meter cannot carry a very heavy current.

When used as a voltmeter, no difficulty in construction is encountered, because the current required is not more than 0.1 A. This amount of current can be brought in and out of the moving coil through the springs. When the electrodynamometer is used as a voltmeter, its internal connections and construction are as shown in Fig. 153a. The fixed coils (a and b) are wound with fine wire, since the current through them will be no more than 0.1 A. They are connected directly in series with the movable coil (c) and the series current-

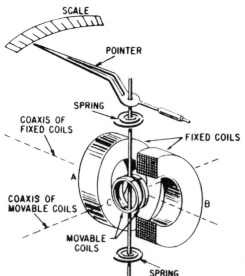

FIGURE 152   Inside construction of an electrodynamometer.

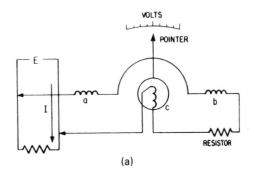

(a)

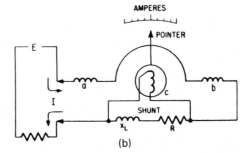

FIGURE 153 Circuit arrangement of
electrode dynamometer-type meter:
(a) voltmeter; (b) ammeter.

(b)

limiter resistance. For ammeter applications, however, a special type of construction must be used, because the large currents that flow through the meter cannot be carried through the moving coils.

In the ammeter, the stationary coils (a and b of Fig. 153b) are generally wound of heavier wire to carry up to 5 A. In parallel with the moving coils is an inductive shunt, which permits only a small part of the total current to flow through the moving coil. This current through the moving coil is directly proportional to the total current through the instrument. The shunt has the same ratio of reactance to resistance as the moving coil. Thus, the instrument will be reasonably correct at all frequencies with which it is designed to be used. The meter is mechanically damped by means of aluminum vanes that move in enclosed air chambers.

Although electrodynamometer meters are very accurate, they do not have the sensitivity of the D'Arsonval-type meter. For this reason, they are not widely used outside the laboratory.

Electroencephalograph: An instrument used for measuring and recording the rhythmically varying potentials produced by the brain. The electroencephaloscope is used to detect brain potentials at many different sections of the brain by means of electrodes, which are applied to the scalp at selected points. These electrodes pick up brain voltages, which, in turn, are received and recorded by the electroencephalograph and displayed on a cathode-ray tube in the form of a graphic wave. The voltages being recorded are of very low level and therefore mandate that the equipment being used for recording display excellent noise rejection. The electroencephalograph is a very valuable instrument and is used to detect brain abnormalities and to isolate areas of brain and nervous disorders. Abbreviated *EEG*.

Electrolysis: The process by which the chemical composition of a material is changed by sending an electrical current through the material. This process is employed in the commercial production of such metallic elements as aluminum, sodium, magnesium, and various gaseous elements such as oxygen, hydrogen, and chlorine. Electrolysis made the production of these elements possible at a substantially less expensive cost than was formerly possible.

Electroplating and electrotyping also employ the process of electrolysis. Several examples are the commercial production of sodium hydroxide (lye), potassium hydroxide (caustic potash), and the purification of copper, silver, gold, and other useful metals.

Electrolysis can occur only when the material, which may be a solid or liquid, through which an electrical current is passed is electrically conducting. Electricity is passed through solids by electric charges called *electrons* and through liquids by electrically charged atoms called *ions*.

Electrolytic conduction: The flow of electricity that occurs when electrodes are immersed in an electrolyte. An example of this method is the electrolysis of water, in which the elements hydrogen and oxygen are prepared in quantity. First, a small amount of a substance (such as sulfuric acid) is added to the water to make it a good conductor of electricity. Two platinum plates called *electrodes* are then placed into the solution, and each electrode is connected to the poles of a battery or other source of direct current. The water is then decomposed into the two gases. Hydrogen collects at the negative electrode (or cathode), and oxygen collects at the positive electrode (or anode).

Electrolytic rectifier: A device that will pass current flow in only one direction. It consists of an aluminum electrode and a lead or carbon electrode, both of which are immersed in a solution of borax or sodium bicarbonate or in a solution of ammonium citrate, ammonium phosphate, and potassium citrate.

Electrolytic resistor: A device that is formed by immersing two conductors in an electrolyte solution. The amount of resistance will be controlled by the quality of the solution. Higher resistances will be produced by weaker electrolyte solutions. Electrolytic resistors are emergency devices that are often made by immersing two conductors in salt water. As more salt is added, the solution becomes stronger and the resistance drops.

Electromagnet: A core of soft iron around which a coil of insulated wire is wound. These devices are used in signal devices such as bells and buzzers, as well as in relays, lifting magnets, electric motors, and generators. They can be made extremely powerful and have the advantage of being magnetized or demagnetized at will simply by turning the coil current on or off.

Electromagnetic frequency: The number of electromagnetic waves transmitted per second (rate of vibration). A rainbow is the perfect example of a means of viewing the entire range of visible electromagnetic frequencies at one time. Rainbows are a form of electromagnetic radiation, as are gamma rays, X rays, ultraviolet rays, infrared rays, and radio waves, all of which travel at the same speed, but differ in wavelength and frequency.

Electromagnetic induction: The induction of a voltage by a magnetic field. A current-carrying conductor always creates a magnetic field around it, but the reverse is also true. A changing magnetic field creates or induces a voltage in a conductor in this field. Electromagnetic induction takes place whenever a change occurs in the number of magnetic lines of force passing through a conductor. Such change is due to a relative movement of the lines and the conductor. The relative movement may be obtained by moving the magnet near a stationary conductor, or by moving the conductor in the stationary magnetic field.

The same effect may also be obtained without moving either the magnet or the conductor if the magnet is an electromagnet and the current in its coil is changed. The changes in the current cause a change in the number of lines cutting the conductor, and an EMF is induced in the conductor. Such an induction is called *mutual induction*. *Self-induction* takes place when changes of current in the coil of an electromagnet induce a voltage in this same coil.

Electromagnetic pickup: A transducer that provides an electrical output that corresponds directly to the changing magnetic field at its input.

Electromagnetic radiation: That which occurs when electrical currents of sufficient amplitude flow in an antenna whose dimensions are about the size of the wavelength of the radiation to be generated. Radiation can be generated by devices that are not supposed to produce radiation, such as receivers or industrial machines. This unwanted radiation can cause interference with radiation that is actually carrying information. When the radiation is produced by a radio transmitter, the pattern of energy depends primarily on the design of the antenna. All of the possible radiation patterns have counterparts in visible patterns that can be seen radiating from sources that are familiar to everyone.

Electromagnetic repulsion: The tendencies of light poles of magnets to repel each other while opposite poles attract. Electromagnetic repulsion is used in some electromechanical relays and solenoids to set up mechanical movements that are electromagnetically controlled by electric current flowing in a conductor.

Electromagnetic switch: A device that is actuated by magnetism produced by a current flow through a coil of wire wound on an iron core. When current is passed, a magnetic effect causes a mechanical reaction that opens and closes two or more switch contacts. Examples of electromagnetic switches are solenoids and electromechanical relays.

Electromagnetic tube: A cathode-ray tube that uses electromagnetic deflection. Most television picture tubes are of this design, and some oscilloscopes may also incorporate this type of deflection.

Electromechanical amplifier: A device that converts an electrical input signal into mechanical motion (vibratory or rotary), which is then converted back into an electrical output signal of higher current, voltage, or power. Examples of this type of amplifier are the amplidyne, the carbon-button amplifier, and the electrocoustic amplifier.

Electromechanical chopper: A device that interrupts the flow of current at some predetermined and usually rapid rate. It is composed of an on/off switch in the dc power lead to any load. When the switch is initially turned on, the current rises from zero to a peak value. When switched off, the current returns from the peak value to zero again. By rapidly engaging and disengaging the switch, a square wave input is seen by the load.

An electromechanical chopper is a vibrator type of interruptor used primarily to chop direct current, converting it into a square wave signal whose amplitude is proportional to current strength. The dc power is applied to a magnetic vibrator that causes a reed contact to swing back and forth at a rapid rate (usually 60 times/s or more). This reed makes and breaks the electrical contact in the same manner as described previously. Here, the human element is removed, and the switching process is accomplished through electromechanical means.

While electromechanical choppers are still seen from time to time, most have been replaced by pure electronic circuits. These use transistors or silicon-controlled rectifiers to electronically switch the power on and off. There are no moving parts using this latter means. Thus, reliability and overall circuit efficiency are greatly increased.

Electromechanical frequency meter: A direct-reading instrument used for measuring frequency in the lower and middle portions of the AF spectrum. It records the mechanical motion resulting from the applied signal. There are two varieties of electromechanical frequency meters—the movable-iron type and the reed type.

In the movable-iron type, a soft-iron vane (to which a pointer is attached) rotates in the magnetic field of two stationary coils mounted perpendicularly to each other and through which current flows (through a resistor in series with one coil and an inductor in series with the other). The direction of the field varies with frequency because of the phase shift introduced by the resistor and inductor. This causes the deflection of the vane to be proportional to frequency. The pointer travels over a scale reading directly in hertz.

In the reed type, the alternating current flows through the coil of a field magnet. Metal reeds cut to vibrate at a different frequency are mounted within the field of the magnet. The reed that vibrates most violently indicates the frequency of the current. A scale under the row of reeds is marked with the resonant frequency of each reed.

Electromechanical modulator: A meter-type modulator that resembles a D'Arsonval movement. An alternating supply current in coils wound on the poles of the movement's permanent magnet induces an ac output signal in the movable coil. This signal is coupled to the output through a transformer and a dc blocking capacitor. A dc input signal is also applied to the movable coil, which is deflected to a position proportional to the current. The ac output voltage is proportional to the position of the coil in the ac field and to the dc input voltage.

Electromechanical oscillator: An oscillator that utilizes electronic excitation control to produce a highly stabilized output. The two general types of electromechanical oscillators are the crystal type and the magnetostriction type. The crystal type uses natural or synthetic crystals vibrating at or near their natural frequency to control the frequency of oscillation. The magnetostriction type uses a magnetic field to excite a metal bar, rod, or tuning fork at the mechanically resonant frequency. The magnetostriction oscillator is generally used at audio frequencies, whereas the crystal type is used at both audio and radio frequencies.

Electromechanical oscilloscope: A galvanometer-type instrument used to display a varying or alternating current or voltage. The signal is applied to a meter movement having a movable coil that swings or vibrates in response to the signal. A tiny mirror cemented to the coil reflects a beam of light to a rotating mirror that sweeps the beam across a translucent screen, on which the image is produced.

Electromechanical transducer: A transducer that translates mechanical signals into electrical ones, or vice versa, without the intermediary of electronic devices (tubes, transistors, etc.). It may also refer to a special triode having a stylus attached to the plate and extending beyond the envelope. Pressure on the tip of the stylus moves the plate, changing the gap between plate, grid, and cathode and altering the electrical characteristics of the tube as an amplifying device. The tube's output for a standard input signal, then, is proportional to the applied pressure. The electromechanical transducer has many uses, among them the checking of surface roughness.

Electrometer: A specially designed electronic voltmeter used to measure extremely low voltages. Typically, these devices use highly sensitive electronic voltage amplifiers to perform their functions, which also include measurement of extremely low currents in the microampere range. Tube-type electrometers use an electrometer tube, which is a specially selected vacuum tube that exhibits a high grid to cathode resistance, negligible photoelectric emission, and low input to circuit leakage. Typically, vacuum tubes are chosen for electrometer applications that have very high current gain.

The amplifier circuitry used in electrometers must be very stable. Therefore, placement of components is critical, and wiring lengths must be kept to a minimum. This circuit must be designed to exhibit a very high gain while inducing minimal noise characteristics.

Electromotive force: A theoretical unit that is used in establishing the relationship between it and other electrical and magnetic theoretical units. Taking a magnet having a field of unit strength (one line of force per square centimeter) and moving a conductor through this field at a uniform rate so that it will cut across it in 1 s, there would be an EMF generated in the conductor. In other words, if a conductor cuts one line of force per second, it will generate a voltage of unit value.

Electron: The fundamental negative electrical charge found in matter. Electrons surround the positively charged nucleus of an atom and help determine the chemical properties of that atom. An electron has approximately 1/1840th the mass of a hydrogen atom (which is the lightest of atoms) equal to 9.107×10^{-28}g. An electrical current can be created by the flow of electrons through a conductor. Methods have been devised to free electrons from metals and control their flow. This knowledge has proven very useful, as it has led to the development of electronic devices and brought into being the fields of radio, television, and radar.

Electron avalanche: The phenomenon in semiconductors that are operated at high inverse bias voltage, whereby carriers acquire sufficient energy to produce new electron-hole pairs. This occurs when the carriers collide with atoms. This action causes the inverse current to increase sharply. The sudden marked increase of reverse current at the bias voltage where avalanche begins is called the *breakdown point.* This action is nondestructive when the current is limited by external means.

Electron avalanche may also be referred to as *electron multiplication,* but this term is more descriptive of the production of additional electrons as a result of collisions between electrons, atoms, and molecules in a gas discharge. However, during true electron avalanche, there is an increased production of electrons in a semiconductor material.

Electron-beam generator: A sophisticated device used for melting, welding, and machining conducting materials. Since the process takes place in an evacuated jar, this approach is particularly convenient for reactive metals such as zirconium and titanium. It also allows for processing refractory metals like tungsten and molybdenum, and high thermal-conductivity metals such as copper and silver that are difficult to process by conventional methods. Such sophisticated equipment is quite expensive, but aeronautic and aerospace industries can hardly do without it.

In an electron-beam machine, a tungsten and molybdenum heater-cathode emits electrons that are focused by a negative-potential conical grid and strongly accelerated toward the work by a high-voltage supply (Fig. 154). The strongly concentrated beam hits the work in a very small impact spot (less than 0.01-in. diameter), where the considerable kinetic energy is dissipated, leading to immediate, intense heating of a minute region of the work and quick fusion of the irradiated matter. Skillfully putting this heat to work, the target can be melted, welded, or cut.

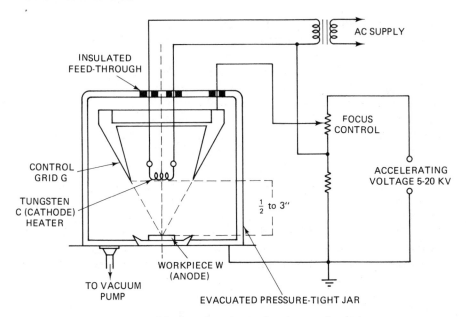

FIGURE 154 Diagram of an electron-beam processing device.

This phenomenon can only take place at a very low pressure (less than 2×10^{-4}/mm/Hg), so the whole device is enclosed in a pressure-tight jar that is permanently connected to a vacuum pump. The need to introduce the materials to be processed into a vacuum chamber of obviously limited dimensions and control the operations from the outside (possibly by a closed-circuit television system) is evidently a shortcoming, but vacuum processing is mandatory for some exotic materials and advantageous for many others. Often, more convenient or less costly processing methods simply do not exist.

The heat quantity Q (in calories per minute) dissipated in the impact spot is

$$Q = 1.4(10^{-2} \, EI)$$

where E is the accelerating voltage and I is the beam current in amperes. In the machine in Fig. 154, the work acts as an anode. Accelerating voltages of 5 to 30 kV are used with powers up to 5 kW; spot diameters range from 0.3 to 3 mm. For highest heat dissipation, the work is positioned in the focal point of the beam.

To obtain more heat concentration in a narrower fusion zone and increase the penetration depth, magnetic focusing and higher acceleration voltages are used. With voltages of up to 100 kV and 1 kW of dissipated power, a beam of less than 0.01-in. diameter (0.25 mm) is obtained, allowing for cutting holes of intricate shapes. Such a machine is much like an electron microscope and costs about five times the price of the above model. Furthermore, the target hit by high-energy electrons emits dangerous X rays and must be screened heavily to provide a safe working environment. Only qualified technicians can operate such a machine.

The self-accelerated gun combines the advantages of the comparatively low and high accelerating voltage systems. The gun in this system not only has a cathode and grid, but also a focusing, diaphragm-like anode and an accelerating anode with a center hole traversed by the pencil-shaped beam (Fig. 155). A focusing coil may be added as shown. Electrons are now accelerated within the gun before they leave by the hole. The workpiece is no longer a part of the focusing system, and there is no electric field between the gun and workpiece. The spot diameter is from 0.8 to 3 mm, but it can be as low as 0.25 mm with magnetic focusing. Accelerating voltages range from 10 to 30 kV with 5 to 10 kW of power.

Electron-coupled oscillator: An oscillator circuit used to produce an RF output of constant amplitude and an extremely constant frequency, usually within the RF range. This circuit is normally used in any type of electronic

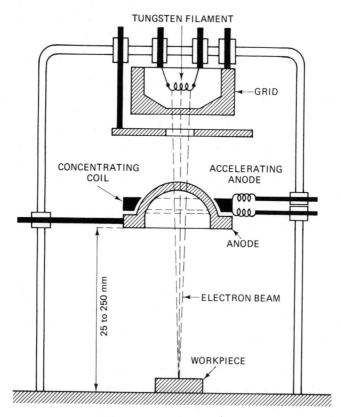

FIGURE 155   Diagram of an improved self-accelerating gun electron-beam heating device.

equipment where good stability is required, and where the output waveshape is not critical, as some distortion of the waveform is normal for this oscillator. The electron-coupled oscillator uses the shielding effect between the plate and screen grid in a tetrode or pentode to isolate the plate load device. While this circuit will operate with other configurations, the LC type of oscillator is used most frequently. The frequency stability of the electron-coupled oscillator is quite good.

Electron emission: That which occurs when an electron leaves the surface of the material that contained it and escapes into the space surrounding the material. Most metals have a large number of free electrons. The free electrons cannot normally escape from the surface of the metal and can only wander about within the lattice structure of the metal.

Electron gun: An assembly composed of the electrodes that form the electron beam in a cathode-ray tube. As shown in Fig. 156, the electron gun assembly is placed in the neck of the tube and connects through internal leads to the pins on the base.

Electron-hole pair: In semiconductor terminology, an electron and its related hole contained within a specially treated crystalline chip. In the conduction band, electrons are contained and each has a counterpart in the valence band. A hole is a vacancy left by the electron after it moves into the conduction band.

Electronic breadboard: An operational mock-up of an experimental circuit in order to test its function. The basic support for such a circuit is often a piece of perforated circuit board that has been fitted with a number of clip-spring contacts. Component leads are fitted through the tightly coiled springs, which take the place of solder connections. Using this procedure, component leads may be easily changed in a short period of time to modify the circuit until the desired operational characteristics are obtained. Once the proper circuit configuration is arrived at, a permanent circuit is made using standard wiring techniques.

Electronic clock: A clock whose motor is driven by a constant-frequency oscillator (crystal or tuning fork type), which is followed by multivib-

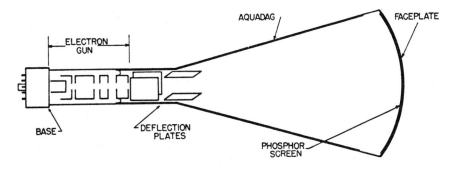

FIGURE 156 Cathode-ray tube assembly.

rators and amplifiers. It may also refer to any electronic timing circuit that produces pulses at predetermined intervals for the purpose of regulating the operation of other circuits, subsystems, or assemblies.

Electronic control: The activation, monitoring, and general control of circuits, devices, and their functions by electronic means. Often, the devices under control are remote from the actual control site. Electronic control often involves the use of sophisticated radio transmitters and receivers to link the control site with the work area. The link equipment is used to broadcast and receive signals, which are modulated with audio tones or coded information. At the receiving end, the demodulated signals are sent by electronic circuits, which provide drive to other circuits, mechanical devices, etc.

Electronic control may also be used to describe a specific component that is electronic in nature and replaces a control that is normally mechanical. A standard volume control used with audio equipment is usually a wire-wound variable resistor. The electronic counterpart of this component might consist of an IC amplifier whose gain may be varied. This device would be called an *electronic volume control.*

As the state of the art progresses, more and more mechanical switches and other control devices are being replaced with electronic controls that offer a higher degree of reliability due to their construction. This involves solid-state circuitry with no moving parts.

Electronic counter: A fully electronic circuit that indicates the number of pulses that have been applied to it. This circuit may be a cascade of flip-flops that keep track of the pulses. Unlike the electromechanical counter, the electronic counter has no moving parts and is therefore capable of extremely high-speed, noiseless operation.

Electronic data processing: Data processing performed largely by electronic equipment.

Electronic divider: An electronic device that is used to perform arithmetic division. In a digital computer, such a divider may be a sequence of flip-flops, each of which produces a single output for every two input pulses. In an analog computer, the output-signal amplitude is equal to the quotient of two input-signal amplitudes. An electronic divider may also be a frequency divider or voltage divider that uses active components rather than resistors.

Electronic microphone: A device that changes sound waves into varying electrical signals. A simple microphone can be found in the mouthpiece of a telephone. Here, a thin metal plate or flexible diaphragm lies over a small metal box, which contains carbon grains through which an electric current flows. This is triggered by tiny sound waves projected by the voice. The frequency of the sound waves causes changes in the frequency of the electrical current.

Microphones that are used in radio transmissions are much more sensitive and can be adapted to receive and convert sound waves produced by orchestras, singing voices, etc. Other examples of microphone applications are hearing aids, intercom systems, and phonograph pickups.

Electronic relay: A switching circuit that employs one or more tubes or transistors to perform the relay function without moving parts. An electronic relay may also refer to an electronic component that is designed to switch on application of appropriate gating signals. The triac, diac, and silicon-controlled rectifier are examples.

Electron microscope: An electronic instrument that allows magnifications of over 100,000 times the size of the object, which is between 20 and 30 times more than is possible with an optical microscope. In the electron microscope, an electromagnetic field, which is manipulated by several focusing coils, bends a stream of electrons that passes through the specimen being observed. Another series of coils bends the beam to form a magnified image on a fluorescent screen. The electron microscope made possible the examination of the cells that compose plant and animal tissues.

Electron motion: The movement of electrons in a conductor, semiconductor, or space as the result of electric or magnetic attraction or repulsion. Electron motion may also be the movement of an electron as a charged mass. In an electric field, this movement simulates that of a free-falling body in a gravitational field.

Electron tube: A device made up of a highly evacuated glass or metal shell that encloses several elements (electrodes). The elements consist of the cathode (emitter), the plate (anode), and sometimes one or more grids. Another element of importance in many tubes is the heater, also called the *filament,* which serves to heat the cathode.

Electron tubes are of many types and designations and perform many functions. They can be made to convert currents and voltages from one waveform to another, amplify weak signals with minimum distortion, and generate frequencies much higher than any conventional ac generator.

Electron-tube parameter abbreviations: Letter symbols used as a form of shorthand notation in technical literature when designating electron-tube operating conditions. These parameters are:

1. Maximum, average, and root-mean-square values are represented by capital (uppercase) letters; i.e., *I.E.P.*
2. Where needed to distinguish between values as above, the maximum value may be represented by the subscript *m*; i.e., E_m, I_m, P_m.
3. Average values may be represented by the subscript *av*; i.e., E_{av}, I_{av}, P_{av}.
4. Instantaneous values of current, voltage, and power that vary with time are represented by the small (lowercase) letter of the proper symbol; i.e., *i, e, p.*
5. External resistance, impedance, etc., in the circuit external to an elec-

tron-tube electrode may be represented by the uppercase symbol with the proper electrode subscripts; i.e., R_g, R_{sc}, Z_g, Z_{sc}.

6. Values of resistance, impedance, etc., inherent within the electron tube are represented by the lowercase symbol with the proper electrode subscripts; e.g., r_g, z_g, r_p, z_p, c_{gp}.

7. The symbols g and p are used as subscripts to identify ac values of electrode currents and voltage; i.e., e_g, e_p, i_g, i_p.

8. The total instantaneous values of electrode currents and voltages (dc plus ac components) are indicated by the lowercase symbol and the subscripts b for plate and c for grid; i.e., i_b, e_c, i_c, e_b.

9. No-signal or static currents and voltages are indicated by the uppercase symbol and lowercase subscripts b for plate and c for grid; e.g., E_c, I_b, E_b, I_c.

10. Maximum values and r.m.s. of a varying component are indicated by the uppercase letter and the subscripts g and p; i.e., E_g, I_p, E_p, I_g.

11. Average values of current and voltage for the with-signal condition are indicated by adding the subscript s to the proper symbol and subscript; i.e., I_{bs}, E_{bs}.

12. Supply voltages are indicated by the uppercase symbol and double subscript bb for plate, cc for grid, ff for filament; i.e., E_{ff}, E_{cc}, E_{bb}.

An alphabetical list of electron-tube symbols is shown in Fig. 157.

Electroplating: The act or process of depositing metal by electric means. This process consists of obtaining an electrodeposit of one metal, used as an anode, upon some metallic article that is connected to form the cathode in

C_{gk}	Grid-cathode capacitance	I_k	Cathode current
C_{gp}	Grid-plate capacitance	i_b	Instantaneous plate current
C_{pk}	Plate-cathode capacitance	i_c	Instantaneous grid current
E_b	Plate voltage, dc value	i_g	Ac component of grid current
E_c	Grid voltage, dc value	i_p	Ac component of plate current
E_{cc}	Grid bias supply voltage	P_g	Grid dissipation power
E_{co}	Negative tube cutoff voltage	P_o	Output power
E_f	Filament voltage	P_p	Plate dissipation power
E_{ff}	Filament supply voltage	R_g	Grid resistance
E_k	Dc cathode voltage	R_L	Load resistance
e_b	Instantaneous plate voltage	R_k	Cathode resistance
e_c	Instantaneous grid voltage	R_p	Plate resistance, dc
e_g	Ac component of grid voltage	R_{sc}	Screen resistance
e_p	Ac component of plate voltage	r_L	Ac load resistance
I_b, I_o	Dc plate current	r_p	Plate resistance, ac
I_c	Dc grid current	t_k	Cathode heating time
I_f	Filament current		

FIGURE 157

an electrolytic bath. That is, the object upon which it is desired to deposit the metal is connected with the negative pole of the source of current, and the metal that is to be deposited is connected with the positive pole.

The chemical nature of the electrolyte employed depends on the kind of plating. For plating with gold or silver, the electrolyte is always alkaline; for plating with nickel or copper, it is usually acid. Substances other than metal can be electroplated by first coating their surfaces with powdered graphite or plumbago, as in the case of electrotyping.

Electroscope: An instrument capable of measuring the existing charge on a body and which is useful in the study of electrostatic phenomena. One form of this device consists of two thin gold leaves attached to a metallic rod, the other end of which is terminated by a small sphere. These leaves are usually enclosed in a glass-walled container to ensure that they are not adversely affected by air currents. The sensitivity of the device will be determined primarily by the thickness and type of material used for the leaves. An example of an electroscope is shown in Fig. 158.

If an object containing a charge is brought near the terminal of an electroscope, the two leaves will diverge. This occurs as a result of the repelling effect existing between the like charges that are forced onto the two leaves by the charged object. The greater the charge, the greater the mutual repulsion between the leaves.

When using a sensitive electroscope, the charge on the leaves should be felt through induction. Otherwise, the leaves may be ripped from the device due to the great repelling force of the similarly charged leaves. However, charging by contact with the metallic sphere may be necessary in the case of less sensitive devices.

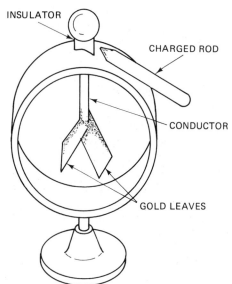

FIGURE 158 The electroscope.

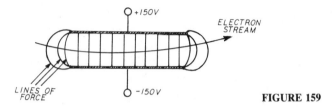

FIGURE 159

Electrostatic deflection: In oscilloscope tubes and some early television picture tubes, deflection of the electron beam by the electrostatic fields between pairs of internal horizontal and vertical deflecting plates. Figure 159 illustrates this action.

Electrostatic generator: Any device that produces a high-voltage electric charge. These devices are used in many chemistry, electricity, and physics experiments and are often constructed from two disks that rotate in opposite directions. As the disks rotate, a static electrical charge is built up. Metal collector brushes that are in physical contact with one plate conduct the electricity to a high value capacitance until discharge occurs.

Electrostatic generators synthesize the buildup of static electricity in the atmosphere, which results in lightning striking. When the charge has reached a sufficient potential, a flashover occurs in lightning-like fashion. Typically, these devices will produce potentials of 200,000 V or more. However, while the potential is high, current is low, typically less than 2 μA.

The same basic principle applies to static electricity buildup on the surface of the human body, which can be created by certain types of clothing when the wearer walks across a carpet. The movement between the shoes and the rug causes the body to charge to a high voltage potential. Discharge occurs when the body comes in contact with a metal object.

Electrostatic hysteresis: The tendency of some dielectrics (especially ferroelectric materials) to saturate and retain a portion of their polarization when an alternating electric field to which they are exposed reverses polarity. This causes the charge to lag behind the charging force.

Electrostatic induction: A means of placing a charge on a body. If a neutral body is suspended as shown in Fig. 160 and a charged body is brought near but not in contact with the neutral body, charging by induction can be demonstrated. If a negatively charged rubber rod is brought near a neutral body (a copper bar), the excess electrons that have accumulated on the surface of the rod repel electrons of the copper toward the end opposite that of the charged body. There is now an accumulation of electrons at one end of the copper bar for as long as the charged body continues to be held nearby. Although electrons become bunched together at the end of the bar, the total charge throughout the material remains unchanged.

If the copper bar is now placed in contact with a much larger neutral body, such as the earth, the electrons that are being repelled along the copper bar are transferred to the larger neutral body. This transfer of electrons takes place

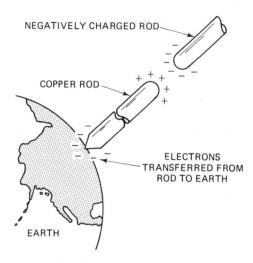

NEGATIVELY CHARGED ROD

COPPER ROD

ELECTRONS
TRANSFERRED FROM
ROD TO EARTH

EARTH

FIGURE 160   Electrostatic induction.

until there is an equalization between their charges or until the contact between
the bodies is broken. Equalization occurs when the force of repulsion from the
negative rubber rod is exactly counterbalanced by the positive force of attraction resulting from the positive charges generated at the unearthed end of the
copper bar.

In the instance when the neutral rod made contact with the earth,
electrons were repelled to the earth by the electrostatic force from the
negative-charged rubber rod. Although many electrons were repelled to the
earth, it is still considered to be neutral. This is permissible because of the vast
quantity of electrons that would be necessary to appreciably change the earth's
charge. Due to the electrons being repelled to the earth from the neutral
copper bar, the copper is left lacking electrons and acquires a detectable
positive charge. Keep in mind the important fact that the negatively charged
rubber rod was near, but not in contact with, the copper bar. Therefore, no
electrons were transferred from the rod to the copper bar. In all cases of
electrostatic induction, the body being charged will acquire a charge opposite
to that of the charging body.

Electrostatic precipitator: A device designed to remove dust, lint,
and other contaminating particles from the air. Also referred to as a *dust
precipitator* or *electronic air filter,* this device consists of a pair of screens or
wires through which the air is drawn. A potential of several thousand volts is
maintained between the two grids. As the air passes, the contaminants acquire
a charge, causing them to be drawn to the oppositely charged screen. They
adhere to this element of the precipitator, which is removed periodically for
cleaning.

Due to the charge attraction, electrostatic precipitators are far more
efficient than mechanical types of air filters. However, most precipitators draw

air through a finely woven screen, cleaning out some of the contaminants before reaching the electronic filter.

Electrostatic printer: A device for printing an optical image on paper in which dark and light areas of the original are represented by electrostatically charged and uncharged areas on the paper. The paper is dusted with particles of finely powdered dry ink, and the particles adhere only to the electrically charged areas.

Electrostatics: The branch of electrical science that deals with the properties of electricity at rest or of frictional electricity (as opposed to electrokinetics, or electricity in motion). An electrostatic field can exist between two bodies that have a difference of potential, wherein the force of attraction between the charged bodies becomes great enough that ionization will occur in the intervening space. This will result in an arc between the two bodies, which then reduces the potential.

Electrostatic shield: A metallic enclosure that is specially designed to contain an electrostatic field. Often, these devices will allow for the passage of electromagnetic energy.

Electrostatic speaker: A loudspeaker whose vibrating diaphragm is one of two plates in an air capacitor, the other being a closely situated metal plate (or plug). An AF voltage applied to the plates causes them to vibrate. This type of device may also be called a *capacitive loudspeaker* or *capacitor loudspeaker*.

Electrostatic unit: A unit of measurement in the electrostatic system of centimeter-gram-second (cgs). Mass is measured in grams, whereas time is measured by mean solar seconds. Electrical units in the cgs system fall into two categories—electrostatic and electromagnetic—with the *esu* denoting the electrostatic portion of this system. In the cgs system, all electrostatic units carry the prefix *stat*. For example, electrostatic current is measured in statamperes, whereas potential difference is measured in statvolts.

Electrostatic voltmeter: An instrument that utilizes the attraction between two closely spaced metal plates (one stationary and the other rotating) to which the unknown voltage is applied. A pointer attached to the rotating plate moves over a voltage scale. A spiral spring returns the rotating plate to rest (and the pointer to zero) when the voltage is removed.

Electrostriction: A mechanical deformation created when an electric field is applied to any dielectric material. The amount of deformation is a product of the square of the applied field. A good example of electrostriction is the contraction of a ceramic plate when a voltage is applied across its parallel faces. Electrostriction is closely aligned with *magnetostriction,* which is the deformation of any dielectric in proportion to the strength of an applied magnetic field. In both cases, the deformation sets up a vibration in the dielectric, which is comparable to piezoelectric vibration in a quartz crystal.

Elements, circuit: Any electronic devices or parts that are connected in such a manner as to form an electronic circuit. Circuit elements include capacitors, resistors, inductors, transistors, and any other component that may be used to make an electronic circuit. The wiring that is used to connect these components is not generally classified as a circuit element, however, as this term applies to the basic components only.

Element spacing, antenna: The physical placement of driven, reflecting, and parasitic elements in relationship to one another to make up a directive system. The actual physical distance is a portion of a full wavelength at a specific operating frequency. For example, the optimum spacing for a two-element array is approximately 0.115 wavelength for a director and 0.135 wavelength for a reflector. Both distances are measured from the driven element.

When a directional antenna contains both a director and reflector to form a three-element array, the optimum spacing is established by the bandwidth over which the antenna will be required to operate efficiently. Generally speaking, greater spacing length will result in a wider bandwidth. Some high efficiency beam antennas will space the director and reflector a full quarter wavelength from the driven element. This results in increased gain and wide bandwidth characteristics.

Emergency communications equipment: Any of a multitude of devices and systems that are designed to provide established communications during anticipated emergency conditions. Emergency communications systems usually consist of a radio transmitter, receiver, antenna, appropriate switching and processing equipment, and emergency power supplies. The power supply or supplies are usually the main difference between communications systems designed for standard operation and those intended for emergency usage. An emergency communications system usually provides for operation from several different power sources. For example, a radio transceiver that will operate from the standard ac supply or from a storage battery could be considered as emergency communications equipment. During a disaster, the main power lines may be inoperative, so the equipment would be operated from an alternate and independent storage battery source.

Emergency communications equipment often includes a means of power generation. This is often a gasoline- or diesel-operated generator whose output will be equivalent to the standard ac line or a value with which the equipment is designed to operate. Other forms of emergency power supplies might include a bank of storage batteries that are charged by means of a gasoline- or diesel-powered generator, solar cells, or even hydroelectric systems.

EMF: The force that causes a flow of electricity in a closed circuit when there is a difference of potential maintained between its ends by a dynamo or an electric battery. The EMF of such a generator is a measure of the potential difference between its terminals when it is an open circuit and there is no current flow because the circuit is not closed. An increase in EMF is accompanied by an increase in the rate of flow of current in an electrical circuit. The unit of measure used to signify electromotive force is the volt.

Emission-type tube tester: A device that is used to rate electron tubes in terms of electron emission. In using this instrument, the tube's electrodes, except the cathode, are joined to act as the plate of a simulated diode. Applying a test voltage results in diode current that indicates tube condition.

Emitter: A body that discharges particles or waves. In a semiconductor device, the area, region, or element from which carriers are injected into the device. In a transistor symbol, the emitter is that electrode shown with an arrowhead. In a punched-card machine, the device that produces signals simulating holes; i.e., nonexistent perforation, is also known as the emitter.

Emitter-base junction: The boundary between the base and emitter regions of the semiconductor material and which makes up a bipolar transistor.

Emitter current: A direct current that flows in the emitter circuit of a transistor. Electrons are emitted into the base and are attracted by the positive charge of the collector material, creating a small bias, or base current.

Emitter follower: A circuit in which the output impedance depends upon the source impedance. That is

$$Z_{out} = \frac{Z_G + h_{IE}}{1 + h_{FE}}$$

where h_{IE} is the input impedance of the transistor in ohms, and h_{FE} is the forward-current transfer ratio of the transistor. Both h_{FE} and h_{IE} may be measured or taken from the transistor manufacturer's specifications. Figure 161 shows an emitter-follower circuit.

Emitter resistance: The value (measured in ohms) of the resistance of the emitter electrode in a bipolar transistor. An emitter resistor may also carry the same designation and is an external device that is connected between the emitter electrode and the positive or negative supply voltage for this device. The emitter resistor serves to limit current flow through the bipolar transistor and is a means of power control. The symbol for emitter resistance is Re.

Emitter voltage: The value of the dc potential at the emitter electrode of a bipolar transistor. This value is often abbreviated with the symbol Ve. The Ve designation may also indicate the maximum dc potential, which may be applied to the emitter of a specific bipolar transistor.

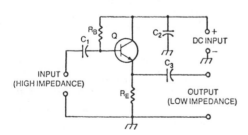

FIGURE 161 Emitter follower.

Emulation: In computer applications, a procedure that is used to imitate one system with another in such a manner that the imitating system accepts the same programs and achieves the same end results as the imitated system. It involves the software techniques used to imitate the original system and can minimize the impact of conversion from one system to another during program development.

Emulation of a number of devices can sometimes be done using a single general-purpose unit. The general-purpose device, adapted to several different configurations through microprogramming, becomes a host serving the more specialized devices. One in-circuit emulator system uses two processors, one to execute commands and control peripherals, and the other to interface directly with the user's prototype system.

Emulation allows custom instructions through microprogramming, which permits software designed for larger machines to run on microprocessors. The user is also allowed to run programs and integrated hardware and software very early in the development cycle.

Enable: To initiate the operation of an electronic circuit or device by applying a pulse or trigger signal. In digital computer applications, enable is a binary pulse that augments a write pulse, which causes a magnetic core to change states.

Encapsulated circuit: A circuit that is composed of two or more components that are completely encased. Component leads are brought through the case for connection to other circuit portions. An integrated circuit is one example of this process, although the term often applies to other circuits that are built from a multitude of components and then are completely encased in epoxy. This encapsulation process is often referred to as *potting*.

Today, many entertainment devices use encapsulated circuits. All components are wired on the circuit board. Once operation has been established, the potting compound is applied. This effectively seals the entire circuit from moisture and makes it mechanically secure. Vibration, heat, and other climatic elements can adversely affect electronic circuits, especially those which are used to establish a critical operating frequency. Through encapsulation, individual components are prevented from moving and frequency tolerances are held more closely in line. The encapsulation process often voids the possibility of servicing the electronic circuit. When devices using this type of construction become defective, the entire circuit is usually removed and replaced.

Individual components are often encapsulated. This includes resistors, capacitors, transformers, and all solid-state devices. Other circuit elements such as coils, variable capacitors, chokes, and a few others are not usually encapsulated, but they may be potted in the overall circuit once installed and operational.

End-fire array: An antenna system in which the principle radiation is off the ends of the array; that is, the radiation is maximum in the plane of the elements. A simple end-fire array employing two vertical half-wave antennas

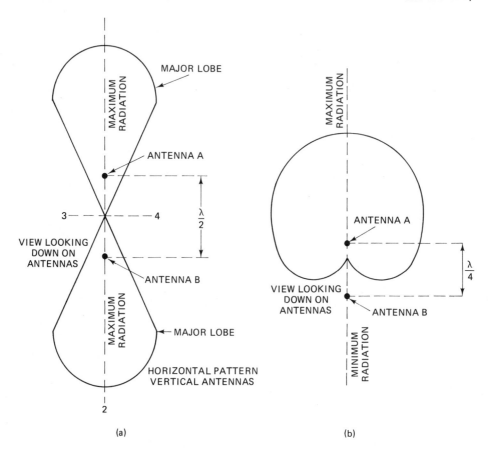

FIGURE 162 End-fire array: (a) half-wave spacing; (b) quarterwave spacing.

spaced λ/2 wavelength apart and fed 180° out of phase is shown in Fig. 162a. Because the two antennas are λ/2 wavelengths apart and fed 180° out of phase, radiated energy will reinforce along line 1-2 in the plane of the antennas. Likewise, radiated energy is cancelled along line 3-4, broadside to the antennas.

If the spacing of the antennas is reduced to λ/4 wavelength and the excitation of A lags that of B by 90°, a cardioid (heart-shaped) pattern is produced. This is illustrated in Fig. 162b. The cancellation of the radiated energy in the direction of B and the reinforcement in the direction of A may be explained as follows: By the time the radio wave leaving antenna B reaches antenna A, the wave leaving antenna A will have the same phase. Therefore, the two waves will add to produce maximum radiation in the direction shown (in the direction of A). However, by the time the radio wave leaving antenna A reaches antenna B, the wave leaving antenna B will be 180° out of phase, causing cancellation in the direction of B. Maximum radiation occurs in the direction of the antenna that lags.

Endoradiosonde: A tiny electronic circuit that consists of a biomedical transducer and a radio transmitter. The transducer is designed to sense physiological conditions in the human stomach and intestines. Its output is in the form of fluctuating electronic signals that are passed to the input of the radio transmitter. Here, the transducer signals are modulated and transmitted to sensing equipment placed outside the physiological system. The detected transmissions are fed to recording instruments that extract the information needed for diagnosis and treatment. The entire circuit is enclosed in a tiny pill that is treated to present nominal rejection by the human body.

Endothermic reaction: A chemical reaction that causes a loss of kinetic energy. Endothermic reactions produce cold, whereas the exothermic equivalent results in an output of heat. In electronics applications, exothermic reactions are far more prevalent.

End plate: In a motor, the component that is fastened to the stator frame by means of screws or bolts and serves mainly to keep the rotor in position. The bore of the end plate, in which the rotor shaft rests, is fitted with either ball bearings or sleeve bearings. These sustain the weight of the rotor, keep it precisely centered within the stator, and permit rotation without allowing the rotor to rub on the stator. An end plate is shown in Fig. 163.

Energy: The ability or capacity for doing work. Energy may be in the form of a charged storage battery, a raised weight, a compressed spring, or a tank of compressed air. It may be mechanical, electrical, magnetic, chemical, thermal, etc. The different kinds of energy may be readily converted from one form to another. However, each conversion results in a loss of some of the useful energy, although the total amount of energy remains the same.

Since the energy of a device represents the total amount of work that it can do, the units for work and energy are the same. The unit of energy most frequently used in electrical work is the joule, which is equal to approximately 0.74 ft.lb.

Epitaxial layer: A crystal layer that is grown or deposited onto the original collector substrate during the diffusion process in the production of transistors (both mesa and planar). This layer is usually of the same basic material as the parent material. The process involves placing the original substrate into a closed container that contains a vapor of the same impurity.

FIGURE 163   *END   PLATE*

Through maintaining the proper temperatures, the atoms of the vapor will be deposited upon the substrate and arrange themselves as directed.

Epitaxial mesa transistor: A transistor that is created by a process that diffuses a thin mesa crystal material over the original substrate material. The original P-type substrate (collector) is placed in a closed container that houses a vapor possessing the same impurity. Upon maintaining proper temperature controls, the atoms of the vapor are forced to deposit themselves upon the original substrate, which results in the epitaxial layer. After this layer is established, the base and emitter regions are formed.

The original P-type substrate in this type of transistor has a higher doping level and therefore less resistance than the epitaxial layer, which results in a low-resistance connection to the collector lead that will reduce the dissipation losses of the transistor.

Epitaxial transistor: A transistor that contains one or more layers of a semiconductor material (usually crystalline) that has been deposited or grown onto the original substrate material.

Epitaxy: The controlled growth of a layer of semiconductor material (usually crystalline) onto a collector substrate, as in the production of epitaxial transistors.

EPROM: A storage media for digital systems in which the code stored in the memory array can be programmed by the user, erased, and then reprogrammed to a different code. An acronym for erasable programmable read-only memory. Special equipment such as ultraviolet light fixtures is necessary to erase the units, which must also be removed from the system to be erased and reprogrammed. EPROMs are useful in many different applications, such as the home computer or small business system. However, the bit density in an EPROM is lower than that of a PROM and usually costs more per bit of storage capacity.

Equalizer: A connection used in large dc armatures to minimize circulating currents. These currents are usually due to uneven air gaps between the field poles and the armature and may be eliminated by connecting commutator bars of equal potential together. The bars to be connected depend on the number of poles in the motor and the number of commutator bars. It should be understood that equalizer connections are used on lap windings only.

Equivalent circuit: A circuit that clarifies the relative connections of resistances and reactances in a simplified visual way and makes the calculations pertaining to transformer operation easier. For calculation of the resistances, currents, and voltages in a transformer, a schematic diagram of the equivalent circuit is helpful. An equivalent circuit of a transformer with a 1 to 1 ratio is shown in Fig. 164. Every winding has a resistance R and an inductive reactance X. They are inherent to each coil and are not physically separated. In a schematic diagram, however, the resistance and reactance of a coil are shown separately, represented by their standard symbols and connected in series. Thus, when showing the equivalent circuit of a transformer (Fig. 164), the

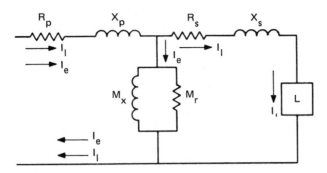

COMPLETE EQUIVALENT CIRCUIT OF TRANSFORMER

FIGURE 164   Complete equivalent circuit of transformer.

resistance R_p of the primary winding and the reactance X_p of the primary winding are shown in series.

When voltage is applied to the primary circuit, the exciting current I_e flows through R_p and X_p. Since this produces losses, these no-load losses are shown in the equivalent circuit as a parallel combination of a fictitious resistance M_r due to the core losses, and a reactance M_x due to the effect of the magnetizing component of the exciting current. The instantaneous direction of the exciting current I_e is shown by arrows through R_p, X_p, the parallel combinations M_r and M_x, and back to the source.

When the load L is connected across the secondary winding, which has a resistance R_s and a reactance X_s, the load current affects the current in the primary. Therefore, the equivalent circuit of a transformer shows the secondary resistance R_s, the reactance X_s, and the load L connected in series with R_p and X_p. The instantaneous direction of the load current I_l is shown by arrows through R_p, X_p, R_s, X_s, and L back to the source. Note that the currents I_l and I_e both flow through the primary, but only I_l flows through the secondary winding.

The ratio of 1 to 1 has been selected in this example because the secondary voltage will be equal to the primary voltage, the load current I_l is the same in both windings, and resistances and reactances may be indicated directly in ohms.

When a transformer has a ratio other than 1 to 1, the voltage ratio causes proportional changes in the secondary current. Therefore, the voltages, the currents, the resistances, and the reactances have to be expressed in percentages, the calculations made with these percentage values according to the equivalent circuit, and the actual values determined on the basis of the obtained percentage values.

Error signal: An output signal that is used for automatic correction purposes in a servo-system. Its value is proportional to the difference between the actual operating quantity of the system and a standard reference quantity.

When the output signal differs from that of the standard reference, the servomotor is activated to bring the two into alignment. When no error signal exists, the system is aligned and the motor is shut down until another deviation occurs.

Error voltage: A signal whose potential difference is sensed for automatically correcting a servo-system.

Esaki diode: A diode that is fabricated by first doping the semiconductor materials that form the PN junction at a level that is one hundred to several thousand times that of a typical semiconductor diode. The Esaki diode was introduced by Dr. Leo Esaki in 1958 and is better known today as the *tunnel diode.*

The depletion region of the diode is very thin, allowing many carriers to tunnel through. The peak voltage is limited to about 600 mV, but the peak current can vary from a few microamperes to several hundred amperes. Due to this characteristic, a tunnel diode can easily be damaged by a simple VOM that possesses a voltage potential that is too high.

The Esaki diode is very stable and can be operated in an environment that exposes it to severe moisture and contamination. It is employed in many areas that subject the equipment to widely varying temperature changes and radiation bombardment. Several devices that utilize the Esaki diode are oscillators, mixers, high-speed switching devices, logic circuitry, and amplifiers.

Esnault-Pelterie formula: A formula used for calculating the total inductance of a single-layer, air-wound coil. The formula reads: $L = 0.0018$ $(a^2N^2)/(l - 0.92a)$. In this formula, L is the inductance of the coil in microhenrys; a is the radius of the coil measured in inches; l is the coil length in inches; and N is the number of turns. Figure 165 shows how this would be applied to an actual coil.

Etched circuit: A circuit that is produced by eating away or etching the metallic coating of a printed circuit board to provide the required pattern of conductors and terminals to which discrete components are soldered. Figure 166 shows an example of an etched circuit board that began as one that was completely coated with a thin copper sheet on one or both sides.

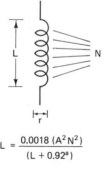

$$L = \frac{0.0018\,(A^2N^2)}{(L + 0.92^a)}$$

FIGURE 165

ETCHED CIRCUIT

FIGURE 166

The designer decides (from a schematic drawing) where each component is to be placed, as well as the number and paths of each conductor. This pattern is drawn on paper and checked for accuracy. When the anticipated pattern is known to be correct, this pattern is superimposed upon the surface of the circuit board.

A pen that contains a nonetchant solution is then used to trace the pattern exactly on the copper surface of the board. The final trace should duplicate the drawn pattern. With a tiny drill bit, component mounting holes are drilled through the board at the desired locations. One final check is made to be certain that connections for every component have been included and that the nonetchant traces are correct and uniform.

The entire board is then immersed in an etchant solution, which is a type of acid that will eat away all of the copper surface that has not been covered by the nonetchant. The tray containing the solution is vibrated periodically to make sure that all loose copper particles are allowed to float free. After the required etching time, the board is removed and a stiff brush is used to clear away any remaining copper particles. The end result is a circuit board that is completely clear of its original copper coating, except at those points that were marked by the antietchant. The circuit is then compared with the original drawing to assure further accuracy.

Etched circuit boards form very reliable bases for electronic components, especially semiconductors, which can be flush-mounted for extreme rigidity. Since external wiring is avoided in many instances, performance reliability is better assured. Equipment that is built from etched circuit boards (also called *cards*) can be subjected to a higher degree of mechanical stress and still continue to perform as intended. Also, etched circuit boards provide compact construction techniques and allow a large, complex circuit to be contained in a small enclosure.

E-transformer: A differential transformer whose windings are placed on an E-shaped core. This is shown in Fig. 167. The core has three legs that share a common base. Typically, the primary winding is attached to the center leg, while the secondaries are located on the outer legs. The center coil serves

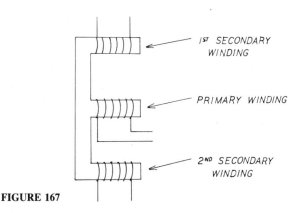

FIGURE 167

as a reference point and indicates the difference or imbalance between it and the two secondary windings, respectively.

Ettinghausen heat: A phenomenon that occurs when a metal strip carrying current longitudinally is placed into a magnetic field perpendicular to the plane of the strip. The corresponding points on the opposite edges of the strip exhibit different temperatures. This effect is similar to the Hall effect.

Even-order harmonic: A signal whose frequency is an even multiple of the fundamental frequency. For example, a complex waveform consists of a fundamental frequency of 100 kHz. The even-order harmonics would be equal to 2, 4, 6, 8, etc., times the fundamental, or 200, 400, 600, 800, etc., kHz.

Radio-frequency transmitting devices usually generate even- and odd-order harmonics. These undesirable signals are greatly suppressed in the output circuitry, but are transmitted to some degree. The Federal Communications Commission has established suppression standards for this type of equipment to avoid undue interference.

Harmonic interference can play havoc with the radio-frequency spectrum. For example, if a transmitter has an output at a fundamental frequency of 6 MHz and the second harmonic is not adequately suppressed, the information will also be transmitted at 12 MHz. Generally, the intensity of fundamental frequencies decreases as the multiplication factors increase.

Except gate: A gate in which the specified combination of pulses producing an output pulse is the presence of a pulse on one or more input lines and the absence of a pulse on one or more other input lines.

Exciter, transmitting: Any device or circuit that supplies the input signal to the output amplifier in a radio transmitter. Circuits that are incorporated to perform this function in a single complex device are often known as *drivers*.

In a simple radio transmitter, a small oscillator circuit may supply the drive signal that is fed directly to the input circuit of the final amplifier, whose output is transferred to the antenna. In moderate- to high-powered transmitters, the oscillator output will not supply adequate drive, so preamplifying stages are incorporated. Often, the oscillator output will be fed to a driver circuit, which is, in itself, a low-powered amplifier. The original signal from the oscillator is increased in magnitude and then fed to the final amplifier. This intermediate stage then becomes the driver.

Many types of transmitters that are capable of effecting communications when used alone are also designed to be operated with outboard amplifiers. These are discrete devices that are complex electronic circuits in themselves. They are designed to receive an excitation signal from the output of the transmitter. This output would be connected to an antenna in a typical discrete operation. Here, the transmitter output is connected to the amplifier input circuit. In this mode of operation, the transmitter becomes the exciter and may be referred to as an *amplifier driver*.

Exciting current: The output current measured in amperes, milliamperes, or microamperes by the exciter of an ac generator. The exciter itself is a small dc generator that supplies direct current to the field winding of an ac generator. In electronic terminology, exciting current is the current drawn at the primary circuit of a transformer whose secondary is not connected. This is the amount of current needed to establish unloaded transformer operation.

Exciting power: The amount of input/signal power required for a full output from a power amplifier. Also called *driving power,* it may also describe the maximum output power that is capable of being produced by an exciter circuit designed to drive a power amplifier. Exciting power is measured in watts.

Exothermic reaction: Any chemical reaction that produces heat.

Exploring coil: A small inductor that consists of a few turns of copper conductor, the ends of which are attached to a metering circuit or a grid dip oscillator. The purpose of the exploring coil is to pick up stray radiation from a transmitter or antenna circuit and couple this energy to a separate metering circuit for evaluation.

The exploring coil is sometimes known as a *measurement length.* Its inductance value is very small in order to have a very minor effect on the normal operation of the circuit under test. While an exploring coil is usually intended for equipment testing, similar devices may be permanently incorporated in radio-frequency equipment to provide RF sampling from a fixed point in the circuit to effect constant metering of operation. When used in this manner, the coil is often called a *fixed link.*

Exploring electrode: A transducer that is sealed in a dielectric casing and is designed to be inserted into a substance for measurement purposes. Many types of exploring electrodes are used in physiological monitoring applications. These are designed for insertion into the human body. See *endoradiosonde.*

In electronic service work, an exploring electrode can be a simple ohmmeter test probe or an electromagnetic pickup that is inserted into tank circuits to sample radio-frequency energy. See *exploring coil.*

Explosion-proof apparatus: Apparatus enclosed in a case that is capable of withstanding an explosion, of a specified gas or vapor, which may occur within it and preventing the ignition of a specified gas or vapor surrounding the enclosure by sparks, flashes, or explosion of the gas or vapor within, and which operates at such an external temperature that a surrounding flammable atmosphere will not be ignited.

Extended class-A amplifier: A push-pull AF amplifier in which a triode and pentode are connected in parallel on each side of the circuit. At low signal levels, the circuit operates entirely with the push-pull triodes, the pentodes being cut off. At high signal levels, however, the AF output is almost entirely that of the push-pull pentodes.

Extended cutoff tube: A vacuum tube in which the amplification does not vary directly with the control-grid bias, and cannot be completely cut off by this bias. Structurally, this tube is characterized by closer spacing of wires at the center of the grid than at the ends. This type of tube may also be called a *remote cutoff tube* or *variable-mu tube.*

Extremely high frequency band: That frequency band that composes a range of 30 to 300 GHz. It lies near the upper limits of the radio-frequency spectrum and falls into the microwave spectrum, which includes all extremely short radio waves, especially those shorter than 0.3 m in wavelength. Exploration and experimentation in the EHF band is still a fairly new science, and most of the communications and control possibilities of this upper spectrum are yet to be discovered.

Extrinsic semiconductor: A semiconductor material to which a controlled amount of impurity has been added. This gives the semiconductor a desired resistivity and polarity. The two chief semiconductor materials used in this type of construction are germanium and silicon.

Facsimile: An exact copy or the process of transmitting printed matter or still pictures by a system of either telephones or telegraph or radio for reproduction. Copy machines accept an original document containing printed and pictorial information and then duplicate this visual pattern on another piece of paper. Facsimile systems do exactly this.

The difference between a facsimile system and an office copier is primarily in the location of the original and the copy. In the facsimile system, these are located remotely some distance from each other; whereas in the office copier, they are located in the same machine. Thus, the basic facsimile system is as shown in Fig. 168.

The document or photograph (source document) to be transmitted is input through the sending unit. This unit performs the conversion from the visual information on the document or film into electrical signals representing

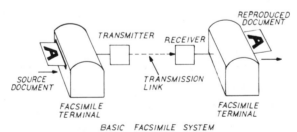

FIGURE 168 Basic facsimile system.

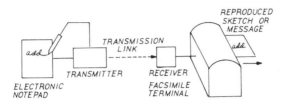

ELECTRONIC   NOTEPAD   SYSTEM

FIGURE 169   Electronic notepad system.

the information. These electrical signals modulate a transmitter so that the information can be sent to the receiver electrically through the transmission lines, such as wires or cables and amplifiers, electromagnetic waves in waveguides, or electromagnetic waves through space. The receiver first converts the transmitted information back to the same sort of electrical signals generated by the sending unit and then converts the patterns of the electrical signals to a copy of the original source document, whether picture or printed page.

A slight variation of the basic facsimile system is shown in Fig. 169. The sending unit is an electronic sketchpad that accepts the handwriting as an input. A facsimile receiving unit generates the output. The pen positions on the sketchpad are encoded as a sequence of electrical signals, which, when decoded by the receiving unit, will duplicate the handwritten sketch or message composed at the transmitting unit.

This blackboard, depending on the type of message being sent, still qualifies as a facsimile system, since a copy on paper of a written document is produced at a remote location. The only difference is that the original has not been prepared beforehand, but is being produced at the same time it is being sent through the system. One obvious use for this type of system would be to send authorization signatures for financial transactions to a remote bank.

Fading: The variation in signal strength that occurs at a receiver during the time a signal is being received. There are several reasons for fading. Some are easily understood, while others are more complicated. One cause is the direct result of interference between single-hop and double-hop transmissions occurring simultaneously from the same source. If the two waves arrive in phase, the signal strength will be increased, but if the two waves arrive in phase opposition (180° out of phase), they will cancel each other and the signal will be weakened.

Interference fading also occurs when the ground wave and sky wave come in contact with each other. This type of fading becomes severe if the two waves are approximately equal in strength. Fluctuations in the sky wave with a steady ground wave can cause worse fading than sky wave transmission alone.

Variations in absorption and the length of the path in the ionosphere are also responsible for fading. Occasionally, sudden disturbances in the ionosphere cause complete absorption of all sky wave radiation. Receivers located near the outer edge of the skip zone are subjected to fading as the sky wave alternately strikes and skips over the area. This type of fading sometimes causes the received signal strength to fall to nearly the zero level.

Fail safe: A system, circuit, network, or component with built-in protective measures that preclude system failure. Fail safe systems usually allow some degradation of performance that does not prevent proper system operation.

Fan-out: The ability of an output to sink current from a number of loads (N) when at a logic 0 voltage level, or to supply current when at a logic 1 voltage level. Normally, each standard output is capable of sinking current or supplying current to 10 loads ($N = 10$). These dc loads are normalized with respect to a standard 5400/7400 gate input, which is considered to represent one load. The fan-out capability of a device is based upon its ability to drive multiples of this normalized input load.

Farad: The capacity of a condenser that would contain a charge of 1 C under 1 V of pressure. The farad is sometimes inconvenient for practical use. The microfarad, which is one millionth of a farad, is more commonly seen.

Faraday: A unit of electrical quantity that is approximately equal to 9.65×10^4 C. This is the quantity of electricity required in the electrolytic process to free 1-g atomic weight of a univalent element.

Faraday effect: The rotation of the plane of polarization of radio waves when they pass through the earth's ionosphere. This produces a loss between linearly polarized antennas. Also called *magneto-optical rotation*, this describes the tendency of a magnetic field to rotate the plane of polarization of light passing through a physical substance.

Faraday shield: A device that blocks electric flux. Also called an *electrostatic screen*, the shield consists of a number of straight rods separated by a very small distance or a number of wires joined at only one end. While blocking electric flux, the Faraday shield has a minimum effect on magnetic flux. These shields are often incorporated into oscilloscope and other CRT circuits that use electrostatic fields for electron beam deflection. The Faraday shield contains the electrostatic flux, confining it to one specific portion of the circuit.

Faraday's law: Two fundamental laws of electrolysis that are the basis of all quantitative calculations. The English scientist and physicist, Michael Faraday, discovered these laws in 1832–1833. The laws may be formulated as follows:

1. The weight of any material deposited on the cathode during electrolysis is directly proportional to the quantity of electric charge passing through the circuit.

2. The passage of 96,500 C of charge (called one Faraday) through an electrolytic cell deposits a weight (in grams) of any chemical element equal to the atomic weight of the element divided by its valence.

The first law states that the weight of a substance deposited on the cathode (or equivalently, liberated at the anode) is proportional to the quantity of electricity. The quantity of charge is usually measured in coulombs, which is the amount of electricity transported by a current of 1 A flowing for 1 s (ampere-second, As). Equivalently, 1 A is a rate of flow of charge of 1 C/s. To obtain the total charge (in coulombs) that has passed through a circuit, the current (in amperes) is multiplied by the time (in seconds).

Sometimes a larger unit than the coulomb, called the *ampere-hour,* is used. An ampere-hour (Ah) is the amount of charge transferred in 1 h when the current is 1 A. Since an hour contains 3600 s, 1 Ah is equal to 3600 As, or 3600 C.

Faraday's second law states that the same quantity of electricity will produce weights of different substances that are proportional to the ratio of the atomic weight to the valence for each substance. This ratio is called the *chemical equivalent.*

Moreover, it states that a charge of 1 faraday (96,500 C) will liberate or deposit the chemical equivalent (atomic weight/valence) of any substance. Note that the atomic weight enters into it, since any substance is deposited atom by atom on the cathode and the number of atoms in a gram depends on the atomic weight.

Each ion of the substance combines with one or more electrons to form a neutral atom of the substance. Thus, the copper ion (Cu^{2+}) with a valence of $+2$ requires two electrons to form a neutral copper atom. The hydrogen ion (H^+) with a valence of $+1$, in contrast, requires only one electron to form a hydrogen atom.

The greater the valence, the more electric charges (electrons) are required to form neutral atoms of the substance deposited. Hence, for a given total charge (total number of electrons), the weight deposited must be inversely proportional to the valence of the substance.

Federal Communications Commission: The U. S. government agency that regulates electronic communications. Established in 1934, this agency succeeded the Federal Radio Commission (FRC), which came into being in 1927. Prior to the establishment of the FRC, the Radio Division of the Bureau of Navigation in the Department of Commerce regulated communications. This was established in 1912.

The FCC controls all communications services that are not owned by the federal government. It licenses individuals who are permitted to operate and maintain radio equipment in many different categories, which include citizens band, commercial broadcasting, and amateur radio. Some licenses require the passing of a comprehensive examination, which is prepared and given by the FCC.

Before 1912, radio-frequency communications were not governed at all. Some farsighted individuals realized the potential of these communications and saw the need for regulation. Today, many millions of people use the air waves for many types of communication and entertainment. The FCC regulations are designed to assure proper and efficient use of the radio spectrum space that is allotted to the United States. The FCC negotiates with the International Telecommunications Union in procuring new frequencies for use in this country. This is a coordinated system that is designed to eliminate interference, both with services in this country and throughout the world.

One of the most important aspects of the FCC is in establishing standards of operation in radio-frequency communications. Signal quality, bandwidth, modes of communication, modulation, and minimum equipment specifications are all covered under many FCC rules and regulations. The FCC has the power to issue nonperformance citations and levy fines against violators. As a government agency, the FCC is charged with responding to the needs of the communications public in the United States and coordinating these desires with services in other parts of the world. The main FCC offices are located in Washington, D.C., with district offices maintained for all 50 states and U.S. territories.

Feedback: The part of a closed loop system that automatically brings back information about the condition under control.

Feedback control system: A type of system control obtained when a portion of the output signal is operated upon and fed back to the input in order to obtain a desired effect.

Feeders: Conductors extending from overcurrent devices in the switchboard to the distribution center. Generally, there would be no connections made to feeders between these two points.

Feedthrough insulator: An insulator that is designed to be mounted through a wall, chassis, or other surface and contains an internal conductor. Shown in Fig. 170, this device is often used for running conductors from the bottom of a chassis to the top. The internal conductor is often a long bolt that protrudes from each end of the cylindrical insulator. The latter is often made from ceramic, but may also be constructed of a flexible dielectric with a stranded internal conductor. This allows it to be bent into many configurations while still retaining its insulating and conducting properties.

Ceramic feedthrough insulators are sometimes used to bring two wire transmission lines through a building for connection to a transmitter. Here, a hole is drilled through the wall for each insulator that is designed to conduct RF current while insulating it from the building structure. This is shown in Fig. 171.

Feedthrough insulators range from a length of an inch or so to over a foot. The diameter of these devices can be anywhere from 0.5 to 1 in. or more. Occasionally, coaxial cable can be used to make a flexible insulator. The plastic cover is removed, along with the outside conductor or braid. This leaves only the internal conductor, which is surrounded by a foam dielectric. This design is often used for radio-transmission applications, but it can serve as a replacement for many standard types of feedthrough insulators at a fraction of the cost.

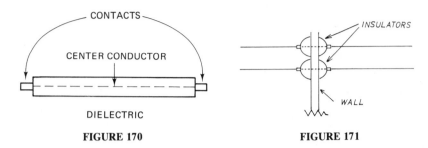

FIGURE 170   FIGURE 171

Felici mutual-inductance balance: An inductive null circuit that is used for determining mutual inductance (Mx) in terms of a standard mutual inductance (Ms). The secondary coils of two mutual-inductance circuits are connected in phase-bucking. The standard mutual inductor, which is variable, is adjusted for null. At null, Mx equals Ms. See Fig. 172.

Ferric oxide: A red oxide of iron that is used in many electronic applications, especially as a magnetic coating for recording tapes. Its formula is Fe_2O_3 and is the result of the contact of the iron material with oxygen. When mixed with other metals, a high-resistance magnetic material is formed and is called *ferrite*. This latter material is used in electronics applications and serves as a core for induction devices.

Ferrite antenna: A compact device that consists of a single coil wound on a ferrite form.

Ferrite bead: A tiny cylinder of ferrite material that is slipped over some current-carrying leads in radio-frequency circuits. When installed, the beads increase circuit Q and serve to prevent stray RF from flowing. Ferrite beads are also used as magnetic storage devices in digital computers. Here, a bead of ferrite powder is fused onto the signal conductors of a memory matrix. Ferrite beads are available in many different sizes and shapes in order to handle a myriad of electronic functions. The quality of the ferrite material is normally rated to operate properly at a specific maximum frequency. As operating frequency is increased, the quality and cost of each bead increase.

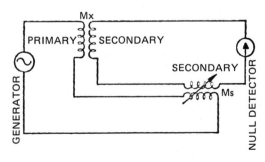

FIGURE 172

Ferrite isolator: A microwave device that permits electromagnetic energy to pass with negligible loss in one direction, while absorbing energy passing in the opposite direction. The ferrite isolator is inserted at one end of a waveguide or coaxial line used to transport microwave energy and acts in a similar manner to a diode, i.e., passing energy in one direction and not in another.

Ferrite-rod antenna: In its most basic form, a circular coil of wire wound upon a circular ferrite form. In some instances, this coil may be flattened into a symmetrical oval for space-saving advantages. The ferrite core increases the inductance of the coil by a multiplication factor that is determined by the quality of the ferrite material. This effectively allows a coil with a high inductance to fill a physical space that is very small compared to that which would be required by an air-wound coil exhibiting the same value.

While the ferrite coil may be used as a transmitting antenna, it is almost always seen in compact receivers, especially those designed for the AM broadcast band. A full-length antenna at these frequencies could be several hundred feet long. The ferrite-rod antenna serves to load the antenna circuit so that it is resonant at the operating frequency. Nearly every AM "pocket radio" contains a ferrite-rod antenna, although some have short, telescoping whips that may be connected to the top end of the ferrite-core coil.

When used for transmitting purposes, the size of the ferrite core must increase in cross-sectional dimensions and in length as power output is increased. For high power levels, the size of the ferrite core, along with its elevating costs, tend to make this type of antenna construction impractical. Circular ferrite cores are often used as baluns in symmetrical antenna systems, but these serve only a matching purpose and are not considered to be discrete antennas.

While ferrite-rod receiving antennas are more efficient than air-wound coils of the same size, a full-length antenna cut for the operating frequency is far more desirable. Due to the small physical size of ferrite-rod antennas, their capacity with the atmosphere is very small, and reception will not be at an optimum.

Ferroelectric: Pertaining to a phenomenon exhibited by certain materials in which the material is polarized in one direction or the other, or reversed in direction by the application of a positive or negative electric field of magnitude greater than a certain amount. The material retains the electric polarization unless it is disturbed. The polarization can be sensed by the fact that a change in the field induces an electromotive force that can cause a current.

Ferroelectric capacitor: Also called a *ferroelectric cell,* a device composed of a crystalline material that serves as the dielectric. The ferroelectric material may be made from barium titanate, barium strontium titanate, potassium dihydrogen phosphate, guanadine aluminum sulfate hexahydrate,

Rochelle salt, triglycene sulfate, and other combinations. This is a nonlinear dielectric material that is capable of producing ferroelectricity. This results when electric polarization occurs in certain crystalline materials.

Ferromagnetic: Pertaining to a phenomenon exhibited by certain materials in which the material is polarized in one direction or the other, or reversed in direction by the application of a positive or negative magnetic field of magnitude greater than a certain amount. The material retains the magnetic polarization unless it is disturbed. The polarization can be sensed by the fact that a change in the field induces an electromotive force, which can cause a current.

Ferromagnetic element: A type of magnetic material that becomes strongly magnetized in the same direction as the external magnetizing field. The atoms of ferromagnetic materials become aligned in a specific direction according to the attraction of the magnetic lines of the external field. Readily magnetized materials include iron, nickel, cobalt, and steel, plus a few of these elements' alloys. Should a maximum number of domains become aligned in one direction, the magnetic materials are said to be *saturated*. After the magnetizing force has been removed, the property of the magnetic material that causes it to hold its magnetism is referred to as its *retentivity*.

Fiber optics: A technique used in electromagnetic wave propagation in which infrared and visible light frequencies are transmitted by an LED (light-emitting diode) or a laser through a low-loss glass fiber. This method is used in VHF (very high frequency) radiation transmissions.

Field-effect transistor: A monolithic semiconductor amplifying device in which a high-impedance gate electrode controls the flow of current carriers through a thin bar of silicon called the *channel*. Ohmic connections made to the ends of the channel constitute source and drain electrodes.

The characteristic curves and high-impedance input of the field-effect transistor closely resemble those of a pentode electron tube. For this reason, circuits using the field-effect transistor closely resemble those used with the pentode tube.

Field-effect transistors are used in a myriad of electronic applications. They are generally small physically and are not designed to handle a high amount of power. They often see service in low-noise RF amplifier circuits designed to be used ahead of a radio receiver. There are two categories of field-effect transistors in use today—the junction field-effect transistor (JFET) and the insulated gate field-effect transistor (IGFET). The junction type uses two closely spaced junctions to provide the gate. These add to or subtract from the current flow through the device. The IGFET design is quite different, in that instead of two junctions, a tiny, solid-state capacitor forms the gate over the channel between the source and drain.

Bipolar transistors form a one-way channel or diode path between contacts. This is not true of the FET, which has a source-to-drain channel made up of a pure resistance that is not polarity conscious. Because of this type of construction, FETs can be readily used as electrically controlled resistors for other than dc signals.

FETs are generally considered to be electronically delicate devices owing to their extremely high-input impedances. They often come from the manufacturer with a special shorting band that temporarily ties all leads together. They must be treated with care to avoid subjecting them to static electricity charges that can quickly destroy them.

Field intensity: The strength of an electric or magnetic field. More specifically, field intensity sometimes refers to the strength of a radio wave, usually expressed in microvolts, millivolts, microvolts per meter, or millivolts per meter, but sometimes in volts or volts per meter.

Field strength: The effective value of the electric field intensity in microvolts or milliwatts per meter produced at a point by radio waves from a particular station. Unless otherwise specified, the measurement is assumed to be in the direction of maximum field intensity. Seldom are the actual operational characteristics of an antenna exactly the same as those determined on the basis of theoretical considerations. In order to determine these differences, various measurements must be made after the antenna is installed and while it is being test-operated. Often, on the basis of these measurements, changes are made in the design or installation of the antenna to improve the radiation pattern.

It is very important to know the direction and intensity of the power being radiated from an antenna. To determine these values, measurements of the field intensity are made at various distances and directions from and around the antenna. In order to determine the field strength or field intensity, it is desirable to use some type of standard antenna as a basis for all comparison. The standard antenna is a wire 1 m long. The magnitude of the single voltage (in microvolts) induced into this antenna is called the *absolute field strength* and is measured in microvolts per meter. Other types of antennas are commonly used, but they are calibrated against a standard 1-m antenna to obtain an absolute field strength measurement. Rod antennas are used to measure ground wave signal intensities. When the signal is reflected from the ionosphere, a loop antenna is commonly used.

The calibrated antenna picks up the radio wave and feeds the induced voltage to a sensitive receiver that is well shielded against extraneous signals. An indicating voltmeter is connected across the output of the receiver. When the receiver is properly calibrated, the absolute value of the field strength in microvolts per meter is indicated on the voltmeter.

If the meter in the available equipment is not calibrated in microvolts per meter, but a signal generator (covering the desired frequency range and whose output is calibrated in microvolts per meter) is available, the calibrated field strength readings may be made without difficulty. The antenna is connected to the receiver, and the meter reading is noted. The antenna is then disconnected, and the calibrated signal generator is connected to the receiver input. The output of the calibrated signal generator is adjusted until the receiver meter reads the same as it did when the antenna was connected. The output of the signal generator is then equal to the output of the standard meter antenna. If the antenna is not 1 m long, an antenna correction factor is used. The field strength in microvolts per meter is computed as

$$\text{Field strength} = \frac{\text{Output of calibrated signal generator}}{\text{Antenna correction factor}}$$

The power of a radiated wave, such as a light wave, falls off as the square of the distance between the source and point of measurement. This same law holds for the field intensity of electromagnetic waves when the power intercepted in a unit area is considered. However, since electric power expressed in terms of the voltage present is proportional to E^2 (because $P = E^2/R$), then the square of the voltage falls off as the square of the distance, or voltage itself falls off as the distance.

Absolute field intensity measurements are not difficult to make. However, the necessary equipment is relatively complex and bulky and must be calibrated very carefully. Often, all that is necessary to know is relative field strength, and simple field-strength meters and a pickup antenna are all that is necessary to make the measurements. The pickup antenna should be polarized in the same manner as the antenna whose field intensity is being measured.

Field-strength meter: An instrument that gives relative measurements of the field strength of an antenna system. It is composed of a pickup antenna, a tuned input circuit, a crystal rectifier, and an indicating device, such as an ammeter. It is very useful in the VHF range and for determining the beam pattern when using directional antennas. When using the field-strength meter, care must be taken to make the field-pattern checks at least several wavelengths away from the antenna at heights corresponding with the desired angle of directivity.

Figure 173 shows a field-strength measuring device especially suitable for measurements in the VHF range. The input or antenna coil shown, which is of the variometer type, may be changed for different frequency ranges. The amount of energy introduced into the tuned circuit is determined by the amount of coupling between the primary and secondary. Capacitor C_1 is tuned for maximum indication of the meter at one particular location of the pickup antenna. Resistors R_1 and R_2 make the response of the crystal more linear with variations in radiated power and also lessen the loading effect of the meter on the tuned circuit.

Filament evaporation: A technique for electrically depositing a film of selected metal on a metallic or nonmetallic surface by means of a heated filament. A filament of the metal to be deposited is heated by electric current in a vacuum chamber. This makes filament particles travel to the nearby object

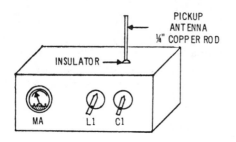

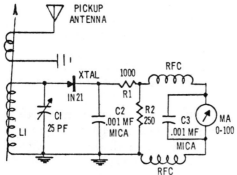

FIGURE 173   Field strength measurement device.

that is to be deposited with the filament film. As the filament particles strike the object to be coated, condensation occurs and a film is formed. Filament evaporation also describes the emission of electrons by a hot cathode in an electron tube.

Filament transformer: A transformer designed primarily to deliver operating current to the filament of an electron tube. These devices are almost always powered from the 115-V ac house current line or from 220-V ac in some applications. A filament transformer can also be referred to as a *step-down transformer,* because it produces a secondary voltage that is usually lower than that of the ac line connected at its primary winding.

Filament transformers are often connected with their secondary leads attached directly to the tube filament contacts, although some circuits may require rectification and filtering of the ac voltage. Most tube circuits, however, are designed so their filaments will accept alternating current. The secondary voltage values of most filament transformers will range from a low of approximately 2.5-V ac to a high of approximately 20 V. The most commonly seen values are 5-, 6.3-, and 12.6-V ac, as these voltages fit the requirements of over 90 percent of the electron tubes in use today. Current ratings of filament transformers will vary widely. Some may deliver only 1 A or less, whereas others may be designed for currents exceeding 20 A or more. The latter are heavy components with large-diameter wiring making up their primary and secondary windings.

When filament transformers are used with rectifier tubes, especially in high-voltage power supplies, the transformer insulation rating is very important, as the high-voltage potential will be seen at the secondary of the filament transformer. A 1500-V supply using rectifier tubes must use a filament transformer rated to withstand the full amount of the supply voltage to prevent a short circuit to ground and avoid arcing between the primary and secondary windings.

Filament, vacuum tube: The thin wire in a vacuum tube that is heated by electric current. In a filament type tube, the filament is the cathode. In indirectly heated tubes, the filament is usually called the *heater* and serves only to bring the cathode sleeve to a temperature that is great enough to cause it to emit electrons. For the most part, electron tubes employ thermionic cathodes that are heated directly or indirectly. A directly heated cathode is one in which the current used to supply the heat flows directly through the emitting material. In an indirectly heated cathode, the heating current does not flow through the emitting material. Figure 174 illustrates the construction of the two types of cathodes.

The directly heated cathode, commonly called a *filament,* has the advantage of being fairly efficient and is capable of emitting large amounts of energy. However, due to the small mass of the filament wire, the filament temperature fluctuates with changes in current flow. If an ac source is used to heat the filament, undesirable hum may be introduced into the circuit. This is especially evident in low-level signal circuits.

A relatively constant rate of emission under mildly fluctuating current conditions may be obtained with an indirectly heated cathode. This type of cathode is in the form of a cylinder, in the center of which is a twisted, electrically insulated wire called the *heater.* The emitting material in this type of cathode remains at a relatively constant temperature, even with an alternating heater current.

Figure 175 illustrates the schematic symbols for directly and indirectly heated cathodes. The cathode or filament is normally placed in the center of the tube plate electrode so that the electrons that are emitted may strike this latter element from all directions. Figure 176 shows cutaway views of directly and indirectly heated cathodes in a typical diode electron tube.

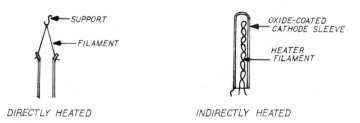

DIRECTLY HEATED   INDIRECTLY HEATED

FIGURE 174

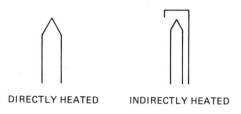

DIRECTLY HEATED   INDIRECTLY HEATED

FIGURE 175

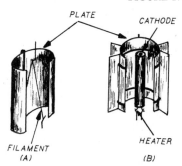

PLATE   CATHODE

HEATER

FILAMENT
(A)   (B)   **FIGURE 176**

Filament voltage: The operating potential of vacuum tube filaments. Typical values range from 2.5 to 22.5 V. However, many other values may be encountered. Filament voltage may either be ac or dc, with the former often preferred because of its tendency to increase active filament life. The most common value of filament voltage is 6.3 V, although many tubes are designed to operate from 12.6 V. When a potential is applied to the filament winding of a vacuum tube, this element is heated and begins to give off electrons that flow to the plate electrode. As filaments begin to age, they give up fewer and fewer electrons and, eventually, all practical emission ceases.

Filament voltage may be either ac or dc. While the longer life expectancy of tubes operating with alternating current is advantageous, some circuits will pick up and amplify the 60-Hz hum from the power line. When this becomes a problem, the filament winding of the transformer is rectified and filtered to provide a pure dc potential. The life of vacuum tube filaments operating at dc potentials can be lengthened by periodically reversing the polarity of the filament power supply. If this is done often, filament life expectancy will be nearly equal to that obtained in ac operation.

Filament winding: A designation that applies to one output of the secondary of a power transformer that is designed to provide operating current for the filaments of vacuum tubes. The filament winding may actually be of almost any value. However, those with output voltages of 2.5-, 5-, 6-, 7.5-, 10-, and 12-V ac are most common. While two filaments will operate from direct current, this requires rectification circuitry. In most instances, the filament winding of a power transformer will be directly connected to the vacuum tube filament leads. The ac operation of vacuum tube filaments tends to lengthen the useful life of the vacuum tube.

Sometimes a power transformer will be wired so that its secondary will provide several values of filament voltage. Here, a 6.3-V winding might be found in addition to one with a 12-V potential. The 5-V filament winding that is common to many older types of power transformers was designed to supply filament voltage to vacuum tube rectifiers, which were most often designed to operate from this potential. Since most electronic circuits of modern design use solid-state rectifiers, these older transformers are often used to supply a high operating potential from the main secondary winding, whereas the 5-V potential from the filament winding is rectified and often passed through a series regulator to provide a stable low-voltage output for powering solid-state circuits.

Current ratings of filament windings may vary from less than 1 to 10 A or more, depending upon the intended application. Some filament windings are center-tapped to allow for dc return of the high-voltage supply in amplifier circuits.

Filter: A selective network of resistors, inductors, or capacitors that helps to isolate or separate signals with different frequencies in order to channel dc, ac, and radio-frequency signals through circuits with minimum distortion and interference. Filters are extensively used in modern communication systems in order to reduce interference among the signals that are transmitted over broadcast channels that are crowded. Other types of electronic devices employ filters to keep individual transfers of information separated that are being transmitted over a common radio carrier. Frequency-selective arrangements are also responsible for the high-fidelity performance obtained from audio equipment.

Filter cutoff: The pointer where a filter circuit begins attenuating. For example, a typical low-pass filter is designed to pass frequencies below a certain point unattenuated. Above this point, all signals are suppressed. If the cutoff frequency of the low-pass filter is 30 MHz, all energy below this frequency will be passed, whereas any frequency above 30 MHz is attenuated.

Filter cutoff frequency is determined by the value of the inductive and capacitive components used in the design. In many instances, multisection design is incorporated to bring about a very sharp cutoff frequency. A broad cutoff frequency is not desirable in many instances, as the pass frequencies will tend to be attenuated to a higher and higher degree as they approach cutoff. Past cutoff, some signals are still passed, with attenuation levels increasing in direct proportion to frequency. A filter with a sharp cutoff exhibits very little attenuation within the entire range of the passband, but the suppression factor becomes sharply significant at the design cutoff and above.

Filter passband: The frequency range that can be efficiently passed by a filter circuit with little or no attenuation. The value of the component with which the filter is built will determine the passband frequency. Highly selective filters exhibit properties of high attenuation above and below the passband frequency. Lower quality filters will exhibit a gradual increase of attenuation to frequencies that lie just outside of the range that is passed unimpeded. Some filters offer very narrow passbands of a few hertz, whereas others may be broad, passing frequencies over a range of several kilohertz or megahertz. The passband encompasses the signals that are to be passed through the filter and on to other portions of an electronic circuit.

Filter tube: A vacuum tube that acts as a replacement for a choke in a power supply filter. A small amount of the ripple in the unfiltered dc is applied to the grid of the tube, which amplifies it. The amplified ripple then appears across the load in the phase that will cancel the ripple in the dc original. See Fig. 177.

Final amplifier: The last amplifier circuit in a cascade of amplifier stages. This is often called the *output amplifier* and is usually descriptive of the final output section of radio-frequency amplifiers. These may be referred to as *finals*. The purpose of a final amplifier is to accept the drive from an exciter stage, increase the input signals in magnitude, and pass them to its output, which is often the antenna section in RF circuitry. At audio-frequency ranges, the final amplifier or amplifiers would be that group of circuits that supplies drive directly to an audio transducer (speaker).

Some final amplifiers may be driven directly from an oscillator (in RF) or transducer, but usually the signals produced by these devices are too low in amplitude and must be increased by smaller amplifier stages. The final output from these intermediate devices will be to the input of the final amplifier, where the low-magnitude signal is produced at a much higher level.

Final voltage: The dc potential on the plate of a final amplifier, typically in a radio transmitter. Final voltage can also describe the output voltage of a power supply or other similar device that is derived from a source with a potential that is either higher or lower than the voltage which the circuit finally delivers. For example, in a 12-V dc power supply which operates from the ac line, the final voltage is 12-V dc. The original voltage is 115 ac. During the rectification process, the original voltage is changed to approximately the dc equivalent of the alternating current output potential of the transformer secondary winding. After filtering, the dc potential may be higher than 12-V dc. After regulation, the final voltage is usually arrived at.

Finned surface: A device or an attachment to an electronic component or circuit that is incorporated to enable heat to discharge into the atmosphere

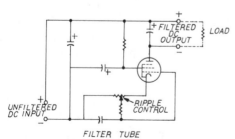

FILTER TUBE

FIGURE 177

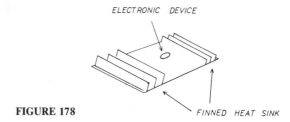

ELECTRONIC DEVICE

FIGURE 178

FINNED HEAT SINK

better. A good example of such a device is the *heat sink,* which is designed to be incorporated with a transistor, diode, or other solid-state device. Shown in Fig. 178, the heat sink becomes part of the device case and increases its radiating surface. Due to the finned construction, air currents can flow more freely across the surfaces and remove unwanted heat.

Some electric and electronic components that are designed to operate at high temperatures may incorporate finned surfaces into their basic construction. Figure 179 shows the plate element of an electron tube that uses a finned surface to convect heat to the atmosphere more efficiently. Some transformers are built with finned surfaces on their cases to accomplish this same function. The actual cooling ability of any finned surface will relate directly to total surface size and the ducting of cooling air across the complex surface.

Firing angle: A portion of the operation of electronic devices and circuits such as magnetic amplifiers, thyratron tubes, and silicon-controlled rectifiers. For a magnetic amplifier, firing angle is the angular distance through which the input voltage vector rotates before the core is driven into saturation. This is abbreviated by the symbol φ. For a thyratron with ac anode voltage, firing angle is the point, as an angle, in degrees in radians along the anode voltage half-cycle at which the tube conducts or fires. The firing angle symbol changes when used to describe thyratron operation to *a.* In a thyratron with dc anode voltage and ac control voltage, firing angle is the point along the voltage control cycle at which the tube fires. The symbol remains the same as for the thyratron with ac anode voltage. For a silicon-controlled rectifier, firing angle is

PLATE

CATHODE

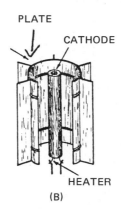

HEATER

FIGURE 179   (B)

the point along the control voltage half-cycle at which the SCR fires. Actual firing angles will vary from device to device and with the extraneous circuits to which they are connected.

Fishbone antenna: An untuned, wideband directional antenna of the general end-fire type, which consists of a number of collector antennas, each loosely capacitively coupled to the resistor-terminated transmission line in collinear pairs. It is so called due to its resemblance to the skeleton of a fish, as shown in Fig. 180.

Fixed electrode: An electronic component that is comprised of electrode elements that are mounted in one position or area and are not designed to be moved. Vacuum tubes, semiconductor devices, many resistors, capacitors, inductors, and most transformers fall into this category. Variable electrode devices are designed so that one or more of their elements may be altered in configuration to result in different physical and/or electric properties. A variable capacitor, for example, usually consists of two plates (in the most basic form). One is fixed in place, while the other can vary its position with the former. In this instance, the stator is the fixed electrode, while the rotor is variable. The action of the variable resistor is quite similar in operation, as one element is stationary, while the other is movable.

Flashover voltage: The sudden discharge of electrical energy between conductors or electrodes. This is normally an undesirable occurrence and may be accompanied by a flash of light and electrode damage. This condition is usually the result of an excessive voltage potential between the electrodes or elements.

Flashover voltage is a rating that indicates the peak voltage at which flashover occurs. This is also the voltage at which disruptive discharge occurs between elements and across the surface of an insulating material. In most instances, the flashover voltage rating is an indication of the minimum potential value, which is to be avoided in circuit operation. When this value is reached or

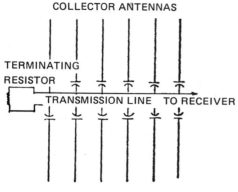

FISHBONE ANTENNA   **FIGURE 180**   Fishbone antenna.

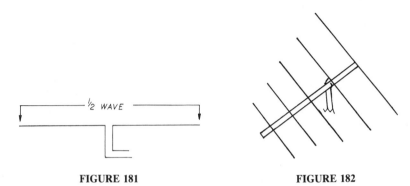

FIGURE 181   FIGURE 182

exceeded, flashover will occur, with the subsequent undesirable effects typical of this condition.

Flat response: A characteristic in which the dependent variable is substantially constant over a specified range of values of the independent variable. In amplifier operation, an output signal whose component fundamental frequencies and their harmonics are in the same proportion as those of the input signal being amplified exemplifies a flat response.

Flattop antenna: Any radiator that has its element or elements all mounted in a horizontal plane. The half-wave dipole antenna is a good example of this type of radiator, in that it is composed of a single horizontal element. This device is shown in Fig. 181. Horizontally polarized yagi antennas are also of flattop design. Here, all elements are mounted in a horizontal plane in relationship to the ground below, as shown in Fig. 182. Some antenna types consist of both vertical and horizontal elements. The cubical quad antenna shown in Fig. 183 represents this type of design. Here, the flattop section of the quad encompasses the two horizontal element portions, which lie at the top of the vertical portions. Many antennas of complex design can be referred to as partial flattops or as having flattop sections.

F layer: That portion of the ionosphere that extends approximately from the 90-mi level to the upper limits of the ionosphere. The ionosphere is found in the rarefied atmosphere approximately 40 to 50 mi above the earth. At different altitudes, it appears to be made up of different densities. At night, only one F layer is present. During the day, especially when the sun is high, this layer often separates into two parts, F_1 and F_2, as shown in Fig. 184.

As a rule, the F_2 layer is at its greatest density during early afternoon hours, but there are many notable exceptions of maximum F density existing

FIGURE 183

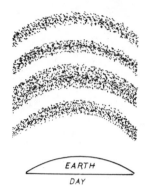

EARTH

DAY   **FIGURE 184**

several hours later. Shortly after sunset, the F_1 and F_2 layers recombine into a single F layer.

Fleming's rules: Rules by which the direction of movement of a current-carrying conductor through a magnetic field can be quickly determined. In applying this rule, the thumb, forefinger, and middle finger of the left hand are extended at right angles to one another. If the forefinger is pointed in the direction of the lines of force (from the north pole to the south pole) and the middle finger is pointed in the direction of current in the conductor (from positive to negative), the thumb will indicate the direction of the conductor motion.

There is a difference in application of Fleming's right-hand and left-hand rules. When generating action is considered and the direction of the voltage induced due to Lenz's law is required, apply Fleming's right-hand rule. Whenever the movement due to motor action is required, apply the left-hand rule. These rules are ilustrated in Fig. 185.

Flexible diode: A specialized semiconductor rectifier that may be altered from a PN junction in one direction to a PN junction in the opposite direction. It may also be so configured as to have no junction at all. In other

FLEMING'S RULES

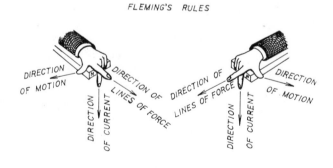

LEFT-HAND RULE   RIGHT-HAND RULE

FIGURE 185

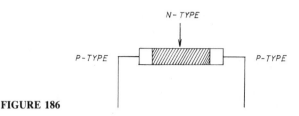

N- TYPE

P-TYPE

P-TYPE

FIGURE 186

words, its PN junction can be changed or reversed without reversing its lead. Its resistance is variable from the forward to the backward resistance value. As shown in Fig. 186, the basic flexible diode consists of a center section of N-type semiconductor material sandwiched between two layers of P-type material.

Flexible resistor: A component that consists of resistance wire wound upon a flexible form and covered with flexible insulation. This device is rarely seen in modern electronics, but it was designed for special applications that required a device that could be bent to unusual configurations while still maintaining a fixed ohmic value.

Floating paraphase inverter: A dual-tube or dual-transistor adaptation of the paraphase inverter. The second tube receives its grid-input signal from a tap on the load resistor of the first tube and provides the additional phase shift that is required.

Floating point calculation: A calculation made with floating point arithmetic.

Floppy disk: A means of bulk storage in a computer system. These disks are made from thin plastic and are coated with a magnetic material, permanently contained within a sleeve that protects it from dust and handling. Head access is provided by a slot in the sleeve, through which a read-write head obtains and deposits information. Floppy disk systems offer fast data transfer rates, easier data modification, and updating than smaller cassette tapes. They also offer relatively large data capacity in a relatively small package.

Flowchart: A graphical representation of the definition or solution to a problem in which symbols are used to represent functions, operations, and flow. A flowchart might contain all of the logic steps in a routine or computer program in order to allow the designer to conceptualize and visualize each step. It defines all the major phases of processing, as well as the path to problem solution.

The flowchart can contain logical operations by using symbolic notation to describe the arithmetic operations in terms of input and output. Functional flowcharts define all operations sequentially, but do not contain enough detail to allow program coding. Detailed charts are derived from the functional flowchart and the command codes, along with the way each command code acts in the system. The detailed charts include every operation in step-by-step form that must be performed during coding. The programmer is only required to know the processor programming language.

Fluorescent lamp: An electrical discharge source in which a mercury arc generates ultraviolet energy, which, in turn, activates the phosphor coatings to produce light. Such lamps take a variety of shapes, the dominant one being a smooth, long tubular shape of various diameters and lengths. Due to their negative resistance characteristic, fluorescent lamps require a ballast to start and limit the current flow through them.

Fluorescent screen: A surface of glass that is coated with a fluorescent material, such as platinocyanide, that emits light on exposure to electron bombardment. Fluorescent screens are commonly used in cathode-ray tubes and television receivers.

Flux density: The means of measuring the amount of flux lines per unit area.

In the cgs system, the gauss (G) is the unit of measurement. The flux density is 1 G when there is 1 Mx/cm^2.

Flux linkage: The passage of lines of force set up by one component through another component. Flux is the quantity of the lines of force that extend in all directions from an electric charge or from a magnetic pole. Figure 187 shows two inductors. The first is excited by a voltage potential and gives off lines of force. These force lines intercept the turns of a second coil and a linkage occurs. An induced electromotive force is present at the output of the second coil, although it is not driven directly by a separate voltage source. This is the principle upon which transformers operate.

Fluxmeter: An instrument used primarily to measure magnetic flux. The measuring element of a fluxmeter consists of a permanent magnet moving-coil element of negligible restoring torque. In such a moving element, the lead-in springs are very light filaments so designed that they exert a minimum force on the moving element. A small voltage applied across the terminals of the coil

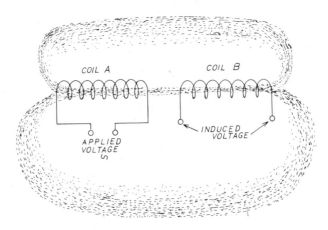

FIGURE 187

will cause the element to deflect continuously within the limits of its range as long as the voltage is applied.

If the coil terminals are connected to a search coil linking the coil to a magnetic field, any change in the linkage will induce a voltage in the coil. The maximum deflection of the instrument will be proportional to the total change of flux caused by this induced voltage. Consequently, if the search coil is introduced into the air gap of a magnet, the instrument will give an indication representing the value of the flux passing through the coil. If the coil is quickly withdrawn, there will be a deflection in the opposite direction.

The flexibility and versatility of the ballistic fluxmeter are due to the fact that the search coil can be wound by the user and made of any desired shape or dimensions most useful for the magnet to be measured. The resistance of the search coil and its leads should be kept within certain specified limits, which are not critical. Of course, the number of turns in the search coil must be known. A multiplying constant supplied by the manufacturer of the instrument is applied to the number of turns, and the scale can be read directly in terms of the strength of the magnetic field.

Where the magnetic circuit is closed and no air gap is included, as in a transformer, the search coil may be wound about the core at a convenient location, and the field strength may be determined from the deflection obtained when the current in the magnetizing winding is turned on or off or is reversed.

Fluxmeters are usually supplied with a few standard coils, but there is no limit to the variety of coils that may be made up by the user. Since the fluxmeter lacks restoring torque, it is customary to provide an auxiliary circuit actuated by a small flashlight cell for returning the pointer to zero after each observation.

Flywheel effect: In an LC tank circuit, the action in which energy continues to oscillate between the capacitor and inductor after an input signal has been applied. The oscillation stops when the tank circuit finally loses the energy absorbed. The lower the inherent resistance of the circuit, the longer the oscillation will continue before dying out.

Folded dipole: A full-wavelength conductor that is folded to form a half-wave element. It consists of a pair of half-wave elements connected together at the ends. The voltage at the ends of each element must be the same. In operation, the field from the driven element induces a current in the second element. This current is the same as the current in the driven element.

An ordinary dipole with a given current (I) produces a certain field intensity in space. Due to this field, there is also a certain power density per square meter in space. This power density is produced by the input power P. The relationship between the input resistances, the current, and the input power is expressed by the equation $R = P/I^2$. When the same current (I) exists in each of the two sections, the field strength in space is doubled. This causes the power density per square meter to increase four times. In turn, the input power must be four times as great.

As long as each section of a dipole has the same diameter, the input resistance is four times that of the simple half-wave dipole. Increasing the diameter of one section makes the increase in impedance still greater. While the input impedance to the driven element of a parasitic array drops to about a fourth of the value of the coaxial impedance, the use of a folded dipole increases the impedance by about four times. In this way, a good impedance match is effected.

Follower: A single-stage amplifier whose output impedance is substantially lower than its input impedance. The maximum theoretical voltage gain of a follower is 1. The follower is useful as a step-down impedance transformer and often serves as a buffer between a voltage source (generator) and a load device that would overload the voltage source. There are three types of followers—the cathode follower (vacuum tube), the emitter follower (bipolar transistor), and the source follower (FET). Figure 188 shows circuits of these devices. No type of follower operating correctly will introduce a phase shift.

Force: Any agent that produces or tends to produce motion. It may be mechanical, electrical, magnetic, or thermal in character. Force does not

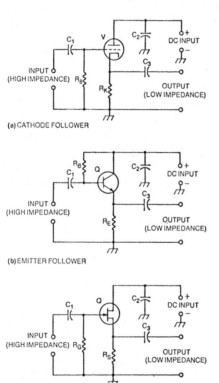

FIGURE 188 Active follower circuits: (a) cathode follower; (b) emitter follower; (c) source follower.

necessarily produce motion, as a relatively small force may fail to move a large body; but it tends to do so. The word *body* refers to any material object such as an electron, rock, wheel, gears, or even air. Force is usually measured in pounds or foot-pounds.

Forced-air cooling: A method of cooling on self-cooled transformers that takes care of peak loads that occur only infrequently, provides reserve capacity in emergency situations, or takes care of high ambient temperatures that may exist during certain seasons. Forced-air cooling can increase the rating of standard power transformers. Some units have been built using a single large blower with air ducts to direct the air to the cooling surfaces. At present, a large number of smaller, propeller-type fans are commonly used, mounted on the radiators. Each of the individually mounted blowers has its own motor and propeller-type fan, which forces the air horizontally past the cooling surfaces of the radiators. The blowers can also be mounted so that the air passes vertically through the transformer.

The use of several blowers has many advantages over the use of a single equivalent blower. These include power consumption, simpler installation, and quieter operation. Furthermore, the use of more than one blower per transformer may be considered very advantageous, because it introduces a greater factor of safety in the equipment, since the failure of a single blower unit does not result in complete loss of auxiliary capacity. Automatic control is usually provided so that the auxiliary air-moving equipment will start and stop under predetermined temperature conditions.

Fork oscillator: An audio-frequency oscillator controlled by a tuning fork. The dimensions of the fork determine its vibration frequency and accordingly, the frequency of the oscillator.

Form factor: Shape factor for a filter or tuned circuit. For a filter, the ratio of bandwidth at high attenuation to that at low attenuation. For a tuned circuit, the ratio of the 60-dB bandwidth to the 6-db bandwidth.

FORTRAN: A programming language designed for problems that can be expressed in algebraic notation, allowing for exponentiation and up to three subscripts. The FORTRAN compiler is a routine for a given machine that accepts a program written in FORTRAN source language and produces a machine language routine object program.

FORTRAN II added considerably to the power of the original language by giving it the ability to define and use almost unlimited hierarchies of subroutines, all sharing a common storage region if desired. Later improvements have added the ability to use Boolean expressions, and some capabilities for inserting symbolic machine language sequences within a source program.

Forward breakover voltage: In a silicon-controlled rectifier, the potential value at which the device abruptly switches on. This voltage is applied between the anode and gate electrode leads.

Forward current: The current that flows across a semiconductor junction when a forward bias is applied. This rating for a PN junction diode is broken down into *forward average current* (I_{fav}) and *forward surge current* (I_{fsm}).

The first rating indicates the average current that the device may safely pass without exceeding design ratings. The forward surge current value may be 50 times higher than the average rating and indicates the maximum surge current that may be handled by the device. The forward surge current occurs for a very short duration, especially upon initial circuit activation. Its time of occurrence is often rated in microseconds or milliseconds.

When purchased, PN junction diodes are usually rated only for forward average current, although surge current ratings as well as other technical information may be available on an enclosed data sheet.

Forward resistance: The ohmic value of a PN junction diode at a specified forward voltage drop or forward current.

Forward scatter: The scattering of a radio wave in the normal direction of propagation to points beyond the skip zone. The phenomenon results from reflections from nonuniform regions in the ionosphere and points beyond the skip zone.

Four-layer diode: A two-terminal switching device that permits a very small current to flow in a forward direction with forward bias until its critical voltage is reached. The diode switches into high conduction when this firing voltage is reached and continues to conduct freely until current is reduced to below a minimum holding value. Here, it switches back to the off state. It differs from a zener because the voltage across it drops to a very low value as soon as conduction begins. A four-layer diode may also be referred to as a *Shockley* or *PNPN diode.*

Four-layer transistor: A solid-state device that has four conductivity regions made from layers of N- and P-type materials.

Franklin oscillator: A dual-terminal AF/RF oscillator circuit that consists of a two-stage, RC-coupled vacuum tube amplifier with a tuned LC tank in the first (input) grid circuit and capacitive feedback from the second plate to the tank. It possesses sufficient loop gait to permit extremely loose coupling to the resonant circuit. See Fig. 189.

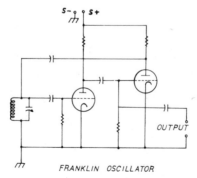

FRANKLIN OSCILLATOR   **FIGURE 189**

Free electrons: Electrons that have become dislodged from the outer shell of an atom. These electrons can exist by themselves outside of the atom, and it is the free electrons that are responsible for most electrical and electronic phenomena. Free electrons carry the current in conductors, as well as in all types of electron tubes and transistors.

Free space loss: Radio transmission loss that occurs as the signal travels through free space. This theoretical condition disregards variable factors such as weather conditions, atmospheric propagation, and the interference of physical objects. Free space loss is measured in decibels and is directly dependent upon operating frequency and the distance between the transmitting and receiving antennas. It becomes a monumental factor in designing transmit and receive stations for satellite communications.

At 1 GHz, the free space loss between a satellite orbiting at 20,000 mi and the earth station will be approximately 180 dB. The loss factor, however, begins to decrease percentage-wise with distance, with the most significant losses occurring a short distance from the transmitting antenna.

Using the previous example, if the satellite were 2000 mi above the earth station, the free space loss would still be approximately 153 dB. At a distance of 2-mi separation, free space loss would be approximately 103 dB. As frequency decreases, so does free space loss. At 10 MHz and a distance of 2 mi between transmit and receive stations, free space loss would be measured at approximately 63 dB.

Free space pattern: A chart of the radiation pattern of a transmitting antenna that is situated in a free space environment. This type of condition is only theoretical and cannot be achieved on the earth. Free space indicates that there are no surrounding objects or conditions that will interfere with the antenna pattern in any way. Free space conditions can be closely approximated by mounting an antenna at least one wavelength above the earth and the same distance from any surrounding objects.

In many instances, this is not practical, and the actual antenna radiation pattern will differ from the calculated free space pattern because of its physical capacity with the earth. Figure 190 shows the free space pattern of a quarter-

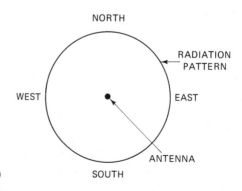

FIGURE 190

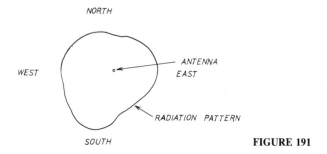

FIGURE 191

wave vertical antenna. This is an overhead view that looks straight down on the antenna element. It can be seen that the pattern is uniform, and an equal amount of radiation is transmitted in all directions. Therefore, the pattern is a circle with the antenna at the exact center.

In practical operation, ground contours and nearby objects can significantly alter the calculated free space pattern. Figure 191 shows what a typical pattern might look like when these other factors are taken into account. It can be seen that the pattern is not symmetrical or uniform in all directions. Ground contours to the left and right of the antenna have caused a reduction in radiated power toward these areas. The signal is now directional to some degree, with the main lobe portions broadcast from the top and bottom of the graph.

If it were possible to mount this antenna one wavelength above the earth and away from all other surrounding objects, the indicated pattern would more nearly conform with the previous free space pattern. The conformity would not be exact, but it would be far closer than the previous example shown in Fig. 191.

Frequency changer: A device that receives alternating current electrical energy at one frequency and delivers it at a different frequency. Frequency changers are used to interconnect power systems of different frequency and provide power to specific loads that require a frequency different from that of the available power system. Frequency changers are generally classified as either static or rotating. Mercury-arc rectifiers and the various controlled solid-state devices are examples of static frequency changers. They are primarily used when the desired output is direct current. They are also used to synthesize current of a required frequency from the available power supply.

Rotating frequency changers include motor-generator sets in which the generator is a synchronous machine and the motor is synchronous, induction, or direct current; and induction frequency changers, which are essentially wound-rotor induction machines with provision for taking the slip frequency power off the rotor winding.

Frequency compensation: The modification of an electronic circuit in such a manner that it responds, passes, or amplifies a signal of a specific frequency or frequency range. For example, an audio amplifier that will normally pass signals in the range of 20 to 20,000 Hz might be frequency compensated to pass only those signals that lie within the range of 300 to 3000 Hz. This latter range encompasses the great majority of frequencies that are generated by normal human speech. All frequencies that lie outside of the passband will be attenuated.

Frequency compensation is often obtained by incorporating LC filters at the input of a circuit. Alternately, electronic filters may be used that involve transistors and/or integrated circuits. The latter type of frequency compensation networks often have the advantage of being adjustable or tunable so that the pass frequency range may be altered by varying certain controls.

All electronic circuits are inadvertently frequency compensated to some degree, in that none can be built that will pass or amplify all frequencies within the range of dc to microwave. Frequency compensation, however, usually describes the specialized tailoring of the circuit so that it responds to a range that is substantially less than typical for similar uncompensated circuits.

Microphones used for communications work are often frequency compensated so that they respond best to the human voice range. Those used for professional recording work may be compensated to respond to the complete range of frequencies used in the production of music. Other may be intended for recording certain instruments and will be compensated to respond best to frequencies that lie in a particular instrument range.

Frequency compensation allows for the rejection of unwanted frequencies that lie either above or below the filter passband. As these frequencies fall farther and farther from the primary range, they are attenuated to a higher degree. Unwanted frequencies are passed to some degree, but at such a reduction from the primary pass frequencies as to be negligible.

Most stereo systems offer adjustable frequency compensation controls that will allow the output amplifier to be adjusted for the listener. A bass boost control is a good basic example of frequency compensation, as is the treble control.

Frequency control: A device, component, or circuit that allows for control of frequency or frequency response. Frequency control in radio transmitters is accomplished by using a quartz crystal that operates at a specific frequency, or by a variable-frequency oscillator that can be changed by the operator. In audio systems, inductors and capacitors (usually variable) are incorporated into the amplifier circuitry to allow for adjustment of the frequency response curve of the entire system.

The term *frequency control* can describe any device or process that establishes the frequency output or response of any circuit. When used as a noun, this usually pinpoints a specific operator-adjusted potentiometer, capacitor, inductor, etc., that is used to establish frequency. As a verb, the term describes the process by which output frequency or frequency response is selected and maintained.

Frequency deviation: The degree to which a frequency changes from a prescribed value. Thus, if the frequency of a 1-kHz oscillator drifts between 990 and 1010 Hz, the deviation is ± 10 Hz. In an FM signal, frequency deviation is the amount of frequency shift above and below the unmodulated carrier frequency.

Frequency modulation: A method of modulation in which the frequency of the carrier voltage is varied with the frequency of the modulating voltage, and the amount of variation is determined by the amplitude of the modulating signal.

Frequency offset: The difference between an actual frequency and the desired frequency. This term is often used in describing radio receivers and transmitters and specifically, in the adjustment and alignment of their calibrated frequency readout panels. For example, if a transmitter frequency control is set for 8 MHz, as indicated on its readout panel, and the actual frequency produced is 7.1 MHz, the frequency offset is 0.1 MHz, or 100,000 Hz. The same would apply to a receiver whose frequency readout indicates it is tuning 7 MHz, but in fact is actually receiving a signal at 7.1 MHz.

Frequency offset may also be used as a descriptive term in transceiver operations where the transmitter sends at one frequency, while the receive section responds to another. For instance, the transmitter section of a transceiver sends at 10 MHz, while the receiver section responds to a frequency of 5 MHz. Here, the transceive offset is 5 MHz.

Frequency response: A measure of the ability of a device to take into account, follow, or act upon a varying condition. As applied to amplifiers, it would be the frequencies at which the gain has fallen to the one-half power point, or to 0.707 of the voltage gain, either at the high or low end of the frequency spectrum. When applied to a mechanical controller, the maximum rate at which changes in condition can be followed and acted upon, since it is implied that the controller can follow slow changes.

Frequency shift: The measurement of the output frequency of a transmitter operating in two different modes and the comparison of those measurements. This is sometimes a desirable feature that is built into transmitting equipment, but it is also an indication that equipment problems exist. An example of undesirable frequency shift would be a transmitter that produces an output at 3 MHz when loaded for a power input of 100 W; but when loaded for an input of 500 W, it shifts frequency to 3.5 MHz. In this instance, the final amplifier is loading the master oscillator to such a degree that the final output frequency is altered. In the above example, the frequency shift is 0.5 MHz. Some transmitters that experience these difficulties will constantly change frequency in direct response to the keying or voice input. The AM or SSB signal then becomes frequency modulated, causing erratic and often illegal operation.

Frequency shift keying: A method of transmission that is often used in radioteletype communications. Using FSK, the carrier has two states, each at a different frequency. During the open or *space* condition, the carrier is usually at a lower frequency than when in the *mark* condition, which is several hundred hertz higher. Some forms of telegraph may also use this keying system.

In radioteletype operation, the mark frequency is typically 850 Hz above the space frequency. Narrow shift radioteletype will have mark and space separations of approximately 120 Hz. While these two shifts are considered standard, many others may be used for specialized applications.

The transmitter is shifted in frequency normally by switching a capacitance or inductance in and out of the frequency-determining oscillator circuit. Frequency shift occurs rapidly and in a coded form that is determined by the radioteletype machine operation.

On the receiving or decoding end, the shift patterns are extracted from the receiver audio channel and converted into baud pulses that drive another radioteletype machine. In this manner, the information that is originally input to the keyboard of the sending machine is printed out on the one used for receive. Frequency shift keying may be used in line communications as well as in radiobroadcasts.

Frequency swing: The degree to which a frequency changes above or below its nominal value. The amount of swing is measured in hertz, kilohertz, and sometimes megahertz. In the majority of cases, frequency swing will be measured over a range of a few cycles per second.

Frequency tolerance: The acceptable amount by which a frequency may vary from its intended value. The tolerance may be specified as ± a percentage range of the stated frequency, ± so many parts per million, or ± a number of frequency units (hertz or fractions thereof); i.e., 1 MHz ± 10 Hz.

Frequency variation: The drifting of a signal of prescribed value to a higher or lower value. This is normally an undesirable characteristic. A method of determining the Q of a tuned circuit by varying the frequency of the applied test voltage from resonance to a high point and to a low point at which the circuit voltage is 0.707 times the resonant voltage is called the *frequency variation method*. This is a test that determines the selectivity of an electronic circuit.

Front-to-back ratio: At a semiconductor junction, a comparison of forward current to reverse current, both being at the same value of voltage. This may also be called *forward/reverse ratio* and compares positive to negative current handling capabilities of the rectifier.

Front-to-back ratio is also commonly used to describe the performance of directional radio-frequency antennas. The forward lobe that is emitted from the director element of the antenna is compared with the lobe emanating from the reflector or the back of the antenna. Directional antenna systems are designed to broadcast a signal in one direction only, while eliminating radiation from the sides and back. However, some radiation can be measured from these last two points. If the radiation from the front of the antenna is 20 times that of the back, the front-to-back ratio will be 20 to 1. Generally, antenna directivity and efficiency is increased by high front-to-back ratios.

Front-to-back ratio plays an important role when the same antenna system is used for receiving. Still assuming a 20 to 1 ratio, signals entering the antenna system from the front will have 20 times the signal strength at the receiver as those that enter at the back. This assumes that both signals in the area of the antenna are of equivalent strength.

Fuch's antenna: A simple antenna that consists of a single-wire radiator without a feeder or transmission line, connected directly to the transmitter. The disadvantage of this type of antenna is that part of its radiated field is often inside the transmitter building.

Function: A variable (often represented by the letter x) that is so related to a second variable (y) that the value of the second is always determined in terms of the first. Algebraically written, the relationship is: $x - fy$. In digital computer terminology, a function is the part of a computer instruction that specifies the operation to be done. This instruction is usually written into the master program or may be input by external means. When the latter method is used, a function character is input to the machine, which is used to start the control of a peripheral. This character is determined by the internal programming of the microprocessor.

Furnace transformer: A transformer that is used to deliver heavy currents at low voltages. Most are of the single-phase, shell-form transformer type, with interleaved disk coils consisting of high-voltage coils and low-voltage coils. The heavy low-voltage leads are made for the large current and extend upward to the bus bars. Heavy bus bars are necessary to carry such currents without overheating. The end frames clamp the core and coils, and the guide pieces engage in the guide rails for tanking and untanking. The core, coils, and cover may be lifted as a unit by means of four lifting hooks.

In order to reduce the loss in the cover and consequent heating, adjacent low-voltage leads are of opposite polarity. This low-voltage winding is arranged in four parallel circuits, with eight bus-bar leads, each lead consisting of two bus bars. When used to supply arc furnaces, the iron-core, current-limiting reactor necessary to secure proper operation of the arc is usually mounted in the same case as the main transformer. The reactor is connected in series with the primary high-voltage winding of the transformer.

Fuse: An inexpensive electrical component that instantly opens a circuit when subjected to excessive current. When using an electrical device, care must be taken that the electric current passing through the device does not become excessive. All electrical apparatus contains a certain amount of resistance, and when electric current passes through a resistance, electric energy is transformed into heat energy. If faulty circuit operation causes an excess amount of current flow through an electrical device, the resulting increased temperature could cause considerable damage.

Figure 192 shows several types of fuses. Fuses are fabricated from wire made of zinc or similar metals having a low resistance value and a low melting point. When the current in a circuit is less than or equal to the current rating of the fuse, the element of the fuse is below its melting temperature. However, as soon as current flow in the circuit exceeds the current rating of the fuse, the fuse element melts rapidly due to the increased heat. Hence, the circuit will open and the circuit components are protected.

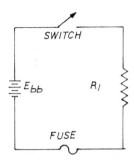

TYPES OF FUSES

FIGURE 192

FIGURE 193

Since a fuse is placed into a circuit for reasons of safety, a blown fuse should always be replaced with another fuse having the same rating as the original fuse. A fuse should never be replaced with one having a high current rating or a piece of wire. Although a circuit may become temporarily operational by improper replacement of a fuse, the circuit is left completely unprotected. Any sudden surge of current will not only damage circuit components, but may also set fire to the surrounding area.

The insertion of a switch, a fuse, or both into a circuit will not effectively change the operation of the circuit, since these electrical devices are constructed so that they will have negligible resistance. Figure 193 shows the proper placement and schematic symbol for a fuse. The fuse must always be placed in series with the components to be protected.

Gain: The ratio between the output signal and the input signal of a device.

Gain control: A component or circuit that has a direct effect on the amplification factor. One example is the volume control found on most audio-frequency devices, such as stereos, television receivers, etc. When the control is turned in one direction, audio-amplifier gain is increased and the output signal becomes louder. When the control is turned in the reverse direction, gain is decreased.

Many types of electronic equipment contain several gain controls, each attached to a different amplifier circuit. For example, a radio receiver has a gain or volume control that allows for adjustment of the audio-amplifier output to a speaker or headphone. At the same time, it may have an RF gain control that allows for increasing or decreasing the gain of the RF amplifier, regulating its sensitivity to the incoming signal from the antenna.

Galena: Also called *galenite* or *lead glance,* a mineral ore that is the primary source of lead. Its composition includes lead sulfide, PbS, and possibly small amounts of zinc, silver, cadmium, antimony, copper, and bismuth. Galena, which is commonly found in veins in limestone or in igneous rocks, is a soft, heavy, brittle mineral that exhibits a bluish-gray luster. It is commonly used in the crystal of a variable crystal detector.

Gallium arsenide semiconductor: A compound of gallium and arsenic used as a semiconductor material. Abbreviated *GaAs,* this material is used in solid-state diodes, varactors, and field-effect transistors (FETs). GaAs FETs are relatively expensive devices that exhibit very low internal noise characteristics in specialized circuits. Recently, these have seen much usage in microwave applications, especially in low-noise amplifiers that are used at the antenna output of microwave receiving stations.

Typically, the signal strength from the antenna is extremely low, and a low-noise amplifier is required that will boost the pure signal while adding as little noise factor as possible. These applications are especially significant to satellite communications systems, such as television receive-only earth stations, which are becoming increasingly popular.

Galvanometer: A type of stationary permanent magnet moving-coil instrument. Shown in Fig. 194, the galvanometer indicates very small amounts (or the relative amounts) of current or voltage and is distinguished from other instruments used for the same purpose, in that the movable coil is suspended by means of metal ribbons instead of a shaft and jewel bearings.

The moving coil (bobbin) of the galvanometer is suspended between the poles of the magnet by means of thin, flat ribbons of phosphor bronze. These ribbons provide the conducting path for the current between the circuit under test and the movable coil. They also provide the restoring force for the coil.

The restoring force, exerted against the driving force of the coil's magnetic field, is balanced in order to obtain measurement of the current intensity. The ribbons thus tend to oppose the motion of the coil and will twist through an angle that is proportional to the force applied to the coil by the action of the coil's magnetic field against the permanent magnet's field. The ribbons thus restrain or provide a counter force for the magnetic force acting on the coil.

When the driving force of the coil current is removed, the restoring force returns the coil to its zero position. In order to determine the amount of current flow, a means must be provided to indicate the amount of coil rotation. Either of two methods may be used—the pointer arrangement or the light-and-mirror arrangement.

In the pointer arrangement, the end of the pointer is fastened to the

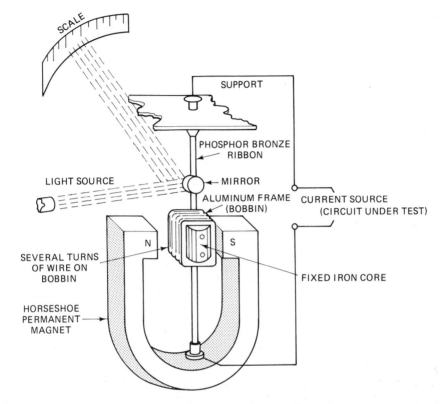

FIGURE 194 Simplified diagram of a galvanometer.

rotation coil. As the coil turns, the pointer turns as well. The other end of the pointer moves across a graduated scale and indicates the amount of current flow. An advantage of this type of arrangement is that it permits overall simplicity. A disadvantage is that it introduces the problem of coil balance, especially if the pointer is long.

The use of a mirror and a beam of light simplifies the problem of coil balance. When this arrangement is used to measure the turning of the coil, a small mirror is mounted on the supporting ribbon. An internal light source is directed to the mirror and then reflected to the scale of the meter. As the moving coil turns, so does the mirror, causing the light reflection to move over the scale of the meter. The movement of the reflection is proportional to the movement of the coil. Thus, the intensity of the current being measured by the meter is indicated.

If a beam of light and mirrors is used, the beam of light is swept to the right or left across a central-zero translucent screen (scaled) having uniform divisions. If a pointer is used, the pointer is moved in a horizontal plane to the right or left across a center-zero scale having uniform divisions. The direction in which the beam of light or the pointer moves depends on the direction of current through the coil.

This instrument is used to measure minute currents, such as in bridge circuits. In modified form, the galvanometer has the highest sensitivity of any of the various types of meters in use today.

Galvanometer shunt: A low-value resistor that is placed in parallel with the input of a galvanometer to increase deflection. Using these shunts, an ammeter with a specific internal resistance may be used to measure wide ranges of current, which would normally lie outside of its fixed metering range. For example, if the meter were designed with an internal impedance that would provide a full-scale deflection of the pointer at 1 A, a shunt resistor with a value equal to the internal resistance could be placed in parallel with the meter terminals to provide a full-scale deflection at 2 A. This extends the useful range of the galvanometer.

In practical applications, galvanometer shunts may be switched across the meter contacts. This allows 1 m to be used to measure several different current values ranging from low to high. Since these meters are most accurate when the current they pass causes a pointer deflection to fall within the top half of the scale, a shunt of the proper value can be chosen to allow for this inherent condition to be used to best advantage.

For example, a 1-A full-scale ammeter might not provide a highly accurate reading when used to measure a current value of 200 mA (lower third of the meter scale). However, a 300-mA meter could be used to measure this low value accurately and then be fitted with a shunt to enable it to also read values of 1 A or more. Using the galvanometer shunt, a single meter may be incorporated into a piece of equipment, along with a switching network. This avoids the cost and complexity of using two or more meters to accomplish the same purpose.

Gamma rays: A form of electromagnetic waves that are of high frequency and that have great penetrating power. The frequencies are generally in the range from three quintillion to six hundred quintillion vibrations per second, differing from ordinary light waves in wavelength and frequency. This term is often used to describe the radiation that is emitted or produced in radioactivity, or from the disintegration of atomic nuclei. An example of such radiation occurs in the nuclear fission in uranium piles and in the atomic bomb. Gamma rays contribute to the devastating flash-burn effects of atomic bomb explosions and are also emitted in the radioactivity that follows such a blast. Gamma rays have proven useful, however, in the treatment of certain cancer cells, as long as the dosage is carefully regulated.

Gas diode: A device that conducts electricity by the use of a small amount of gas (usually mercury vapor, argon, helium, or neon) placed within a tube envelope with an anode and cathode. As the potential of the anode plate is increased, the electrons flowing to it release electrons from the gas molecules, thus ionizing them. The resulting current is much greater than that produced by a vacuum tube and is limited only by the external circuit or current rating of the tube.

There are two major types of gas diodes—the hot cathode, which has a filament, and the cold cathode, which employs no heating element. The cold-cathode gas diode is most commonly used in neon signs. Other uses for the gas diode include voltage-regulation circuits, rectifiers, voltage-reference circuits, and fluorescent lamps.

Gas-filled tubes: An electron tube that contains a small amount of gas at low pressure. During normal tube operation, the gas ionizes and can serve to neutralize the negative space charge that exists between tube elements. The space charge of electrons around the cathode of a diode will cause the plate-cathode voltage drop to be directly related to the current being carried between the cathode and the plate. In standard vacuum tubes, this voltage drop can be very high when large currents are being passed. This causes a large amount of energy loss, which shows up as increased plate dissipation.

The negative space charge can be neutralized by injecting a small amount of gas between the electrodes. This provides a supply of positive ion. When the voltage drop across the tube reaches a potential to ionize the gas, the gas molecules will form positive ions. They tend to neutralize the space charge in the direct vicinity of the cathode. When this occurs, the voltage drop across the tube tends to remain fairly constant up to a current drain that is equal to the maximum emission capability of the cathode.

Gas tubes include gas diodes, thyratrons, and mercury-vapor tubes. The latter were quite popular in high-voltage power supplies of several decades ago. Their main disadvantage is the fact that they must be operated within a specific temperature range in order for the mercury-vapor pressure in the glass envelope to reach operating potential. Pressure increases in direct relationship to temperature. When the temperature is too high, the vapor pressure rises also, and the flashover voltage is lowered to a point where destruction of the tube can take place under load. If operating temperature is too low, internal vapor pressure is also lowered, and internal voltage drop will increase substantially.

In most modern applications, gas-filled tubes of these types are obsolete and have been replaced with solid-state equivalents. The latter are more rugged and reliable and do not require the stringent operating conditions needed by these earlier devices.

Gate: A circuit that yields an output signal that is dependent on some function of its present or past input signals.

Gate circuit: A circuit designed to operate in such a manner as to have no output until it is triggered into operation by one or more enabling signals. Digital computers used specialized gate circuits in determining the logic of an operation. A gate circuit is also the combination of components that are connected at the gate electrode of a silicon-controlled rectifier or triac. Connections are normally made between the anode and gate to cause the

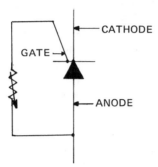

FIGURE 195

device to fire or conduct electricity. Figure 195 shows an SCR gate circuit that is composed of a variable resistance. When the value of the resistor is above a certain point (in ohmic value), insufficient current flows in the gate circuit and the SCR is cut off (nonconducting). As the resistance is lowered, current begins to flow and the device eventually fires.

Gate contact: The physical connection to the gate of a silicon-controlled rectifier or triac. These four-layer semiconductor devices are diodes or rectifiers, which conduct only when sufficient current is flowing in the gate circuit. An electrical circuit is often formed by placing a resistance (often variable) between the anode contact and the gate.

Gauss: A unit of measurement that represents flux density, or the concentration of the magnetic field being measured. This unit of density is stated in lines per square centimeter, or cgs. One gauss is equal to one line per square centimeter.

Gauss meter: A moving-coil assembly, together with stationary soft-iron pole pieces, which is placed in the field of the magnet to be measured. With a known current flowing in the moving coil, the deflection of the instrument becomes a measure of the strength of the magnet. For mechanical reasons, this principle is restricted in its application. A more common method of obtaining field-strength measurements employs the fluxmeter.

Gearmotors: A self-contained combination of a motor and an enclosed speed-reducing gear built as an integral unit. It is a more compact and readily adaptable unit than that obtained by using a motor coupled to a gear reducer. A motor used in a gearmotor combination can be any type of dc or ac motor. Gearmotors are available in sizes up to 75 hp. They have output shaft speeds from about 4 to 1430 rpm, making it possible to couple or to connect them by gear or chain to nearly any machine.

Generally, high-speed motors with 1800 rpm at 60 cycles are used in gearmotors, thus obtaining the advantages of high power factor and efficiency of the motor. The gearing efficiency is also high, usually about 98 percent, for a single reducer of the helical or spur types. It has a 2 percent loss for one reduction or a 4 percent loss for a double reduction. Consequently, the overall performance of a gearmotor is much higher than that of a combination of open bearing, belting, countershaft, or other arrangement.

Gearmotors are extensively used to drive belt and chain conveyors, agitators, mixers, feeding devices, and numerous other slow-speed drives. Besides being more efficient than the other combination drives and saving space, gearmotors provide the important advantage of reducing operating hazards and maintenance.

Germanium: A semiconductor material used in the manufacture of transistors. Germanium is a grayish-white metallic element. In some of its properties, it resembles carbon; whereas in others, it resembles tin. It is found in the fourth column of the periodic table and has an atomic weight of 72.60. Metallic germanium is secured by reduction (in a hydrogen or helium atmosphere) of germanium dioxide, a gray powder. The dioxide is obtained in commercial quantities in the United States as a flue residue in zinc smelting, and in England is a component of the chimney soot from gas works.

After purified germanium has been produced by electronic manufacturers and doped to specifications, comparatively large single crystals of it are drawn from a melted mass of the metal by dipping in a seed crystal of germanium and withdrawing it slowly under rotation. The melted germanium adheres and follows the seed to be pulled out of the melt in single-crystal form. During the process, temperature is controlled closely, and air is excluded. The tiny germanium wafers used in diodes and transistors are later cut out of this single crystal.

The advantages of single-crystal material are uniformity and reproducible electrical characteristics, such as resistivity. When, on the contrary, a germanium sample is composed of numerous intimately bonded, separate crystals, wafers sliced from this material might cut through crystal interfaces and exhibit nonuniformity of characteristics due to separate crystal properties.

While the germanium is in the molten state, impurities of the proper kind and amount are added to make it either N or P type, as required. Without controlled doping, pure germanium would behave like an insulator. Later, during the single-crystal drawing, impurities may be added at proper times during the withdrawal to produce separate N and P layers in the same crystal. Most general-purpose germanium is prepared to be N type.

Germanium diode: A semiconductor diode composed of the metallic element germanium (Ge). Germanium transmits a current only in one direction and is therefore able to rectify an alternating current. At high frequencies such as those employed in radar, television, and frequency modulation radio-broadcasting, germanium crystals prove to be more efficient than electronic radio tubes or vacuum tubes. Due to the electron arrangement of germanium atoms, stable crystalline structures can be formed from the element, which acts as an insulator.

Germicidal lamp: A source of radiant energy at the wavelength of 2537 A that is not contained in natural sunlight. These waves are lethal to bacteria. In germicidal lamps with no fluorescent powders or phosphors within the bulb and with a bulb made of special clear glass, this 2537-A radiation is transmitted and directly released for air irradiation and the killing of airborne bacteria and mold spores. The use of germicidal radiant energy is a science in itself, and much information is available as to its application.

Direct exposure to the rays of these lamps must be avoided. Neither the eyes nor the skin should be exposed to these short, ultraviolet rays. If a germicidal lamp is grasped, a burn will result, even though the lamp is cool to the touch.

Germicidal lamps are operated from the same ballasts as equivalent fluorescent lamp sizes. They are supplied in only a few sizes, from 4 to 40 W, and are sufficient for use in air ducts and wall fixtures for upper air radiation.

Getter: A small rectangular, square, or circular section of metal that is housed in the envelope of a vacuum tube to absorb gases during the evacuation process. These devices are most often constructed of magnesium and are flashed by an external radio-frequency field to aid the absorption process.

Ghost: A slightly displaced image that appears on a television screen simultaneously with its twin image. This condition is objectionable and is created by out-of-phase signals entering the receiver front end. The main image is received in one time segment, while the ghost image is received at a later time. The time differential is extremely small, being measured in microseconds.

Out-of-phase reception results when the broadcast from the television station travels along two or more different paths to the receiver. The ghost signal travels a slightly farther distance and is therefore received in a later time segment. Television signals travel at the speed of light, so distance covered will have a direct bearing on the exact time the transmission is actually received.

Nearly every transmission will contain a certain amount of signal portion that lags the main one. However, in most instances, the amplitude of the lagging signal is so small that it is not detected at the receiver front end. Ghost images result when the lagging signal is comparatively high in amplitude and is readily detected by the receiver. While both images were transmitted during the same time frame, they are not received simultaneously.

Gigaelectronvolt: Abbreviated *GeV*, an extremely large unit of voltage equal to 10^9 V ac or dc. Giga is a prefix that means one billion, so a gigaelectronvolt is equivalent to one billion volts. This term is used for scientific and theoretical evaluation purposes, as potentials of this magnitude are not encountered in any electronic applications.

Gigohm: A measure of resistance of a material with regard to how easily the electrons move through the material. One gigohm is equal to 1000 MΩ, or 10^9 Ω.

Gilbert: A csg (centimeter, gram, second) unit of measurement of the electromagnetic force that is required to produce 1 Mx of magnetic flux. One gilbert is equal to approximately 0.7958 ampere-turn.

Glow discharge: The luminous electrical discharge that results when electrical current is passed through ionized gas in a partially evacuated tube. Light is produced by the ionization of gas as electrons flow between two internal electrodes. The color of the glow is determined by the particular gas used in the tube or globe. Fluorescent light and neon bulbs are good examples of the glow discharge phenomenon.

Governors: Devices that maintain constant engine speed under various load conditions. They must have provisions for adjustment of speed (which controls generator frequency) and speed drop, from no load to full load. Average speed drop, from no load to full load, is about 2 cycles (60 cps for an 1800-rpm generating plant). A slightly greater drop is sometimes necessary to prevent *hunting* of paralleled generating plants.

Graded-junction transistor: A transistor composed of a grown junction semiconductor in which the temperature of the melt and the rate at which the crystal was pulled from it are closely controlled as the N and P layers are formed. The tolerances used to design and manufacture these types of transistors are quite critical, and microscopy is used to measure the various occurrences during the growth process and to monitor the quality of the finished product.

Graphite resistor: An electric component made of graphite that resists the flow of electric current. Graphite is a soft form of carbon and is widely used in electronics and is the basic element of attenuators, contacts, brushes, cathode-ray tube coatings, and others.

Graphite resistor also describes an emergency, makeshift resistor that is quickly made by drawing a pencil line on a piece of paper. The resistance between the start and end of the line will vary in relation to line length and width. The heavier line will produce a lower resistance than a thin line of the same length.

Again, this is an emergency resistor that might be used to test the operation of a circuit when a standard carbon resistor is not available. The length of the line is determined by the desired resistance. An ohmmeter is used to arrive at the correct value.

Grid: An element that is inserted between the anode and cathode of a gas-filled tube to regulate the amount of current that it handles. This tube is called a *triode* because it has three electrodes. The filament is not counted as an electrode when it is used to heat the cathode. The grid consists of a wire fence encircling the cathode so that it lies between cathode and anode. It is mesh-constructed so that electrons can readily get through to the anode from the cathode. However, if the grid is made sufficiently negative, the electrons leaving the cathode will be repelled and will never reach the anode.

Even though the anode is positive, no electrons will reach it, because the negatively charged grid repels all electrons. The amount of grid voltage necessary to cut off the plate current to zero will depend on the amount of anode voltage. The greater the anode voltage, the greater the grid voltage necessary to reduce the electron flow to zero.

If the grid voltage or *bias* is reduced, the number of electrons reaching the anode will be increased. The smaller the grid bias, the greater the current in the anode circuit. This can be accomplished by connecting a potentiometer across the grid-bias battery and varying the grid potential by moving a sliding contact.

The value of the triode is that a comparatively small voltage between grid and cathode has the same effect on the current in the anode circuit as a large voltage between the anode and the cathode. The triode is useful as an amplifier.

Grid modulation: Amplitude modulation that is accomplished by varying the dc control-grid bias of an RF amplifier at an audio rate. While this is a form of amplitude modulation, it is quite inefficient when compared to the more standard plate modulation used almost exclusively by commercial AM broadcast stations. Class C grid modulation induces a slightly higher distortion factor than linear amplification, but this can be kept to within tolerable limits for communications work. Using this system, a higher plate voltage is required on the modulated stage if maximum output is to be had. The plate potential is normally run about 50 percent higher than for maximum output with plate modulation.

A comparatively small amount of audio power is required to modulate the amplifier stage to 100 percent than when using plate modulation. For example, an audio amplifier having a 20-W output will be sufficient to modulate an RF amplifier with a 1-kW input. This compares with a 500-W amplifier requirement for plate modulation.

Grid-bias modulation was quite popular among radio amateurs several decades ago when amplitude modulation was used almost exclusively. This preceded the single sideband era. Today, amplitude-modulated systems are used by commercial and military operations, and the more efficient plate modulation systems are usually incorporated.

Ground absorption: The loss of radiant energy that is absorbed into the earth. This is especially applicable to energy losses associated with the transmission of radio waves that travel along the ground. Ground absorption will vary depending on the conductivity of the earth. This is especially critical within the immediate vicinity of the antenna installation.

Ground absorption is more accurately earth absorption, in that certain losses are incurred over water as well. Due to the higher conductivity of water as compared to soil, the absorption loss is not nearly as high. This type of loss is especially high when mountains are in close proximity to the antenna. As the electromagnetic energy leaves the prime radiator, it travels on an angle toward the horizon. As the wave travels farther and farther from the original point of transmission, its proximity with the earth decreases and ground absorption is lessened. When high terrain surrounds the antenna, greater absorption takes place because of the close proximity of the radiated wave to the earth.

Ground-plane antenna: A quarter-wave vertical antenna that carries its own artificial ground system. Shown in Fig. 196, this device is an effective long-angle radiator consisting of a vertical element surrounded by a grouping

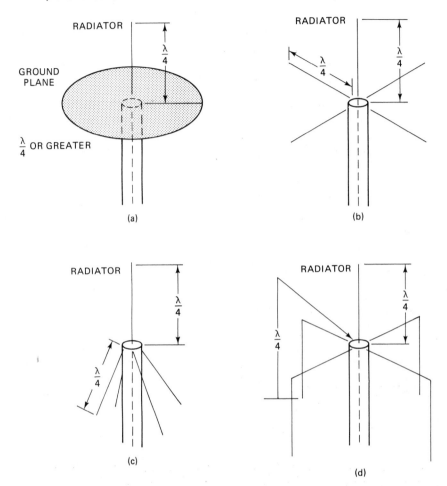

FIGURE 196   Ground-plane antennas.

of radial elements that serve as a ground system. The ground-plane antenna may be mounted high above the earth proper and is not affected by soil conditions in its immediate vicinity.

The ground-plane antenna exhibits properties similar to grounded quarter-wave antenna, but the angle of radiation is usually lower. Base impedance of the ground plane is on the order of 40 Ω, which makes this antenna a good match for 50-Ω coaxial cable, this being a standard transmission line for this type of system. The ground-plane elements may be trimmed slightly for a better match to the cable, or they may be angled about 45° from the horizontal (125° from the vertical) element.

The ground-plane antenna is often used for communications work, especially at VHF frequencies where element lengths are of practical proportions. It is an omnidirectional radiator, transmitting and receiving equally well in all directions. For high-frequency applications, the ground-plane antenna is often decreased in size by installing an inductor in the vertical element. Depending on the value of this latter component, the vertical element may be shortened from 10 to 40 percent. Sometimes, similar inductors are installed within the radials to allow them to be shortened as well.

Ground-plane antennas are often constructed of lightweight aluminum tubing instead of copper wires. This provides a sturdy structure that presents fewer mounting difficulties due to its self-supporting design. The entire assembly is then mounted to a telephone pole or aluminum mast and extended above the earth by 20 ft or more.

The vertical element is cut for an electrical quarter-wavelength, as are the radials. Most commercial systems for VHF operation contain three radial elements, but more may be added for a slight increase in radiation efficiency.

Ground wave: A wave that is used for both short-range communications at high frequencies with low power and for long-range communications at low frequencies with very high power. When a radio wave leaves a vertical antenna, the field pattern of the wave resembles a huge doughnut lying on the center of the ground with the antenna in the hole at the center, as shown in Fig. 197. Part of the wave moves outward in contact with the ground to form the ground wave. Daytime reception from most commercial stations is carried by the ground wave.

As it passes over and through the ground, this wave induces a voltage in the earth, setting up *eddy currents*. The energy used to establish these currents is absorbed from the ground wave, thereby weakening it as it moves away from the transmitting antenna. Increasing the frequency rapidly increases the attenuation so that ground wave transmission is limited to relatively low frequencies.

Since the electrical properties of the earth along which the surface wave travels are relatively constant, the signal strength from a given station at a given point is nearly constant. This holds essentially true in all localities except those having distinct rainy and dry seasons. Here, the difference in the amount of moisture causes the conductivity of the soil to change.

The conductivity of salt water is 5000 times as great as that of dry soil.

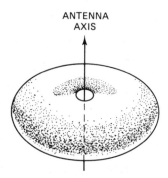

ANTENNA
AXIS

FIGURE 197   Vertical antenna field pattern.

High-power, low-frequency transmitters are placed as close to the edge of the ocean as practical because of the superiority of surface wave conduction by salt water.

Gunn diode: A specially designed semiconductor junction diode for operation at microwave frequencies. Named after its inventor, this component consists of a single section of semiconductor material such as N-type gallium arsenide. At a specific potential, this device develops a negative resistance region and begins to generate microwave signals.

Gunn oscillator: A solid-state (gallium arsenide crystal) bulk-effect source of microwave energy. The discovery that microwaves could be generated by applying a steady voltage was made in 1963 by J. B. Gunn. The operation of this device results from the excitation of electrons in the crystal to energy states higher than those they normally occupy. In a gallium arsenide semiconductor, there exist empty electron valence bands, higher than those occupied by electrons. These higher valence bands have the property that electrons occupying them are less mobile under the influence of an electric field than when they are in their normal state at a lower valence band.

To simplify the explanation of this effect, assume that electrons in the higher valence band have essentially no mobility. If an electric field is applied to the gallium arsenide semiconductor, the current that flows will increase with an increase in voltage, provided the voltage is low. However, if the voltage is made high enough, it may be possible to excite electrons from their initial band to the higher band, where they become immobile. If the rate at which electrons are removed is high enough, the current will decrease even though the electric field is being increased.

If a voltage is applied across an unevenly doped N-type gallium arsenide crystal, the crystal will break up into regions with different-intensity electric fields across them. In particular, a small domain will form within which the field will be very strong, whereas in the rest of the crystal outside this domain, the electric field will be weak.

It is not difficult to see that such a domain is unstable. Consider the result of momentarily disturbing the electron density in such a crystal. Assume there is a sudden increase in electron density at some point in the crystal, which tends to reduce the electric field to the left of the disturbance, while increasing the electric field to the right. If the material has a positive resistance (an increase in current with an increase in voltage), the decreasing electric field to the left will result in a decreasing current flowing into the region from the left, and the increasing field to the right will result in an increasing current flowing out of the disturbed region. The excess electrons will drain away. Thus, a disturbance caused by a temporary local increase in electron density will be dissipated as a result of the changing pattern of currents.

The situation is quite different in a negative-resistance material. In a negative-resistance material, the decreasing field to the left of the disturbance will cause an increase in current flowing into the disturbed region, whereas the increase in the field to the right will tend to lower the current outside this region. This current pattern will have the effect of building up the charge disturbance even more. Hence, the situation will become unstable and will result in a redistribution of the electric field within the crystal.

This concept may be more familiar in a somewhat different connection. It is possible to obtain negative resistance by accelerating electrons to such a velocity that when they collide with atoms in the system, they produce more free electrons. Once this happens, the voltage necessary to produce a given current declines. If voltage is applied across the material, different regions of the material may conduct different quantities of current. In fact, filaments form across the material, each containing a different current.

The extreme example of this situation is an electric spark in a gas, which consists of narrow filaments of high current, while the rest of the gas in the region is transporting much smaller currents. The spark is, in some sense, a current domain, whereas this discussion concerns electric field domains.

The domains formed in the gallium arsenide crystal will not be stationary, since the electric field acting on the electron energy will cause the domain to move across the crystal. This is illustrated in Fig. 198. The domain will travel

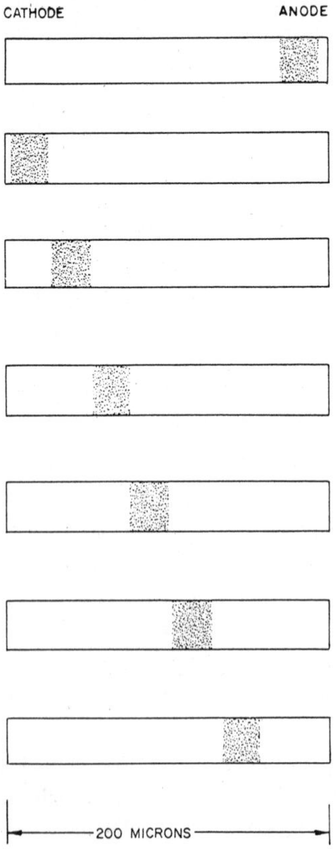

FIGURE 198 Gallium arsenide crystal domain.

across the crystal from one electrode to the other. As it disappears at the anode, a new domain will form near the cathode.

The Gunn oscillator will have a frequency inversely proportional to the time required for a domain to cross the crystal. This time is proportional to the length of the crystal and, to some degree, to the potential applied. Each domain results in a pulse of current at the output. Hence, the output of the Gunn oscillator is a microwave frequency, which is determined, for the most part, by the physical length of the chip.

The Gunn oscillator has delivered power outputs of 65 mW at 2 GHz (continuous operation) and up to 200 W in pulsed operation. The power output capability of this device is limited by the difficulty of removing heat from the small chip. It is conceivable that much higher power outputs may be achieved by using many wafers of gallium arsenide as a single source.

The advantages of the Gunn oscillator are its small size, ruggedness, low cost of manufacture, lack of vacuum or filaments, and relatively good efficiency. These advantages open a wide range of application for this device in all phases of microwave transmissions. This and other solid-state bulk-effect microwave devices are still new, and much research is being carried on in this area. It is conceivable that these devices will rival the conventional electron tube microwave devices in the future.

Guy insulator: A means of insulating a guy wire in order to prevent the lower end from becoming live if a line wire should accidentally come in contact with the upper end of the guy wire. The insulator should preferably be 6 ft out from the pole and cut 8 ft or more above ground. Such insulators should have a strength of at least equal to that of the guy wire. They should be made of wet-process porcelain, which will not be punctured by lightning discharges down the guy wire.

Figure 199 shows the most commonly used interlocking type of guy insulator. The two ends of the wire which it joins are interlocked, so that they will not separate if the insulator should break. Interlocking insulators are also sometimes used for dead-ending low-voltage lines.

Guy wire: Wire that is made of stranded steel, either galvanized, copper covered, or aluminum covered. The most common sizes and strengths are listed in Fig. 200. All the guy wires in this table are seven-strand wires, except the 3M, which is a three-strand wire. The utility strength (UTS) wires are somewhat more flexible and of somewhat larger diameter than the extra-high-strength (EHS) wires of comparable strength. Guy clamps are suitable for use with the UTS, whereas preformed grips are normally used with the EHS.

The computed load on the guy wire consists of the total loaded tension of the wires to be held, plus any wind pressure on pole or wires that may be

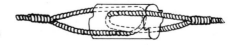

FIGURE 199

GUY WIRES

Specification	Strength, in Pounds
3M utility strength (UTS)	3,000
6M utility strength (UTS)	6,000
10M utility strength (UTS)	11,500
16M utility strength (UTS)	18,000
¼ in. extra-high strength (EHS)	6,650
$\frac{5}{16}$ in. extra-high strength (EHS)	11,200

FIGURE 200

combined with it, increased by an allowance for the vertical angle at which the guy runs. A method of estimating the load in pounds on the guy wire is to multiply the load in pounds to be held by the length of the guy in feet, and to divide by its horizontal span in feet. The guy should be horizontally in line with the load if possible. Otherwise, the estimated load on it must also be increased in the calculations to allow for any angle.

Half-adder: A circuit having two output points, *S* and *C*, representing sum and carry, and two input points, *A* and *B*, representing addend and augend, such that the output is related to the input according to the following table.

Input		Output	
A	B	S	C
0	0	0	0
0	1	1	0
1	0	1	0
1	1	0	1

A and *B* are arbitrary input pulses, and *S* and *C* are sum without carry and carry, respectively. Two half-adders, properly connected, may be used for performing binary addition and form a full serial adder.

Half-wave: An electronic term that means half of a complete wave at any frequency. A half-wave of an alternating current cycle encompasses the complete rise and fall of current in one polarity direction only. A half-wave rectifier operates from only one-half of the ac sine wave. A half-wave antenna is equal in length to the distance a radio wave will travel (at the speed of light) during one-half of an ac cycle at a specific operating frequency.

Half-wave antenna: A transmitting and/or receiving antenna consisting of a length of conductor whose physical length is equal to one half of the distance a radio wave will travel in free space during a half cycle. Radio waves travel at the speed of light, which is equivalent to 3×10^8, so a full wavelength is obtained by the formula

$$\text{Wavelength} = (3 \times 10^8)/f$$

where f is the design frequency given in hertz (cycles per second).

For example, the wavelength of a 1,000,000 Hz (1MHz) frequency is found by the formula

$$(3 \times 10^8)/1,000,000 = 300 \text{ m}$$

A half-wavelength, then, is equal to 150 m, and a half-wave antenna would have an element 150 m in physical length. In practical usage, a half-wave antenna will be about 5 percent shorter than an actual half-wavelength due to ground capacitance effects. A standard, practical formula for determining physical dimensions of a half-wave antenna is

$$\text{Half-wavelength} = 464/f\text{MHz}$$

where fMHz is the design frequency in megahertz.

Shown in Fig. 201, the entire antenna system consists of the wire element and a feed system or transmission line that connects the element to the transmitter or receiver. This line becomes an active part of the overall system.

Half-wave antennas may be connected to the feedline at their centers, off-center, or at either end. The impedance at the center point is approximately 70 Ω and increases to several thousand ohms at either end. In order for these impedances to be obtained and for the formulas to work properly, the antenna must be mounted at least a quarter-wavelength (half its physical length) above the ground.

The half-wave antenna has been a standard of the amateur radio service since its early inception and is a simple and inexpensive structure to build and mount, providing an adequate amount of lateral space is afforded. Sometimes, portions of the wire element may be bent to enable the antenna to fit into

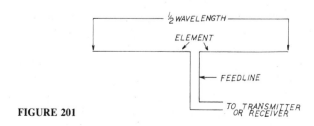

FIGURE 201

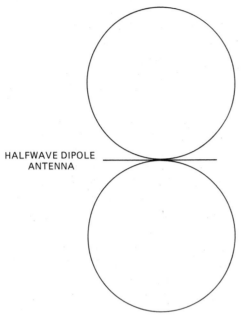

HALFWAVE DIPOLE
ANTENNA

FIGURE 202

minimum space allotments. This will have an effect on the physical length requirements for the element to become resonant at the design frequency.

The radiation pattern of a horizontal half-wave antenna is generally bidirectional. Figure 202 shows an example of an ideal pattern when the antenna is mounted in free space and is devoid of all capacitance effects from the earth. The two main lobes are equal in radiated power and are broadside to the wire element. Thus, a half-wave antenna that is mounted with its ends pointing east and west will transmit signals best to the north and south. Likewise, it will be most sensitive to receiving transmitted signals that intersect its element from northerly and southerly directions.

Half-wave rectifier: A circuit used to convert alternating current into pulsating direct current that acts on only one-half of the sine wave of the ac current. It offers a half-cycle of dc output for every other half-cycle of an ac input. This means that the successive dc half-cycles are 180° apart, all having the same polarity.

Figure 203 shows the ac sine wave on the left, while on the right is the

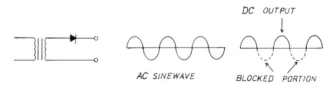

FIGURE 203

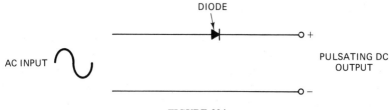

FIGURE 204

resultant dc output from a half-wave rectifier circuit. Note that the bottom or negative portion of the cycle has been blocked from the output side of the circuit. Every positive cycle portion is passed through to the output.

Half-wave rectifier circuits are used today for simple dc power supplies, which receive input normally from the 60-Hz house current line. Figure 204 shows a typical circuit, which consists of a single diode rectifier. The output from this circuit is pulsating dc and is usually filtered through a capacitor and/or choke-capacitor circuit. The resultant dc output is relatively pure, without most of the original ac component or ripple frequency.

After filtering has taken place, the pure dc output will usually be near the peak value of the ac line. This is roughly equivalent to 1.4 times the average value of the line. For example, a step-down transformer is used to supply alternating current to a half-wave rectifier circuit, as shown in Fig. 205. The transformer secondary winding is designed to deliver an ac output of 6 V, which is the average value. Once this output has been rectified and filtered, the dc output will be 1.4 times the 6-V value under conditions of zero loading (no current drain from the circuit). Since 1.4×6.0 equals 8.4 V dc, the output from this circuit under no-load conditions will be about 8.5 V. When current is drawn, the output voltage level will begin to drop toward the average ac value and below to approximately $0.5\ t \times$ the rms value.

Half-wave rectifier circuits require more filtering than do full-wave circuits. This is due mainly to the pulsating or ripple frequency of the dc output. A half-wave rectifier presents a ripple frequency equivalent to the ac frequency. In most cases, this will be 60 Hz. Low ripple frequencies require more filter capacitance values than are necessary to accomplish the same amount of filtering in full-wave circuits whose outputs exhibit ripple frequencies of twice the ac frequency.

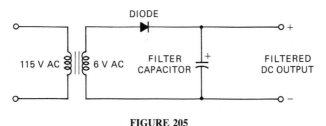

FIGURE 205

The PRV (peak reverse voltage) a rectifier must be rated to withstand in a half-wave circuit will vary with the load. The PRV is the voltage that the rectifier must be able to block when it is not conducting. This would be during the negative portion of the ac wave for the circuits previously discussed. When the load is resistive, the PRV is equal to the peak value of the ac voltage, or 1.4 × rms value. When the load is capacitive and under no-load conditions, the PRV can rise to 2.8 × rms value.

A major disadvantage of the half-wave rectifier is in the needed power ratings of transformers used with this circuit. A considerably higher power rating is required when compared to that of full-wave circuits to obtain the same dc power output. A value of approximately 45 percent has been set as the recommended increase percentage when half-wave circuits are used.

Half-wave rectifiers are extremely simple and inexpensive, but are relegated to noncritical applications. Many of the low-current power supplies used to provide operating current for transistor radios, tape recorders, and other general hobby devices may use this type of circuit to convert 115-V ac house current to 6-, 9-, or 12-V dc for line operation. The commonly seen ac-to-dc converter packs often depend on the half-wave rectifier to accomplish this conversion in as small a package as possible.

Half-wave transmission line: A line that may be used between an antenna and a transmitter or receiver in order to couple energy between the two devices. The physical length of such a transmission line will be equal to the distance a radio-frequency wave at the operating frequency of the system will travel (at the speed of light) during one-half of one operational cycle. Wavelength is usually measured in meters, with one wavelength being equivalent to 300,000,000 m (the speed of light) divided by the frequency in hertz. For example, a frequency of 4 MHz has a wavelength of 75 m. A half-wave transmission line for this frequency would be 37.5 m in electrical length.

Hall coefficient: The constant relationship between the transverse electric field of a conductor carrying current in a magnetic field and the magnetic flux density. This is tied in closely with the *Hall effect,* which is a phenomenon observed in thin strips of metal and in some semiconductors.

When a strip carrying current longitudinally is placed in a magnetic field that is perpendicular to the plane of the strip, a voltage appears between opposite edges of the strip that will force a current through an external circuit. These induced voltages are quite low in magnitude, but are capable of establishing a separate electronic flow or circuit.

Hall effect: That phenomenon in which an electrical current can be controlled by a magnetic field, because the magnetic field changes the resistances of some elements with which it comes in contact. When a bubble passes the Hall effect elements, a change in current of the circuit will create, say, a *one.* If there is a *zero,* there will be no bubble. In this time appearance slot, there will be no current change in the output circuit. The read-in device would be an opposite effect, wherein the Hall device creates a magnetic field when

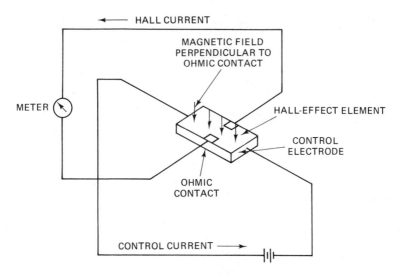

HALL CURRENT

MAGNETIC FIELD
PERPENDICULAR TO
OHMIC CONTACT

METER

HALL-EFFECT ELEMENT

CONTROL
ELECTRODE

OHMIC
CONTACT

CONTROL CURRENT

FIGURE 206   Hall generator.

supplied with a pulse of current. This, in turn, creates a little domain. A magnetic bubble is then created.

Hall generator: A semiconductor device that exhibits the Hall effect. It is a thin wafer or film of indium antimonide or indium arsenide with leads on opposite edges. This device is used primarily for measuring ac power and magnetic field strength, with its output voltage being proportional to the current passing through it times the magnetic field, which is perpendicular to it. See Fig. 206.

Handshaking: In a computer system, that which occurs when a system and an external device tell each other that the data are ready to transfer and that the receiver is ready to receive the data. For example, if the system is outputting eight parallel bits to a printer, the system loads the port and then tells the printer that the data are ready. The printer reads the port and tells the system it is ready for more data. The system loads the port with the next data word, and the process is repeated. Handshaking is accomplished using one bit of an input port and one bit of an output port.

The devices that require handshaking are those that transfer several data words and/or operate on the data. For example, if a printer did not print the word from the system before the next word appears, some data would be lost. If the printer processed data faster than the system provided the data, the same character might be printed twice.

Harmonic: A frequency that is a multiple of the fundamental frequency.

Harmonic filter: A filter that is designed to remove unwanted harmonics from a generated frequency. These are actually bandpass filters with cutoff frequencies just above that of the main frequency. Low-pass filters are often used with radio transmitters and will conduct all energy below a certain cutoff frequency. Above this point, all energy is suppressed.

A radio transmitter with an output at 1 MHz will also broadcast a second harmonic at 2 MHz, a third at 3 MHz, etc. A harmonic filter with a cutoff frequency of 1.5 MHz would effectively pass the prime frequency (1 MHz) and suppress the harmonics.

Harmonic loss: That loss derived by subtracting the amount of power contained in the harmonic output of any signal generator or transmitter from the power content of the prime signal. For example, if a radio transmitter is designed to produce a 3 MHz output signal, this is the main frequency, and all other outputs are harmonics. If the second harmonic (6 MHz) is 3 dB in amplitude below the strength of the 3 MHz signal, it is being transmitted at half the power of the main output.

If the measured output of the 3 MHz signal is 100 W, the 6-MHz transmission is being transmitted at 50 W. The harmonic loss to the prime signal is 50 W, as this amount of power is *robbed* from the prime signal.

Hash noise: Electrical interference that is evidenced in many different circuits with audio-frequency output. Hash noise is recognized by a frying or sizzling sound emanating from the output speakers. It can be created from the discharge in gas tubes and mercury-vapor tubes. It may also be generated by a defective silicon-controlled rectifier, which allows a varying amount of current to flow while in a partially conducting state.

Nearly every electronic device or electrical component is capable of producing hash noise due to defective operation. Simple light switches can lose their insulation properties due to age and pass small amounts of electric current intermittently. Hash noise may then be heard from the speakers of stereo systems, radios, and television sets. Some types of electrical and electronic equipment incorporate built-in hash filters, which are usually comprised of an inductor and a shielded capacitor.

Head: A device that reads, records, or erases information in a storage medium. It is usually a small electromagnet, which has access to a magnetic tape or drum. Under specific directions generated by the system or operator, a head is capable of reading, writing, or erasing information found on the tape or drum.

Headphone: An audio transducer that is designed to be worn in direct contact with the ear or ears. For stereophonic applications, a headphone will usually consist of two high-quality miniature speakers, each housed in a protective cushion and connected by a headband. When placed over the head, the cushions encircle the ears and serve to protect from outside noises. The left speaker will be connected to the left amplifier channel, while the other is connected to the right channel. The use of stereophonic headphones often brings about a more realistic reproduction as sensed by the listener.

Other types of headphones are designed for communications work and may be constructed from magnetic diaphragm transducers rather than true speakers. Some may even contain crystal elements, but most of these are designed to be inserted into the ear channel and are called earphones.

Communications headphones are used in areas where high noise exists. They serve to channel the output of the receiver directly to the human ear, increasing the relative volume to the listener above that of environmental noise. Communications headphones are often used by aircraft pilots. These devices usually contain a miniature boom that supports a small microphone element near the pilot's mouth. Using this arrangement, two-way communications can be maintained without manual controls. Communications headsets normally contain only one transducer, leaving the operator's uncovered ear free to hear and respond to fellow operators.

Heat: A form of energy that is produced in the agitation of the molecules of matter. The energy expended in agitating these molecules is transformed into heat. Heat is measured in calories or *British thermal units* (abbreviated Btu). A calorie is the amount of heat necessary to raise the temperature of 1 g of water from 0 to 1°C. A Btu is 1/180 of the heat required to raise 1 lb of water from 32 to 212°F.

The eminent English physicist, James Prescott Joule, worked for more than 40 years in establishing the relationship between heat and mechanical work. He stated the doctrine of the conservation of energy and discovered the law for determining the relationship between the heat, current pressure, and time in an electric circuit. This is known as *Joule's law*.

Heat sink: A heat exchanger in the form of a heavy, metallic mounting base or a set of radiating fins. It removes heat from such devices as tubes, power transistors, or heavy-duty resistors, and radiates the heat into the surrounding air.

Helical antenna: An antenna consisting of a coil wound around a cylindrical form. Often, a reflecting ground plane is used in conjunction with this type of antenna to make it highly directional and broadbanded as well. The radiation from a helical antenna is circularly polarized and will be most sensitive (in the receive mode) to transmitted signals that are broadcast using the same polarization.

A simple helical antenna is shown in Fig. 207. A typical eight-turn helix

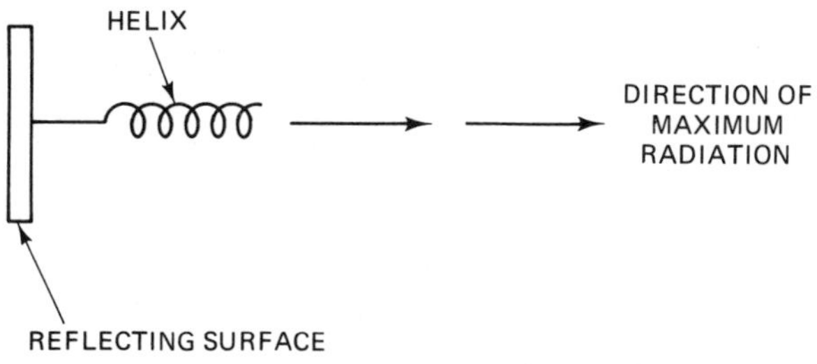

FIGURE 207

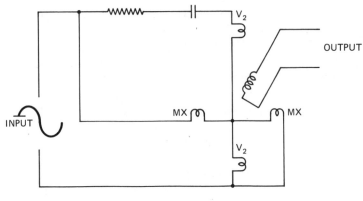

FIGURE 208

can produce a directional antenna with a 12-dB gain. Helical antennas may be stacked by mounting them within a half-wavelength of a common reflector. Typically, the diameter of each coil in the helix will be one quarter-wavelength, with a spacing between turns of one eighth- and one quarter-wavelength.

Helmholtz coil: A device consisting of two crossed-field primary windings in which an inductively coupled secondary winding rotates. The primary windings carry currents that differ in phase by 90°. Rotating the secondary coil provides 360° of continuously variable phase shift. See Fig. 208.

Henry: A unit of inductance that measures the variation of a current at the rate of 1 As and induces an electromotive force of 1 V. Abbreviated *H*.

Heptode: A vacuum tube that has seven internal elements.

These will include the cathode (or filament), anode, control grid, screen grid, and 2 auxiliary electrodes that are also normally grids.

Hertz: The unit of frequency of periodic phenomena such as alternating or pulsating current equivalent to one complete cycle in 1 s. A sine wave that takes exactly 1 s to rise from zero to a peak positive value, falls back to zero and attains a peak value, and then returns to zero again has a frequency of 1 Hz. This process occurs 60 times in 1 s in standard household current. Therefore, the frequency of the household current is 60 Hz (60 complete cycles in 1 s). Abbreviated *Hz*.

Hertz antenna: Any antenna that is one half-wavelength long, or any even or odd multiple thereof, and may be mounted either vertically or horizontally. A distinguishing feature of all hertz antennas is that they need not be connected conductively to the ground. At the low and medium frequencies, these antennas are rather long.

Vertical half-wave and five-eighths-wave antennas are widely used with AM broadcasting stations and have been built to heights of 1000 ft or more for the lower broadcast frequencies. At the medium and high frequencies, they are used extensively in fixed service when operation is not required at a large number of frequencies.

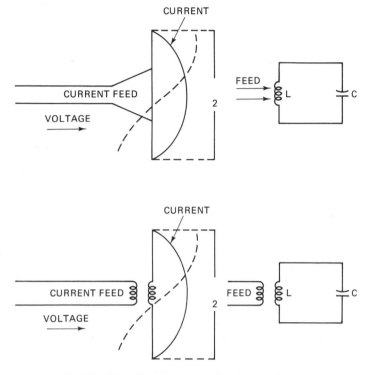

FIGURE 209 Hertz antenna and equivalent circuits.

This type of antenna is not particularly suited to services where a large number of different and unrelated frequencies must be transmitted using the same antenna. Half-wave antennas showing two different methods of connecting the feedline together with the equivalent resonant circuits are shown in Fig. 209. For a half-wave dipole, the effective current is maximum at the center and minimum at the ends, while the effective voltage is minimum at the center and maximum at the ends. The voltage and current relationships are similar to those of simple dipoles.

Heterodyne: The process of combining two or more frequencies in a nonlinear device and producing new frequencies. This may also be referred to as mixing, modulating, beating, or frequency conversion. The stage that this process is taking place in may be called a *mixer, converter, translator,* or *first detector.*

The principle of heterodyning is not related to electronics alone. The basic principles are also related to physics. The production of an audible beat note is a phenomenon that is easily demonstrated. For example, if two adjoining piano keys are struck simultaneously, a note will be produced that rises and falls in intensity at regular intervals. This action results from the fact that the rarefactions and compressions produced by the vibrating strings will

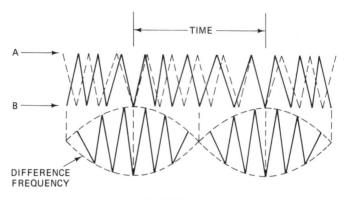

FIGURE 210

gradually approach a condition in which they reinforce each other. This occurs at regular intervals of time with an accompanying increase in the intensity of the sound. Likewise, at equal intervals of time, the compressions and rarefactions gradually approach a condition in which they counteract with each other, and the intensity is periodically reduced. Figure 210 illustrates graphically how the resultant difference vibration appears.

If wave A were 98 Hz and wave B were 100 Hz, both of the same amplitude, the resultant vibration would rise and fall at a difference frequency of 2 Hz. It is important to note that the mere existence of this amplitude variation does not directly indicate the presence of a difference frequency component. In the diagram shown, the difference frequency is observed, but is not necessarily detected until the ear itself acts as a nonlinear device or mixer. To obtain a difference frequency component, it is necessary to apply the original frequencies to a nonlinear device. Since the transistor and the tube, like the ear, display nonlinear characteristics, the simultaneous application of various frequencies will result in the reproduction of the original frequencies, plus the production of various new frequencies.

The necessity for the use of a nonlinear device to produce the heterodyning process can best be demonstrated by the use of a response curve. The response curve shown in Fig. 211 is a graph of current versus voltage. This curve, a dynamic transfer curve, can represent $V_{BE} - I_C$ of a transistor or $E_g - I_p$ of an electron tube. Simultaneous application of two different frequencies to a linear device will produce an output containing only the original frequencies, while simultaneous application of the same two frequencies to a nonlinear device will produce not only the original frequencies, but also the sum and difference of two original frequencies.

Referring to Fig. 211, waveforms A and B are used to form the composite waveform C. Although C appears to contain a modulation component at the difference frequency of A and B, the average value of wave C at any constant is zero. Thus, no useful energy exists at the difference frequency. However, the application of waveform C to a nonlinear device (point X) causes a heterodyn-

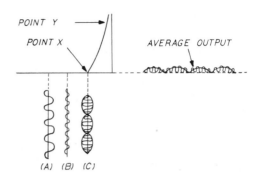

FIGURE 211   *CURRENT VERSUS VOLTAGE*

ing action between the two original frequencies. This causes the average of the output wave D to be other than zero and to vary at the difference frequency. Although not shown graphically, energy is also present at the sum frequency.

If waveform C were applied at point Y rather than point X and the output waveform graphed, it would be noticed that (due to the linear characteristics of the curve in this region), the variations of I would be nearly symmetrical. Thus, very little heterodyning action would take place and the average output would be very nearly zero.

Heterodyne repeater: A type of repeater used in microwave systems in which the incoming signals are heterodyned (entire received signal shifted in frequency as a block) to another band, amplified, and retransmitted. In microwave communications, the length of a single link is determined by the terrain. Most systems are composed of links of 30 mi or less, except where especially favorable sites can be found. Repeater stations may be used to connect one link to another to form long chains, thereby setting up long paths for many voice channels where needed. For example, chains of more than 40 lines cross the United States carrying voice and television signals. With proper engineering, excellent quality may be preserved through many sequential links.

In an RF heterodyne repeater, the shifting is done directly to the desired transmitting channel in one step. Amplification will normally be used both before and after the frequency changing process. In an IF heterodyne repeater, the incoming signals will be shifted to a relatively low, intermediate frequency at which most of the amplification is done. They are then shifted again to the desired channel for transmission.

Hexode: A six-element tube which has all the elements of the pentode, with one auxiliary grid. These devices are often used in high-amplification circuits.

High-C circuit: A resonant circuit that exhibits a high capacitance and low inductance at a given frequency. These tuned circuits are characterized by high selectivity and low-voltage operation. No specific value of capacitance is required to form a high-C circuit as long as the measured capacitance value is very high in relationship to the actual inductive component.

High fidelity: A system that is capable of reproducing sounds with a minimum of distortion and interference. The study and improved use of acoustics, better microphones, more precise cutting tools, and improved needles and electronic parts contributed to modern high-fidelity recording systems used in radio transmissions and phonograph play.

FM broadcast stations usually transmit high-fidelity program material that is received by FM console models with high-fidelity FM tuners and high-fidelity audio amplifiers. The receivers are constructed to provide stability, low-noise operation, and high-power audio output. A squelch system is usually included in the FM receiver to reduce the annoyance of high-output pulses when tuning between stations. Additional metering and indicating circuits are used for proper tuning and as an aid in switching among various receiver modes.

The audio system contains several audio-amplifier stages, including a preamplifier stage for controlling and assuring adequate acoustic power, proper circuits for the frequency response of the audio system, and loudness and volume levels.

High-level language: A programming language developed to allow for shorter, more understandable instructions to be used by the system programmers. These languages use the alphabet and decimal system in their coding procedures and greatly reduce the steps that would be necessary using machine language. Computer systems are designed to use a simple and universal machine language that consists of binary-coded digits by which it is able to transfer information and instructions internally. However, machine language is tedious and time-consuming. Therefore, it is not adequate for communications between the computer and user. Hence, high-level languages were developed to make communications easier.

Two very useful and versatile examples of high-level language are COBOL and FORTRAN. FORTRAN is generally used in scientific applications, and COBOL is used extensively in business situations.

High-Q: A relative term that indicates the Q of a tuned circuit, which is of a higher value than standard or normal. Q is a measurement of the factor-of-merit of a capacitor inductor. It is this factor that determines the sharpness of resonance of a tuned circuit. A high-Q for one type of circuit could be an extremely low-Q for another. The description of high-Q generally applies to circuits that are extremely selective, having very narrow passbands.

Hole conduction: The unfilled tracks of a moving electron. When an electron is freed in a block of pure semiconductor material, it creates a hole that acts as a positively charged current carrier. This electron liberation creates two currents, known as *electron current* and *hole current*.

Holes and electrons do not necessarily travel at the same rate. When an electric field is applied, they are accelerated in opposite directions. The life

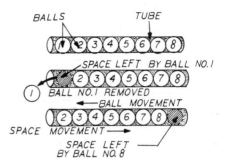

FIGURE 212   ANALOGY OF HOLE MOVEMENT

spans (time until recombination) of a hole and free electron in a given semiconductor sample are not ordinarily the same.

Because the hole is a region of net positive charge, the apparent motion is like the flow of particles having a positive charge. An analogy of hole motion is the movement of balls through a tube (Fig. 212). When ball 1 is removed from the tube, a space is left. This space is then filled by ball 2. Ball 3 then moves into the space left by ball 2. This action continues until all the balls have moved one space to the left, at which time there is a space left by ball 8 at the right-hand end of the tube.

A pure specimen of semiconductor material will have an equal number of free electrons and holes, the number depending on the temperature of the material and the type and size of the specimen. Such a specimen is called an *intrinsic semiconductor,* and the current that is borne equally by hole conduction and electron conduction is called *intrinsic conduction.*

If a suitable impurity is added to the semiconductor, the resulting mixture can be made to have either an excess of electrons, causing more electron current, or an excess of holes, causing more hole current. An impure specimen of semiconductor material is known as an *extrinsic semiconductor.*

Hole, semiconductor: In a semiconductor atom, the vacancy resulting from the loss of an electron. When an electron is lost, the negative charge is lost as well, leaving a hole that exhibits an equivalent positive charge, which, like the electron, can apparently migrate as a current carrier.

Horn: An acoustical device that is used to radiate or receive acoustic waves. Air is constricted in an area called the *throat,* which is considerably smaller than the diaphragm area closely coupled to it. Therefore, a greater portion of the drive energy is devoted to moving air than to drive the apparatus itself, as is the case in a cone acoustical device. The gradually expanding horn then changes this movement from a high-pressure wave in a constricted space to a low-pressure wave in the air of the room.

An advantage of the horn design is that the diaphragm is more uniformly loaded at all frequencies that the horn handles, yielding a much smoother frequency response than is easily obtained from a direct-radiator type. It is also more efficient for this same reason.

Horns have been made with conversion efficiencies in excess of 50 percent. Direct radiators seldom reach or exceed 10 percent efficiency, and many are less than 1 percent efficient.

Horn antenna: An antenna that provides a method of matching the impedance of waveguides to free space and at the same time provides the desired beam characteristics. Normally, the radiation pattern of an open-ended waveguide is broad in both the vertical and the horizontal planes. It is also an inefficient radiator system because of the mismatch between the waveguide antenna and free space.

The evolution of the horn came about as a result of the effort to minimize the reflections that occur when a straight piece of waveguides radiates into space. By flaring the open end of the waveguide, the RF energy in the guide does not encounter such an abrupt impedance change when it reaches the end of the guide. Hence, reflections and standing waves are minimized. If the waveguide is flared out at a gradual angle, a horn antenna is obtained. The flare must be gradual so as to permit a better match between antenna and free space. At the same time, a sharper beam is obtained when a flared horn is used.

The horn is very practical at microwave frequencies, since its physical size is not prohibitive. Since a resonant element is not involved, this type of antenna is capable of wideband operation. Horn antennas used in aircraft installations produce beam patterns that are not highly directive. In order to produce a highly directive beam at lower frequencies, the antenna would have to be longer than practical for aircraft.

The horn antenna may be used with either circular or rectangular waveguides, and the same considerations will apply; i.e., the longer the flare of the horn, the higher the gain and the narrower the beam width. For both circular and rectangular horns, the flare angle for the narrowest beam is between 40 and 60°. Generally speaking, the gain of a horn is less than that of a parabolic antenna. In order for the horn to have the same gain as a parabolic antenna, the horn must be much longer, which soon proves to be a big disadvantage.

A sectoral-type horn antenna is illustrated in Fig. 213. Its field distribution is spherical in shape and depends on the mode of operation of the waveguide feeding it and the flare angle of the horn. The spherical shape of the wavefront becomes flatter (elliptical) as the flare angle of the horn decreases. This horn has the greatest degree of directivity in the plane of the widest dimension.

Two other types of horns to which the principles just discussed apply are shown in Figs. 214 and 215. The pyramidal horn in Fig. 214 has equal directivity in both the vertical and horizontal plane. The conical horn in Fig. 215 produces a conical beam pattern.

For horn antennas (as well as for dipole arrays and reflectors), a rule that

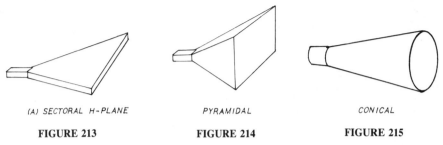

(A) SECTORAL H-PLANE	*PYRAMIDAL*	*CONICAL*
FIGURE 213	**FIGURE 214**	**FIGURE 215**

may be kept in mind is that the larger the aperture of the waveguide or the larger the area of the array, the greater the gain and directivity. Since a large aperture is desired at the mouth, the length of the horn may become unwieldy if the gradual flare requirement is to be incorporated in the design. However, a development in antenna design has made possible the use of horn antennas having a much larger equivalent size than formerly practicable. This type of antenna is called the *jelly-roll antenna* due to its configuration.

Effectively, it is the same as a conventional horn antenna rolled from the feed end toward the aperture end like a jelly roll. The resulting cylindrical shape uses less space than the conventional horn with the same size feed, the same angle of flare, and the same size aperture. Therefore, this savings in size could be used to good advantage by increasing the size of the aperture, with the resultant increase in directivity and gain.

Another method of improving the directivity and gain of a horn antenna is by the use of electromagnetic lenses to focus the beam in much the same manner that the light beams are focused by an optical lens.

Horn-type loudspeaker: A type of loudspeaker that is widely used in high-fidelity sound equipment due to its efficiency in the reproduction of audio waves of high frequency. A vibrating diaphragm and sound chamber are mounted in the neck of the horn, which is rigid and does not vibrate. The horn expands the diaphragm vibration, resulting in a substantial sound pressure variation at the opening. The diaphragm is small, enabling it to respond to high-frequency changes. The shape of the horn controls the sound characteristics, such as frequency response, sound dispersion, and efficiency.

Horsepower: The equivalent of the work necessary to raise 33,000 lb 1 ft in 1 min or 550 lb in 1 s. This may be expressed as the work necessary to raise 330 lb 100 ft in 1 min. Horsepower was defined by James Watt as the result of his experiments with draft horses. This unit was adopted by Watt in the rating of his steam engines and has been used since then as a unit for the rating of power equipment.

Since the horsepower equals 33,000 ft·lb·min, the rate of doing work of any device expressed in horsepower equals ft·lb·min/33,000.

Hunting: A swinging motion of the rotor in a synchronous motor due to a sudden change in load. The rate of swinging is called the *natural frequency of the rotor*. If a coil spring, suspended at one end, has a weight attached to the lower end, the spring will deflect a certain amount, depending on the stiffness of the spring (the spring constant) and the amount of weight. In the synchronous motor, the load angle corresponds to the deflection of the spring and is a measure of the motor spring constant. The flywheel effect of the motor and its load correspond to the weight attached to the spring.

If the weight is pulled down and then released, it will oscillate about its original position and finally come to rest owing to friction. The frequency of oscillation is determined by the spring constant and the amount of weight. The swinning of the rotor induces a current in the amortisseur winding owing to the flux moving back and forth across the pole face. This current causes a loss in the winding, which, if the change in load is not repeated, will stop the hunting, just as the friction losses stop the weight from oscillating.

If the weight, when oscillating, is pulled downward each time it starts to move downward, the amplitude of oscillation will increase and may become sufficient to break the spring. A similar condition may exist in a motor driving a reciprocating load, such as a plunger pump or compressor. Such equipment has a variable load. For example, the load of a single-cylinder equipment varies once each revolution if single-acting and twice each revolution if double-acting. If the rate of load change is too close to the natural frequency of the rotor, excessive swinging or hunting will result.

When the rotor of the synchronous motor swings in the manner described, each swing produces a variation in line current (or current pulsations) approximately proportional to the amount of swing. If the changes in line current are excessive, they may be a source of line disturbance, causing lights to flicker and possibly causing the motor to drop out of synchronism. The possibility of such disturbances is carefully considered when using synchronous motors for the applications mentioned, or for similar applications, and a limit of current swing not to exceed 66 percent of normal full-load motor current is usually required.

Since the natural frequency of the rotor depends on the spring constant of the motor and the flywheel effect, either or both of these factors may be changed to avoid excessive hunting and, hence, excessive current variations. Where a relatively small increase in flywheel effect is required, ballast rings or flywheels may be bolted to the rotor spider rims. In other cases, large flywheels are mounted on the shaft.

Hybrid circuit: A circuit that is formed by a combination of the film and monolithic IC techniques. The multichip IC employs either process to form the various components, which are then interconnected on an insulating substrate material and packaged in the same container. The more sophisticated hybrid IC is created by first forming the active device within a semiconductor wafer, which is then covered with an insulating layer. The process employs film techniques to form the passive elements on the insulating layer surface, after which connections from the film to the monolithic structure are made through openings cut in the layer.

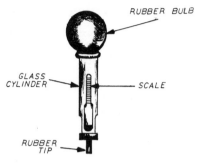

FIGURE 216   *HYDROMETER*

Hybrid circuit may also refer to a hybrid equivalent circuit in which two-wire and four-wire circuits are interconnected through a differential balance or bridge circuit in which the two sides of the four-wire circuit form conjugate arms.

Hydrometer: A device that measures the specific gravity of an electrolyte. In the syringe-type hydrometer (Fig. 216), part of the battery electrolyte is drawn up into a glass cylinder by means of a rubber bulb at the top. The hydrometer float consists of a small hollow glass tube weighted at one end and sealed at both ends. A scale calibrated in specific gravity is laid off axially along the body (stem) of the float.

The hydrometer float is inside the glass cylinder. The electrolyte to be tested is drawn up into the cylinder, thus immersing the hydrometer float in the solution. When the syringe is held in a vertical position, the hydrometer float will sink to a certain level in the electrolyte, depending on the specific gravity of the solution. The reading on the scale at the surface of the liquid is the specific gravity of the electrolyte in the syringe.

Hysteresis loop: A graphic loop that represents a power loss determined by the area within the loop. In ferromagnetic material of low retentivity, the magnetic domains are not completely elastic and will not realign to their initial positions after a magnetizing force is decreased or removed. The flux lag that occurs out of step with the increase or decrease of the magnetizing force is called *hysteresis.*

This characteristic can be graphically shown beginning at the point when the magnetizing force reaches the saturation level. When the magnetizing force is decreased to zero, the flux density fails to decrease accordingly due to the property of hysteresis. A coercive force, which is a magnetizing force with an opposite polarity to the magnetizing force, must be applied.

As this coercive force increases the magnetizing force in the opposite polarity, the flux density again increases toward saturation with a magnetization of polarity opposite to the original saturation. As the magnetizing force reaches zero, the flux density declines only to a certain point. To reduce the flux density to zero, it is necessary to reverse the magnetizing force again. This is done by passing a current through the coil in the original direction. The flux density falls to zero as the magnetizing force rises, only to rise again as the magnetizing force increases. This cycle constitutes the hysteresis loop.

I

IEEE: A professional organization that resulted from the merger of the IRE (Institute of Radio Engineers) and the AIEE (American Institute of Electrical Engineers). It is composed mainly of scientists and engineers whose purpose is the advancement of electrical engineering, electronics, and the related branches of engineering and science.

Ignitron: A cold-cathode-controlled, unidirectional device that has much greater power-handling capabilities than thyratrons or thyristors. An ignitron comprises a water-cooled envelope containing a pool of mercury that forms the cathode, a graphite anode, and a conical tip made of boron carbide extending slightly into the mercury pool. In spite of its high efficiency, the high power losses induce considerable heat dissipation. Since a pair of ignitrons can handle up to 10,000 A, the cathode and anode connections are obviously very sturdy.

Like a thermionic tube, an ignitron can conduct only if its anode is positive with respect to its cathode. During its conductive half-cycle, it can be fired by connecting the ignitor to a positive voltage point, such as the anode. Since there is considerable resistance between the boron carbide tip and the nonwetting mercury pool, a very tiny arc builds up between them, releasing a sufficient quantity of electrons from the mercury to immediately ionize the cathode-anode space and start tube conduction.

The discharge stops at the end of the half-cycle and is started again by the ignitor at the next positive half-cycle. For a very short time (microseconds), the ignitor passes a rather heavy current (about 40 A), but the anode then takes over and the ignitor current becomes negligible. Just as in a thyratron or thyristor, the discharge-initiating electrode (ignitor) loses control as soon as the device is conducting.

Illumination: Intensity, in candelas, divided by the square of the distance of the source from the surface, or

$$E = \frac{I}{D}$$

Illumination is measured in footcandles. As the area covered by a given solid angle gets larger with distance from the source, the included light flux remains the same. The illumination, therefore, decreases as the square of the area. This equation is true only if the receiving surface is perpendicular to the source direction. If light is incident at some other angle, the equation is

$$E - \frac{I}{D} \cos \varnothing$$

where E is illumination in footcandles (fc), I is intensity in candelas (cd), D is distance in feet, and $\varnothing$ is the angle of incidence.

Image converter: An optoelectronic device that is capable of changing the characteristics of an input image and relating it in another spectral angle. The image converter does this by responding to one form of light energy and producing an output that is the equivalent of the original at a different light frequency. An example of this is an infrared detector, which is used to display the insulation characteristics of a dwelling visually. The energy given off by escaping heat is displayed on a screen in such a manner that the heat image can be seen by the human eye.

An image converter is actually a form of electronic transducer that accepts energy from one system and transfers it to another. Often, the output image is transmitted so that it may be visually inspected.

Image frequency: An interfering transmitted signal, the frequency of which always differs from the desired station frequency by twice the intermediate frequency (IF). In other words, image frequency = station frequency + (2 × intermediate frequency). The plus sign of the formula is used if the local oscillator frequency tracks (operates) above the station frequency. The oscillator tracking below the station frequency is generally used for the higher frequency bands, and the oscillator tracking above the station frequency is generally used for the lower frequency bands, such as the broadcast band.

Image impedance: For a network that connects a generator to a load, the total impedance of the generator and matching network, which is the same as the characteristic impedance of the generator. With respect to the generator, the image impedance is the total impedance of the matching network and load, which is the same as the characteristic impedance of the load. For example, with the proper load impedance attached to the network at one end, the generator sees an image impedance that is equal to the generator impedance. With the proper generator impedance attached, the load sees an image impedance that is equal to the load impedance.

Image intensifier: A system whereby the sensor response to a radiation pattern or image is increased by interposing active elements between the sensor and the image after power is supplied to the active element. This can be accomplished by focusing the scene to be imaged on the photocathode of the tube, which produces a photoelectron pattern corresponding to the optical image. This pattern is then accelerated and focused onto a material (usually phosphor) that emits light to reproduce a visual image of the object or scene.

Image orthicon tube: A sensitive television or camera tube that uses low-velocity electrons in scanning, enabling it to pick up scenes under all lighting conditions or by infrared radiations. In this device, a photoemitting surface produces an electron image and focuses it on one side of a separate storage target. Low-velocity electrons are then used to scan the opposite side of the target to produce the output.

Impedance: The total opposition to the flow of current in a circuit. Impedance, like other oppositions to the flow of current, is expressed in ohms. When the resistance and inductive reactance are known in a circuit, the impedance may be found by the equation

$$Z = \sqrt{R^2 + X_L{}^2}$$

In ac circuits, impedance is used in Ohm's law just like resistance

$$E = IZ$$

$$I = \frac{E}{Z}$$

$$Z = \frac{E}{I}$$

In circuits with both inductive reactance and capacitive reactance, the equation for finding impedance is

$$Z = \sqrt{R^2 + (X_L - X_C)}$$

where Z equals impedance, R equals resistance, X_L equals inductive reactance, and X_C equals capacitive reactance.

Impulse: An electrical current that begins and ends within a very short time and is often regarded mathematically as infinitesimal. The result it produces in the medium is generally of a finite value. An impulse may occur as an abrupt change in voltage, either positive or negative, which conveys information to a circuit.

Impulse generator: An electrical device that produces high-voltage surges for testing insulators and for other purposes. It may also refer to any electronic device that is capable of producing a broad energy spectrum by means of a very narrow impulse. This spectrum is usually generated by the discharge of a short coaxial or waveguide transmission line with the pulses being discrete and regularly spaced.

These pulses usually have a repetition rate from a few pulses per second to a few thousand pulses per second. The output of such a generator can be described or calculated as the r.m.s. equivalent of the peak voltage in decibels above 1 μV/MHz.

Incandescent lamp: An electric light source that emits a yellow-red to infrared radiation. In incandescent lamps, a flow of current heats a metal filament. Because of the use of filaments, incandescent lamps are sometimes called *filament lamps* or *bulbs.* An incandescent lamp emits light by virtue of a filament that is heated to incandescence by an electric current passing through it. The filament is placed inside a bulb from which every particle of air has been exhausted. Otherwise, the filament would burn out almost instantly. As the lamp is burned, the filament slowly evaporates and becomes thinner and thinner, until it finally breaks or burns out.

Inductance: The property of a circuit to oppose any change in the current, regardless of the means used to produce the change. It is measured in units called the *henry.*

Induction heating: The heating of a nominally conducting material due to its own I^2R losses when the material is placed in a varying electromagnetic field.

Induction motor: A motor commonly consisting of a primary to which alternating current is applied and a secondary to which no electric current is supplied from an outside source. The primary is usually stationary and is called the *stator.* The secondary is usually the rotating part and is known as the *rotor.* The inductor motor stator, or *primary,* has a laminated iron core with slots that contain coils. The frame of the stator holds a connection box for connecting the terminals to the line.

The thin core laminations are held in place by dovetails that slip into grooves machined in the inner surface of the stator frame. The laminations are pressed tight and held by a flange keyed in the stator frame. Outside space blocks or fingers, placed between the iron core and flange, are used in the larger motors. They prevent the laminations from flaring out at the ends. Stators of large motors, such as the one shown in Fig. 217, have spacers inserted during the stacking of the laminations in order to provide ventilating ducts.

Inductive divider: A circuit composed of two or more inductors and has a single input and two or more outputs. A good example of an inductive divider is a common power transformer with a multiple secondary winding. Shown schematically in Fig. 218, an electromotive force is applied at the input or primary winding, and induced electromotive forces appear at the secondary outputs. The input has been divided into two outputs in this case.

Inductive dividers are also used in RF circuits to receive the output from a transmitter, dividing it into two outputs for feed to separate antennas. Inductive dividers, when used in reverse, are called *inductive combiners.* These are frequently seen in transistorized RF power amplifiers, as shown in Fig. 219. Here, the outputs from four separate amplifier modules are fed to the combiner circuit, which has a multiinductor input and a single output. The outputs from the four amplifier modules are combined into a single output, which is passed on to the tuning network/antenna.

Inductive dividers may consist of air-wound coils or inductors having iron

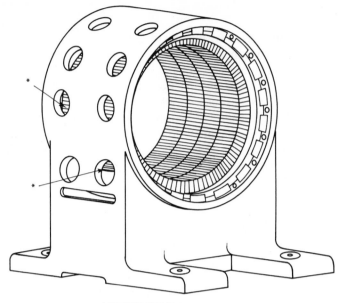

* VENTILATING DUCTS

LARGE STATOR CORE WITH SPACERS

FIGURE 217   Alternating current motors.

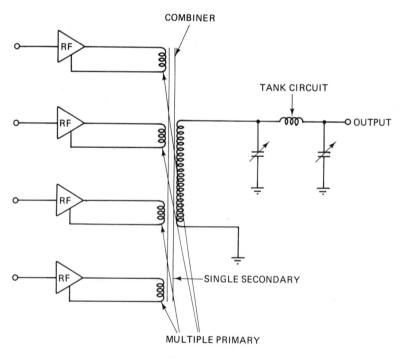

FIGURE 218

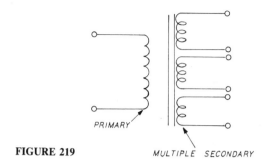

FIGURE 219

PRIMARY

MULTIPLE SECONDARY

or ferrite cores. Their construction will depend on their intended frequency of operation, with iron being used at audio frequencies and ferrite at radio frequencies. Air-wound coils are generally used only for low-frequency ac power applications.

Inductive reactance: The opposition that inductance presents to the flow of alternating current. It is designated by the sumbol X_L. In any circuit, the amount of inductive reactance depends on the value of inductance and the frequency of the alternating current. Thus, the value of inductive reactance increases with an increase in inductance and an increase in the frequency of current alternations. The value of inductive reactance in ohms may be found from the formula

$$X_L = 2\pi\, fL$$

where π equals 3.14, f equals the frequency of current in cps, and L equals the value of inductance in henrys.

In a circuit containing only inductance (resistance is negligible), the only opposition to current flow is the EMF of self-induction. As a result, the voltage applied to the circuit will be equal and opposite to the EMF of self-induction. The value of self-induced EMF will be maximum when change in current (and hence, change in flux) is maximum. This will be when the current wave is going through its zero point. The value of self-induced EMF will be zero when the rate of current (and again, flux) change is zero. This latter condition occurs at the maximum value points in the current wave.

From all of these conditions, it is found that the current lags the applied voltage by 90° in a purely inductive circuit. If there is any resistance in a circuit containing inductance (usually the case), the angle of current lag will be less than 90°, depending on the amount of resistance and inductance.

Inductor: In its simplest form, a coil of wire wound according to a prearranged design, with or without a core of magnetic metal, to concentrate the magnetic field in order to exhibit a higher self-inductance than a straight wire. Figure 220 shows examples of two types of inductors and their schematic symbols. The air-core-type inductor is most frequently used in circuits that are above the audio range. The iron-core type is widely used in the audio range

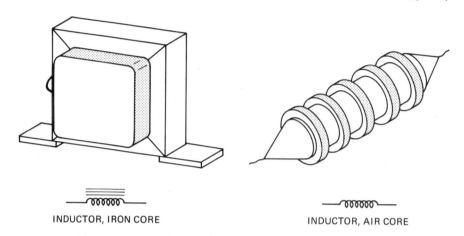

INDUCTOR, IRON CORE   INDUCTOR, AIR CORE

FIGURE 220   Inductor types and schematic symbols.

(below 20 KC). The iron-core type is usually made of laminated sheets of iron to reduce core losses.

Infrared heating lamp: A lamp in which the heat in an incandescent lamp is used as a main product. Infrared lamps use reflecting materials such as electrolytic gold, silver, Alzak aluminum, and copper to guide the heat in the desired direction. Infrared lamps are very efficient in producing heat, but very inefficient as producers of light, because the light they give off is a by-product; in many cases, this light is undesirable. In fact, one type of infrared reflector lamp has a red-lacquer face to absorb most of the visible light. Although ordinary incandescent lamps could be used for heating, infrared lamps are more practical because, since the light is not necessary, the infrared lamps are operated at lower temperatures and thus have a longer life. The rated life of infrared lamps is 5000 h or more.

An advantage of infrared heating is its speed, because the radiation is immediately absorbed and converted into heat at the areas where it is required. In most instances, drying time is reduced to a few minutes. High product temperatures are quickly obtainable. The simplicity and flexibility of infrared lamps adapt them to conveyor production lines. Ovens and lamp banks are easily assembled from lamp-socket and reflector sections made by the infrared equipment manufacturers. The assembly is usually in the form of complete ovens engineered to the desired specific heating cycle.

Insertion loss: The ratio of power received at a load before and after the insertion of an apparatus in series with the load and its power source. Insertion losses are created by impedance mismatch, increased resistance, and reactive components between devices or lines. For example, if a coaxial transmission line is placed between the output of a transmitter and the input to its antenna, a certain loss will be incurred. If the length of this transmission line is doubled, the insertion loss is doubled.

The insertion loss of the short line would be the difference between the power output of the transmitter and the power delivered to the antenna terminals. The same would apply to the longer line, although its loss could be compared by measuring the power delivered to the antenna with the power delivered to the same point by the shorter line. This latter insertion loss measurement would state the difference in efficiency between the long line and the short one.

Insulated conductor: A conductor encased within material of composition and thickness that is recognized by the National Electrical Code as electrical insulation.

Insulation resistance test: A test that is conducted by applying a usually high potential between the winding of a machine and its frame or between the active element of a circuit and ground. For example, tests on the insulation resistance of a motor may be made by applying 500-V dc between the winding of the machine and the frame. The current that this pressure forces through or over the insulation to the frame is measured by a sensitive instrument, the scale of which is usually calibrated to read in megohms. The 500-V dc may be developed by a hand-operated generator, as in the megger, or it may be supplied from an ac source by a rectifier-filter combination, as shown in Fig. 221.

Insulation resistance tests are necessary, since the quality of insulating materials used on any electrical machine deteriorates with age due to the action

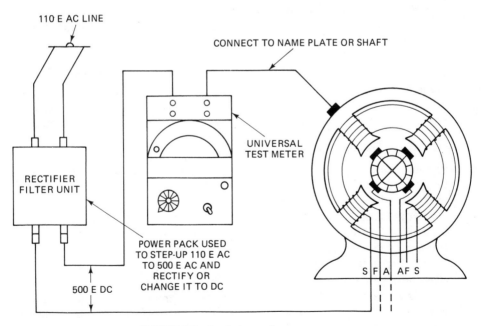

FIGURE 221 Insulation resistance test.

of moisture, dirt, oil, acids, etc. It is necessary to test the electrical resistance of the insulation periodically, so that weaknesses may be detected and corrected before they result in complete failure.

The readings obtained on any given machine will vary greatly with the temperature of the insulation, a 10°C rise in temperature reducing the insulation resistance by as much as 50 percent. The dampness of the location and the amount of oil, dust, and dirt on the winding will also materially affect the readings. Wherever possible, the test should be made when the insulation is at the maximum operating temperature. The minimum safe insulation resistance at maximum operating frequency should not be lower than 1 MΩ for equipment having a voltage rating below 1000 V.

Insulator material: Substances in which it is very difficult to cause a flow of current with any electrical force that may be applied. If two copper wires or other conductors should touch each other while carrying electric currents, electricity from one conductor would pass into the other and escape from the path it follows through the first conductor. To prevent the escape of electric current from conductors into other conductors, all current-carrying conductors should be surrounded and isolated or supported by materials that are not conductors. Any material that is not a conductor is called a *nonconductor* or an *insulator*. Among the insulators that are most useful in electrical work are porcelain, glass, mica, hard rubber, soft rubber, paper, Bakelite and similar compounds, cotton, linen, silk, various oils and waxes, and air.

Integrated circuit: A chip made up of a number of components contained in a single package. One semiconductor chip can contain two or more transistors, several resistors and capacitors, and many individual diodes or other components. Figure 222 shows a pictorial and schematic of a simple integrated circuit.

A single integrated circuit may be used to replace many discrete components. In addition to the space-saving features of ICs, the fact that they receive almost identical processing enables them to be closely matched in characteristics. The closer such circuit elements are matched, the greater the reliability of the circuit.

Intensity: Sometimes called *candlepower,* the amount of light in a unit of solid angle, assuming a point source of light. The candela is the unit of intensity (I) and may be compared to pressure in a hydraulic system. Intensity in a given direction is constant regardless of distance.

Interchange, inadvertent: The act of directly substituting one component for another component of the same kind, such as substituting one transistor for another, one capacitor for another, etc.

Interlock: To arrange the control of machines or devices so that their operation is interdependent in order to assure their proper coordination.

Inverted-L antenna: A grounded antenna so constructed that a portion of it is mounted horizontally and takes the form of an inverted L. In this type of antenna, a fairly long horizontal portion, or flattop, is used. The vertical

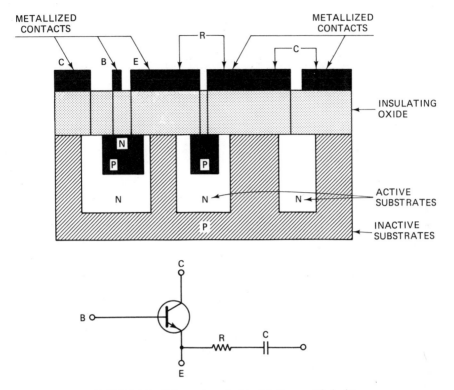

FIGURE 222 Schematic and pictorial diagram IC device.

down lead, which forms an important part of the radiating system, is connected to one end of the flattop. The length of the antenna is measured from the far end of the flattop to the point at which the down lead is connected to the transmitter.

Figure 223a shows an inverted-L, single-wire vertical antenna. The length of the straight vertical portion is one quarter-wavelength; the length of the horizontal portion is also a quarter-wavelength. Thus, the current loop is at the topmost part of the straight vertical portion, with resultant increased radiation efficiency. However, considerable energy angles in a direction opposite the free end due to current flow in the horizontal section.

Figure 223b shows an inverted-L antenna with a multiple-wire flattop arrangement. The efficiency of this inverted L is somewhat better than that of the single-wire type. Although it has good ground wave propagation characteristics, it also radiates considerable energy at a high radiation angle. Since a high-voltage loop exists at the input end, a parallel-tuned coupling arrangement is suitable.

Figure 223c shows the method for obtaining some cancellation of the fields in the vertical plane due to the out-of-phase currents in the flattop conductors. Figures 223d and e show the arrangement of horizontal conductors

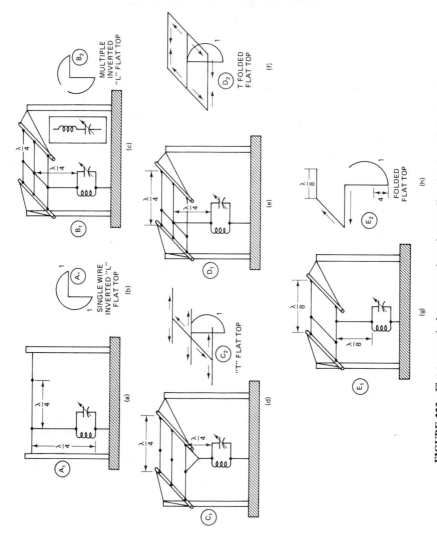

FIGURE 223 Flat-top vertical antennas show the coupling circuits and the current distribution.

to effect a maximum cancellation of the fields due to current flow in the flattop. The folded top arrangement offers the best possibilities for minimizing high-angle radiation, especially in the low-frequency operating range.

Inverter: A circuit that takes in a positive pulse and puts out a negative one, or takes in a negative pulse and puts out a positive one. The physical meaning of positive and negative depends on the specific circuit and the conventions established for it.

Ion: An atom that has become electrically unbalanced by the loss or gain of one or more electrons. An atom that has lost an electron is called a *positive ion,* while an atom that has gained an electron is known as a *negative ion.*

In other words, when an atom loses an electron, its remaining orbital electrons no longer balance the positive charge of the nucleus and the atom acquires a charge of $+1$. Similarly, when an atom gains an electron, it acquires an excess negative charge of -1. The process of producing ions is called *ionization.*

Ionosphere: That portion of the rarefied atmosphere located approximately 40 to 350 mi above the earth. The ionosphere differs from the atmosphere, in that it contains a much higher number of positive and negative ions. The negative ions are believed to be free electrons. The ions are produced by the ultraviolet and particle radiations from the sun.

The rotation of the earth on its axis, the annual course of the earth around the sun, and the development of sunspots all affect the number of ions present in the ionosphere, and these in turn affect the quality and distance of radio transmissions. The ionosphere is constantly changing. Some of the ions are recombining to form neutral atoms, while other atoms are being ionized by the removal of electrons from their outer orbits. The rate of formation and recombination of ions depends on the amount of air present and the strength of radiation from the sun.

At altitudes above 350 mi, the particles of air are too sparse to permit large-scale ion formation. Below about 40-mi altitude, only a few ions are present because the rate or recombination is so high. Ultraviolet radiations from the sun are absorbed in passage through the upper layers of the ionosphere, so that below an elevation of 40 mi too few ions exist to affect materially sky wave communications.

Densities of ionization at different heights make the ionosphere appear to have layers. Actually, there is thought to be no sharp dividing line between layers, but for the purpose of discussion, a sharp demarcation is indicated. These layers are called the *D layer,* the *E layer,* and the *F layer.*

Iron-vane meter: An ac meter whose movable element, a soft-iron vane, carries the pointer and pivots near a similar stationary vane. The vanes are mounted in a multi-turn coil of wire. The current to be measured flows through the coil, with the resulting magnetic field magnetizing the vanes. Because the magnetic poles of the vanes are identical, they repel each other. The movable vane is deflected against the torque of returning springs over an arc proportional to the scale. Although the iron-vane meter is a current meter, it may be converted into a voltmeter by connecting a suitable multiplier resistor in series with the coil.

Jar: The name given to certain types of battery containers. For example, nickel-iron batteries (alkaline) contain a jar that is made of nickel-plated sheet steel. Since the jar on this type of battery is an electrical conductor, it must be insulated internally from the element and also from the battery posts where they pass through the top of the container. The insulation from the element is accomplished by hard rubber or plastic sheeting, and hard rubber bushings encircle the posts.

JFET: A bar of doped silicon, called a *channel,* that behaves as a resistor. The doping may be N- or P-type, creating either an N-channel or P-channel JFET. There is a terminal at each end of the channel. One terminal is called the *source,* and the other is called the *drain.* Current flow between source and drain for a given drain source voltage is dependent on the resistance of the channel. This resistance is controlled by a *gate.* The gate consists of P-type regions diffused into an N-type channel or N-type regions diffused into a P-type channel.

As with any PN junction, a depletion region or an electric field surrounds the junction when reverse biased. As the reverse voltage is increased, the electric fields spread into the channel until they meet, creating an almost infinite resistance between source and drain.

J-K flip-flop: A configuration readily adapted to integrated circuit applications that permits set/reset, toggle, and gated functions to be performed utilizing a single basic circuit. Figure 224 is the presently accepted logic symbol for such a flip-flop. As indicated by the diagram, the *J-K* flip-flop will usually have a minimum of five inputs and two outputs. The inputs are *S,* the preset connection; *R,* the reset connection; *J,* the set input; *K,* the clear input; and *C,*

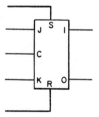

FIGURE 224 Logic symbol for J-K flip-flop.

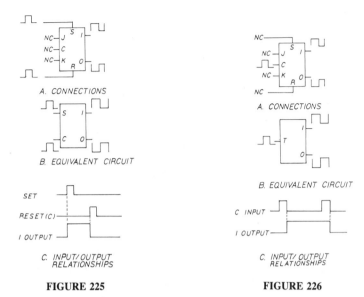

A. CONNECTIONS

B. EQUIVALENT CIRCUIT

SET

RESET (C)

I OUTPUT

C. INPUT/OUTPUT RELATIONSHIPS

FIGURE 225

A. CONNECTIONS

B. EQUIVALENT CIRCUIT

C INPUT

I OUTPUT

C. INPUT/OUTPUT RELATIONSHIPS

FIGURE 226

the clock or toggle input. The outputs are L, the set output; and O, the clear output.

If the J-K flip-flop is to perform the set/reset function, the S and R connections will be utilized, as indicated in Fig. 225. For the toggle function, the C input will be utilized, as in Fig. 226. For the gated function, the J, K, and C inputs will be utilized, as in Fig. 227. The C input is utilized in this case as a clock input. Depending on its design, a J-K flip-flop will change states on either the leading or trailing edge of the clock or toggle pulse.

Joule: A unit of electrical measurement representing the work done by a difference of potential of 1 V (E) while moving 1 C (ϕ) of charge (6.28×10^{18} electrons). The equation can be stated as follows

$$J = E \times Q$$

Joule may also refer to a measurement of work that is done by a force of 1 N acting through a distance of 1 m.

Joule heat: Also called *joule effect*, the thermal effect that results when electrical current flows through a resistance. It is measured in watts. When a current of 1 A flows through a resistance of 1 Ω, the joule heat given off is equivalent to 1 W. This is explained by Ohm's law for power, which reads

$$P - I^2R$$

where P is power in watts, I is current in amperes, and R is resistance in ohms.

Using this formula, when 2 A of current flow through a resistance of 1 Ω, the total power dissipation is 4 W.

Joule's law: The heat generated in a conductor by an electric current is proportional to the resistance of the conductor, the time during which the

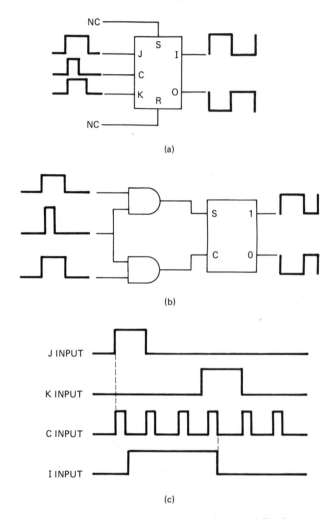

FIGURE 227   J-K flip-flop connected as gated flip-flop.

current flows, and the square of the strength of the current. The quantity of heat in calories may be calculated by the use of the equation

$$\text{Calories per second} = \text{Volts} \times \text{Ampere} \times 0.24$$

The total number of calories or heat developed in seconds will be given by

$$\text{Heat} = \text{Volts} \times \text{Amperes} \times \text{Seconds} \times 0.24$$

Joystick: A manual input control device for graphic display consoles that allows the user to control the coordinates of a point of light on the screen. By driving the cursor around the screen, freehand drawings may be made in electronic graphic form. Graphics computers and many video games incorporate joysticks. For example, in a TV tennis game, each opponent controls a joystick, which, in turn, aligns the electronic cursors at different coordinates in order to repel the automatically controlled cursor, which corresponds to the tennis ball. The joystick cursors repel the ball.

Joysticks normally consist of a small lever that moves vertically and horizontally (and sometimes diagonally). The cursor on the visual display screen moves in accordance with the physical position of the joystick.

Junction diode: A semiconductor device created by joining an N-type region and P-type region of a crystalline material, such as germanium or silicon. The junction diode has four important ratings that must be taken into consideration. These are the maximum average forward current, maximum repetitive reverse voltage, maximum surge current, and maximum repetitive forward current. These ratings are important when it becomes necessary to troubleshoot a circuit or to select junction diodes for replacement when the desired one is not readily available.

The maximum forward current is the maximum amount of average current that can be permitted to flow in the forward direction. This rating is usually given for a specified ambient temperature and should not be exceeded for any length of time, as damage to the diode will occur. The maximum repetitive reverse voltage is that value of reverse bias voltage that can be applied to the diode without causing it to break down.

The maximum surge current is that amount of current allowed to flow in the forward direction in nonrepetitive pulses. Current should not be allowed to exceed this value at any time and should only equal this value for a period not to exceed one cycle of the input. The maximum repetitive forward current is the maximum value of current that may flow in the forward direction in repetitive pulses.

All of the ratings mentioned are subject to change with temperature variations. If the temperature increases, the ratings given on the specification sheet should all be lowered, or damage to the diode will result.

Karnaugh map: A logic chart showing switching-function relationships in digital computers. It is used in computer logic analysis to determine speedily the simplest form of logic circuit to use for a given function. The Karnaugh map is sometimes regarded as a tabular form of the more conventional Venn diagram.

Kelvin double bridge: A special bridge that is used for measuring very low resistance (0.1 or less). The arrangement of the bridge reduces the effects

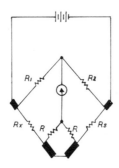

KELVIN DOUBLE BRIDGE   **FIGURE 228**

of contact resistance, which causes significant error when such low resistances are connected to conventional resistance bridges. See Fig. 228.

Kelvin's temperature: Any temperature in the absolute scale. This is a temperature scale with its zero point at -273.1°C, or absolute zero. The unit of thermodynamic temperature is the *kelvin,* and its symbol is K.

Keying, transmitter: A means of causing an RF signal to be radiated only when the key contacts are closed. When the key is open, the transmitter does not radiate energy. Keying is accomplished in either the oscillator or amplifier stages of a transmitter. A number of different keying systems may be used. In some transmitters, the hand telegraph key is at low potential with respect to ground. The keying bar is usually grounded to protect the operator. Generally, a keying relay with its contacts in the center tap lead of the filament transformer is used to key the equipment. Because one or more stages use the same filament transformer, these stages are also keyed. The class-C final amplifier, when operated with fixed bias, is usually not keyed, because with no excitation applied, no current flows. Hence, keying the final amplifier along with the other stages is not necessary.

Two methods of oscillator keying are shown in Fig. 229. In part a, the grid circuit is closed at all times, and the key opens and closes the negative side of the plate circuit. This system is called *plate keying.* When the key is open, no plate current can flow and the circuit does not oscillate. In part b, the cathode circuit is open when the key is open, and neither grid current nor plate current can flow. Both circuits are closed when the key is closed. This system is called *cathode keying.* Although both circuits may be used to key amplifiers, other keying methods are generally employed because of the larger values of plate current and voltage encountered.

Two methods of blocked grid keying are shown in Fig. 230. The key in part a shorts cathode resistor R_1, allowing normal plate current to flow. With the key open, reduced plate current flows up through resistor R_1, making the end connected to grid resistor R_g negative. If R_1 has a high enough value, the bias developed is sufficient to cause cutoff of plate current. Depressing the key short circuits R_1, thus increasing the bias above cutoff and allowing the normal

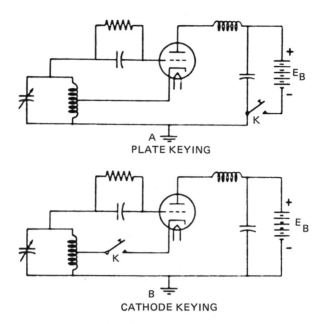

A
PLATE KEYING

B
CATHODE KEYING

FIGURE 229   Oscillator keying.

flow of plate current. Grid resistor R_g is the usual grid leak resistor for normal bias. This method of keying is applied to the buffer stage in a transmitter.

The blocked grid keying method shown in part b affords complete cutoff of plate current and is one of the best methods for keying amplifier stages in transmitters. In the voltage divider with the key open, two-thirds of 1000 V, or 667 V, is developed across the 200-kΩ resistor and one-third of 1000 V, or 333 V, is developed across the 100-kΩ resistor. The grid bias is the sum of -100 and -333 V, or -433 V. Because this is below cutoff, no plate current flows. The plate voltage is 667 V. With the key closed, the 100-kΩ resistor is shorted out, and the voltage across the 200-kΩ resistor is increased to 1000 V. Thus, the plate voltage becomes 1000 V at the same time the grid bias becomes -100 V. Grid bias is now above cutoff, and the amplifier triode conducts. Normal amplifier action follows.

Where greater frequency stability is required, the oscillator should remain in operation continuously while the transmitter is in use. This procedure keeps the oscillator tube at normal operating temperature and offers less chance for frequency variation to occur each time the key is closed. If the oscillator is to operate continuously and the keying is to be accomplished in an amplifier stage following the oscillator, the oscillator circuit must be carefully shielded to prevent radiation and interference to the operator, while receiving.

Keypunch: A special device to record information in cards or tape by punching holes in the cards or tape to represent letters, digits, and special characters.

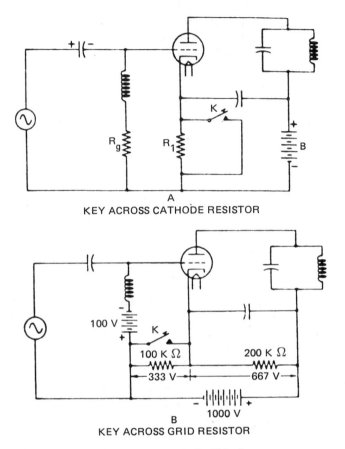

KEY ACROSS CATHODE RESISTOR

KEY ACROSS GRID RESISTOR

FIGURE 230 Blocked grid keying.

Kilo: A prefix representing 1000. A kilohertz is equal to 1000 Hz, while a kilowatt is the equivalent of 1000 W. Abbreviated *k*.

Kilohertz: A unit of frequency that is equivalent to 1000 Hz, or 1000 cycles/s. Abbreviated *kHz*.

Kinescope: Any device containing a cathode-ray tube used to form pictures or visual displays from an electronic signal. This is an old term that is rarely used in modern terminology.

Kinetic energy: Energy contained by an object due to its motion. Whenever work is accomplished on an object, energy is consumed (changed from one kind to another). If no energy is available, no work can be performed. Thus, energy is the ability to do work. One form of energy is that which is contained by an object in motion. In driving a nail into a block of wood, a hammer is set in motion in the direction of the nail. As the hammer strikes the nail, the energy, or motion of the hammer, is converted into work as the nail is driven into the wood. This energy is called *kinetic energy.*

Kirchhoff's current law: The law that states that the algebraic sum of the currents entering and leaving a junction of conductors is equal to zero. That is

$$I_1 + I_2 + I_3 + \ldots = 0$$

where I_1, I_2, I_3, etc., are the currents entering and leaving the junction. Currents entering the junction are assumed to be positive, while currents leaving the junction are negative. When solving a problem using this equation, the currents must be placed into the equation with the proper polarity signs attached.

Klystron amplifier: An amplifier that is used mainly as a power amplifier and that has application in many facets of microwave transmission. The basic theory of a klystron amplifier is quite simple. Figure 231 shows a cutaway representation of a basic klystron amplifier. It consists of three separate sections—the electron gun, the RF section, and the collector.

The electron gun structure consists of a heater, cathode, control grid, and anode. Electrons are emitted by the cathode and drawn toward the anode, which is operated at a positive potential with respect to the cathode. The electrons are formed into a narrow beam by either electrostatic or magnetic focusing techniques. The control grid is used to control the number of electrons that reach the anode region. It may also be used to turn the tube completely off or in certain pulsed amplifier applications.

The electron beam is well formed by the time it reaches the anode. The beam passes through a hole in the anode and onto the RF section of the tube,

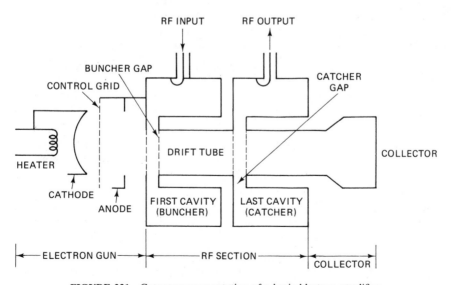

FIGURE 231 Cutaway representation of a basic klystron amplifier.

eventually striking the collector. The electrons are returned to the cathode through an external power supply.

Referring to the RF section of the klystron amplifier, the resonant circuits are reentrant cavities. Electrons pass through the cavity gaps in each of the resonators, as well as the cylindrical metal tube between the gap. These metal tubes are called *drift tubes*. The low-level RF input signal is coupled to the first resonator, which is called the *buncher cavity*. The signal may be coupled through either a waveguide or a coaxial connection. If the cavity is tuned to the frequency of the RF input, it will be excited into oscillation. An electric field will exist across the buncher gap, alternating at the input frequency. For half a cycle, the electric field will be in a direction that will cause the field to increase the velocity of electrons flowing through the gap. On the other half-cycle, the field will be in a direction that will cause the field to decrease electron velocity. This effect is called *velocity modulation* and is illustrated in Fig. 232. Note that when the voltage across the cavity gap is negative, electrons will decelerate; when the voltage is zero, the electrons will be unaffected; when the voltage is positive, the electrons will accelerate.

After leaving the buncher gap (Fig. 231), the electrons proceed through the drift tube region toward the collector. In the drift tube region, electrons that have been speeded up by the electric field in the buncher gap will tend to overtake electrons that have been slowed down. Due to this action, bunches of electrons will begin to form in the drift tube region and will be completely formed by the time they reach the gap of the last cavity. This last cavity is called the *catcher cavity*. Bunches of electrons periodically flow through the gap of this catcher cavity; and during the time between bunches, relatively few electrons flow through the gap. The time between arrival of electron bunches is equal to the period of one cycle of the RF input signal.

The initial bunch of electrons flowing through the catcher cavity will cause the cavity to oscillate at its resonant frequency. This sets up an alternating electric field across the catcher cavity gap, as illustrated in Fig. 233. With proper design and operating potentials, a bunch of electrons will arrive in the catcher cavity gap at the proper time to be retarded by the RF field. Thus, energy will be given up to the catcher cavity. This action is illustrated in Fig. 234.

The RF power in the catcher cavity will be much greater than that applied in the buncher cavity. This is due to the ability of the concentrated bunches of electrons to deliver great amounts of energy to the catcher cavity. Since the electron beam delivers some of its energy to the output cavity, it arrives at the collector with less total energy than it had when it passed through the input cavity. This difference in beam energy is approximately equal to the energy delivered to the output cavity.

It is appropriate to mention that velocity modulation does not form perfect bunches of electrons. There are some electrons that come through out of phase. These electrons show up in the output cavity gap between bunches.

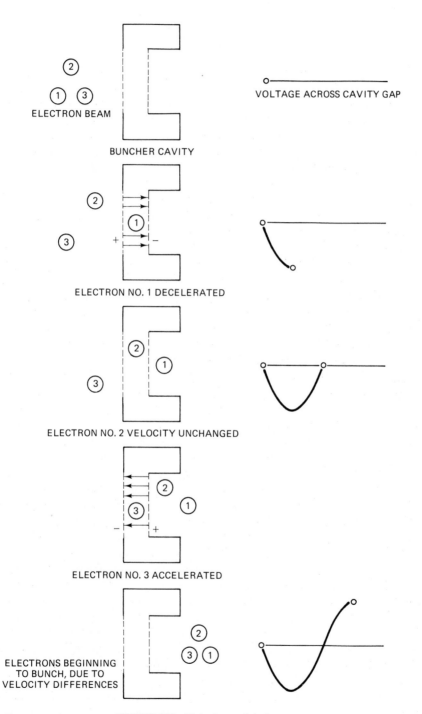

FIGURE 232 Velocity modulation.

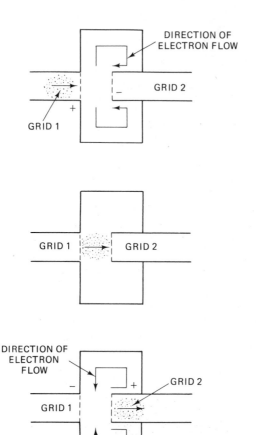

FIGURE 233 Catcher cavity gap alternating electric field.

The electric field across the gap, at the time these out-of-phase electrons come through, is in a direction to accelerate them. This causes some energy to be taken from the cavity. However, much more energy will be contributed to the output cavity by the concentrated bunches of electrons than will be withdrawn from it by the small number of out-of-phase electrons.

Klystron tube: A specially constructed electron tube using the properties of transit time and velocity modulation of the electron beam to produce microwave frequency operation. It can be used as an oscillator or an amplifier. The amplifier employs two or more cavities to produce the proper bunching of electrons, upon which its function and amplifying properties are based. The amplifier type of klystron can produce a large amount of power (up to megawatts) and can be used as an oscillator if proper feedback arrangements are made.

However, the reflect klystron offers a simpler type of feedback arrangement and performs specifically as a special tube designed for oscillator opera-

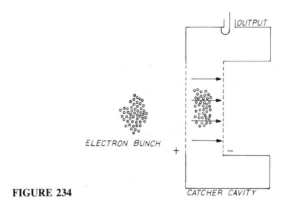

ELECTRON BUNCH

FIGURE 234

CATCHER CAVITY

tion alone. Although the power output of the reflex klystron is limited, it is adequate for receiving, test equipment functions, and for low-power transmitters.

Where high power is required, it can be achieved by using the reflex klystron as a master oscillator and the conventional amplifier-type klystron as a power amplifier. Since microwave radiation is limited to line-of-sight distances, the reflect klystron usually furnishes sufficient power for these relatively short RF transmission paths.

The operation of the klystron is based on the development of velocity modulation of the electron beam; that is, the velocity of the electron beam is controlled to produce a grouping or bunching of electrons. These bunches of electrons are then passed through grids or cavities to produce oscillations at the desired frequency by direct excitation of the cavities.

A basic klystron (not the reflex type) is shown in Fig. 235. The electrons from the cathode are attracted to the accelerator grid (1), which is at a positive potential with respect to the cathode. The accelerator grid may be a grid structure or an annular ring (cylinder or sleeve through which the electrons pass unhindered). Assume that this attraction produces a constant-velocity electron beam, which is further attracted to the next electrode, the buncher grids (or cavity), and then to the next electrode, the catcher grids (or cavity), also at a higher positive potential. If the output from the catcher is fed back to the buncher and the proper phase and energy relations are maintained between the buncher and the catcher, the tube will operate as an oscillator. The collector plate, which is also at a positive potential, serves only to collect the electrons that pass the catcher. Successful operation requires that the energy needed for bunching is less than that delivered to the catcher. Amplifying action is obtained because the electrons pass through the buncher in a continuous stream and are effectively grouped so that they pass through the catcher in definite bunches or groups.

Bunching is produced by applying an alternating voltage to the buncher grids (produced by excitation of the buncher resonator by the passing electron

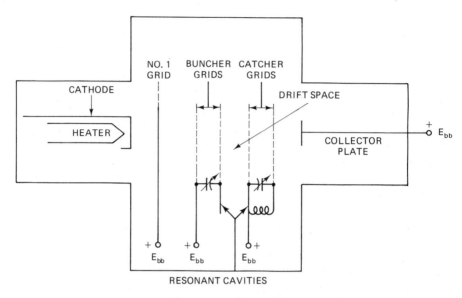

FIGURE 235 Basic klystron.

beam). Assuming that a sine-wave voltage is produced and applied between the buncher grids, it is evident that on the positive alternation, the buncher grid nearest the catcher effectively has its positive potential increased and therefore further accelerates the electron flow. On the negative alternation, the same buncher grid voltage is made less positive and the electron stream is slowed down. Since a continuous stream of electrons enters the bunching grids, the number of electrodes accelerated by the alternating field between the buncher grids on one half-cycle of operation is equaled almost exactly by the number of electrons decelerated on the negative half-cycle. Therefore, the net energy exchange between the electron stream and the buncher is zero over a complete cycle of alternation, except for the losses that occur in the tuned circuit (cavity) of the buncher.

After passing through the buncher grids, the electrons move through the drift space in the tube with velocities that have been determined by their speed through the buncher grids. Since in a conventional klystron, the drift space is free of any fields, at some point in this drift space, the electrons that were accelerated will catch up with those that were previously decelerated (in a prior passage) to form a bunch. The catcher grids are placed at this point of bunching (determined by frequency and transit time) to extract RF energy from the bunched electrons.

At the catcher, a different situation exists. Since the electrons are traveling in bunches, spaced so that they enter the catcher field only when the oscillating circuit is in its decelerating half-cycle, more energy is delivered to

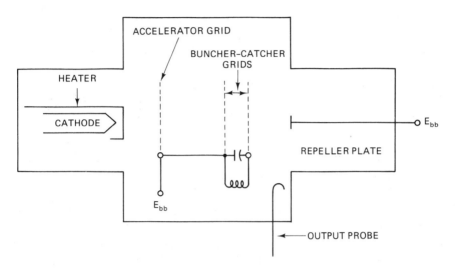

FIGURE 236   Reflex klystron.

the catcher than is taken from it. The remaining electrons in the beam pass through the grid and travel to the collector plate, where they are absorbed.

In the reflex klystron, the catcher grids are replaced by a repeller plate, to which a negative potential is applied, as shown in Fig. 236. In this type of klystron, the electron beam is also velocity-modulated, and by proper adjustment of the negative voltage on the repeller plate, the electrons that have passed the bunching field may be made to pass through the resonator again (in the reverse direction) at the proper time to deliver energy to this circuit. Thus, the feedback necessary to produce oscillation is obtained, and the tube construction is simplified. Spent electrons are removed from the tube by the positive accelerator grid (when used), or by the grids of the positive buncher cavity. Energy is coupled out of the cavity by means of a one-turn coupling loop. The operating frequency can be varied over a small range by changing the negative potential applied to the repeller, because this potential determined the transit time of the electrons between their first and second passages through the resonator. Since maximum output depends on the fact that the electrons must return through the resonator at just the time when they are bunched and at exactly the decelerating half-cycle of oscillating resonator grid voltage, the output is more dependent than the frequency upon the repeller potential. Therefore, the amount of tuning provided by varying the repeller voltage is limited.

Usually, the volume of the resonator cavity is changed (by mechanical tuning) to make a coarse adjustment of the oscillator frequency, and the repeller voltage is varied over a narrow range to make a fine adjustment of the frequency (electronic tuning) consistent with good output. As the same grids

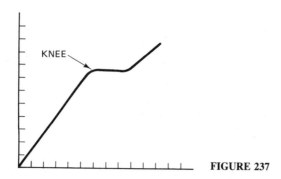

FIGURE 237

perform the dual function of bunching and catching in the reflect klystron, they are frequently referred to as the *buncher-catcher grids*. Because of the variation of frequency with accelerating voltage, it is difficult to achieve linear amplitude modulation with a klystron. Frequency modulation may be readily accomplished, however, by introducing a small modulating voltage at the cathode or repeller. The tuned circuit used in the reflect klystron is a cavity resonator which has a very high Q. Depending on the tube type, the cavity may be an integral cavity (built into the tube) or an external cavity (clamped around the tube).

Several methods are used to tune cavity resonators. Capacitive tuning is provided by mechanically varying the grid gap spacing. Inductive tuning is provided by moving screw plugs in or out of the cavity. This changes the volume, making it either smaller and tuned to a higher frequency, or larger and tuned to a lower frequency. In some instances, thermal tuning is used by applying heat to the grid gap to produce capacitive variations. Several types of output couplings are used, but the coupling loop is the most popular type.

Knee: The bend in a response curve that is most often an indication of the onset of saturation or cutoff. The point on the curve that represents the knee marks an abrupt change, as shown in Fig. 237.

Labyrinth loudspeaker: A loudspeaker whose enclosure (wooden cubicle) includes a folded pipe or acoustic transmission line behind the speaker. The inner walls are lined with a sound-absorbent material. When the pipe, which is open ended, is half as long as the wavelength of the frequency being reproduced, the sound emerging from the open end is in phase with that radiated by the front of the speaker and therefore reinforces it. Because there is no sudden change in pressure as the sound leaves the pipe, the pipe produces no antiresonance.

Language, problem-oriented: A language designed for convenience of program specification in a general problem area rather than for easy conversion to machine instruction code. The components of such a language may bear little resemblance to machine instructions. It is a machine-independent language in which one needs only to state the problem, not the how of the solution.

L antenna: An antenna composed of a horizontal radiator and a vertical feeder or lead-in attached to one end of the element, resembling the letter L. L antennas are often single-wire designs that are bent at a 90° angle near the center in order to be accommodated in a limited mounting space. These antennas may be a half-wave or more in length, but are often cut to a quarter-wavelength and tuned against ground. L antennas are used for RF communications purposes, in both receive and transmit, and are quite popular for military shipboard use. They are also seen at amateur radio stations and in modified form at commercial broadcast installations.

Lap winding: A parallel winding that has an even number of segments and slots. The number of bars, however, is equal to or a multiple of the number of slots. It is known as a parallel winding because there are as many circuits in parallel as there are poles. Thus, a four-pole motor has four circuits in parallel, and a six-pole motor has six circuits in parallel through the armature. This requires that there be as many brushes as there are poles for the motor.

Since there are so many paths in parallel, a larger current can pass through the winding because it has several paths to travel. This, in turn, reduces the amount of voltage needed in the armature to push the current through so many paths.

The lead connections for a lap winding come from each side of the coil as top and bottom leads and connect adjacent to each other on the commutator. The position that they connect on the commutator cannot be set down to a hard-and-fast theoretical rule, but is rather at the discretion of the manufacturer who designs the motor.

Laser: An optoelectronic circuit that generates intense coherent light. The term *laser* is an abbreviation for *light amplification by simulated emission of radiation.* In recent years, lasers have been used for industrial applications involving the fine cutting of metals and diamonds. The intense concentration of light energy can be electrically controlled. For this reason, it is used in some types of delicate eye surgery.

In communications, lasers may be amplitude-modulated and used to carry speech information that is received by a light-beam detector. When coupled with sophisticated processing equipment, laser beams may be used to provide highly accurate remote measuring of distant objects, such as the surface of the moon.

Laser diode: A semiconductor diode that emits coherent light when a voltage is applied to its terminals. It is usually constructed of gallium arsenide. The laser diode is the prime component of the gallium-arsenide injection laser that uses electric energy directly, pushing electrons to high energy states by injecting them across a PN junction. This type of laser is of special interest to home experimenters, because it is readily available and quite inexpensive, as compared to other types of lasers.

LCD: An acronym for *liquid crystal display,* a display technique that uses segments of a liquid crystal solution in a sandwich of glass plates. The light-reflecting properties of the solution are controlled by an electric field. See *liquid crystal display.*

Lead-acid battery: The most widely used type of storage battery, which has an EMF of 2.2 V per cell. In its charged condition, the active materials in the lead-acid battery are lead dioxide (sometimes referred to as lead peroxide) and spongy lead. The lead dioxide is used as the positive plate, while the spongy lead forms the negative plate.

The electrolyte is a mixture of sulfuric acid and water. The strength (acidity) of the electrolyte is measured in terms of its specific gravity. Specific gravity is the ratio of the weight of a given volume of electrolyte to an equal volume of pure water. Concentrated sulfuric acid has a specific gravity of about 1.830. Pure water has a specific gravity of 1.000. The acid and water are mixed in a proportion to give the specific gravity desired. For example, an electrolyte with a specific gravity of 1.210 requires roughly one part of concentrated acid to four parts of water. As a storage battery discharges, the sulfuric acid is depleted and the electrolyte is gradually converted into water. This action provides a guide in determining the state of discharge of the lead-acid cell. The electrolyte that is usually placed in a lead-acid battery has a specific gravity of 1.350 or less. Generally, the specific gravity of the electrolyte in portable batteries is adjusted between 1.210 and 1.220.

In a fully charged battery, the positive plates are pure lead dioxide and the negative plates are pure lead. Also, all of the acid is in the electrolyte, so the specific gravity is at its maximum value. The active materials of both the positive and negative plates are porous and have absorptive qualities similar to a sponge.

The pores of the plates are filled with the battery solution (electrolyte) in which they are immersed. As the battery discharges, the acid in contact with the plates separates from the electrolyte. It forms a chemical combination with the plate's active material, changing it to lead sulfate. Thus, as the discharge continues, lead sulfate forms on the plates and more acid is taken from the electrolyte. The water content of the electrolyte becomes progressively higher; that is, the ratio of water to acid increases. As a result, the specific gravity of the electrolyte will gradually decrease during discharge.

When the battery is being charged, the reverse action takes place. The acid held in the sulfated plate material is driven back into the electrolyte. When fully charged, the material of the positive plates is again pure lead dioxide, and that of the negative plates is pure lead.

Electrical energy is derived from a cell when the plates react with the electrolyte. As a molecule of sulfuric acid separates, part of it combines with the spongy lead plates. This makes the spongy lead plates negative and forms lead sulfate at the same time. The remainder of the sulfuric acid molecule, lacking electrons, has thus become a positive ion. The positive ions move through the electrolyte to the opposite (lead dioxide) plates and take electrons from them. This action makes the lead dioxide plates positive and neutralizes the positive sulfuric acid ions. Lead sulfate and water are formed in the process.

In the charged condition, the positive plate contains lead dioxide, the negative plate is composed of spongy lead, and the solution contains sulfuric acid. In the discharged condition, both plates contain lead sulfate and the solution contains water. As the discharge progresses, the acid content of the electrolyte decreases because it is used in forming lead sulfate, and the specific gravity of the electrolyte decreases.

A point is reached where so much of the active material has been converted into lead sulfate that the cell can no longer produce sufficient current to be of practical value. At this point, the cell is said to be discharged. Since the amount of sulfuric acid combining with the plates at any time during discharge is in direct proportion to the ampere-hours (product of current in amperes and time in hours) of discharge, the specific gravity of the electrolyte is used as a guide in determining the state of discharge of the lead-acid cell.

If the discharged cell is properly connected to a dc source, the voltage of which is slightly higher than that of the cell, current will flow through the cell in the direction opposite of that of discharge. The cell is then said to be *charging*. The effect of the current will be to change the lead sulfate on both the positive and negative plates back to its original form of lead dioxide and spongy lead, respectively. At the same time, the sulfate is restored to the electrolyte, with the result that the specific gravity will again be maximum. The cell is then fully charged and ready for use again.

Note that adding sulfuric acid to a discharged lead-acid cell does not recharge the cell. Adding acid only increases the specific gravity of the electrolyte and does not convert the lead sulfate on the plates back into active material (spongy lead and lead dioxide). Consequently, this does not bring the cell back to a charged condition. To recharge a cell, a charging current must be passed through the cell.

As a cell becomes nearly charged, hydrogen gas is liberated at the positive plate. This action occurs because the charging current has become greater than the amount necessary to reduce the remaining amount of lead sulfate on the plate. Thus, the excess current ionizes the water in the electrolyte. This action is necessary to develop a full charge in the cell.

Individual plates are formed into positive and negative groups. When these groups are assembled, they become a cell element (Fig. 238). The

POSITIVE PLATE
GROUP
NEGATIVE PLATE
GROUP
SEPARATOR

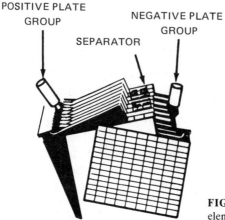

FIGURE 238 Partly assembled cell element.

number of negative plates is always one more than the number of positive plates, so both sides of each positive plate are acted upon chemically. The active material on the positive plates expands and contracts as the battery is charged and discharged. The expansion and contraction must be kept the same on both sides of the plate to prevent buckling.

Separators of wood, rubber, or glass are placed between the positive and negative plates to act as insulators (Fig. 238). These separators are grooved vertically on one side and smooth on the other. The grooved side is placed next to the positive plate to permit free circulation of the electrolyte around the active material.

An assembled lead-acid cell with the positive and negative terminals projecting through the cell is shown in Fig. 239. A hole fitted with a filler cap is provided in each cell cover to permit filling and testing. The filler cap has a vent hole to allow the gas that forms in the cell during charge to escape.

The ordinary 6-V portable storage battery consists of three cells assembled in a molded hard rubber (monobloc) case. Metal cannot be used because of the acid electrolyte. Each cell is contained in an acid-proof compartment within the case. The cells are connected in series by means of lead-alloy connectors that are attached to the terminal posts of adjacent cells.

CELL COVER

FIGURE 239 Assembled lead-acid cell.

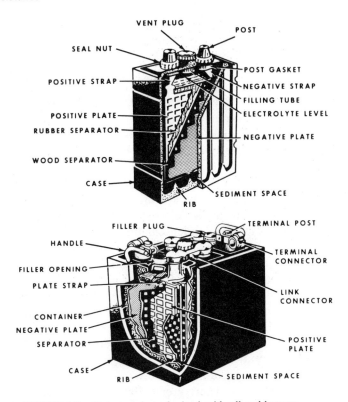

FIGURE 240 — VENT PLUG, POST, SEAL NUT, POST GASKET, POSITIVE STRAP, NEGATIVE STRAP, FILLING TUBE, POSITIVE PLATE, ELECTROLYTE LEVEL, RUBBER SEPARATOR, NEGATIVE PLATE, WOOD SEPARATOR, CASE, SEDIMENT SPACE, RIB

FILLER PLUG, TERMINAL POST, HANDLE, TERMINAL CONNECTOR, FILLER OPENING, PLATE STRAP, LINK CONNECTOR, CONTAINER, NEGATIVE PLATE, SEPARATOR, POSITIVE PLATE, CASE, RIB, SEDIMENT SPACE

FIGURE 240 Cutaway view of a lead-acid cell and battery.

The space between the case and the edges of the cell covers is filled with an acid-proof battery sealing compound or pitch. This compound is a blend of bituminous materials that are processed so that they remain solid at high temperatures.

Figure 240 shows a cutaway view of a lead-acid cell and battery.

Leakage resistance: The ohmic value of the path between two electrodes that are insulated from each other. Figure 241 shows a basic capacitor with both elements separated from each other by a dielectric. In this case, the leakage resistance would be the total ohmic value measured from one point on one plate to a similar point on the other. The leakage resistance would be the same as the dielectric resistance. If the two electrodes were separated by

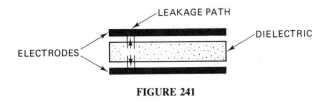

LEAKAGE PATH, DIELECTRIC, ELECTRODES

FIGURE 241

air, the leakage resistance would be measured through this medium. In almost every case, leakage resistance is extremely high and is measured in megohms or gigohms.

Lecher wire: A circuit segment that consists of two parallel wires or rods that are joined by a coupling loop on one end, with the other end open. A short-circuiting bar is moved along the wires to vary the effective length of the circuit. RF energy is inductively coupled into the system through the loop, and the bar is slid along to various response points, as shown by a meter or lamp coupled to the wires. The frequency may be determined by measuring the distance between adjacent response points.

Lenz's law: That which states that whenever the value of an electric current is changed in a circuit, it creates an electromotive force by virtue of the energy stored up in its magnetic field, which opposes the change.

Leyden jar: A form of condenser generally used in making experiments on static electricity. It consists of a glass jar coated inside and out to a certain height with tinfoil, having a brass rod terminating in a knob passed through a wooden stopper, and connected to the inner coat by a loose chain.

Light: Radiant energy lying within a wavelength that spreads from 100 to 10,000 nm. The human eye can perceive light radiations in a frequency range of between 450 and 700 nm. The color of light is determined by its wavelength. Energy at the short-wave end of the visible spectrum, from 380 to about 450 nm, produces the sensation of violet. The longest visible waves, from approximately 630 to 760 nm, appear as red. Between these lie the wavelengths that the eye sees as blue (450–490 nm), green (490–560 nm), yellow (560–590 nm), and orange (590–630 nm), the colors of the rainbow.

The region of the spectrum adjoining the long wavelength end of the visible band is known as the infrared (below the red); adjoining the short-wavelength end of the visible is the ultraviolet (beyond the violet). Neither the infrared nor the ultraviolet is visible to the human eye, but both have applications in the electronics field.

The spectrum of a light source may be continuous, including all the visible wavelengths, or it may be a line or a band spectrum, containing only one or a few separated groups of wavelengths. A tungsten filament has a continuous spectrum; a mercury arc, a line spectrum. An equal energy spectrum (all the visible wavelengths in equal quantities) produces the sensation of white light. Noon sunlight approximates an equal-energy spectrum.

Light-emitting diode: A light-sensitive, solid-state device that glows when it passes current. Abbreviated *LED,* this device contains a semiconductor material that has been treated with a chemical known as gallium arsenide. When current is passed through a light-emitting diode, it emits light in the infrared as well as in the visible ranges, depending on the type of materials used and the manner in which they have been treated. LEDs come in many different sizes, shapes, and forms and are used mainly as indicators for electronic devices. LEDs take the place of the old panel lamps, which drew much more current.

The glow that emanates from an LED is a cool light and is not produced through heating effects. Because of this, LEDs are much more efficient with regard to energy consumption than other types of artificial light. LEDs are often available in integrated circuit forms that will display numbers and letters when current is fed to the proper contacts.

Lighting outlet: An outlet intended for the direct connection of a lampholder, a lighting fixture, or a pendant cord terminating in a lampholder.

Lightning: The discharge of enormous charges of static electricity accumulated on clouds. These charges are formed by air currents striking the face of clouds and causing condensation of the moisture in them. When the wind strikes the cloud, these small particles of moisture are blown upward, carrying negative charges to the top of the cloud and leaving the bottom with positive charges. As very heavy rains or other forms of heavy condensation fall through a part of the cloud, one side of the cloud becomes charged positively and the other side negatively, with many millions of volts difference in potential.

When clouds (under the condition described) come near enough to the ground or to another cloud with opposite charges, they will discharge to the ground or to another cloud with explosive violence. Since there is a strong tendency for lightning discharges to strike trees, structures, and other objects and travel on any metal parts that extend in the general direction of the discharge, lightning rods and properly grounded electrical systems can prevent much of the damage.

Lightning rod: A device used to protect a structure from lightning. Lightning rods should be placed on upward projections such as chimneys, towers, and the like. On flat roofs, rods should be placed 50 ft on center, and on edges of flat roofs and ridges of pitched roofs, about 25 ft on center. The rods should project from 10 to 60 in. above flat roofs and ridges, and from 10 to 14 in. above upward projections.

Light quantity: A measure of the total light emitted by a source of light falling on a surface. It may be expressed in lumens or watts. If all the light were of green-yellow color, at the peak of the spectral luminosity curve, 1 W would be equal to 681 lm. This gives some idea of the relative luminous efficiency of common light sources. For instance, a 100-W incandescent lamp emits about 1600 lm, only 2.4 W as light and the balance as heat. A 40-W fluorescent lamp emits about 3100 lm, 4.5 W as light and the balance as heat. From this, it can be seen that the fluorescent lamp gives four to five times as much light as the incandescent on a watt-for-watt basis.

Light relay: A photoelectric device that triggers a relay in accordance with fluctuations in the intensity of a light beam. These may be purely electronic in nature or electromechanical, depending on the individual circuit.

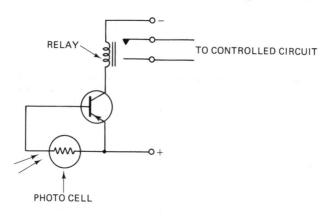

FIGURE 242

Often, light relays use an active or passive photoelectric device in the base circuit of a transistor, which is connected in series with an electromechanical relay. When light strikes the photocell, the transistor is driven to saturation and conducts current through the relay coil, causing it to change states. Such a circuit is shown in Fig. 242 using a photovoltaic cell.

Light sensors, semiconductor: Optoelectronic devices that convert light into electric current. Optoelectronic semiconductor devices are specially made diodes and transistors that interact with light to a useful extent. All diodes and transistors interact with light to some extent (one of the functions of their packages is to shut out light), but optoelectronic devices are designed to make efficient use of this phenomenon.

There are two important categories of optoelectronic devices—*light sensors* and *light emitters*. Figure 243 shows the symbols for these devices. The photodiode and phototransistor are examples of light sensors, and the light-emitting diode (LED) is a light emitter. All optoelectronic devices are based on one simple principle. Whenever light strikes the semiconductor material, it tends to knock bound electrons out of their sockets, so to speak, creating free electrons and holes. Conversely, when an electron falls into a hole, it tends to create a particle of light, or a *photon*.

Figure 244 shows a photodiode (a light-sensing semiconductor) being used to control a very small dc motor. The object is to operate the motor only when light strikes the photodiode. At that time, the stronger the light, the

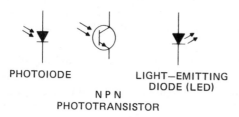

PHOTOIODE

LIGHT—EMITTING
DIODE (LED)

N P N
PHOTOTRANSISTOR   **FIGURE 243**

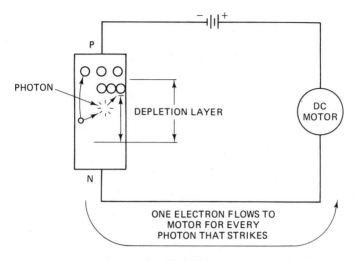

FIGURE 244

faster the motor will operate. The photodiode is much like an ordinary diode, except that it has a window or lens that lets light fall onto the PN junction. In this circuit, the battery is attempting to pump electrons from P to N through the diode. But this is the reverse direction for electrons. The free electrons and holes are forced apart, leaving a depletion layer around the junction devoid of holes and free electrons. The only current that flows is leakage current, and it is negligible.

Assume that a photon of light shoots into the semiconductor element within the area of the depletion layer. A free electron and hole are then created in the depletion layer. This assumes that the diode is properly designed, uses the right kind of semiconductor material, usually silicon, has enough of the right dopants, and has a chip shaped and placed properly to receive the light.

Immediately, the free electron is forced by the battery into the N region and out the cathode terminal, while the hole is driven in the opposite direction. The net result is that one electron passes to P to N through the circuit and the motor. Multiply this sequence by the countless millions of photons in a strong beam of light, and the result is a considerable current to drive the tiny motor. Thus, the function of a photodiode is to switch and regulate a working current under the control of light striking the device.

Light shining on a semiconductor junction greatly increases the reverse leakage current. This applies to all semiconductor devices. Light sensors are merely diodes and transistors in which this effect is enhanced and put to efficient use. In the case of photodiodes, there is one qualification to the rule that light increases the leakage current. It is true that light falling on a PN junction always tends to make electrons flow in the reverse direction from P to N. But the fact is that an external power supply is not always required to make current flow. In the circuit of Fig. 243, if the battery is removed so that the circuit consists of only the photodiode and the motor, a small amount of current will flow when light strikes the photodiode, generated entirely by light.

To put it simply, every photodiode is theoretically capable of converting light energy into electrical energy. Solar cells in artificial satellites and the sensors in many light meters are simply photodiodes specially constructed to enhance this ability of generating electric current. When photodiodes are used along with a separate power supply as in Fig. 244 they are called *photoconductive,* which means they conduct current when illuminated and block current when dark. But photodiodes used to generate current without the assistance of another power supply are called *photovoltaic,* because they actually produce a voltage pressure in the reverse direction.

The phototransistor, which is another light-sensing semiconductor, is also a photoconductive device. Figure 245 shows an NPN (a PNP version is possible too) device. It functions to switch and regulate current as the power supply attempts to pump electrons from emitter to collector. When there is no light, no current flows because there is no base control current. But when light strikes the base-collector junction, reverse leakage current flows.

This, in effect, constitutes a current of electrons being withdrawn from the base, which is what is needed to turn on an NPN transistor. In the phototransistor, unlike the diode, a much larger working current flows from emitter to collector. Thus, the phototransistor works like the photodiode; but in addition, it amplifies the tiny current produced by the light. With the diode, only one electron of current is required for each photon of light. With the phototransistor, however, each electron leaving the P region allows perhaps a hundred electrons to pass from emitter to collector and through the working circuit.

Light wave: A stream of electromagnetic energy that falls into the light spectrum. This lies between 100 and 10,000 nm. Visible light (that which can be seen by the human eye) lies within a very narrow bandwidth of this spectrum at a frequency range of about 500 to 600 nm.

Line: The projection of an electron beam upon a screen in a horizontal

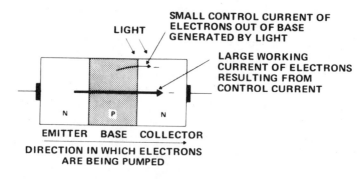

FIGURE 245

boundary, as found in cathode-ray tubes and television sets. The standard used in the United States designates a complete picture based on 525 lines. Line may also refer, in general, to any conductor of electrical energy.

Linear amplifier: An amplifier whose output is linearly proportional to its input. Linear amplifiers are used for audio reproduction and in high-frequency applications requiring a linear response. They are often used to boost the output of single-sideband transmitters and accept the modulated signal at their inputs and then output the equivalent at a much higher amplitude.

A linear amplifier may be operated class A or AB single-ended or B push-pull. Class A RF power amplifiers, due to their low efficiency, are primarily used in low-power applications. Class B power amplifiers provide greater power output with increased efficiency but require well-regulated bias supply voltages. Class AB amplifiers represent a compromise in power and efficiency between class A and class B. Linear amplifiers are used to increase levels of power or voltage in cases such as amplitude-modulated carriers or single-sideband signals.

Linearity: The degree to which performance or response approaches the condition of being linear. Linear response is indicated when one quantity varies directly with another. By definition, linear means in a straight line. When one quantity varies directly with another, the graph of the response is a straight line.

Linearity may also describe the characteristic of a signal that is a replica of another. For example, an audio tone is fed to the input of an amplifier at a frequency of 1 kHz. If the output frequency of the amplifier is also 1 kHz, the frequency response is said to be linear. A linear amplitude response from the same amplifier would be had if reduction of the input amplitude by 3 dB results in a reduction of the output amplitude by the same quantity.

Linear resistance: A trait whereby ohmic variation is directly proportional to shaft rotation. This means that a component with a maximum resistance of 1000 Ω would present an ohmic value of 250 Ω when its shaft was rotated a distance corresponding to a movement of 25 percent of the distance of full rotation. When the shaft is rotated to the halfway point, the measured resistance would be 500 Ω, or one-half of the total resistance.

Line trap: A filter circuit that is designed to be placed in series with an ac power line to prevent radio interference. In certain situations, ac power lines may be of the correct length to resonate at RF frequencies. When a radio transmitter is operated in close proximity to these lines, they conduct some of the transmitted energy to appliances and other devices which receive power from these lines. This causes interference or disruption of the devices, especially those which produce audio and video outputs.

Shown in Fig. 246, the ac line filter is designed to conduct low-frequency ac current to the device, but presents a high resistance to the flow of RF energy. Instead, this latter energy is conducted to ground, where it is eliminated. The

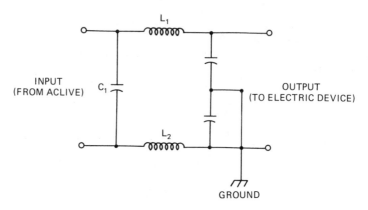

FIGURE 246

size of the conductors used to wind the coils for a line trap will depend on the amount of current the device to which it is attached normally draws. It is usually necessary to install line traps in the power cords to every device being affected by interference.

Link coupling: A specialized form of transformer coupling. The type shown in Fig. 247 requires the use of two tuned circuits, one in the output circuit of the driver and another in the input circuit of the power amplifier. A low-impedance RF transmission line having a coil of one or two turns at each end is used to couple the plate and grid-tank circuits. The coupling links or loops are coupled to each tuned circuit at its cold end (point of minimum RF potential).

Circuits that are cold near one end are called *unbalanced circuits*. Link-coupling systems normally are used where the two stages to be coupled are separated by a considerable distance. One side of the link is grounded in cases where capacitive coupling between stages must be eliminated or where harmonic elimination is important.

Link coupling is a very versatile interstage coupling system. It is used in transmitters when the equipment is sufficiently large to permit the coupled coils to be so positioned that there is no stray capacitive coupling between them. Link circuits are designed to have low impedance so that RF power losses are low. Coupling between links and their associated tuned circuits can be varied without complex mechanical problems.

Liquid crystal display: A type of seven-segment indicator that is instrumental in the decimal display of numerals. Liquid crystals become opaque when an EMF is applied, in contrast to other indicators that use light-emitting diodes (LEDs). The operating current of liquid crystals is extremely low, making them ideal for digital clocks and watches, where continuous display with the use of batteries is necessary. However, they do require sufficient surrounding light for proper visible projection, but they are capable of incorporating front or rear illumination for operation in poor lighting conditions. Abbreviated *LCD*.

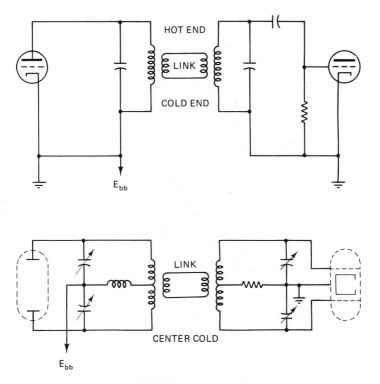

FIGURE 247 Link coupling.

Lissajous figures: The patterns that are produced on the screen of a cathode-ray tube when sine-wave signal voltages are introduced to the circuits. These sine-wave voltages are of various amplitude and phase relationships and are applied to the horizontal and vertical deflection circuits at the same time, resulting in rather geometrically shaped patterns of varying symmetrical shapes.

L network: An impedance-matching circuit, filter, or attenuator that resembles the letter L. It closely resembles a standard resonant circuit with the load resistance in either series or parallel. It is the simplest form of impedance-matching circuit. Shown in Fig. 248, the L network consists of two elements, a capacitor, and an inductor. The capacitor may be placed at the input to the circuit or at the output, as shown.

L networks are often used as external impedance-matching circuits

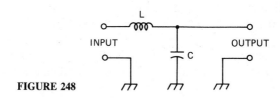

FIGURE 248

between radio transmitters and antennas. It normally follows an internal pi network tank circuit. Here, the L network serves to extend the tuning range of the transmitter output, matching it to the input impedance of the transmission line.

Load:

1. To put data into a register or storage.
2. To put a magnetic tape onto a tape drive, or to put cards into a card reader.
3. Name given to the amount of current being drawn from an electrical circuit.

For example, a 1200-W appliance connected to a 120-V circuit will load the circuit to an ampacity of 10 A or 1200 W. When a 15-A circuit is connected to a load of, say, 18 A, the circuit is said to be overloaded, or carrying more current than it was designed to carry.

Load factor: The ratio of the average load to the maximum load over a certain period. The time may be either the normal number of operating hours per day or 24 h, as generally used by the power companies. The average load is equal to the kilowatt-hours used in the specified time, as measured by a watthour meter, divided by the number of hours. The maximum load is the highest load at one time, as measured by some form of maximum-demand or curve-drawing watthour meter.

Local oscillator: That stage in a receiver whose function is to produce a constant-amplitude sine wave of a frequency that differs from the desired station frequency by an amount equal to the intermediate frequency of the receiver. The operation of an oscillator may be either above or below the station frequency. In most broadcast band receivers, the oscillator is operated above the station frequency. In order to allow selection of any frequency within the range of the receiver, the tuned circuits of the RF stage and the local oscillator are variable. By using a common shaft or ganged tuning for the variable component of the tuned circuits, both circuits may be tuned in such a manner as to maintain the difference between the local oscillator frequency and the incoming station frequency equal to the receiver intermediate frequency.

Log-periodic antenna: A type of directional antenna that uses geometric iteration in order to achieve its wideband properties. The ratio of element length to element spacing remains constant due to the fact that the radiating element and the spacing between elements have dimensions that logarithmically increase from one end of the array to the other.

Long-wire antenna: Long single wires, longer than a half-wavelength, in which the current in adjacent half-wave sections flows in opposite directions. Such antennas have two basic advantages—increased gain and directivity. If the length of a long-wire antenna is such that two or more half-waves of energy are distributed along it, it is often referred to as a harmonic antenna. Consider the half-wave antenna shown in Fig. 249a. At a given instant, the polarity of the RF generator connected to the center of the antenna is positive at its left-hand terminal. As a result, current in the left half of the antenna flows away from the generator. In both halves of the half-wave antenna, current flows in the same direction, from left to right, as shown by the wave of current above the antenna wire.

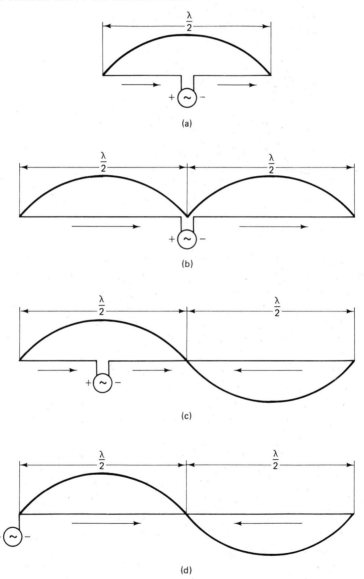

FIGURE 249 Harmonic and nonharmonic antennas.

Assume that the antenna just discussed is increased until it is two half-wavelengths, as shown in Fig. 249b. With the RF generator still connected at the center and the same instantaneous polarities as in Fig. 249a, current in the left side of the antenna must flow toward the generator and the current in the right side must flow away from the generator.

Since the antenna is now two half-wavelengths, two half-waves of current can be accommodated on the antenna, and the current polarity is the same in both halves of the antenna. It is important to note that this is not a true long-wire or harmonically operated antenna, since there is no reversal of current flow in adjacent half-wave sections. Instead, this arrangement is simply two half-wave antennas operating in phase at their fundamental frequency. Such an arrangement is called a *driven colinear array* and has characteristics quite different from those of the true harmonically operated or long-wire antenna.

The antenna in Fig. 249b can be converted into a true long-wire, harmonically operated antenna by simply moving the generator to a current loop, as shown in Fig. 249c. With the RF generator polarity as shown, current flows from left to right in the half-wave section of the antenna. The direction of current flow is then reversed in the second half-wave section. If the generator is moved to the extreme end of the antenna, as shown in Fig. 249d, the antenna is also a long-wire antenna, and the current distribution on the antenna is exactly the same as in Fig. 249c. The harmonically operated antenna, therefore, must be fed either at a current loop or at its end for proper operation.

As the length of the antenna is increased, it is natural to expect a change in the radiation pattern produced by the antenna. A long-wire antenna can be considered to be made up of a number of half-wave sections fed 180° out of phase and spaced a half-wavelength apart. As a result, there is no longer zero radiation off the ends of the antenna, but considerable radiation occurs in the direction of the long wire as a result of the combined fields produced by the individual half-wave sections. In addition, radiation also occurs broadside to the long wire. Consequently, the resultant maximum radiation is neither completely at right angles to the long wire nor completely along the line of the long wire. Instead, the maximum radiation occurs at some acute angle with respect to the wire, the exact angle being determined by the length of the antenna.

As shown in Fig. 250, as the length of a long-wire antenna is increased, characteristic changes occur. First, the gain of the antenna increases considerably compared with that of the basic half-wave antenna, especially when the long wire is many wavelengths. Second, the direction along which maximum radiation occurs makes a smaller angle with respect to the wire itself. Consequently, as the antenna is made longer, its major lobe of radiation lies closer to the direction of the wire itself. Third, more minor lobes are produced as the antenna length is increased.

Loop antenna: A small, portable antenna that is configured as a single coil of wire. This antenna design is often used for bidirectional tracking

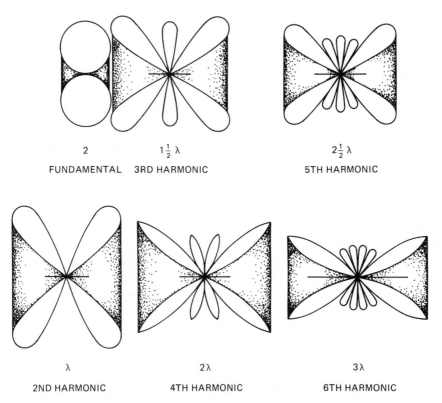

2	$1\frac{1}{2}\lambda$	$2\frac{1}{2}\lambda$
FUNDAMENTAL	3RD HARMONIC	5TH HARMONIC

λ	2λ	3λ
2ND HARMONIC	4TH HARMONIC	6TH HARMONIC

FIGURE 250 Radiation patterns of harmonic antennas.

purposes, as its two main areas of response are through the center of the loop, as shown in Fig. 251. For omnidirectional tracking purposes, the loop antenna may be shielded on one side and possibly on either end. In this arrangement, the antenna responds best to radio signals that approach from the nonshielded side of the element.

Loop antennas are usually not highly efficient, either in receive or transmit modes. However, they are quite simple to construct, respond to radio signals that are both horizontally and vertically polarized, and can be easily rotated to home in on signals transmitted from an unknown site.

When used for communications purposes, loop antennas may contain a coil of wire with several turns. In this configuration, it takes on highly directional properties, especially when used with a reflector. This is often

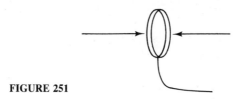

FIGURE 251

known as a *helical antenna*. Here, each coil is made from a length of conductor that is an electrical wavelength at the operating frequency.

Cubical quad antennas are quite similar to loop designs, using a full-wavelength element. However, a square coil is used for mechanical mounting ease.

LORAN: An acronym for *long-range navigation,* a system that provides a means of obtaining accurate navigational fixes from pulsed radio signals radiated by shore-based transmitters. Depending on the mode of LORAN operation and the time of day or night, fixes are possible at distances up to 3000 nautical mi from the transmitting stations. A LORAN set aboard a ship or airborne is a receiving set and indicator that displays the pulses from LORAN transmitting stations on shore.

Loudspeaker: An audio transducer that outputs acoustical energy from an electrical source. The purpose of audio-reproduction devices such as loudspeakers and headphones is to convert electrical audio signals to sound power. Figure 252 shows a diagram of a loudspeaker called a permanent magnet speaker. This speaker consists of a permanent magnet that is mounted on soft-iron pole pieces, a voice coil that will act as an electromagnet, and a loudspeaker cone that is connected to the voice coil.

The audio signal has been previously amplified (in terms of both voltage and power) and is applied to the voice coil. The voice coil is mounted on the center portion of the soft-iron pole pieces in an air gap so that it is mechanically free to move. It is also connected to the loudspeaker cone. As it moves, the cone will move as well.

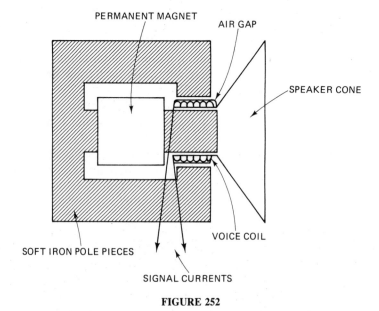

FIGURE 252

When audio currents flow through the voice coil, the coil is moved back and forth proportionally to the applied ac current. As the cone (diaphragm) is attached to the voice coil, the coil is moved back and forth proportionally to the applied ac current. As the cone (diaphragm) is attached to the voice coil, it also moves in accordance with the signal currents and in so doing, periodically compresses and rarefies the air and thus produces sound waves.

Most speakers of this type receive their input by means of transformer coupling. This is necessary because of the normally low impedance of the voice coil. Standard impedance values for such speakers are 4, 8, 16, and 32 Ω. Other impedance values may be obtained, but these are the most common.

While permanent magnet speakers perform reasonably well in the audio range, they nevertheless have inherent limitations. When the speaker is constructed, only a limited number of turns may be built into the voice coil. Therefore, it has a fixed inductance. At low frequencies, the inductive reactance of the voice coil will be relatively low, and large audio currents will flow. This provides a strong magnetic field around the voice coil and a strong interaction with the field of the permanent magnet. Low-frequency response is therefore excellent.

At mid-band frequencies, the inductive reactance has increased, and less current can flow in the voice coil. This produces less magnetic field and less interaction. Mid-band response, however, is still acceptable in a properly designed speaker.

At high audio frequencies, inductive reactance is quite high, and very little current flows in the voice coil. This results in a greatly reduced voice coil field and very little interaction with the permanent magnet field. Also, at high frequencies, the interwinding capacities of the voice coil tend to shunt some of the high audio frequencies and further reduce the response.

The frequency response of most permanent magnet speakers will fall off at the higher audio frequencies. This problem is normally overcome either by the procurement of an expensive, specially designed speaker, or through the use of two speakers, one of which is designed to operate well at the higher audio frequencies.

As shown in Fig. 253, an electromagnet may be used in place of the permanent magnet to form an electromagnetic dynamic speaker. However, in this instance, sufficient dc power must be available to energize the field electromagnet. The operation is otherwise much the same as that of the permanent magnet type.

Figure 254 shows a diagram of representative headphones as used in conjunction with this type of equipment. The device consists of a permanent magnet and two small electromagnets, through which the signal currents pass. A soft-iron diaphragm is used to convert the electrical effects of the device into sound power. When no signal currents are present, the permanent magnet exerts a steady pull on the soft-iron diaphragm. Signal current flowing through the coils mounted on the soft-iron pole pieces develops a magnetomotive force

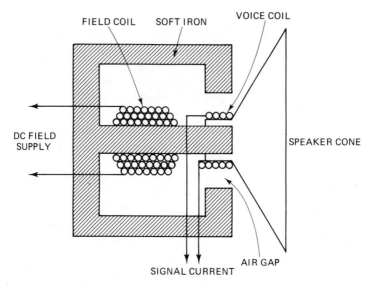

FIGURE 253 Electromagnetic speaker.

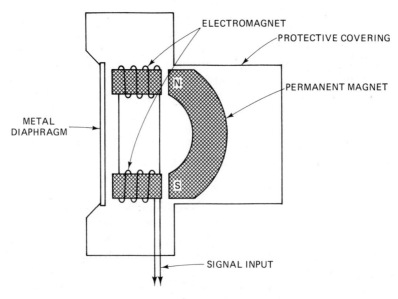

FIGURE 254 Headphone.

that either adds to or subtracts from the field of the permanent magnet. The diaphragm thus moves in or out according to the resultant field. Sound waves are then reproduced that have an amplitude and frequency (within the mechanical capability of the reproducer) similar to the amplitude and frequency of the signal currents.

As compared to the permanent magnet speaker, standard headphones are considered to be high-impedance devices. Headphone elecromagnets are normally wound of small wire with many turns providing the larger impedance. Because of the physically small size and inflexibility of the metal diaphragm, the headphones often give poor response to the lower audio frequencies. In the voice range of audio, however, most headphones are adequate reproducers.

LSI: An acronym for *large-scale integration,* the combining of electrical components and circuits into a piece of solid-state material to produce subsystems containing 100 or more gates. The hand-held calculator was one of the first devices to employ large-scale integration. The introduction of such sophisticated circuits changed the field of digital electronics extensively.

Much more information can be placed on a single silicon chip without significant increase in its size. The cost of making such chips has not risen much, because manufacturers have improved the ways complex integrated circuits are made. Also, because there are so many gates per chip, the cost per gate has been drastically reduced.

Luminaire: A complete lighting unit that includes the lamp, sockets, and equipment for controlling light, such as reflectors and diffusers. On electric discharge lighting, the luminaire also includes a ballast. The common term used for luminaire is *lighting fixture,* or in some cases, simple fixture.

Luminance: Intensity in a given direction divided by projected area, as intercepted by the air. It is subjective intensity and ranges from very dim to very bright. Luminance is expressed as candelas per square inch in a certain direction. Candelas per square inch may be put into more convenient form by multiplying by 452, giving luminance in footlamberts. Another way of looking at luminance is in relation to illumination and the reflection factor. For a nonspecular surface

$$Luminance = Illumination \times Reflection\ factor$$

or

$$L = E \times R$$

where E equals footcandles, R equals reflection factor, and L equals footlamberts.

To illustrate, if $E = 100$ fc and $R = 50\%$, then $L - 100 \times 0.50 = 50$ fL.

M

Machine language: A language designed for interpretation and use by a machine without translation. Also, a system for expressing information that is intelligible to a specific machine; e.g., a computer or class of computers. Such a language may include instructions that define and direct machine operations and information to be recorded by or acted upon by these machine operations. Also, the set of instructions expressed in the number system basic to a computer, together with symbolic operation codes with absolute addresses, relative addresses, or symbolic addresses.

Magnet: A piece of iron or steel that has the ability to attract and hold other pieces of iron and steel, and that is attracted and held in certain positions by another magnet. Magnets are put to practical use in magnetic screwdrivers, compasses, and many, many other devices. Natural magnets are lumps of iron core or oxide that have the power of attracting small pieces of iron. Oblong pieces of natural magnets, if suspended on a thread, will always turn to a position with its length north and south. This type of natural magnet that was used in the first crude compasses was often called *lodestone,* meaning leading stone.

Artificial magnets are made of steel and iron in various forms and are usually much more powerful than natural magnets or lodestones. Artificial magnets can be made by stroking a bar of steel in a certain way with a lodestone or some other magnet. The most common method, however, is to pass an electric current through a coil around the bar.

All magnets, whether natural or artificial, usually have their strongest pull or effects at their ends, which are called poles. If a magnet is placed under a piece of glass or paper covered with iron filings, and the glass or paper is then tapped, the filings will arrange themselves as shown in Fig. 255. This is the shape and direction of the lines of force acting around a magnet. The lines of force around a magnet are called *magnetic flux.* The area they occupy is called the *field* of the magnet.

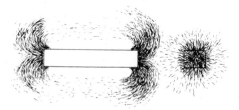

MAGNETS   **FIGURE 255**

Magnetic amplifier: A device used to reproduce an applied signal at an increased amplitude. Although all amplifiers have this capability, the method by which this is accomplished is vastly different when a magnetic amplifier is used.

The vacuum tube circuit amplifies because small changes in potential difference between grid and cathode produce relatively large changes in plate current, which, in turn, can be converted into large plate voltage changes. The transistor circuit amplifies because the input signal applied across the low-resistance emitter-base junction controls the current through the high-resistance collector-base junction. The increases in magnitude that occur in magnetic amplifier circuits are produced by the variations in magnetism and inductance within the unit.

Magnetic amplifiers are used as regulators, relays, amplifiers, motor starters, timing pulse generators, automatic stabilizers, and automatic pilots. They are also widely used in servo-systems as converters and computers and in many other applications. The magnetic amplifier is known for its dependability, ruggedness, high efficiency, and ability to withstand high temperatures.

A basic magnetic amplifier is shown in Fig. 256. In this circuit, the saturable reactor is used in conjunction with a dry-disc rectifier, CR_1, to produce a controllable dc voltage across the load resistance, R_L.

The reactor contains a three-legged core, and since the two load coils produce opposing fields, the reactor is nonpolarized. The magnitude of the voltage across R_L is dependent upon the level of the dc control voltage. First, the operation of the circuit with a control winding (N_C) voltage of zero will be considered. When the top of the ac supply is positive, current will flow through the load windings (N_L), CR_1, R_L, and back to the source. The absence of a dc potential to saturate the core results in high-core permeability, which means that inductance and inductive reactance will also be high. The current flow through the circuit will be very small and the voltage across R_L will be low.

When a dc potential is applied across the control winding, current will flow through the coil and magnetize the core. If the current flow is of sufficient magnitude, the core will become saturated. The inductive reactance of the load windings will approach zero and current will be high. Hence, the voltage drop across R_L will be large.

An increase in control current beyond the point of core saturation would

FIGURE 256

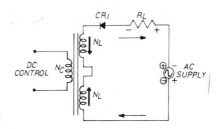

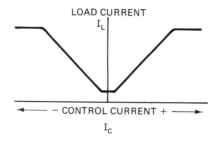

FIGURE 257 Dynamic characteristic curve of a nonpolarized magnetic amplifier.

result in an insignificant increase in load current. In practice, the increase in current beyond this point is considered to be zero. If the current is decreased to values below saturation, the μ of the core will increase, the load current will decrease, and the voltage across R_L will decrease.

A dynamic characteristic curve can be plotted showing the effects of varying control current in any magnetic amplifier. Figure 257 plots load current against control current. It should be observed that an increase in control current (up to a point of saturation) results in an increase in load current in a nonpolarized amplifier, and that the direction of control current is unimportant.

An increase in control current in one direction results in the same increase in load current, as does an increase in control current in the opposite direction. The load current never reaches zero, but will only reach a minimum value. The minimum value of load current is called the *no-signal* or *quiescent current,* and it occurs when the control current is zero.

The use of the term *no-signal current* may seem strange, since the control current is dc. However, the control signal is analogous to the input signal applied between the control grid and cathode of a vacuum tube. Amplification is achieved in the vacuum tube circuit because a small input potential causes relatively large changes in plate current. In the magnetic amplifier, a small change in control current causes large changes in load current.

The load current in a magnetic amplifier is also affected by the size of the load resistor, R_L. Figure 258 shows an output characteristic curve when load current, I_L, is plotted against control current, I_C, for two different values of load resistance. This curve illustrates the fact that the load current is much greater at saturation in the magnetic amplifier, which has the smaller load resistor. It should be noted that an increase in control current when operating on the steep portion of the curve will produce a large change in the load current. Consequently, it can be seen how amplification is achieved.

Magnetic blowout coil: A device that provides a strong magnetic field to extinguish the arc drawn when the circuit is broken. It consists of a few turns of heavy wire wound on an iron core that has its poles placed on either side of the contacts where the circuit is broken. See Fig. 259. This arrangement provides a powerful magnetic field where the circuit is broken.

The arc is a conductor and has a magnetic field set up around it. This field

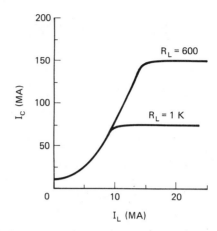

FIGURE 258 Characteristic curves for different load resistors.

will be reacted upon by the flux of the blowout coil distorting the arc so that it is quickly broken or extinguished. This prevents the arc from burning the contacts. Magnetic blowout coils are connected in series with the line, or in series with the contacts being protected.

Magnetic circuit: A circuit that includes the entire path around which the magnetic lines of force flow, just as the electric circuit includes the entire path through which the current flows. Just as an electric circuit must include a source of electromotive force that causes current to flow, so must the magnetic circuit include a source of the force that causes magnetic lines to move around the circuit. In a magnetic circuit, this source is a permanent magnet or an electromagnet. Just as there is resistance or opposition to flow of electric

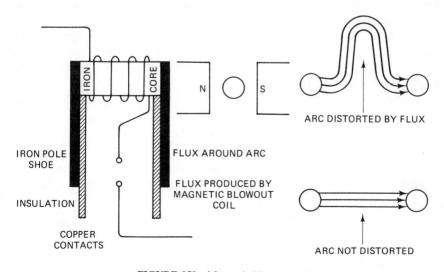

FIGURE 259 Magnetic blowout coil.

current in an electric circuit, so there is opposition to flow of magnetic lines of force in the steel and other parts of the magnetic circuit.

The strength of an electromagnet depends on the number of turns in its coil and the amperes or current flowing through them. The term used is *ampere-turns.* The ampere-turns are the product obtained when the amperes are multiplied by the number of turns. For example, a coil of 100 turns carrying 2 A has 200 ampere-turns. A coil of 400 turns carrying 0.5 A has 200 ampere-turns.

As the ampere-turns are increased in a magnet, the lines of force in its core become more and more dense and numerous until a certain point is reached where additional turns do not cause much increase of flux in the core. This is called the *saturation point.*

Good magnetic iron or steel can carry about 100,000 lines/in.² before reaching the practical saturation point.

Magnetic data: Any data stored by the use of devices that utilize the magnetic properties of materials. Such data may be stored on magnetic tape, disc, or drum, or in magnetic core storage. In the case of a magnetic tape, bits of tiny invisible magnetized spots are stored as rows and columns that have been created in a thin film of iron oxide on the surface of a plastic tape. The writing and reading of bits signifying data is accomplished by using a tiny electromagnet (called a *head*) that the surface passes under.

Magnetic field: The force formed when current flows through the running and starting windings of a motor. The magnetic field rotates and induces a current in the rotor winding, which in turn causes another magnetic field. These magnetic fields combine in such a manner as to cause rotation of the rotor. The starting winding is necessary at the start in order to produce the rotating field. After the motor is running, the starting winding is no longer needed and is cut out of the circuit by means of a centrifugal switch.

Magnetic field intensity: The quantity designated H and sometimes called *magnetic intensity,* which is directly related to the force exerted by a magnetic field. The unit used in measuring field intensity is the oersted, one oersted being equal to the strength necessary to exert a force of one dyne per unit magnetic pole. This relationship may be expressed mathematically as

$$H = \frac{f}{m}$$

where H equals field intensity in oersteds, f equals the force acting upon a magnetic pole in dynes, and m equals the strength of the magnetic pole in unit poles.

For a bar magnet or an electromagnet whose cross-sectional area is small compared to its length, field intensity is directly proportional to the magnetomotive force and inversely proportional to the length of the magnet. The formula then becomes

$$H = \frac{MMF}{cm} = \frac{1.257\ NI}{cm}$$

where H equals the field intensity in oersteds, MMF equals the magnetomotive force in gilberts, and cm equals the length in centimeters.

Magnetic polarization: The quality of a material by which it possesses opposite or contrasting poles. In an atom, the electron has a negative charge, while the nucleus has a positive charge that is carried by the protons. The basic law of electric charges states that like charges repel and unlike charges attract. The positive charge of the nucleus of an atom attracts the electrons. However, the electrons are able to maintain their orbital paths because of their speed and energy. Since the forces of the atoms keep the electrical charges in balance, the atom remains stable and neutral.

Electrons orbiting about an atom, however, can determine the charge of that atom. Atoms that have either lost or gained outside electrons (known as *valence electrons*) can be negative or positive in charge, depending on the number of electrons involved in the transfer. They are then known as *ions*.

When two bodies, whether they be atoms or larger units of matter, have unlike charges, an electric stress is created between the two bodies if they are placed close to each other, and they are said to repel. Oppositely charged bodies attract and tend to combine and neutralize.

The practical unit of measurement of electrical charge is the *coulomb*. The difference between charges or potentials is measured by the *volt.*

Magnetic tape: One of the most widely used mediums for storing data. Rows of tiny bits of invisible magnetized spots are stored in the form of rows and columns on magnetic tape. These bits are covered or created within a thin film of iron oxide on the surface of a plastic ribbon. The writing and reading of these bits is performed by a tiny electromagnet called a *head,* which is situated above a read-write station through which the tape passes. Magnetic tapes can be erased and written upon time and time again.

Magnetizing current: The component of the current in the primary of a transformer that magnetizes the core and produces the required magnetic field. This current lags the applied voltage by 90° but it is in phase with the alternating flux it generates. The magnetizing force that produces the actual magnetic strength is measured in ampere-turns. It is obtained by multiplying the magnetizing current and the number of turns in the windings. If the number of turns is constant, the flux changes depend directly on the changes in magnetizing current.

Magnetostriction effect: An effect that is similar to the piezoelectric effect found in crystal oscillators. Instead of using electric charges, however, it operates by the effect of a changing magnetic field. When an iron alloy is placed within a magnetic field, there is a change in length due to the strain placed on the rod by the magnetic field. This compressional strain in effect squeezes the rod and makes it longer. When the field is removed, the rod returns to nearly its former length.

Similarly, when a rod located within a magnetic field changes its length, it also induces a change in the magnetic field, increasing or decreasing the field. The induced charge is dependent upon the original direction of the magnetic field and the polarization of the metal rod and its composition. When the change in the length of the bar is performed at the resonant (mechanical) frequency of the bar and the induced change is properly phased to enhance the field that produces the strain, mechanical oscillations are set up at the fundamental frequency of the bar. When clamped in a fixed position at the middle, the bar will vibrate with a flexural motion similar to that of a tuning fork.

The metal composition of the bar determines its efficiency and effectiveness as a resonator. Temperature effects will cause changes in the bar length and thus the frequency of operation; hence, for extreme stability, the alloy must have a small temperature coefficient or temperature control must be used. Operation in this respect is similar to the operation of crystal oscillators.

Magnetron: A power oscillator that is used because conventional tubes are not practical at the frequencies and power levels required for radar (microwave) applications. The magnetron is an oscillator that is a self-contained unit. It produces a microwave frequency output within its enclosure without the use of external components such as crystals, inductors, capacitors, etc.

Basically, the magnetron is a diode and has no grid. A magnetic field in the space between the plate (anode) and the cathode serves as a grid. The plate of a magnetron does not have the same physical appearance as the plate of an ordinary electron tube. Since conventional LC networks become impractical at microwave frequencies, the plate is fabricated into a cylindrical copper block containing resonant cavities that serve as tuned circuits. The magnetron base differs greatly from the conventional base. It has short, large-diameter leads that are carefully sealed into the tube and shielded, as shown in Fig. 260.

The cathode and filament are at the center of the tube. It is supported by the filament leads that are large and rigid enough to keep the cathode and filament structure fixed in position. The output lead is usually a probe or loop extending into one of the tuned cavities and coupled into a waveguide or coaxial line. The plate structure, as shown in Fig. 261, is a solid block of copper. The cylindrical holes around its circumference are resonant cavities. A narrow slot runs from each cavity into the central portion of the tube and divides the inner structure into as many segments as there are cavities. Alternate segments are strapped together to put the cavities in parallel with regard to the output. These cavities control the output frequency. The straps are circular metal bands that are placed across the top of the block at the entrance slots to the cavities. Since the cathode must operate at high power, it must be fairly large and able to withstand high operating temperatures. It must also have good emission characteristics, particularly under back bombardment, because much

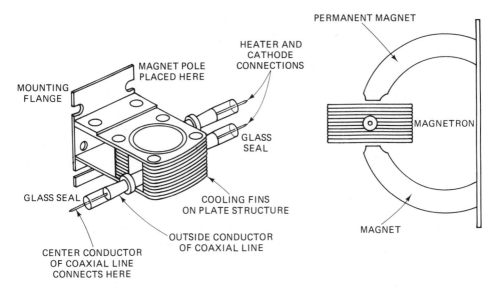

FIGURE 260 Magnetron.

of the output power is derived from the large number of electrons emitted when high-velocity electrons return to strike the cathode. The cathode is indirectly heated and is constructed of a high-emission material. The open space between the plate and the cathode is called *interaction space,* because it is in this space that the electric and magnetic fields interact to exert force upon the electrons.

The magnetic field is usually provided by a strong permanent magnet mounted around the magnetron so that the magnetic field is parallel with the axis of the cathode. The cathode is mounted in the center of the interaction

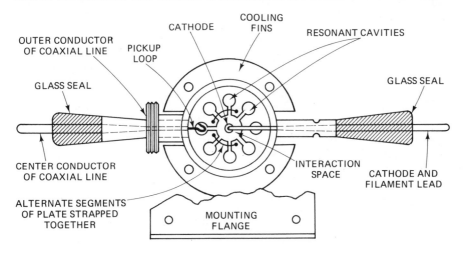

FIGURE 261 Cutaway view of a magnetron.

space. The theory of operation of the magnetron is based on the motion of electrons under the influence of combined electric and magnetic fields. The following laws govern this motion.

The direction of an electric field is from the positive electrode to the negative electrode. The law governing the motion of an electron in an electric or E field states that the force exerted by an electric field on an electron is proportional to the strength of the field. Electrons tend to move from a point of negative potential toward a positive potential as shown in Fig. 262. In other words, electrons tend to move against the E field. When an electron is being accelerated by an E field, as shown in Fig. 262, energy is taken from the field by the electrons.

The law of motion of an electron in a magnetic or H field states that the force exerted on an electron in a magnetic field is at right angles to both the field and the path of the electron. The direction of the force is such that the electron trajectories are clockwise when viewed in the direction of the magnetic field, as shown in Fig. 262.

In this figure, it is assumed that a south pole is below the paper, and a north pole is above the paper, so that the magnetic field is going into the paper. When an electron is moving in space, a magnetic field is built around the electron, just as there would be a magnetic field around a wire when electrons are flowing through a wire. Note in this figure that the magnetic field around the moving electron adds to the permanent magnetic field on the left side of the electron path and subtracts from the permanent magnetic field on the right side of the electron path, thus weakening the field on that side. Therefore, the electron path bends to the right (clockwise). If the permanent magnetic field strength is increased, the electron path will bend sharper. Likewise, if the velocity of the electron increases, the field around it increases and its path will bend more sharply.

A schematic diagram of a basic magnetron is shown in Fig. 263a. The tube

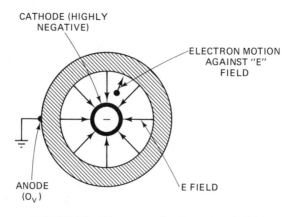

FIGURE 262   Electron motion in an electric field.

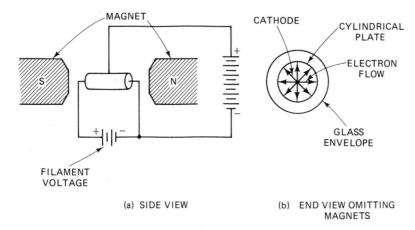

(a) SIDE VIEW   (b) END VIEW OMITTING
                                   MAGNETS

FIGURE 263   Basic magnetron.

consists of a cylindrical plate with a cathode placed coaxially with it. The tuned circuit in which oscillations take place are cavities physically located in the plate. When no magnetic field exists, heating the cathode results in a uniform and direct movement in the field from the cathode to the plate, as illustrated in Fig. 263b. However, as the magnetic field surrounding the tube is increased, a single electron is affected, as shown in Fig. 264. In Fig. 264a, the magnetic field has been increased to a point where the electron proceeds to the plate in a curve rather than a direct path.

In Fig. 264b, the magnetic field has reached a value great enough to cause the electron to just miss the plate and return to the filament in a circular orbit. This value is the critical value of field strength. In Fig. 264c, the value of the field strength has been increased to a point beyond the critical value, and the electron is made to travel to the cathode in a circular path of smaller diameter.

Figure 264d shows how the magnetron plate current varies under the influence of the varying magnetic field. In Fig. 264a, the electron flow reaches the plate so that there is a large amount of plate current flowing. However, when the critical field value is reached, as shown in Fig. 264b, the electrons are deflected away from the plate, and the plate current drops abruptly to a very

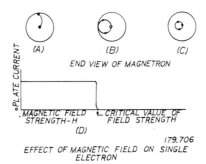

FIGURE 264

small value. When the field strength is made still larger, as shown in Fig. 264c, the plate current drops to zero.

When the magnetron is adjusted to the plate current cutoff or critical value and the electrons just fail to reach the plate in their circular motion, the magnetron can produce oscillations at microwave frequency by virtue of the currents induced electrostatically by the moving electrons. This frequency is determined by the time it takes the electrons to travel from the cathode toward the plate and back again. A transfer of microwave frequency energy to a load is made possible by connecting an external circuit between the cathode and plate of the magnetron. Magnetron oscillators are divided into two classes— negative-resistance and electron-resonance magnetron oscillators.

A negative-resistance magnetron oscillator operates by reason of a static negative resistance between its electrodes and has a frequency equal to the natural period of the tuned circuit connected to the tube. An electron- resonance magnetron oscillator operates by reason of the electron transit time characteristics of an electron tube; that is, the time it takes electrons to travel from cathode to plate. This oscillator is capable of generating very large peak power outputs at frequencies in the thousands of megahertz. Although its average power output over a period of time is low, it can put out very high- power oscillations in short bursts of pulses.

Marconi antenna: A quarter-wave grounded vertical antenna. Figure 265 illustrates the quarter-wave grounded antenna. Note the amplitude of the standing waves of current and voltage on this type of antenna. Note also the similarity to the half-wave dipole when the image is included. The ground is a

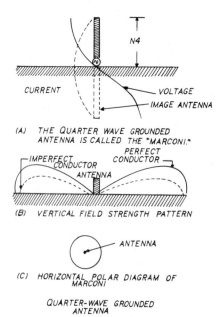

(A) THE QUARTER WAVE GROUNDED
      ANTENNA IS CALLED THE "MARCONI."

(B) VERTICAL FIELD STRENGTH PATTERN

(C) HORIZONTAL POLAR DIAGRAM OF
      MARCONI

QUARTER-WAVE GROUNDED
        ANTENNA   **FIGURE 265**

fairly good conductor for medium and low frequencies and acts as a large mirror for the radiated energy. This results in the ground's reflecting a large amount of energy that is radiated downward from an antenna mounted over it. It is just as though a mirror image of the antenna is produced with the image being located the same distance below the surface of the ground as the actual antenna is located above it. Even in the high-frequency range and higher, many ground reflections occur, especially if the antenna is erected over highly conducting earth, salt water, or a grounded screen.

Utilizing these characteristics of the ground, an antenna only a quarter-wavelength long can be made into the equivalent of a half-wave antenna. If such an antenna is erected vertically and its lower end is connected electrically to the ground, the quarter-wave antenna behaves like a half-wave antenna. Here, the ground takes the place of the missing quarter-wavelength, and the reflections supply that part of the radiated energy that normally would be supplied by the lower half of an ungrounded half-wave antenna.

Master timer, radar: The heart of a radar set, which produces accurately timed pulses that are applied to both the transmitter and indicators. This unit must be very stable, since it determines both the pulse repetition frequency (PRF) and pulse repetition time (PRT) of the radar set. The block diagram of a typical timer unit is illustrated in Fig. 266. The oscillator section generates a steady output at a fixed frequency. The frequency of operation is generally less than 1000 cps and will be determined by the purpose and type of radar set used. A phase-shift or Wein bridge oscillator is usually used to meet the requirements of stability and low-frequency operation. The output of the oscillator section is a sine wave of constant amplitude and frequency.

The output of the oscillator is applied to an overdriven amplifier that produces a square-wave output. This square wave should have steep leading and trailing edges to preserve accurately the time relationships produced by the oscillator. The square-wave output of the overdriven amplifier is applied to an RC differentiating circuit. The differentiator produces two narrow pulses—a positive pulse corresponding to the 0° point of the square wave, and a negative pulse corresponding to the 180° point of the square wave.

The output is applied to a diode clipper that usually removes the negative pulse, but in some cases, positive clipping may be employed. The clipper output is applied to a cathode follower stage. The cathode follower stage is generally used to isolate the timer unit from other circuits in the set and to provide impedance matching between the timer unit and the interconnecting

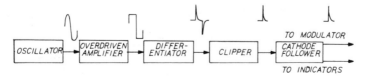

FIGURE 266

coaxial cables. The output of the timer unit consists of very narrow pulses, usually 30 to 60 V in amplitude, which correspond to the 0° point of each cycle produced by the timer oscillator.

Mechanical filter: A filter that consists of an input transducer, bias magnets, resonant metal disks, nickel coupling rods, external capacitors, and an output transducer. SSB transmitters and receivers require very selective bandpass filters in the region of 100 to 600 kHz. The filters used must have very steep skirt characteristics and flat passband characteristics.

Basic crystal filters and filters made up of inductors, resistors, and capacitors are often used, but the mechanical filter is also used to satisfy the requirements of SSB equipment. Mechanical filters offer many advantages over LC and crystal filters. They are small, have excellent rejection characteristics, and are comparatively rugged. The Q obtainable with a mechanical filter is much greater than that obtained with an LC filter.

Figure 267 illustrates a basic mechanical filter. The input transducer converts electrical energy to mechanical energy by utilizing the magnetostriction effect. The mechanical oscillations of the input transducer are transmitted to the coupling rods and metal disks. Each disk acts as a mechanical series resonant circuit. The disks are designed to resonate at the center of the filter's passband. The number of disks determines the skirt selectivity. Skirt selectivity is specified as shape factor, which is the ratio of the passband 60 dB below peak to the passband 6 dB below peak. Practical manufacturing presently limits the number of disks to eight. A six-disk filter has a shape factor of approximately 2.2; a seven-disk filter, 1.85; and an eight-disk filter, 1.5.

The passband of the filter is primarily determined by the area of the coupling rods. Passband may be increased by using more or larger coupling rods. Mechanical filters with bandwidths as narrow as 0.5 kHz and as wide as 35 kHz are practical in the 100 to 600 kHz range.

A terminal disk vibrates the output transducer rod, which induces, by means of generation action, a current in the output transducer coil. The

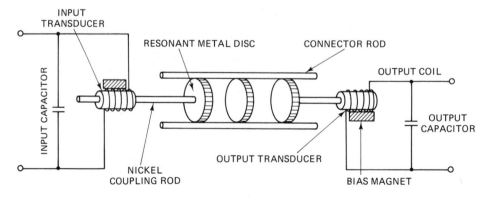

FIGURE 267   Mechanical filter.

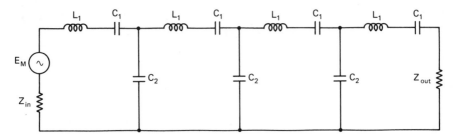

FIGURE 268 Equivalent circuit of the mechanical filter shown in Fig. 267.

external capacitors are used to form parallel resonant circuits with the input and output transducer coils at the filter frequency. The equivalent circuit for the filter in Fig. 267 is shown in Fig. 268. C_1 and L_1 represent a resonant metal disk; C_2 represents the coupling rods.

Figure 269 illustrates a mechanical filter frequency-response curve. Although an ideal filter would have a flat peak or passband, practical limitations prevent the ideal from being obtained. The term *ripple amplitude* or *peak-to-*

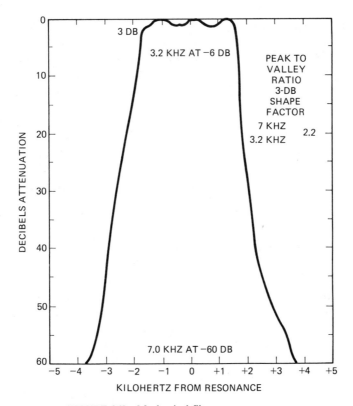

FIGURE 269 Mechanical filter response curve.

valley ratio is used to specify the peak characteristics of the filter. The peak-to-valley ratio is the ratio of the maximum to minimum output level across the useful frequency range of the filter. Mechanical filters with a peak-to-valley ratio of 1 dB may be produced with accurate adjustments of filter elements.

Mechanical filters other than the disk type are presently in use. All mechanical filters are similar, in that they employ mechanical resonance. Mechanical filters differ in that they employ various methods of mechanical oscillation to achieve their purpose.

Megahertz: A unit of frequency that is equivalent to 1,000,000 cycles/s, or 1,000,000 Hz. It is also equivalent to 1000 kHz.

Megger: A test instrument frequently used for measuring resistance. The name comes from the fact that the instrument is commonly used to measure resistances of millions of ohms, and a million ohms of resistance is called a *megohm*. Most meggers consist of a small hand- or battery-driven dc generator and one or more meter elements enclosed in a housing. When the crank is turned or the generator motor is activated, the generator will produce from 100 to 1000-V dc depending on the speed at which the generator is rotated and the number of coil turns in its winding.

Normal operating voltage is from 300 to 500 V, and this is marked on the meter scale. Some of these instruments are manufactured with a built-in voltohmmeter to show the generator voltage and indicate the insulation resistance of the circuit under test. One terminal of the instrument can be connected to the terminal of a motor, for example, and the other to the motor frame. When the crank is turned or the generator is activated by an electric motor, the insulation resistance in megohms can be read directly from the scale.

Memory: That area in a computer system in which programs or instructions that tell the central processing unit what to do are stored. Memory is also sometimes referred to as *storage*. Memory also holds data awaiting processing by the CPU and processed information which is to be delivered at a specified output device or stored for future use.

Memory storage capacity can be greatly expanded by the use of separate peripheral equipment, such as magnetic tapes, disks, and drum units. These auxiliary storage devices are often used to store programs and data to be kept on file for later use and to provide a means of storing large amounts of data during processing.

Mercury lamp: A lamp that produces light with a predominance of yellow and green rays, combined with a small percentage of violet and blue. When lighted, this type of lamp appears to emit white light (red, blue, and green), but the red color is absent. Therefore, red objects appear black or dark brown under mercury lamps. This color distortion has in the past prevented its use in many applications. However, it has now been overcome to a certain extent by the use of red light-generating chemicals within the bulb. Consequently, this type of lamp is now finding its way indoors for more and more commercial lighting applications.

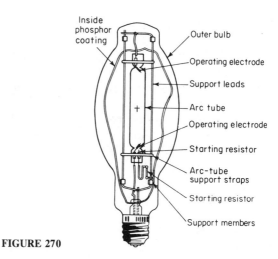

Inside
phosphor
coating

Outer bulb

Operating electrode

Support leads

Arc tube

Operating electrode

Starting resistor

Arc-tube
support straps

Starting resistor

Support members

FIGURE 270

Figure 270 shows typical mercury lamp components, which include an arc tube made of quartz to withstand the high temperatures resulting when the lamp builds up to normal wattage, two main operating electrodes located at opposite ends of the tube, a starting electrode connected in series with a starting resistor that is connected to the lead wire of the operating electrode, tube leads and supports, and an outer phosphor-coated bulb that helps to stabilize lamp operation and prevent oxidation of metal parts.

Mercury lamps for general lighting purposes are available in wattages from 40 to 2000 W, in clear and various type of color-improved lamps. All types of mercury lamps require their own specially designed transformers or ballasts for proper starting and operating performance. While these ballasts are usually located outside the lamps, self-ballasted lamps with built-in filament-type ballasts are available.

One disadvantage of mercury lamps is that they are extinguished in the event of current interruption or excessively low voltages. They will not restart until they have cooled down and the internal vapor pressure has been reduced to the point of restarting the arc with the available voltage. The restarting time interval is between 4 and 7 min.

In addition to being used to produce visible light, mercury lamps have been used to produce ultraviolet radiation as a source of black light. This type of light requires that all visible light be screened out by filters while the ultraviolet energy passes through. Black light lamps are used for decorative and theatrical effects, industrial inspection, medical and chemical analysis, criminal investigations, mineralogy, and varied military applications.

Mercury-zinc battery: A source of power that employs a zinc-powder anode and a cathode comprised of a mercuric-oxide power and a graphite power. It is one of the most expensive types of power supplies. The mercury-zinc battery has extremely uniform output throughout its life, characterized by a high-current capacity in a relatively small space and a continuous-current drain. The service life is usually in excess of 50 h, with a continuous drain of 250 mA. A typical mercury D cell would have a capacity of about 14 Ah.

Metal-clad cable (BX): A type of cable that is widely used for wiring in both residential and small commercial buildings. It is installed in new work in much the same manner as nonmetallic cable. Its flexibility and compactness make it very useful for modernization and alteration work where the cable must be fished inside finished partitions. Metal-clad cable is a fabricated assembly of two, three, or four conductors, ranging in size from No. 14 AWG to a maximum of 500 MCM, in a flexible metallic enclosed, as shown in Fig. 271. Over the years, the trade name BX has been used to identify this type of cable.

The wires in modern metal-clad cable are covered with thermoplastic (TW) insulation, while the steel armor has a corrosion-resistance coating. Type ACL, a similar cable, contains lead-covered conductors for use where the cable is exposed to the weather or to continuous moisture.

A common BX installation fault is to bend the cable too short. This breaks the armor and exposes the conductor to damage or corrosion. Bends should have a radius of not less than seven times the diameter of metal-clad cable, nor five times the diameter of AC or ACL cable.

When connections are made with BX cable to outlets or other points of termination, from 6 to 8 in. of sheathing is stripped from the end of the cable. Special tools are available for stripping BX cable, but the most conventional way is to use a hacksaw and partially saw through two of the sheathing rings. Hand pressure is used to snap the sheathing in two. The loose piece may then

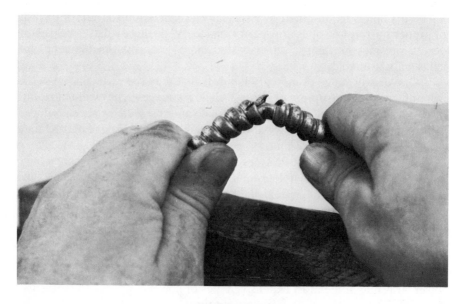

FIGURE 271

be easily pulled from the conductors. An insulating bushing, sometimes called a *redhead,* is used to protect the insulated conductors from the sharp edges of the metal jacket (sheathing) where the cut was made.

Metal-oxide semiconductor: A device that has been fabricated through a technique whereby a metallized electrode or gate is insulated from a semiconductor substrate material by using a dielectric to form a unipolar area. MOS devices have been ushered in by the advancements made in integrated circuit technology, which made possible the placement of large shift registers on a single chip. Elaborate MOS devices may contain as many as 5000 gates on a single chip measuring 4 mm^2. MOS logic is used for pocket calculator chips, as well as for IC computer or microprocessor devices and random-access memories.

Metal-oxide semiconductor field-effect transistor: A field-effect transistor (FET) in which the gate electrode consists of a thin metal film insulated from the semiconductor channel by a thin oxide film. Often abbreviated *MOSFET,* these devices fall into two categories—the depletion type and the enhancement type. Charge carriers are present in the channel of the depletion type when no bias voltage is applied to the gate. In the enhancement type, a forward-biased gate is necessary to produce active carriers. Each type is available with N-channel or P-channel polarization.

Mho: The unit of conductance. Mho is *ohm* spelled backward, as conductance is the reciprocal of resistance, which is measured in ohms.

Mica capacitor: A capacitor consisting of alternate layers of mica and plate material. The capacitance is of a small value, usually in the picofarad range. Although small in physical size, mica capacitors have a high voltage-handling capacity. Figure 272 shows a cutaway view of a mica capacitor.

Microcode instructions: Microprocessor operational instructions that cause the device to respond to user-programmed instructions. A single user-programmed instruction may involve many microcode instructions, and each microcode instruction occurs during a microcycle. Some microcodes are user programmable, and this allows tailoring a microprocessor to fit a particular language, thus making it competitive with larger computers that do not allow modification of their basic structure.

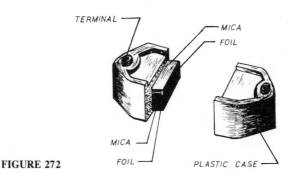

FIGURE 272

FIGURE 273   Microphone schematic symbol.

Microphone: An energy converter that changes sound energy into electrical energy. The diaphragm of the microphone moves in and out in accordance with the compression and rarefaction of the atmosphere, which is known as sound waves. The diaphragm is connected to a device that causes current flow in proportion to the instantaneous pressure delivered to it. Many devices can perform this function. The particular device used in a given application depends on the characteristics desired, such as sensitivity, frequency response, impedance, power requirements, and ruggedness.

The sensitivity or efficiency of a microphone is usually expressed in terms of the electrical power level, which the microphone delivers to a matched impedance in dB (decibels). It is important that the sensitivity be as high as possible. A high microphone output requires less gain in the amplifiers used in conjunction with the microphone. This provides a greater margin over thermal noise, amplifier hum, and noise pickup.

For good quality sound reproduction, the electrical signal from the microphone must correspond in frequency content to the original sound waves. The microphone response should be uniform or flat within its frequency range and free from the electrical or mechanical generation of new frequencies.

The impedance of a microphone is important in that it must be matched to the microphone cable between the microphone and amplifier input as well as to the amplifier input load. Exact matching is not always possible, especially in the case where the impedance of the microphone increases with an increase in frequency. A long microphone cable tends to attenuate the high frequencies seriously if the microphone impedance is high. This attenuation is due to the increased capacitive appearance of the line at higher frequencies. If the microphone has a low impedance, a lower voltage drop will occur in the microphone, and more voltage will be available at the load. The schematic symbol used to represent a microphone is shown in Fig. 273. The schematic symbol does not identify the type of microphone used or its characteristics.

Microphone, carbon: An audio transducer whose operation is based on varying the resistance of a pile of carbon granules by varying the pressure on the pile. The insulated cup, called the *button,* which holds the loosely piled granules, is so mounted that it is in constant contact with the thin metal diaphragm, as shown in Fig. 274a.

Sound waves striking the diaphragm vary the pressure on the button and thus vary the pressure on the pile of carbon granules. The dc resistance of the carbon granule pile is varied by this pressure. This varying resistance is in series with a battery and the primary of a transformer. The changing resistance of the carbon pile produces a corresponding change in the current of the circuit. The varying current in the transformer primary produces an alternating voltage in the secondary. The transformer steps up the voltage, as well as matches the low

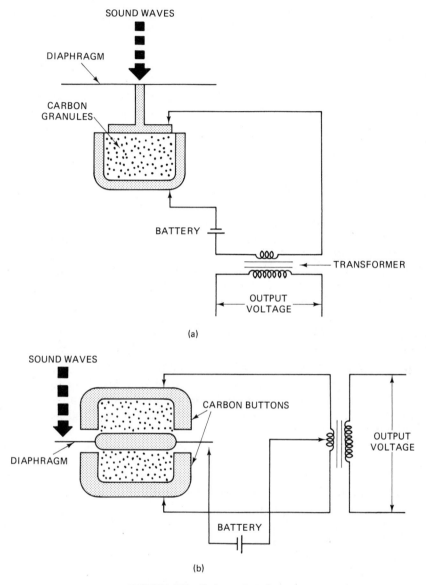

FIGURE 274 Carbon microphones.

impedance of the microphone to the high impedance of the first AF amplifier. The voltage across the secondary may be as high as 25-V peak. The impedance of this type of microphone varies from 50 to 200 Ω.

The double-button carbon microphone is shown in Fig. 274b. Here, one button is positioned on each side of the diaphragm so that an increase in pressure occurs on one side, while an increase in resistance occurs on the other side. Each button is in series with the battery and one half of the transformer primary. The decreasing current in one half of the primary and the increasing current in the other half produce an output voltage in the secondary that is proportional to the sum of the primary signal components. This action is similar to that of push-pull amplifiers.

One disadvantage of the carbon microphone is a constant background hiss, which results from random changes in the resistance between individual carbon granules. Another disadvantage is the reduced sensitivity and distortion that may result from the granules packing or sticking together. Sometimes, this may be cured by tapping the microphone. The carbon microphone also has a limited frequency response. Still another disadvantage is the requirement for an external voltage source.

The disadvantages, however, are offset by advantages. It is lightweight and rugged and can produce an extremely high output when its frequency response is limited to those frequencies that contribute most to intelligibility.

Microphone, crystal: A microphone that consists of a diaphragm that may be cemented directly on one surface of the crystal (Fig. 275a), or in some cases, may be connected to the crystal element through a coupling member (Fig. 275b). A metal plate or electrode is attached to the other surface of the crystal. When sound waves strike the diaphragm, the vibrations of the diaphragm produce a varying pressure on the surface of the crystal. Therefore, an EMF is induced across the electrodes. This EMF has essentially the same waveform as that of the sound waves striking the diaphragm.

A large percentage of crystal microphones employ some form of the bimorph cell. In this type of cell, two crystals, so cut and oriented that their voltages will be additive in the output, are cemented together and used in place of the single crystal.

This type of microphone has high impedance (several hundred thousand ohms), is light in weight, requires no battery, is nondirectional, and has an output on the order of -70 dB. However, the crystal microphone is sensitive to high temperature, humidity, and rough handling. Therefore, its use is restricted where these conditions prevail. Nevertheless, it is used extensively in broadcast work where its relatively high output is an advantage.

Microphone, dynamic: An audio transducer that consists of a coil of fine wire mounted on the back of a diaphragm and located in the magnetic field of a permanent magnet.

Figure 276 shows the typical operation of a dynamic microphone. When sound waves strike the diaphragm, the coil moves back and forth, cutting the magnetic lines of force. This induces a voltage in the coil that is an electrical representation of the sound waves.

The sensitivity of the dynamic microphone is almost as high as that of the carbon type. It is light in weight and requires no external voltage. The dynamic microphone is rugged and practically immune to the effects of vibration, temperature, and moisture. The microphone has a uniform response over a

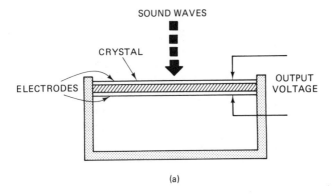

(a)

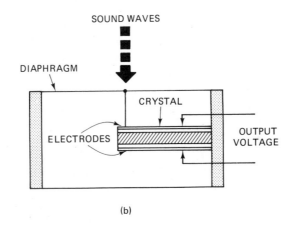

(b)

FIGURE 275 Crystal microphones.

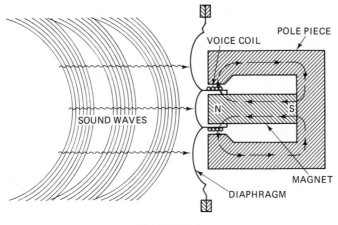

FIGURE 276

frequency range that extends from 40 to 15,000 Hz. The impedance is very low (generally 50 Ω or less). Therefore, a transformer is required to match it to the input of an AF amplifier.

Microphone, magnetic: A microphone consisting of a permanent magnetic and a coil of wire enclosing a small armature. Figure 277 shows the action of a magnetic microphone. Sound waves impinging on the diaphragm cause the diaphragm to vibrate. This vibration is transmitted through the drive rod to the armature, which vibrates in a magnetic field, thus changing the magnetic flux through the armature and consequently through the coil. When the armature is in its normal position midway between the two poles, the magnetic flux is established across the air gap, and there is no resultant flux in the armature.

When a compression wave strikes the diaphragm, the armature is deflected to the right. Although a considerable amount of the flux continues to move in the direction of the arrows, some of it now flows from the north pole of the magnet across the reduced gap at the upper right, down through the armature, and around to the south pole of the magnet. The amount of flux flowing down the left-hand pole place is reduced by this amount.

When a rarefaction wave strikes the diaphragm, the armature is deflected to the left. Some of the flux is now directed from the north pole of the magnet, up through the armature, through the reduced gap at the upper left, and back to the south pole. The amount of flux now moving up through the right-hand pole piece is reduced by this amount.

Thus, the vibrations of the diaphragm cause an alternating flux in the

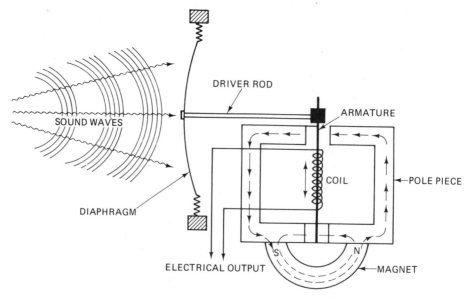

FIGURE 277   Action of a magnetic microphone.

armature. The alternating flux cuts the stationary coil wound around the armature and induces an alternating voltage in the coil. This voltage has essentially the same waveform as that of the sound waves striking the diaphragm.

The magnetic microphone is the type most widely used in communication systems, because it is more resistant to vibration, shock, and rough handling than other types of microphones.

Microphone, omnidirectional: An audio transducer whose sensing element is designed so that it responds equally well to acoustic signals arriving from any direction.

Figure 278 shows a pattern chart for an omnidirectional microphone. The pickup area is uniform and acoustic signals that are broadcast at the same intensity will be detected with equal efficiency regardless of the direction from which they are transmitted as long as the distance from the microphone remains the same.

Microprocessor: A digital integrated circuit or set of such circuits, referred to as a *chip,* which is actually a complete central processing unit capable of processing information and coordinating operations.

Microwave: A form of electromagnetic radiation that has frequencies of one billion hertz. These high-frequency bands of energy are used extensively for radar and wideband communications. More recently, they have been employed in such devices as microwave ovens. The radiation of microwaves can be directed into very narrow beams of energy, which makes these ranges very efficient in the utilization of transmitter energy. Interference between communication systems is also minimized. The extremely small wavelengths

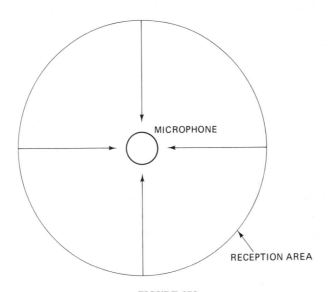

FIGURE 278

require relatively smaller antennas than other forms of transmitted energy. However, because of their shorter wavelength, microwaves are more susceptible to weather effects, such as raindrops, which become small antennas that absorb the energy and cause it to be dissipated before it reaches its destination.

Microwave radio: A system in which long-distance telephone calls and television broadcast programs are relayed by means of microwave-radiated energy bands. Since microwaves are susceptible to weather interference, repeater or relay stations are placed at various places that receive the signals from various other stations, boost their power, and transmit the signals onto other relay stations.

Mixer: That portion of a receiver that performs frequency conversion. The input to the mixer consists of two signals—the modulated RF signal and the unmodulated local oscillator signal. The mixer then combines or mixes these two signals. As a result of this mixing action, the output of the mixer will contain four major frequencies, plus many minor frequencies. The four major frequencies are the original signal frequency, the local oscillator frequency, the sum of the signal and oscillator frequencies, and the difference of the signal and oscillator frequencies.

The addition frequencies present are produced by combinations of the fundamentals and harmonics of the signal and oscillator frequencies. Of the frequencies present in the output of the mixer, only the difference frequency is used in amplitude-modulated broadcast band receivers. The output circuit of the mixer stage contains a tuned circuit that is resonated at the difference frequency.

Modem: An electronic device that performs the modulation and demodulation functions required for communications. A modem can be used to connect computers and terminals over telephone circuits. On the transmission end, the modulator converts the signals to the correct codes for transmission over the communications lines. At the receiving end, the demodulator reconverts the signals for communication to the computer using the computer interface unit. A modem is also sometimes referred to as a *data set.*

Modulate: To change the characteristics of a high-frequency wave, such as its amplitude, frequency, or phase, by impressing one wave on another wave of constant properties. The wave that is produced is a composite of the modulation and the higher-frequency wave.

Modulator: A device that varies a repetitive phenomenon in accordance with some predetermined scheme usually introduced as a signal.

Monostable multivibrator: A multivibrator that produces one output pulse for each input pulse. Monostable multivibrators are used in computer circuits and in counters for one-time control based upon input pulses. Such circuits are often called one-shot multivibrators, as opposed to free-running types, which need only a single trigger pulse to produce a continuous output of pulses.

The emitter-cathode-coupled monostable multivibrator circuit (Fig. 279)

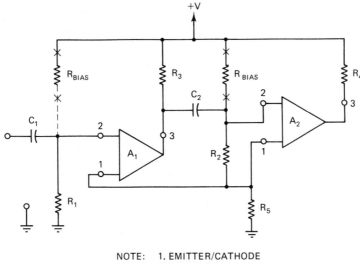

NOTE: 1. EMITTER/CATHODE
         2. BASE/GRID
         3. COLLECTOR/PLATE
         X = TRANSISTOR/CIRCUITS ONLY

FIGURE 279   Monostable multivibrator: (1) emitter/cathode; (2) base/grid; (3) collector/plate. X = transistor/circuits only.

has only one stable state. In this state, one active device conducts while the other active device is normally cut off. The circuit will function for only one complete cycle of operation upon the application of one trigger pulse. During this cycle, the circuit goes to an unstable state, in which the active devices reverse their condition of conduction or nonconduction. The time duration of the unstable state is determined by the circuit constants.

Morse code: A system of dot and dash signals that is used in sending telegraph messages by wire. This code, devised by the inventor of the tele-graph, Samuel F. B. Morse, uses the simplest signals to denote letters, numbers, and punctuation marks. The Morse code was used widely for many years for the transmission of all telegraph messages in the United States and Canada. Today, the code is still in use by some railroads, and for sending overseas telegraph messages by cable and in all wireless telegraph communica-tion. Most telegraph messages sent today, however, originate from a tele-typewriter, which is faster and provides a written copy of the message.

Motorboating: In an audio-amplifier system with two or more stages, a high-frequency howling or a low-frequency staccato noise. This effect is brought about when a small amount of signal fed back from the output to the input sets up self-oscillations in the amplifier, often by way of the supply voltage line that permits a common power supply voltage to be used by each stage of the amplifier. Motorboating can be prevented between stages by the use of decoupling filters. The time constant of a decoupling filter is typically of

0.005 to 0.02 s, which is usually large enough to give good filtering at the lowest frequency in the circuit.

Motor, shaded pole: A single-phase induction motor provided with an uninsulated and permanently short-circuited auxiliary winding displaced in magnetic position from the main winding. The auxiliary winding is known as the shading coil and usually surrounds from one-third to one-half of the pole. The main winding surrounds the entire pole and may consist of one or more coils per pole.

In the unshaded section of the pole, the magnetic flux produced by the main winding is in phase with the main winding current, whereas the flux produced by the shading coil is out of phase with the main flux. Thus, the shading coil acts as a phase-splitting device to produce the rotating field that is essential to the self-starting of all straight induction motors.

As the movement of the flux across the pole face is always from the unshaded to the shaded section of the pole, the direction of rotation can be determined on the normally nonreversible motor by noting the position of the shading coil with respect to the pole itself. This type can be reversed by removing the stator from the frame, turning it through 180°, and replacing it.

The starting torque will not exceed 80 percent of full-load torque at the instant of starting, increases to 120 percent at 90 percent of full speed, and decreases to normal at normal speed. This type of motor operates at low efficiency and is constructed in sizes generally not exceeding 0.05 hp.

Applications of this motor are for fans, timing devices, relays, radio dials, or in general, any constant speed load not requiring high starting torque.

The shaded-pole motor in Fig. 280 has two main windings. This type is externally reversible by means of a single-pole double-throw switch, as shown in this figure. Note that only one set of shading coils is used.

MSI: An acronym for *medium-scale integration,* which is an arbitrary subdivision in integrated circuit terminology that refers to a single chip with

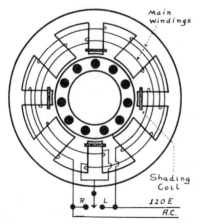

FIGURE 280

between 15 and 100 logic elements. Early integrated circuits contained only a few complete logic operations. However, improvements in monolithic techniques have enabled producers of integrated circuits to greatly increase the number of gates and other logic elements that can be fabricated on a single crystalline chip.

Multioutlet assembly: A type of surface or flush raceway designed to hold conductors and receptacles assembled in the field or at the factory.

Multiplex: The transmission of a number of different signals simultaneously over a single circuit. It is also a process of utilizing a single device for several similar purposes or using several devices for the same purpose. In working with digital electronic systems, a number of analog signals arriving from different sources may be multiplexed into a single digital channel. Conversely, data from a single digital channel may be taken and multiplexed into a number of analog channels.

Multiplier: A device that generates a product from two numbers. A digital multiplier generates the product from two digital numbers by addition of the multiplicand and in accordance with the value of the digits in the multiplier. It then shifts the multiplicand and adds it to the product if the multiplier digit is a 1 or shifts without adding if the digit is 0. This is done for each successive digit of the multiplier.

Mutual induction: The ability of a circuit or device to transfer energy to another electrically isolated circuit or device. Faraday's and Lenz's laws express the concept involved in the process of mutual induction. Faraday discovered that induced EMF was proportional to the rate of cutting of lines of force. Faraday also found that an EMF was induced only when there was a change in the flux linkages. Lenz's law states the same ideas in a more refined manner, but one fact is present in both cases. In order for there to be an induced EMF or a transfer of energy, there must be a change of flux linkage.

Figure 281 illustrates the transferring of energy from a circuit containing a source to an electrically isolated circuit. When the switch is closed, there will be a momentary expansion of flux lines as the field builds up around the wire of the circuit on the left. These flux lines will cut across the conductor of the electrically isolated circuit on the right and produce a momentary EMF. If a steady state of current flow is reached, there will be no more expansion of the field and thus, no more flux lines will be cut.

Since no flux lines are being cut, there will be no CEMF produced. This does not mean the flux field has disappeared. It merely means that as long as there is a steady current flow in the source circuit (primary), the field will exist around the conductors of the primary and the isolated circuit (secondary), but there will be no relative motion.

In order to again induce energy in the secondary, there must be a change in the field (flux linkage). This can be accomplished in one of two ways. The current in the primary could be increased, causing the field to expand further. Additional flux lines would be cut as they passed through the secondary wire.

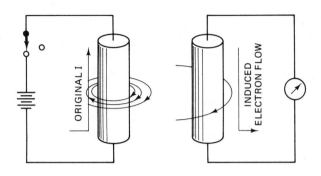

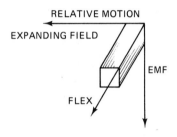

FIGURE 281

All actions would be the same as the initial buildup of the field. The primary current could also be decreased (by opening the switch). Since there is no longer a current in the primary to sustain it, the field will collapse back into the primary wire. In the process of collapsing, the flux lines will have to pass through the secondary wire, again producing a CEMF.

Closer examination of Fig. 281 will show that when the flux lines collapse, they will pass through the secondary in the opposite direction to that which they took when expanding. Applying the left-hand generator rule will show the relative motion of the secondary as being in the opposite direction and the polarity of the CEMF opposite to that shown when the flux was expanding.

A field does not necessarily have to increase from zero to maximum and back to zero in order to produce CEMF. It is now possible to see how a CEMF can be produced by a field that never drops to zero, but simply increases and decreases in density about some average value.

NAND circuit: Also called a *NOT-AND circuit,* a circuit that delivers a zero signal output only when two or more input signals are coincident ones. The performance of a NAND circuit is the inverse of the AND circuit. Two NAND circuits are shown in Fig. 282. Both combine the AND function and the NOT function.

In Fig. 282a, the PNP transistors are biased for saturation conduction in the absence of signals. If positive signals are applied simultaneously to both bases, both transistors will be cut off (the AND function). When cutoff occurs for both transistors, the collector voltage (output) goes relatively negative. This signal inversion is the NOT function.

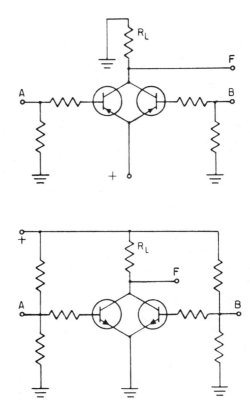

FIGURE 282 NAND circuits.

In Fig. 282b, the NPN transistors are also biased for saturation conduction. When both transistors are cut off by coincident negative signals, the output voltage from the collector is relatively positive. This is a negative-logic NAND circuit.

National Electrical Code (NEC): A set of rules governing the selection of materials, quality of workmanship, and precautions for safety in the installation of electrical wiring. The NEC, originally prepared in 1897, is frequently revised to meet changing conditions, such as improved equipment and materials and new fire hazards. The code is the result of the best efforts of electrical engineers, manufacturers of electrical equipment, insurance underwriters, fire fighters, and other concerned experts through the country.

The NEC is published by the NFPA (National Fire Protection Association) in Boston, Massachusetts. It contains specific rules and regulations intended to help in "the practical safeguarding of persons and property from hazards arising from the use of electricity." The NEC contains provisions considered necessary for safety. Compliance therewith and proper maintenance will result in an installation essentially free from hazard, but not necessarily efficient, convenient, or adequate for good service for future expansion of electrical use.

Natural convection: The circulation of a gas or liquid due to difference in density resulting from temperature differences. A familiar example is the free movement of water contained in a nonforced water-heating system. Another example is the free movement of warm air throughout a room to provide heat.

Natural gas: A form of energy used by power plants to generate electricity. Natural gas varies widely in composition and contains many undesirable materials in its natural state. Some of these undesirables include water and sulfur compounds, all of which must be removed prior to transmission or use. Various chemical processes are used to refine natural gas. In this process, several by-products are produced and marketed, such as propane, butane, and elemental sulfur.

Neutral conductor: A grounded conductor in an electrical system that does not carry current until the system is unbalanced. Neutral conductors must have sufficient capacity for the current that they might have to carry under certain conditions. In a single-phase three-wire system, the middle leg or neutral wire carries no current when the loads on each side between neutral are equal or balanced. If, however, the loads on the outside wires become unequal, the difference in current flows over the neutral wire.

Where a 120/240-V single-phase service is used, it is highly desirable for the 120-V loads to be balanced across both sides of the service. The neutral wire can then be smaller than the two hot or ungrounded wires. The NE code permits the reduction of the neutral to the size that will carry the maximum unbalanced load between the neutral and any one ungrounded conductor.

A general rule of thumb is to reduce the neutral by not more than two standard wire sizes. A further demand factor of 70 percent may be applied in reducing the neutral wire for that portion of the unbalanced load that is in excess of 200 A. However, if 50 percent or more of the load consists of electric-discharge lamp ballasts, the neutral will be the same size as the ungrounded conductors.

In a three-phase, four-wire system, a three-wire branch circuit consisting of two phase wires and one neutral wire has a neutral or grounded wire that carries approximately the same current as the phase conductors. Therefore, it should be the same size as the phase conductors. However, three-phase, four-wire systems generally supply a mixed load of lamps, motors, and other appliances. Motors and similar three-phase loads connected only to the phase wires cannot throw any load onto the neutral, and such three-phase loads can be disregarded in calculating the necessary capacity for the neutral conductor.

Neutralizing capacitor: A capacitor usually employed in a radio receiving or transmitting circuit to neutralize a charge that has been placed across the capacitor plates. This is accomplished by feeding back a portion of the signal voltage equal to one-half of the total charge from the plate circuit to the grid circuit.

Neutron: One of the basic building blocks of an atom. The neutron is located in the central mass, or nucleus, of the atom and possesses no electrical charge. Due to this, the neutron is not deflected by magnetic or electric fields, and its interaction with matter is mainly by collision.

Newton: A unit of measurement that expresses the force of attraction or repulsion between two charged bodies. This force varies directly with the product of the individual charges, and at the same time, varies inversely with the square of the distance between charges.

Nixie tube: An assembly of 10 numeral-shaped wire cathodes numbered from 0 to 9. The cathodes are superimposed, mutually insulated, and arranged before a common anode. On the application of about 100 V between the anode (tied to positive) and one of the cathodes, the corresponding wire numeral stands out glow surrounded, whether it is in the first or in the tenth plane. Since the cathodes are superimposed, all numerals appear at the same place.

The Nixie tubes composing a readout are lined up to provide for convenient in-line reading. Several manufacturers have undertaken the production of complete glow-type display units containing a number of Nixie electrode packs arranged side-by-side in a common rectangular glass envelope, thus reducing the spacing between numerals and simplifying assembly operations.

No-load current: The current that flows through a device or circuit when the same device or circuit is delivering zero-output current. For example, when the primary of a transformer is connected to an ac source, a voltage appears at the secondary winding even when it is not connected to a load. While there is no current flow from the output in this condition, a small amount of current flows in the primary winding to sustain transformer operation. Here, the no-load current is the measured drain from the primary input voltage source.

The same applies to amplifier circuits when their outputs are not connected to a load. A certain amount of input current is required to maintain operation. This is often known as *idle current*.

Non-metallic-sheathed (NM) cable: A type of cable that is popular for use in residential and small commercial wiring systems. In general, it may be used for both exposed and concealed work in normally dry locations.

Type NM cable must not be installed where exposed to corrosive fumes or vapors, nor embedded in masonry, concrete, fill, or plaster, nor run in shallow chase in masonry or concrete and finished or covered with plaster or similar finish. This cable must not be used as a service entrance cable, in commercial garages, theaters and assembly halls, motion picture studios, storage battery rooms, hoistways, hazardous locations, or embedded in poured cement, concrete, or aggregate.

For use in wood structures, holes are bored through wood studs and joists, and the cable is then pulled through these holes to the various outlets. The holes normally give sufficient support, providing they are not over 4 ft on center. When no stud or joist support is available, staples or some similar supports are required for the cable. The supports must not exceed 4.5 ft and must be within 12 in. of each outlet box or other termination point.

Proper tools facilitate the running of branch circuit cables and include

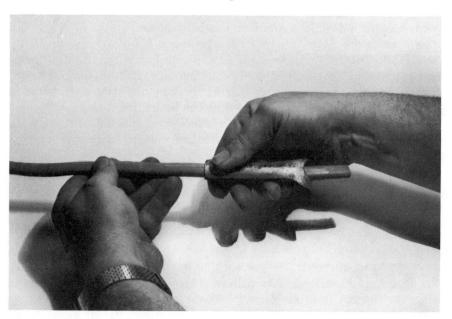

FIGURE 283

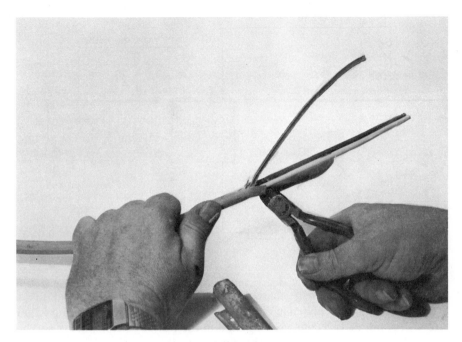

FIGURE 284

sheathing strippers, as shown in Fig. 283. The stripper cuts the sheathing and the wire may then be stripped, as shown in Fig. 284.

NPN transistor: A bipolar transistor that is made from semiconductor materials that form an NPN junction. Three separate sections of semiconductor material are sandwiched together so that the P-type semiconductor lies between the N-type sections, as shown in Fig. 285. The emitter and collector of the transistor are attached respectively to the two N-type materials, while the transistor base lead is connected to the P-type material. NPN transistors may be used to replace similar PNP types, as long as the circuit polarity is reversed.

Nuclear equipment and systems: Any system consisting of a controlled fission heat source, a heat exchanger to remove and transfer the heat produced, and other equipment to convert the thermal energy to electrical power.

The thermal energy is removed from the reactor core by contacting the

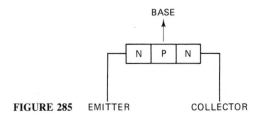

FIGURE 285   EMITTER   COLLECTOR

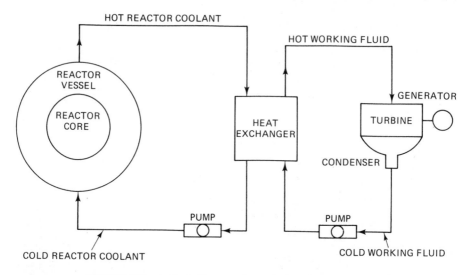

FIGURE 286   A block diagram of a typical nuclear power system.

fuel with a coolant circulated in the reactor vessel and which can be used either directly or indirectly as the working fluid in the power-conversion cycle. The hot working fluid is then used to drive a turbine generator to produce electrical power. A block diagram of such a system is shown in Fig. 286.

Nuclear power systems have a number of advantages. Nuclear fuel has a relatively long life, measured in months or years, whereas fossil fuels must be replenished on a continuous basis. Also, the use of nuclear fuel does not require combustion air, which means that air pollution is kept to a minimum and no stack losses are encountered.

This type of system has disadvantages as well. Nuclear fuel requires a great deal of processing to make it usable in a precise fabricated form, whereas most fossil fuels are essentially raw materials used with comparatively little processing. Burned nuclear fuel is radioactive, thus requiring special handling, processing, and disposal. Also, many areas of a nuclear plant are radioactive during and after operation. This requires special precautions for operation and maintenance of these areas.

Nuclear fuel materials consist essentially of uranium ores, which are widely scattered throughout the world. Besides those deposits available for mining, some uranium is obtained as a by-product of mining operations for other minerals. Uranium ores, however, need to be highly processed before they can be used as a nuclear fuel.

In general, fission occurs when a neutron collides with the nucleus of certain heavy atoms, causing the original nucleus to split into two or more unequal fragments that carry off most of the energy of fission as kinetic energy. The process is accompanied by the emission of neutrons and gamma rays, which release an energy that makes fission possible. The fragments resulting from the fission process are radioactive and decay by emission of beta particles, gamma rays, and, to a lesser extent, alpha particles and neutrons. The fission chain reaction can be easily controlled because some neutrons are emitted after fission. These are called *delayed neutrons.*

Most of the reactors in use today are called *thermal reactors* because they depend on neutrons that are in or near thermal equilibrium with their surroundings to cause the bulk of fissions. Materials called *moderators* are used to decelerate the fast neutrons to thermal energy levels.

The heat, radiation, and radioactive by-products of the fission process in a thermal reactor can be utilized for purposes other than generating electrical power. Exhaust steam, for example, can be utilized to evaporate seawater to produce fresh water. The ionizing radiations emitted during fission can also be used to produce chemicals. Radioactive fission products can be used for small remote power sources, for sterilizing foods, or for polymerizing certain chemicals.

Number system: 1. A systematic method for representing numerical quantities in which any quantity is represented as the sequence of coefficients of the successive powers of a particular base with an appropriate point. Each succeeding coefficient from right to left is associated with and usually multiplies the next higher power of the base. The first coefficient to the left of the point is associated with the zero power of the base.

2. The following are names of the number systems with bases 2 through 20: 2, binary; 3, ternary; 4, quaternary; 5, quinary; 6, senary; 7, septenary; 8, octal or octonary; 9, novenary; 10, decimal; 11, undecimal; 12, duodecimal; 13, terdenary; 14, quaterdenary; 15, quindenary; 16, sexadecimal or hexadecimal; 17, septendecimal; 18, octodenary; 19, novemdenary; 20, vicenary.

Nuvistor tube: A triode that contains a cantilever-supported cylindrical electrode that eliminates the necessity of supplying mica supports. The nuvistor was created in response to the need for a high-vacuum device with increased frequency range to allow the use of the open-circuit approximation. This type of tube can withstand comparatively higher levels of shock and vibration and a much wider frequency range than its forerunners. These characteristics, along with its small size, make the nuvistor suitable for instrumentation equipment, communication systems, and audio and video equipment.

Odd-order harmonic: A signal in a complex waveform that is an odd multiple of the fundamental frequency.

Off-frequency: A term used to describe the operation of a radio transmitter or receiver in relationship to a desired assigned or indicated frequency. If a transmitter frequency indicator reads 7 MHz while the externally measured output is known to be 7.5 MHz, then the indicator is off-frequency by 0.5 MHz.

Ohm: The unit of resistance.

Ohmmeter: A measuring device used to measure resistance and which consists of a dc milliammeter, a dc source of potential, and one or more resistors, none of which is variable. A simple ohmmeter circuit is shown in Fig. 287. The ohmmeter's pointer deflection is controlled by the amount of battery current passing through the moving coil. Before measuring the resistance of an unknown resistor or electrical circuit, the ohmmeter must first be calibrated. If the value of resistance to be measured can be estimated within reasonable limits, a range is selected that will give approximately half-scale deflection when this resistance is inserted between the probes. If the resistance is unknown, the selector switch is set on the highest scale. Whatever range is selected, the meter must be calibrated to read zero before the unknown resistance is measured.

Calibration is accomplished by first shorting the test leads together as shown in Fig. 287. With the test leads shorted, there will be a complete series circuit consisting of the 3-V source, the resistance of the meter coil, R_m, the resistance of the zero-adjust potentiometer, and the series-multiplying resistor, R_s. Current will flow and the meter pointer will be deflected. The zero point on the ohmmeter scale (as opposed to the zero for voltage and current) is located at the extreme right side of the scale. With the test leads shorted, the zero-adjust potentiometer is set so that the pointer rests on the zero mark. Therefore, full-scale deflection indicates zero resistance between the test leads.

If the range is changed, the meter must be zeroed again to obtain an accurate reading. When the test leads of an ohmmeter are separated, the pointer of the meter will return to the left side of the scale due to the spring tension acting on the movable coil assembly. This reading indicates infinite resistance.

After the ohmmeter is adjusted for zero reading, it may be connected in a circuit to measure resistance. A typical circuit and ohmmeter arrangement is shown in Fig. 288. The power switch of the circuit to be measured should always be in the off position. This prevents the circuit's source voltage from

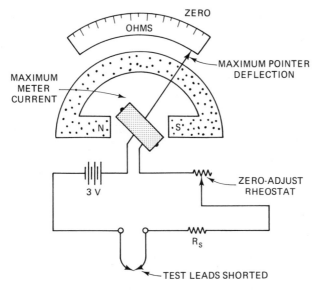

FIGURE 287 Simple ohmmeter circuit.

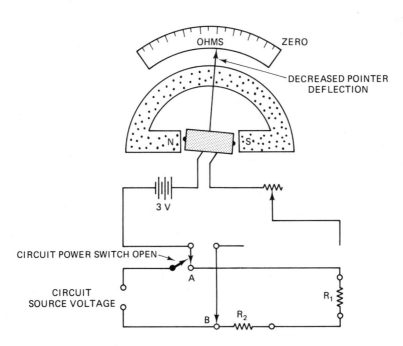

FIGURE 288 Measuring circuit resistance with an ohmmeter.

being applied across the meter, which could cause damage to the meter movement.

As indicated, the ohmmeter is an open circuit when the test leads are separated. In order to be capable of taking a resistance reading, the patch for current produced by the meter's battery must be completed. In Fig. 288, this is accomplished by connecting the meter at points A and B (putting the resistors R_1 and R_2 in series with the resistance of the meter coil, zero-adjust rheostat, and the series multiplying resistor). Since the meter has been preadjusted (zeroed), the amount of coil movement now depends solely on the resistance of R_1 and R_2. The inclusion of R_1 and R_2 raised the total series resistance, decreased the current, and thus decreased the pointer deflection. The pointer will now come to rest at a scale figure indicating the combined resistance of R_1 and R_2. If R_1 or R_2 or both were replaced with a resistor having a larger ohmic value, the current flow in the moving coil of the meter would be decreased still more. The deflection would also be further decreased and the scale indication would read a still higher circuit resistance. Movement of the moving coil is proportional to the intensity of current flow. The scale reading of the meter in ohms is inversely proportional to current flow in the moving coil.

The amount of circuit resistance to be measured may vary over a wide range. In some cases, it may be only a few ohms; in others, it may be as great as 1 MΩ. To enable the meter to indicate any value being measured with the least error, scale multiplication features are incorporated in most ohmmeters. They will be equipped with a selector switch for selecting the multiplication factor desired. For example, a typical meter may have a six-position switch, marked as: RX1, RX10, RX100, RX1000, RX10,000, and RX100,000. This is shown in Fig. 289.

The range to be used in measuring any particular unknown resistance (R_x in Fig. 289) depends on the approximate ohmic value of the unknown resistance. For instance, assume the ohmmeter scale of Fig. 289 is calibrated in divisions from 0 to 1000. If R_x is greater than 1000 Ω and the RX1 range is being used, the ohmmeter cannot measure it. This occurs because the combined series resistor RX1 and R_x is too great to allow sufficient battery current to flow to deflect the pointer away from infinity (00). The switch would have to be turned to the next range, RX10.

When this is done, assume the pointer deflects to indicate 375 Ω. This would indicate that R_x has 375 $\times$ 10 = 3750 Ω of resistance. The change of range caused the deflection because resistor RX10 has only one-tenth the resistance of resistor RX1. Thus, selecting the smaller series resistance allowed a battery current of sufficient value to cause a useful pointer deflection. If the RX100 range were used to measure the same 3750-Ω resistor, the pointer would deflect still further to the 37.5-Ω position. This increased deflection would occur because resistor RX100 has only one-tenth the resistance of resistor RX10.

The foregoing circuit arrangement allows the same amount of current to flow through the meter moving coil, whether the meter measures 10,000 Ω on

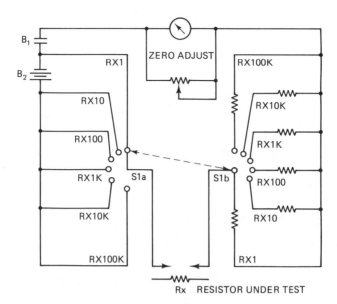

FIGURE 289   Ohmmeter with multiplication switch.

the RX1 scale or 100,000 Ω on the RX10 scale, or 1 MΩ on the RX100 scale. It always takes the same amount of current to deflect the pointer to a certain position on the scale (midscale position, for example), regardless of the multiplication factor being used. Since the multiplier resistors are of different values, it is necessary to always zero-adjust the meter for each multiplication scale desired. The operator of the ohmmeter should select the range that will result in the pointer coming to rest as near midpoint of the scale as possible. This enables the operator to read the resistance more accurately, because the scale readings are more easily interpreted at or near midpoint.

Ohm's law: That law that states that if a voltage of 1 V applied across a resistance causes a current of 1 A through the resistance, the resistance is 1 Ω. This relationship between amperes, ohms, and volts is one of the most important basic electrical laws, because its application helps to solve more different kinds of electrical problems than any other one rule or law.

Ohm's law states that the current in amperes increases and decreases directly with the increase or the decrease of the pressure difference in volts. It further states that doubling the resistance will permit only half as much current to pass; and reducing the resistance to half will permit twice as much current to pass. In other words, the current increases proportionately with every decrease in resistance, and the current decreases proportionately with any increase in resistance, provided the voltage remains the same throughout.

Oil-filled capacitor: A paper capacitor that has been immersed in oil. They are often used in radio transmitters where high output power is desired. The oil-impregnated paper has a high dielectric constant, which lends itself well to the production of capacitors that have a high value. Many capacitors will use

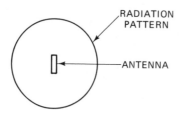

FIGURE 290

oil with another dielectric material to prevent arcing between the plates. If an arc should occur between the plates of an oil-filled capacitor, the oil will tend to reseal the hole caused by the arc. These types are often called self-healing capacitors.

Omnidirectional antenna: An antenna that receives signals with equal sensitivity arriving at the element from all compass points. An omnidirectional antenna, when used for transmitting, will radiate radio-frequency energy equally to all compass points.

Figure 290 shows the radiation pattern for an omnidirectional antenna. It can be seen that this forms a perfect circle. The strength of the radiated signal will be identical at a point 1 mi south of the antenna to that which is measured at a point 1 mi north, east, or west of the radiating element.

The most basic omnidirectional antenna is the half-wave vertical, which is shown in Fig. 291. It is generally not selective in directional frequency response, although this is not exactly true in most practical installations because of earth contours and capacitance effects with nearby objects. To be truly omnidirectional, this antenna would have to be mounted in free space, which is far removed from the earth and any interfering objects.

As a practical definition, an omnidirectional antenna is one that generally tends to radiate and receive signals equally from all compass points. The actual

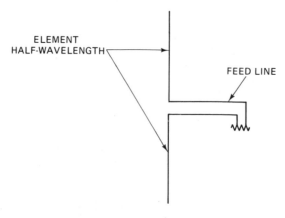

FIGURE 291

mounting environment, however, will usually alter the perfect radiation pattern, so that some signals are better received or transmitted from and in specific directions.

On-frequency: A term most applicable to radio-communications work, indicating that a transmitter or receiver is functioning on an assigned or desired frequency. If a transmission is to be made at 7 MHz and the measured transmitter output corresponds to this frequency, the operation is said to be on-frequency. This term can also be used to describe the proper operation of frequency-indicating dials or readouts that are part of radio-frequency transmitters and receivers. If a transmitter output is measured at 7 MHz by an external frequency meter and the digital readout on the transmitter indicates 7 MHz, the readout is said to be on-frequency. The same terminology applies to receiver operations.

Open-wire line: A transmission line that consists of two or more straight, parallel wires. Typically, low-loss insulating material is used between the wires to keep them uniformly separated throughout the entire line length. Figure 292 shows a common type of two-wire line that is used for connection between radio antennas and a receiver or transmitter.

Parallel lines are often called *tuned feeders* because they are resonant at the operating frequency. Line impedance will depend on the cross-sectional size of the conductors and their spacing with one another.

In radio-frequency communications work, open-wire lines are not as popular as they were in the 1950s. Their use necessitates spacing them away from metallic objects that can affect tuning, and they are quite expensive. Today, coaxial cable, a nonresonant line, is often used for communications work. This is a shielded cable that does not require the stringent mounting considerations of open-wire line.

Open-wire line is still widely employed in television-reception applications where 300-Ω ribbon cable is used to connect the television antenna to the receiver. The spacing element between the two conductors is a solid plastic dielectric. This material also surrounds the conductors, insulating them from moisture.

Operational amplifier: A very high-gain differential amplifier that provides a stabilized voltage gain by using voltage feedback. It is basically a

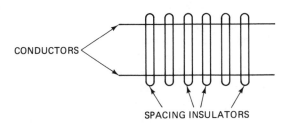

FIGURE 292

differential amplifier, which has a very high open-loop gain in addition to high-input impedance and low-output impedance. Operational amplifiers are generally used for scale changing, in analog computer operations, and in many phase shift, oscillator, and instrumentational circuits.

Optical character reader: A high-speed process by which words, letters, symbols, and numbers are recognized and translated into computer-processable information. The data are machine readable, and at the same time, readable by humans. Modern methods of optical character reading use flying spot electronic and laser scanners for very reliable character recognition. Converting a character into machine representation involves passing it through two systems. The scanning phase determines the presence or absence of a mark or stroke by sensing the amount of light that is reflected from the area it is scanning. The resulting signals are passed onto the recognition system, which uses either hardware or software routines to analyze and decode the output.

Optical character reader procedures are used to enable original documents to be analyzed and coded with the data fed directly into a computer system, eliminating the need for transcription and other keying procedures. However, OCR is not practical in cases where OCR documents are subjected to unclean environmental conditions, extensive handling, or crumpling.

Optoelectronics: An area of electronics that deals with photoelectricity, lasers, and the amplification of light waves. Optoelectric devices normally consist of a light source that is used to trigger a phototransistor, LASCR, light-activated triac, etc. The light source is triggered by current flow in one circuit, while the light-receiver component controls another circuit or simply switches it on or off.

For example, optoelectronic couplers are used to control high-voltage circuits that are connected to the light receiver by means of a low-voltage, low-current source that drives the light source. Both the source and the receiver are located in a single package, and a very high degree of electrical isolation of the two circuits is derived because there is no dc connection between the two.

Optoisolator: An optoelectronic device produced from the combination of an LED and a phototransistor. It allows the transfer of signals from one circuit to another with complete electrical isolation, and an insulation resistance measured in kilovolts between the LED and the phototransistor is usually guaranteed.

Oscillator: A device that electronically produces a pulsating or alternating current. A basic oscillator can be broken down into three main sections. The frequency-determining device is usually an LC tank circuit. While the tank circuit is normally found in the input circuit of the oscillator (both transistor and electron tube), it should not be considered out of the ordinary if it appears in the output circuit of a transistor oscillator.

The differences in magnitude of collector and plate currents and shunting impedances are partly responsible for this. In both types of circuits (solid-state and tube), oscillations take place in the tuned circuit. Both the transistor and tube function primarily as an electric valve that amplifies and allows the

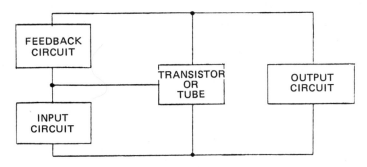

FIGURE 293   Basic oscillator circuit.

feedback network to deliver the proper amount of energy automatically to the input circuit to sustain oscillations.

In both transistor and tube oscillators, the feedback circuit couples energy of the proper amount and phase from the output to the input circuit in order to sustain oscillations. A basic block diagram is shown in Fig. 293. The circuit is essentially a closed loop utilizing dc power to maintain ac oscillations.

Oscillator, backward wave: An oscillator that uses a microwave oscillator tube. Figure 294 shows the backward wave oscillator. It contains a helical transmission line that is surrounded by focusing coils. An electron gun produces a beam in which electron bunching results from interaction of the RF field. Reflection takes place at the collector. Due to this arrangement, the wave moves backward from collector to cathode. Oscillation is sustained because the backward wave is in phase with the input. The oscillator output is taken from the cathode end of the helical transmission line. Backward wave oscillators are tuned by varying the dc voltage potential at the helix. This affords a typical 2 to 1 bandwidth in the 1 to 40 GHz range.

Oscillator, clapp: An oscillator that uses the stabilizing effect of a series-resonant tuned tank circuit coupled to the feedback loop to provide good stability relatively independent of transistor parameters. It also offers capacitive tuning using only one capacitor, without affecting the feedback ration. A shunt-fed common-emitter clapp oscillator is shown in Fig. 295. One voltage supply is used with fixed bias supplied by resistors R_1 and R_B. Emitter swamping resistor R_E, bypassed by C_1, is used for temperature stabilization.

The collector is shunt-fed through choke RFC to keep RF out of the

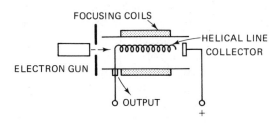

FIGURE 294

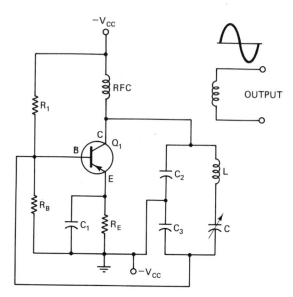

FIGURE 295   Transistor shunt-fed clapp oscillator.

power supply and avoid power supply shunting effects. Note that series feed cannot be used with the clapp circuit, because tuning capacitor C is in series with the tank inductance. Thus, a blocking capacitor in either the base or the collector lead is not required for this circuit. When the circuit is energized, class A bias is supplied for starting by the bias voltage divider consisting of R_1 and R_B, and feedback to the base is applied through feedback divider capacitor C_3. Grounding the common connection between the feedback divider capacitors provides the 180° phase reversal necessary to provide positive feedback from collector to base. The emitter resistor R_E also acts as a thermal stabilizer for collector current temperature variations.

The output may be taken capacitively or by inductance coupling to the tuned tank circuit. With loose inductive coupling, the output obtained is very stable and relatively free of the detuning effects of loading, since the tank circuit is loosely coupled to the feedback loop and is relatively independent of any change in transistor parameters.

Oscillator, overtone: An oscillator that is used in receivers, converters, test equipment, and low-power tansmitters. Its use is generally restricted to the high-frequency range (in excess of 20 MHz) where crystal operation at the fundamental frequency is impractical. Overtone oscillators are crystal-controlled oscillators that make use of crystals cut to have maximum activity at the overtone (normally an odd harmonic of the crystal's fundamental frequency) to maintain stable high-frequency operation.

Overtone oscillators are never untuned. For proper operation, they will make use of one or more tuned circuits that operate at the overtone frequency.

Also, in order to produce an output rich in harmonics, they must be operated similar to class C.

At the present time, the Butler cathode-coupled two-stage oscillator is probably the most widely used oscillator of this type. This is because of its simplicity, versatility, frequency stability, and comparatively great reliability. This circuit seems to be the least critical as to design and adjustments for operation. Its schematic is shown in Fig. 296.

The balanced circuit, plus the fact that twin triodes within a single envelope can be used, contributes to a saving in space and cost and provides for short leads. The problems that exist are that the power output is less than that of a Miller circuit for the same crystal excitation and the broad bandwidth of operation without tuning is not possible.

V_2 is a grounded-grid amplifier whose output is fed back through V_1 and the series-connected crystal. Cathode bias is provided for the tubes by R_1 and R_2. Although both tubes are identical, circuit design is such that V_1 normally conducts harder than V_2. The feedback voltage is coupled capacitively through C_4 to the grid of V_1. The tuned tank circuit (L_1 and C_3) in the plate circuit of V_2 offers maximum impedance at the frequency to which it is tuned. The maximum output voltage (and feedback) occurs at this point, neglecting crystal operation. Resistor R_4 and capacitor C_2 are a conventional plate dropping resistor and decoupling network for V_2. Resistor R_3 and capacitor C1 perform a similar function for V_1. The output of the circuit is normally taken from the plate of B_2, but it could be taken from V_1 or the cathodes without necessarily changing circuit operation.

When power is first applied to the circuit, heavy currents will flow. This is

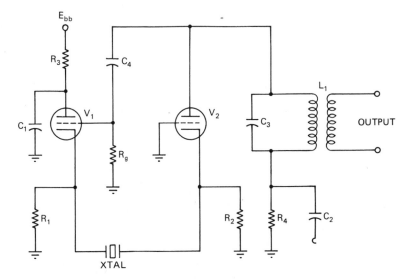

FIGURE 296 Butler cathode-coupled crystal oscillator.

due to capacitor charging and a lack of bias. The difference in conduction levels of V_1 and V_2 will produce a potential across the crystal; crystal current will flow, causing stress; and the resulting stresses will shock it into oscillation.

Assume that the crystal is oscillating in such a manner as to place a positive-going signal on the cathode of V_2 (equivalent to a negative-going signal on the grid). Plate current in V_2 will decrease, and plate voltage will increase. The positive-going signal on the plate of V_2 is coupled through C_4 to the grid of V_1. V_1 will conduct harder, E_{R1} will become more positive than before, and the crystal will couple this positive-going signal to the cathode of V_2. Note that this is a regenerative effect. Observation should make it clear that regeneration will occur on the negative half-cycle of oscillation. Thus, with the crystal providing voltages that will control the conduction of V_2, the variations in the plate voltage of V_2 can then supply the necessary pulsating energy to the tuned tank circuit to maintain oscillations. It should be noted that this circuit will oscillate without the crystal if it is replaced with a resistor. Accurate frequency control, however, is possible only with the crystal in place. Failure analysis of this type of circuit is similar to that of other crystal oscillators.

Oscillogram: The image produced on the screen of an oscilloscope. In some instances, this may be photographed and kept as a permanent record. The photographic print is also called an *oscillogram*.

Oscillograph: An instrument that makes a permanent record of a rapidly varying electrical quantity. These devices may be photographic in nature or may use inked styli to produce a hard image on graph paper. Most chart recorders fall into this category.

Oscilloscope: An instrument that provides a visual display of the electrical characteristics of a device or circuit. The principle components of a basic oscilloscope include a cathode-ray tube (CRT), a sweep generator, deflection amplifiers (horizontal and vertical), a power supply, and suitable controls, switches, and input connectors for proper operation. Figure 297 shows a block diagram of a basic oscilloscope.

Output:

1. The information transferred from the internal storage of a computer to secondary or external storage, or to any device outside of the computer.
2. The routines that direct (1).
3. The device or collective set of devices necessary for (1).
4. To transfer from internal storage onto external media.

Overload: Operation of equipment in excess of normal full-load rating, or of a conductor in excess of rated ampacity, which, when it persists for a sufficient length of time, would cause damage or dangerous overheating. A fault, such as a short circuit or ground fault, is not an overload.

Overmodulation: An amplitude modulation, driving the transmitter to a level greater than 100 percent modulation. Overmodulation is an undesir-

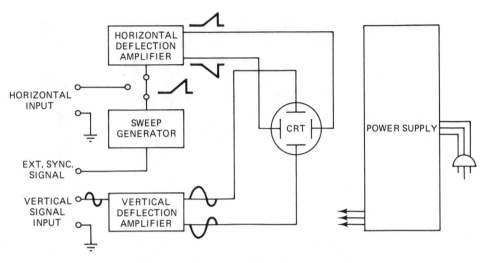

FIGURE 297 Block diagram of a basic oscilloscope.

able characteristic and results in extreme distortion and splatter of the received signal. Most commercially made AM transmitters designed for broadcast purposes use auxiliary equipment to limit the modulation level to 100 percent or less.

Overmodulation indicators are often used to trigger when percentage levels rise above 100 percent. These often consist of a neon bulb, magic-eye tube, light-emitting diode, or incandescent lamp, which is adapted to give an alarm when desired modulation levels are exceeded. Sometimes, an RF relay will be used in place of these other devices to trigger discrete alarm systems.

P

Pacemaker: An electronic device that is used to help control the rhythmic movements of the heart, and also to initiate cardiac beatings when the heart fails to do so on its own. These devices operate by sending a stimulating current to electrodes that have been placed in contact with various heart muscles. This action triggers the beat, producing contractions at or near the normal heart rate.

Panelboard: A single panel or group of panel units designed for assembly in the form of a single panel, including buses, automatic overcurrent devices, and with or without switches for the control of light, heat, or power circuits; designed to be placed in a cabinet or cutout box placed in or against a

FIGURE 298

wall or partition and accessible only from the front. Figure 298 shows a panelboard.

Paper capacitor: A capacitor that uses paper as its dielectric. The construction of a typical paper capacitor is shown in Fig. 299. It consists of flat thin strips of metal foil conductors separated by the dielectric material. In this capacitor, the dielectric used is waxed paper.

Paper capacitors usually range in value from about 300 pF to about 4 μF. Normally, the voltage limit across the plates rarely exceeds 600 V. Paper capacitors are sealed with wax to prevent the harmful effects of moisture from damaging the component.

Parabola: A plane curve that is the locus of points that are equidistant from a fixed point (the focus) and a fixed straight line (the directrix). The parabola is also a conic section and is the curve formed by the intersection of a cone by a plane parallel to the latter's axis.

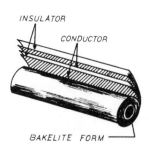

FIGURE 299

Parabolic antenna: An antenna that provides a means of changing the spherical wavefront from the antenna into a plane wavefront for a sharply defined radar beam. A spherical wavefront spreads out as it travels. This produces a pattern that is not very sharp or directive. On the other hand, a plane wavefront does not spread out because all of the wavefront moves forward in the same direction.

Radio waves behave similarly to light waves. Microwaves travel in straight lines, as do light rays. They may be focused and/or reflected just as light rays can. In Fig. 300, a point radiation source is placed at the focal point F. The field leaves this antenna with a spherical wavefront.

As each part of the wavefront reaches the reflecting surface, it is shifted 180° in phase and sent outward at angles that cause all parts of the field to travel in parallel paths. Because of the shape of a parabolic surface, all paths from F to the reflector and back to line XY are the same length. Therefore, all parts of the field arrive at line XY at the same time after reflection.

If a dipole is used as the source of radiation, there will be radiation from the antenna into space, as well as toward the reflector. Energy that is not directed toward the paraboloid has a wide beam characteristic that would destroy the narrow pattern from the parabolic reflector.

To prevent this occurrence, a hemispherical shield is used to direct most radiation toward the parabolic surface. By this means, direct radiation is eliminated, the beam is made sharper, and power is concentrated in the beam. Without the shield, some of the radiated field would leave the radiator directly. Since it would not be reflected, it would not become a part of the main beam and thus could serve no useful purpose. Another means of accomplishing this is through the use of a parasitic array that directs the radiated field back to the reflector.

The radiation of a parabola contains a major lobe that is directed along the axis of revolution and several minor lobes, as shown in Fig. 30l. Very narrow beams are possible with this type of reflector. Figure 302a illustrates the paraboloid reflector.

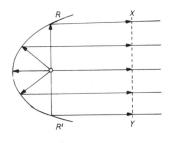

PARABOLIC REFLECTOR RADIATION

FIGURE 300

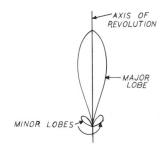

PARABOLIC RADIATION PATTERN

FIGURE 301

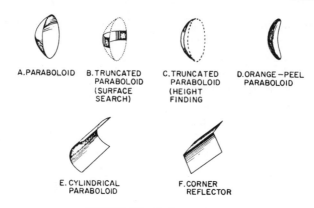

A.PARABOLOID   B.TRUNCATED   C.TRUNCATED   D.ORANGE—PEEL
                PARABOLOID   PARABOLOID   PARABOLOID
                (SURFACE   (HEIGHT
                SEARCH)   FINDING

        E. CYLINDRICAL   F.CORNER
           PARABOLOID   REFLECTOR

FIGURE 302   Reflector shapes.

Figure 302b shows a horizontally truncated paraboloid. Since the reflector is parabolic in the horizontal plane, the energy is focused into a narrow beam. With the reflector truncated or cut so that it is shortened vertically, the beam spreads out vertically instead of being focused. Such a fan-shaped beam is used to determine the azimuth accurately. Since the beam is wide vertically, it will detect aircraft at different altitudes without changing the tilt of the antenna.

The truncated paraboloid reflector may be used in height-finding systems if the reflector is rotated 90° (Fig. 302c). Since the reflector is now parabolic in the vertical plane, the energy is focused into a narrow beam vertically. With the reflector truncated or cut so that it is shortened horizontally, the beam spreads out horizontally instead of being focused. Such a fan-shaped beam is used to determine elevation very accurately.

A section of a complete circular paraboloid, often called an orange-peel reflector because of its shape, is shown in Fig. 302d. Since the reflector is narrow in the horizontal plane and wide in the vertical, it produces a beam that is wide in the horizontal plane and narrow in the vertical. In shape, the beam resembles a huge beaver tail. The RF energy is sent into the parabolic reflector by a horn radiator fed by a waveguide. The horn nearly covers the shape of the reflector, so almost all of the RF energy illuminates the reflector, with very little escaping at the sides. This type of antenna system is generally used in height-finding equipment.

When a beam of radiated energy noticeably wider in one cross-sectional dimension than in the other is desired, a cylindrical paraboloidal section approximating a rectangle can be used. Figure 302e illustrates this antenna. A parabolic cylinder has a parabolic cross section in one dimension only. Therefore, the reflector is directive in one plane only. The cylindrical paraboloid reflector is either fed by a linear array of dipoles, a slit in the side of a waveguide, or by a thin waveguide radiator. Rather than a single focal point, this type of reflector has a series of focal points forming a straight line. Placing the radiator or radiators along this focal line produces a directed beam of energy. As the width of the parabolic section is changed, different beam shapes are obtained. This type of antenna system is used in search and ground control systems.

The corner reflector antenna consists of two flat conducting sheets that meet at an angle to form a corner, as shown in Fig. 302f. This type of reflector is normally driven by a half-wave radiator located on a line that bisects the angle formed by the sheet reflectors.

Parallel circuit: A circuit in which the components are connected across each other so that the circuit could be shown schematically with component leads bridging common conductors as rungs across a ladder. The primary difference that exists between the ideal parallel resonant circuit and the practical parallel resonant circuit is that the practical parallel LC circuit contains resistance. This resistance exists throughout the circuit. However, most of it is located in the inductive branch of the circuit. For purposes of analysis, all of the circuit resistance will be represented by a single resistor placed in parallel with the inductive branch. This resistance will be assumed to account for all of the circuit losses, both ac and dc.

The schematic diagram of a practical parallel LC circuit is shown in Fig. 303a. In this circuit, the impedance of the capacitive branch is equal to X_C, while the impedance of the inductive branch is equal to the vector sum of X_L and R. If the source is adjusted to the frequency at which X_L is equal to X_C, the current through the inductive branch will be smaller than the current through the capacitive branch. The resulting vector cancellation of I_C and I_L yields a low value of I_C (the total amount being determined by the size of the resistance R). Therefore, the total current will lead the applied voltage by a small angle, making the circuit appear slightly capacitive to the source.

If the applied frequency is reduced slightly, the current through the capacitive branch decreases and the current through the inductive branch increases. The current through the inductive branch will cancel a greater percentage of the capacitive branch current, and the total current will attain its minimum value.

Due to the energy losses that occur in a practical parallel LC circuit, some current must be drawn from the source. Because of this current, the circuit will have a finite impedance at the resonant frequency.

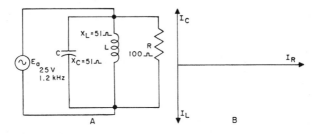

FIGURE 303 Parallel resonant circuit with parallel resistance.

Parasitic array: An antenna system that consists of two or more elements in which not all of the elements are driven. The other elements are excited by induction and radiation fields that are produced by the driven element. With parasitic arrays, it is possible to obtain highly directional patterns.

The action in a parasitic array is analogous to the action in a transformer in which the primary induces a current in the secondary and the current in the secondary produces a magnetic field. This, in turn, induces current back into the primary. The phase relationship between elements in an array varies according to the spacing between elements.

In the two-element array shown in Fig. 304, the driven element is cut at the center for connecting a low-impedance feed line. The length of the driven element is a half-wavelength. This makes it self-resonant. A parasitic element, called the *reflector,* is located 15 percent of a wavelength in space from the driven element and is about 5 percent longer than the driven element.

If the polar diagram is plotted for two elements spaced 0.15 wavelength apart and excited out of phase, the curve shown at the right of Fig. 304 results. This shows that most of the radiation is on the side of the driven element, which is away from the parasitic element, while very little occurs on the side of the reflector.

The field passing from the reflector cuts the antenna (driven element) and induces a voltage in it. This voltage changes the input current. The input impedance, which is a function of this current, is about 50 Ω as compared to 73 Ω for the antenna alone.

A parasitic element becomes a director when it is made shorter than the antenna element. In a director, most of the energy is sent in a direction from the antenna element to the parasitic element. To see what takes place, note the radiation pattern and arrangement of the array in the diagram in Fig. 305. The director is usually about 5 percent shorter and about 0.1 wavelength from the antenna. Sometimes, the impedance is reduced to 20 Ω at the driven element in this array.

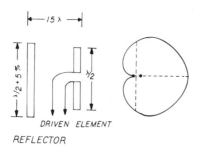

ANTENNA WITH REFLECTOR

FIGURE 304

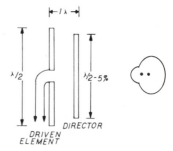

ANTENNA AND DIRECTOR

FIGURE 305

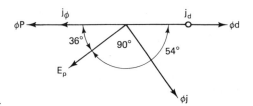

FIGURE 306 Phase relationships.

Figure 306 shows the phase relationship in the parasitic array vertically. Vector **id,** which represents the current in the driven element, is in phase with the H field. The part of the H field that cuts the parasitic element lags the field that leaves the driven element by 0.15 of a cycle. This is the time lapse during the travel between elements. The lag is equal to 54°. The flux at the reflector is shown by vector **oi.** It lags **id** by 54°. The voltage induced by the field is 90° out of phase with the field. This voltage is represented by the vector **Ep.** If the parasitic element were resonant, the current in it would be in phase with **Ep,** but the reflector is longer than the resonant half-wavelength. A long antenna is inductive and the current in this element will lag the voltage by 36° if it is approximately 5 percent longer. The radiated field will be in phase with this current. In summary, the field starting out from the parasitic element will be 180° out of phase with the field leaving the driven element.

Parity bit: A check bit that indicates whether the total number of binary 1 parity bits indicates an odd number of 1 digits and a 0 bit indicates an even number. If the total number of 1 bits, including the parity bit, is always even, the system is called an even parity system. In an odd parity system, the total number of 1 bits, including the parity bit, is always odd.

Partition noise: A type of noise that is present in multigrid tubes. Partition noise is caused by some of the electrons that leave the cathode and move through the control grid, toward the plate, reaching the plate while others strike one of the additional grids and do not arrive at the plate at all. Therefore, a random distribution of electrons between plate and other positive elements will occur and produce random variations in plate current.

Partition noise can be most troublesome. Since this type of noise is caused by grid structures, tubes containing many grids are not used in situations where noise is critical.

Peak limiting: The automatic limiting of the magnitude of an output signal to approximate a predetermined maximum value. Peak limiting is accomplished electronically by reducing amplification when the instantaneous signal magnitude exceeds the predetermined value. Peak limiting may also be called *clipping* because the output waveform presents a clipped appearance on the wave peaks. It may be accomplished with diodes or amplifying devices (transistors or tubes). Diode limiters or clippers may be classified according to the manner in which they are connected (series or parallel). A positive lobe circuit abolishes either part or all of that portion of a waveform that is positive in respect to some reference level. Conversely, the negative lobe limiter affects a waveform's negative portion.

Pentode vacuum tube: A five-element tube that includes a suppressor grid inserted between the screen grid and the plate for the purpose of preventing the screen from attracting secondary electrons from the plate. The five elements are the cathode, control grid, screen grid, suppressor grid, and plate. Figure 307 shows a schematic diagram of a pentode vacuum tube.

In the pentode, the suppressor grid (usually internally connected to the cathode) serves to repel or suppress secondary electrons from the plate. It also serves to slow down the primary electrons from the cathode as they approach the suppressor. This action does not interfere with the flow of electrons from cathode to plate, but serves to prevent any interchange of secondary electrons between screen and plate. The suppressor thus eliminates the negative resistance effect that appears in the tetrode in the region where plate voltage falls below that of the screen. Thus, plate current rises smoothly from zero up to its saturation point as plate voltage is increased uniformly with grid voltage held constant. Typical pentode i_p-e_p characteristic curves are shown in Fig. 308. The amplification factor of pentodes is very high in comparison with triodes or tetrodes.

In the RF pentode, the chief purpose of the screen grid is to eliminate the effects of interelectrode capacitance coupling between control grid and plate circuits. In the power pentode, at audio frequencies, the screen permits the output signal plate voltage variation to be relatively large without the degenerative action occurring as it does in the triode.

Plate current is substantially independent of plate voltage in the power pentode, since the screen voltage is the principal factor influencing plate current. With the addition of the suppressor, the allowable output voltage variation is larger than that of the tetrode, and the distortion effects shown in the tetrode in Fig. 308 are eliminated.

Thus, an audio-frequency power pentode has an allowable output voltage variation in which the plate voltage can fall a large amount below that of the screen voltage on the positive half-cycle of input signal without clipping the plate signal current. Thus, the ratio of output power to grid driving voltage is relatively large.

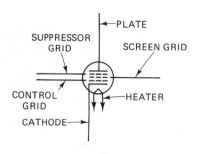

FIGURE 307

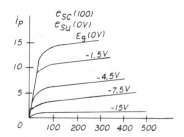

FIGURE 308

Permanent magnet moving-coil meter: The basic movement used in most measuring instruments for servicing electrical equipment. The basic movement consists of a stationary permanent magnet and a movable coil. When current flows through the coil, the resulting magnetic field reacts with the magnetic field of the permanent magnet and causes the coil to rotate. The greater the intensity of current flow through the coil, the stronger the magnetic field produced; and the stronger the magnetic field, the greater the rotation of the coil.

Permanent magnet speaker: A loudspeaker that converts electrical audio signals to sound power and consists of a permanent magnet on mounted soft-iron pole pieces, a voice coil that acts as an electromagnet, and a loudspeaker cone connected to the voice coil.

Figure 309 shows a diagram of a permanent magnet speaker. The audio signal has been previously amplified (in terms of both voltage and power) and is applied to the voice coil. The voice coil is mounted on the center portion of the soft-iron pole pieces in an air gap so that it is mechanically free to move. It is also connected to the loudspeaker cone. As it moves, the cone will also move.

When audio currents flow through the voice coil, the coil is moved back and forth proportionally to the applied ac current. As the cone (diaphragm) is attached to the voice coil, it also moves in accordance with the signal currents. In so doing, it periodically compresses and rarefies the air, thus producing sound waves.

Most speakers of this type receive their input by means of transformer coupling. This is necessary because of the normally low impedance of the voice coil. Standard impedance values for such speakers are 4, 8, 16, and 32 Ω. Other impedance values may be obtained, but these are by far the most common.

While permanent magnet speakers perform reasonably well in the audio range, they nevertheless have inherent limitations. When the speaker is constructed, only a limited number of turns may be built into the voice coil. Therefore, it has a fixed inductance. At low frequencies, the inductive reactance of the voice coil will be relatively low and large audio currents will flow. This provides a strong magnetic field around the voice coil and a strong

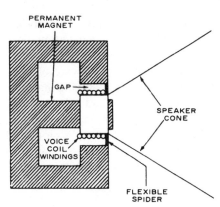

FIGURE 309   Construction of permanent magnet dynamic loudspeaker.

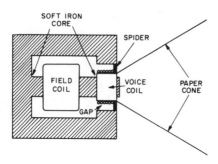

FIGURE 310   Electromagnetic type of speaker.

interaction with the field of the permanent magnet. Low-frequency response is therefore excellent. At mid-band frequencies, the inductive reactance has increased and less current can flow in the voice coil. This produces less magnetic field and less interaction. Mid-band response, however, is still acceptable in a properly designed speaker. At high audio frequencies, however, inductive reactance is quite high and very little current flows in the voice coil. This results in a greatly reduced voice coil field and very little interaction with the permanent magnet field. Also, at high frequencies, the interwinding capacities of the voice coil tend to shunt some of the high audio frequencies and further reduce response.

It can be said that the frequency response of most permanent magnet speakers will fall off at the higher audio frequencies. This problem is typically overcome either by the procurement of an expensive, specially designed speaker, or through the use of two speakers, one of which is designed to operate well at the higher audio frequencies.

As shown in Fig. 310, an electromagnet may be used in place of the permanent magnet to form an electromagnetic dynamic speaker. However, in this instance, sufficient dc power must be available to energize the field electromagnets. The operation is otherwise much the same as that of the permanent magnet speaker.

Permeability: The ability of a core of magnetic material to conduct lines of force. Some core materials have higher permeabilities than others and thus, used with a given winding, they provide inductors having greater inductance values. More precisely, the permeability of a certain magnetic material is the ratio of the flux produced with that magnetic material as the core to the flux produced with air as the core.

It is important to be aware of the fact that the permeability of a given magnetic material varies with the magnetizing force in ampere-turns per unit of core length that is applied to the core. The permeability also depends on the amount and direction of any magnetic flux that might already exist.

Permeability curve: A plot which shows the properties of a magnetic material.

The magnetizing force (H) is plotted along the horizontal axis, while flux density (D) is plotted along the vertical axis. The BH curves of ferromagnetic

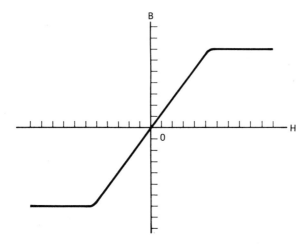

FIGURE 311

materials usually differ from each other. When the field intensity or magnetizing force is increased to a certain point, the material becomes saturated and the curve flattens out past this point. The saturation point is called the *knee*.

Past the saturation point, the magnetizing force has a nominal effect on the value of flux density. The permeability of the material is the ratio of *B* to *H* and is normally measured at the point where saturation is established. Figure 311 shows a typical permeability or *BH* curve.

Permeability tuning: A highly stable method of tuning a radio-frequency receiver or other such circuit by moving a magnetic core into or out of the master oscillator coil to vary its inductance. Most high-frequency receiver tuning involves a fixed inductor in parallel or in series with a variable capacitor. Variable capacitors cannot be manufactured to provide as stable an operation as most adjustable inductors, so permeability tuning is often incorporated in sophisticated receivers whose applications demand rigid frequency tolerance.

Permeability tuning may also be used to establish the output frequency of radio transmitters in the same manner.

Phantrastron circuit: A relaxation oscillator similar to the multivibrator in operation. Whereas the multivibrator derives its timing waveform from an RC circuit, the phantrastron uses a basic Miller sweep generator to generate a linear timing waveform rather than the exponential waveform developed by the RC circuit of the multivibrator. Thus, the output waveform is a linear function of the input (control) voltage and the timing stability is improved.

The phantrastron circuit is used to generate a rectangular waveform or linear sweep whose duration is almost directly proportional to a control voltage. Because of its extreme linearity and accuracy, this waveform is used as a delayed timing pulse, usually in radar or display equipment. It is also used to produce time-delayed trigger pulses for synchronizing purposes and movable marker signals for display. For example, it is used as a time-modulated pulse to indicate antenna position at any instant of rotation, or as a range strobe or delay marker.

Phase angle: A measure of unit, degree, or radian. Its symbol is O.

Phase converter: A device that will permit the operation of a three-phase induction motor from a single-phase power source. Since most commercial suppliers of power place limits on the size of single-phase motors they can serve, such limitations may also apply to three-phase motors supplied through a phase converter. The power supplier should be consulted regarding the size of converter that can be operated in a particular location.

Application of phase converters to easy-starting loads usually involves no particular problems. Such loads would include large fans and centrifugal and turbine-type irrigation pumps. For equipment that requires high starting torque or is subject to wide load fluctuation, application of a phase converter should be made only after consulting the manufacturer of the three-phase motor. Such equipment would include compressors, pumps, and barn cleaners that start under load and feed grinders and blowers on which load may vary due to uneven feeding.

Phase converters may be divided into two general types—the static converter and the rotating transformer converter. Static converters are subdivided into several types, among which are autotransformer converters, series-winding converters, and multimotor converters. Some of these must be matched in horsepower rating to the motor to be driven; that is, a 5-hp converter for a 5-hp motor, etc. The multimotor converter, as its name implies, will operate two or more motors.

The rotating transformer-type converter should have a horsepower rating as large as that of the largest motor to be driven. Additional smaller motors may be supplied by the same converter.

The starting current of a converter (three-phase motor combination) is likely to be less than for a comparable single-phase motor. By the same token, the starting torque of the motor-converter combination is likely to be less than for a similar-sized single-phase motor or for a three-phase motor operated from a three-phase supply. Likewise, the motor connected through a converter may have very little short-time overload capacity.

Overcurrent protection may be difficult to provide because of the longer starting period and the unbalanced currents that occur in the motor windings under overload conditions. Power factor is likely to be near 100 percent at rated load and slightly leading when idling on the line.

Phase protection: A means of preventing damage in an electric motor through overheating in the event a fuse blows or a wire breaks when the motor is running. This is necessary because a motor will continue to operate on single phase even if some part of the device is damaged. In order to provide this protection, phase-failure and -reversal relays are used. Phase-failure relays are available in a number of designs, some of which are quite complex devices.

One type of relay utilizes coils that are connected in two lines of the three-phase supply. The currents in the coils cause a rotating magnetic field to be set up to turn a copper disc clockwise. This movement, which is known as *torque,* results from one polyphase torque turning the disc clockwise and one single-phase torque turning the disc counterclockwise. The disc is kept from turning by a projection resting against a stop.

In the event that the disc rotates counterclockwise, the projecting arm causes a toggle mechanism to open line contacts, thus removing the motor from the line and preventing any damage. In the event that one line opens, the polyphase torque disappears, the single-phase torque rotates the disc counter-clockwise, and the motor is again removed from the line, thus preventing overheating.

Some types of phase-failure relays are quite complex. These are commonly used in situations where it is necessary to protect not only a device, but also any persons or machines from the dangers involved in open-phase or reversed-phase sequence conditions. In some cases, a phase-failure relay may consist of a static, current-sensitive network that is connected in series with the line and a switching relay connected in the coil circuit of the starter. This sensing network serves to monitor the line current in the motor. In the event one phase opens, the sensing network detects it, causing the relay to open the starter coil current, which disconnects the motor from the line. This type of phase-failure relay usually contains a built-in delay of five cycles, which prevents nuisance drops that are caused by transient line fluctuations.

Phasor: A straight line that has a definite length or magnitude and a definite direction within a given plane. It represents a sine wave of voltage or current. A phasor representing the sine wave in Fig. 312 is shown in Fig. 313.

When phasors are used instead of sine waves, the problems of adding and subtracting voltage and current waves are simplified. There is no need for trigonometric formulas or complicated algebraic computations. It is sufficient to know how to draw phasors, understand their meaning, and measure their lengths and the angles between them.

Phone plug: A type of plug that was originally designed for patching telephone circuits, but is now widely used in electronics and instrumentation. In conventional form, the phone plug has a rod-shaped neck that serves as one contact and a ball on the tip of the neck, but insulated from it, which serves as the other contact. A typical phone plug is shown in Fig. 314.

Phonon: A unit of energy resulting from vibration, as in a piezoelectric crystal. When this lattice vibration occurs, a small amount of quantum energy is released.

Phosphor dot: A thin, circular layer of phosphorous that is applied in individual dots on the inside face of television receiver picture tubes in order to respond to the electron-beam trace. Color picture tubes use red, green, and

(a)

(b)

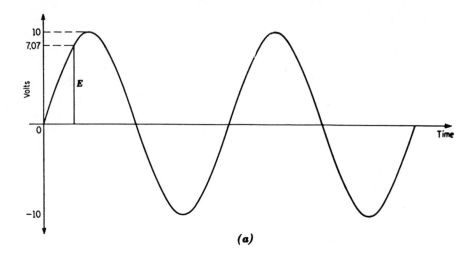

(*a*) Sine wave
(*b*) Phasor

FIGURE 312

(a)

(b)

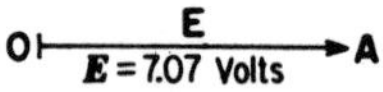

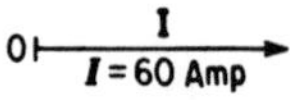

(*a*) Phasor
(*b*) Sine wave

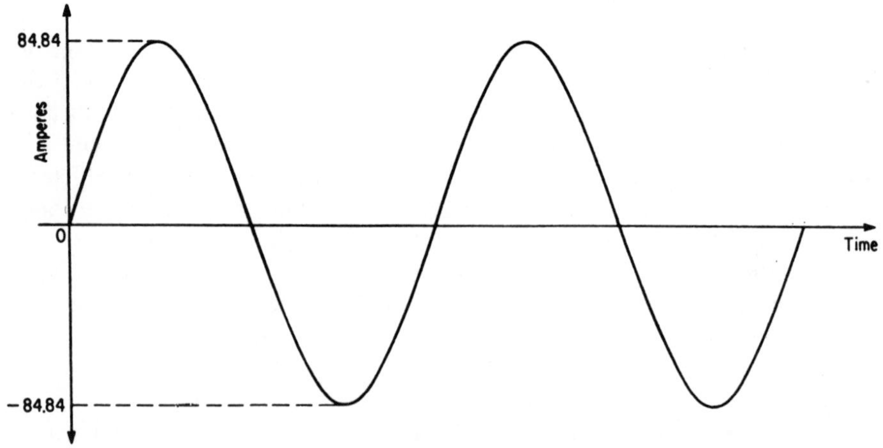

FIGURE 313

FIGURE 314   *PHONE PLUG*

blue phosphor dots to make up the color-forming network. High-contrast color picture tubes separate the dots by a black background that helps to display the differences in high and low picture-tube light levels better.

Phosphorescence: The property of some materials that ordinarily fluoresce to continue to glow after the stimulus has been removed. In most cases, the stimulus will be a light beam or a narrow stream of electrons. Many different types of phosphors are used in electronic applications. These materials are normally painted in a thin layer on the inside of the face of a cathode-ray tube. When the electron beam strikes the sensitive surface, a trace is left that can be seen after the electron beam is focused to a different area of the face.

Cathode-ray tube phosphors are normally rated in P numbers. The lower numbers (P-1, P-3, P-4) offer short persistence. This means the glow continues for only a second or so after the electron beam is removed. High-persistence cathode-ray tubes are also available, and the trace lines will continue to glow for 5 s or more after electron beam removal. These tubes (often P-7 varieties) were used in the early days of amateur slow-scan television (SSTV) experimentation.

Photoflood lamp: A high-efficiency light source designed to produce high levels of illumination in the picture-taking area over a relatively short period of time. These lamps, because of their high filament temperature, produce about twice the lumens produced by similar wattages of general service lamps and have three times the photographic effectiveness of these lamps.

Relatively small bulb sizes are employed so that these lamps may be conveniently used in less bulky reflecting equipment or for certain effects in ordinary residential or commercial fixtures.

Photoresistor: A semiconductor resistor that changes its internal resistance in direct proportion to the amount of ambient light present. They may also be called *photocells, solar resistors,* or *cadmium-sulfide cells,* depending on the structure of the semiconductor material.

The photoresistor is a passive device and does not convert light into electrical energy. Photoresistors operate as they do because of the photoelectric effect, which is the phenomenon whereby temporary changes occur in the atoms of certain substances under the influence of light. Some of these materials undergo a change in electrical resistance, which is lowered as light levels increase.

Photoresistors are often used in photographic light meters and as triggers for electronic relays that are activated by ambient light levels. These are normally small devices that will exhibit a maximum resistance of several megohms when in complete darkness. When subjected to intense light, their internal resistance will drop, often to below 100 Ω.

Phototube: A device that responds to light and is the basis of many electronic controls. This tube is essentially a diode; and like all diodes, it has two electrodes, an anode, and a cathode. Current will flow when the anode is positive with respect to the cathode, provided that the cathode is illuminated.

In rectifier tubes, the cathode releases electrons when heated. In the phototube, the cathode will release electrons when light strikes it. Thus, there are two requirements for the phototube action: The anode must be positive, and the cathode must have light falling on it. The more light that falls on the phototube cathode, the greater the current flow through the tube. At best, however, this current will be very small, about 20 millionths of an ampere. It is so small that it cannot do much work and must be used in conjunction with an amplifying triode to close a relay, which, in turn, can start or stop a motor.

Pierce crystal oscillator: An oscillator usually designed to operate in the lower and medium radio-frequency ranges. It produces an output that is an approximation of a sine wave having a relatively constant frequency and is used in applications that require an output of moderate power and good stability at a specific frequency. It is usually interchangeable with a Miller oscillator in low- and medium-frequency applications, but is not often used in high-frequency applications, primarily due to its low-output power.

The Pierce oscillator utilizes the piezoelectric effect of a crystal to control its frequency of operation. In this circuit, the crystal is connected between the grid and plate of the tube. It does not require an LC tank for operation, although a tank may be used if desirable. The RF feedback occurs only through the crystal and it normally operates with class B or C self-bias, although it may be operated class A, or with a combination of fixed and self-bias. The frequency stability of the Pierce oscillator is excellent with or without temperature compensation, and its output amplitude is relatively constant.

The simplicity of the Pierce oscillator, its lack of tuned plate tank circuit, and its ability to oscillate easily over a broad range of frequencies with the use of different crystals make it popular for use in crystal calibrators, receivers, and test equipment, and in transmitters that do not require much driving power.

The basic Pierce oscillator is shown in Fig. 315. Conventional grid leak bias is obtained through C_g and R_g. The crystal, which is connected between grid and plate, offers a high Q. The plate load is resistor R_1. C_2 is the conventional plate bypass capacitor used in series plate feed arrangements.

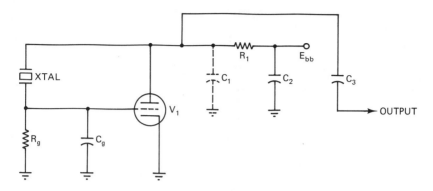

FIGURE 315

The use of R_1 in the plate circuit provides a relatively flat response over a wide range of frequencies. Various crystals may be substituted for operation on other frequencies without any tuning being required. However, when frequency operation is anticipated, R_1 is replaced with an RFC. The choke eliminates the dc power lost in the resistor and provides a high-RF impedance for proper circuit operation. Since this would increase the plate voltage, the output power is correspondingly increased.

Although the Pierce oscillator is normally used without a plate tank circuit, this is not always so. Where the output waveform is important, use of a selective, tuned circuit in the plate circuit minimizes the distortion in the output, thus providing a waveform of greater purity than is produced by a resistive plate load. Selection of the circuit with a resistor plate load for use in a crystal calibrator to supply harmonics of 200 to 300 times that of the fundamental proves particularly advantageous. On the other hand, when harmonic (frequency-mulitplication) operation is desired, the Pierce circuit must use a tuned tank circuit to select the desired harmonic if a useful and practical output is to be obtained.

Pincushion distortion: In television receivers, that effect that is evident when each side of the raster sags toward the center of the screen. This effect is shown in Fig. 316. This type of distortion is normally created by internal defects within the television receiver and is often evidenced when voltage levels drop below minimums.

Pi network: A filter circuit often used to match a radio transmitter to a wide variety of antenna impedances. Pi networks form the output tank circuits in many transmitters of modern design. This circuit is valuable for use as an impedance transformer over a wide ratio of transformation values.

The pi network derives its name from its resemblance to the Greek symbol π. C_1 is the input capacitor, while C_2 is located at the output. The inductor, L_1, can be thought of as two inductors in series. A pi network is actually a combination of two L networks.

The peak voltage rating of C_1 must be high enough to withstand the plate

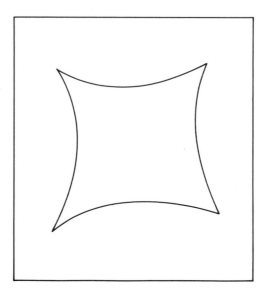

FIGURE 316

voltage used at the plate at the output amplifier tube (or the collector voltage in solid-state circuits). The voltage rating of C_2 can be substantially lower and will depend on the actual power output to the antenna.

Since pi networks often make up the tank circuit of RF power amplifiers, it is quite common to see them built with variable components. Normally, C_1 and C_2 are transmitting variable capacitors, while the inductor is tapped at several locations to provide band-changing capability. The harmonic attenuation of the network is quite good. In well-designed circuits, the second harmonic will be suppressed approximately 45 dB. External pi networks are often coupled to the outputs of RF transmitters to provide a wider tuning range and increase harmonic suppression.

Pink noise: Interfering noise that occurs over a small specific frequency range. This often occurs in radio receivers that are operated near a defective ac power line. The noise is frequency selective and will be tuned at one frequency and not at any others.

Planar array antenna: A two-dimensional configuration of elements arranged to lie in a plane. It is the most versatile of all radar antennas. Radar beams can be rapidly scanned without any mechanical movement of antennas. Also, several independent beams can be generated simultaneously from the same antenna structure. When the array is mounted as a broadside array, the radiation pattern will be bidirectional and perpendicular to the plane of the array. With a reflector screen located a quarter-wavelength behind the elements, the radiation pattern is unidirectional. When the array is mounted as an end-fire array, the radiation pattern will be bidirectional and parallel to the plane of the array.

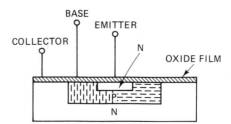

FIGURE 317   Planar transistor.

A rectangular aperture can produce a fan-shaped beam. A square or a circle aperture produces a pencil beam. The array can be made to generate simultaneously many search and/or tracking beams with the same aperture (with the use of phase shifting devices).

Planar transistor: A transistor in which the emitter, base, and collector elements terminate on the same face (plane) of the silicon wafer. A thin film of silicon dioxide is grown on top of the wafer to insulate the exposed junctions after the leads have been attached; i.e., the transistor is passivated. (See Fig. 317.)

Plan position indicator: A type of scan developed for the purpose of presenting range and bearing information. It may also be considered as a modified type of B scan in which the rectangular coordinates are replaced with polar coordinates, thus presenting a polar map-like presentation of the area searched. Electromagnetic deflection is employed where a PPI-scan presentation is used. See Fig. 318.

The antenna for the PPI-scan presentation is rotated uniformly about the vertical axis, thus searching or covering an area in the horizontal plane. The beam radiated by the antenna is usually narrow in azimuth and broad in elevation to give great slant range covered during surface and air search. However, where this type of presentation is used, the antenna can rotate through 360° or scan a pie-shaped sector ahead of the aircraft. The antenna may be caused to scan back and forth through a selected sector by use of controls available to the operator. The scanning rate is quite rapid. Therefore, the area scanned is covered frequently and the observed targets are presented distinctly on the screen of the cathode-ray tube (CRT).

The PPI scan is developed by causing the sweep or trace on the screen to rotate in synchronism with the antenna about the center axis of the CRT. Thus, each sweep is an indication of the instantaneous position of the antenna in azimuth. The rotation of the sweep is accomplished by varying the amplitude of the current through the vertical and horizontal deflection coils of the deflection yoke. The radial sweep is the time base and the range of a target is measured from the center of the screen (or from the sweep's starting point if depressed center is used) outward along the race or sweep. Most PPI scans utilize range markers to calibrate the screen. This is done by the timing system, which causes the trace to be intensified at specific intervals. Thus, equally spaced concentric

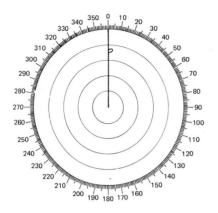

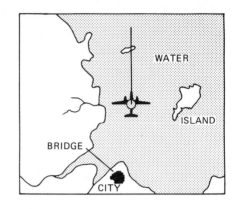

ANTENNA POINTING DIRECTLY AHEAD (0°)

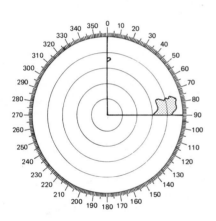

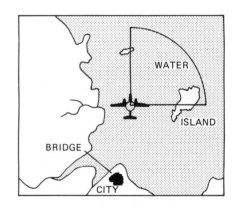

ANTENNA POINTING TO 90°

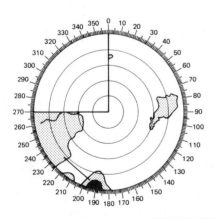

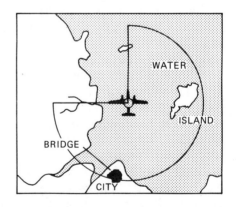

ANTENNA POINTING TO 270°

FIGURE 318 PPI presentation.

circles or arcs are produced to calibrate the screen in multiples of 1, 5, 10, or 20 mi, or any other markers that are available for selection.

When a target echo is received, the intensity of the sweep increases, indicating the position of the target. The target may be presented in some detail, giving an indication of its size and shape. The target's range and relative position may be read directly from the screen with the aid of the range markers and an azimuth scale around the screen of the CRT.

Plate: Another name for the anode of an electron tube. This term may also describe one of the electrodes of a primary or secondary battery cell or one of the electrodes of a capacitor. Often, plate identifies the positive electrode of a number of different electronic components or devices.

Plate current: The direct current that flows in the plate circuit of an electron tube. Abbreviated I_p, when this value is multiplied by the plate voltage potential, the product is equivalent to the total plate power input. Plate current, along with plate voltage, must be monitored in many types of radio-frequency amplifiers according to Federal Communications Commission regulations. These are usually the sole determining factors of total power input, which must be kept to a specific level in many commercial, business, and amateur radio applications.

Plate modulation: A method of superimposing information on a radio-frequency carrier wave by varying the plate current of the RF amplifier at an audio rate. Figure 319 illustrates a plate-modulated RF power amplifier. The V_1 driver circuit is an audio-frequency voltage amplifier whose output is transformer coupled to the V2 modulator stage. The modulator is an audio-

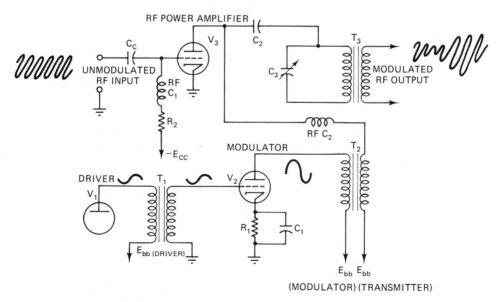

FIGURE 319   Class C Plate modulated RF amplifier.

frequency power amplifier whose output is transformer coupled to the RF power amplifier plate circuit. The AF driver and modulator circuits must be biased class A to minimize audio-modulating frequency distortion. The RF power amplifier is usually operated class C to obtain the necessary nonlinearity for good modulation and maximum efficiency.

Plate resistance: Abbreviated r_p, the total resistance of the internal plate circuit of an electron tube. Alternately, it may be the resistance of the internal plate and all external circuitry attached to this electrode. Plate resistance is measured in ohms. The static value of plate resistance is equal to the resting plate voltage divided by the resting plate current. The dynamic resistance value is equal to maximum plate voltage divided by maximum plate current. Plate resistance may also be a resistor connected in series with the tube plate and the plate power supply, although this component is more often referred to as a plate resistor.

Plate voltage: The dc potential applied to the plate electrode of an electron tube. It is abbreviated E_p and is normally the highest voltage required of any element within the tube proper. Plate voltage potential can range from less than 25 V in some low-power applications to 15,000 V or more in extremely high-powered radio and television transmitters.

Plate voltage power supplies nearly always produce a dc output, although some specialized electronic circuits may dictate an ac potential. In amplifier circuits, the stability of the voltage value of the plate supply is important, especially when high to moderate current is being drawn. In frequency-determining circuits, a change in the plate voltage supply can bring about a subsequent change in frequency. The term *plate voltage* applies only to electronic circuits that use vacuum tubes. This voltage may also be referred to as B +, which is a throwback to the older days of electronics when storage batteries were often used for portable equipment.

In many applications, plate voltage values are high enough to be deadly. Radio-frequency amplifiers with output powers of 500 W or more typically require plate supplies in excess of 1000 V. The equivalent of plate voltage in transistorized circuits is called *collector voltage* and is normally at a potential of less than 100 V and typically 25 V or less. Using solid-state circuitry, the same voltage hazards are not often encountered as in circuits that use vacuum tubes.

Playback head: In audio- or videotape recording, a transducer that picks up signals from magnetic tape and converts them to electrical pulses.

Plugging: The process of braking in an electric motor in which connections are reversed. This reversal causes the motor to develop a counter torque, which results in the exertion of a retarding force. Plugging is used to secure both rapid stop and quick reversal.

Because it is possible for motor connections to be reversed when the motor is running, control circuits should be designed specifically to prevent this from occurring when it is undesirable. However, there are a number of factors that must be considered and investigated thoroughly when it is desired to have this type of operation. It may be necessary to have methods of limiting maximum permissible currents, particularly in situations with repeated operations and also with dc motors. The machine under consideration should be carefully investigated in order to ensure that this type of action will not do damage over an extended period of time.

PN junction diode: A semiconductor diode that is made by taking a single crystal (germanium, for example) and adding a donor impurity to one region and an acceptor impurity to the other. This gives a single crystal with an N section and a P section. The point where the two sections meet is called a *junction*. The result is a simple PN junction diode.

One portion of the crystal is P-type material. This is the portion containing the acceptor impurity. The other portion is the N-type material. This region contains the donor impurity. The end contacts are large surfaces that make a good connection with the crystal. If the connections were not good, there might be rectifying properties where they come in contact with the crystal.

Figure 320 is a pictorial representation of a PN junction. An isolated piece of N-type material is electrically neutral; that is, for every free electron in the conduction band, there is a positively ionized donor atom in the crystal lattice structure. Thus, while there is an abundance of free negative charges, each one is balanced by a fixed positive charge and the overall charge of the crystal is zero.

An isolated piece of P-type material is also electrically neutral; that is, for every hole (positive ionized germanium or silicon atom), there exists a negatively ionized acceptor atom. Thus, the overall charge of the P-type crystal is zero. Figure 321 shows an electrical representation of an isolated N- and P-type material with balanced charges.

Figure 321 is used to represent the even distribution of carriers and ions throughout the crystals. The carriers are placed beside each ion to indicate the balancing of positive and negative charges. While there are many varied methods of combining a P- and N-type material into a PN junction, there is one

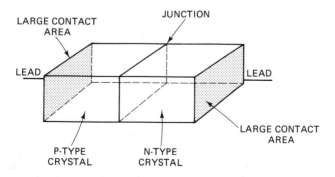

FIGURE 320 PN junction pictorial diagram.

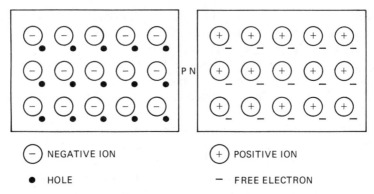

(−) NEGATIVE ION   (+) POSITIVE ION

● HOLE   − FREE ELECTRON

FIGURE 321   Isolated P- and N-type materials.

qualification that must be met. The junction, when completed, must have the properties of a single crystal.

PNP transistor: A bipolar junction transistor that utilizes a PNP junction. Two sections of P-type semiconductor material are sandwiched around a section of N-type material. The P-type semiconductor slabs form the collector and emitter connections, while the N-type material is the base. PNP transistors of similar types may be substituted for NPN transistors in most circuits as long as circuit polarity is reversed.

Point-by-point method: A means of computing the level of illumination in footcandles at any given point in a lighting installation. This is accomplished by summing up all the illumination contributions, except surface reflection, to that point from every fixture individually. Since reflection from walls, ceilings, floors, etc., is not taken into consideration by this method, it is especially useful for calculations dealing with very large areas, outdoor lighting, and areas where the room surfaces are dark or dirty.

With the aid of a candlepower distribution curve, footcandle values for specific points may be calculated as follows

$$Fc = \frac{\text{Candlepower} \times H}{d^3} \quad \text{or} \quad \frac{\text{Candlepower} \times \cos^3 0}{H^2}$$

For vertical surfaces, the equations are

$$Fc = \frac{\text{Candlepower} \times R}{d^3} \quad \text{or} \quad \frac{\text{Candlepower} \times \cos^3 0 \times \sin 0}{h^2}$$

In using the point-by-point method, a specific point is selected at which it is desired to know the illumination level, as, for example, point *P* in Fig. 322. Once the seeing task or point has been determined, the illumination level at the point can be calculated.

It is obvious that the illumination at point *P* in Fig. 322, or at any point in the area, is due to light coming from all of the lighting fixtures. In this case, the

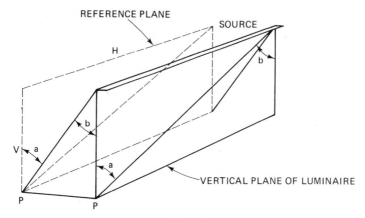

REFERENCE PLANE

SOURCE

H

b

VERTICAL PLANE OF LUMINAIRE

V a

b

a

P   P

FIGURE 322

calculations must be repeated to determine the amount of light each fixture contributes to the point. The total amount is the sum of all the contributing values.

Before attempting any actual calculations using the point-by-point method, a knowledge of candlepower distribution curves and a review of trigonometric functions are necessary. A candlepower distribution curve or graph consists of lines plotted on a polar diagram, which graphically shows the distribution of the light flux in some given plane around the actual light source. It also shows the apparent candlepower intensities in various directions about the light source. Figure 323 illustrates a typical candlepower distribution curve.

A table of trigonometric functions (Fig. 324) will be helpful in determining the degrees of the angle from the light source to the point in question. This is necessary to pick off the candlepower from the photometric distribution curve and also to use in the equations.

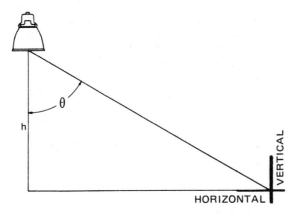

θ

h

VERTICAL

HORIZONTAL

FIGURE 323

Trigonometric Functions

$\theta°$	$\sin\theta$	$\cos\theta$	$\cos^2\theta$	$\cos^3\theta$	$\tan\theta$	$\theta°$	$\sin\theta$	$\cos\theta$	$\cos^2\theta$	$\cos^3\theta$	$\tan\theta$
0	0.0	1.000	1.000	1.000	0.0	46	0.719	0.695	0.483	0.335	1.035
1	.0175	1.000	1.000	1.000	.0174	47	.731	.682	.465	.317	1.072
2	.0349	0.999	0.999	0.998	.0349	48	.743	.669	.448	.300	1.110
3	.0523	.999	.997	.996	.0524	49	.755	.656	.430	.282	1.150
4	.0698	.998	.995	.993	.0699	50	.766	.643	.413	.266	1.191
5	.0872	.996	.992	.989	.0874	51	.777	.629	.396	.249	1.234
6	.105	.995	.989	.984	.1051	52	.788	.616	.379	.233	1.279
7	.122	.993	.985	.978	.1227	53	.799	.602	.362	.218	1.327
8	.139	.990	.981	.971	.1405	54	.809	.588	.345	.203	1.376
9	.156	.988	.976	.964	.1583	55	.819	.574	.329	.189	1.428
10	.174	.985	.970	.955	.1763	56	.829	.559	.313	.175	1.482
11	.191	.982	.964	.946	.1943	57	.839	.545	.297	.162	1.539
12	.208	.978	.957	.936	.2125	58	.848	.530	.281	.149	1.600
13	.225	.974	.949	.925	.2308	59	.857	.515	.265	.137	1.664
14	.242	.970	.941	.913	.2493	60	.866	.500	.250	.125	1.732
15	.259	.966	.933	.901	.2679	61	.875	.485	.235	.114	1.804
16	.276	.961	.924	.888	.2867	62	.883	.470	.220	.103	1.880
17	.292	.956	.915	.875	.3057	63	.891	.454	.206	.0936	1.962
18	.309	.951	.905	.860	.3249	64	.899	.438	.192	.0842	2.050
19	.326	.946	.894	.845	.3443	65	.906	.423	.179	.0755	2.144
20	.342	.940	.883	.830	.3639	66	.914	.407	.165	.0673	2.246
21	.358	.934	.872	.814	.3838	67	.921	.391	.153	.0597	2.355
22	.375	.927	.860	.797	.4040	68	.927	.375	.140	.0526	2.475
23	.391	.921	.847	.780	.4244	69	.934	.358	.128	.0460	2.605
24	.407	.914	.835	.762	.4452	70	.940	.342	.117	.0400	2.747
25	.423	.906	.821	.744	.4663	71	.946	.326	.106	.0347	2.904
26	.438	.899	.808	.726	.4877	72	.951	.309	.0955	.0295	3.077
27	.454	.891	.794	.707	.5095	73	.956	.292	.0855	.0250	3.270
28	.470	.883	.780	.688	.5317	74	.961	.276	.0762	.0211	3.487
29	.485	.875	.765	.669	.5543	75	.966	.259	.0670	.0173	3.732
30	.500	.866	.750	.650	.5773	76	.970	.242	.0585	.0142	4.010
31	.515	.857	.735	.630	.6008	77	.974	.225	.0506	.0114	4.331
32	.530	.848	.719	.610	.6248	78	.978	.208	.0432	.0090	4.704
33	.545	.839	.703	.590	.6494	79	.982	.191	.0364	.0070	5.144
34	.559	.829	.687	.570	.6745	80	.985	.174	.0302	.0052	5.671
35	.574	.819	.671	.550	.7002	81	.988	.156	.0245	.0038	6.313
36	.588	.809	.655	.530	.7265	82	.990	.139	.0194	.0027	7.115
37	.602	.799	.638	.509	.7535	83	.993	.122	.0149	.0018	8.144
38	.616	.788	.621	.489	.7812	84	.995	.105	.0109	.0011	9.514
39	.629	.777	.604	.469	.8097	85	.996	.0872	.0076	.0007	11.430
40	.643	.766	.587	.450	.8391	86	.9976	.0698	.0048	.0003	14.300
41	.656	.755	.570	.430	.8692	87	.9986	.0523	.0027	.0001	19.080
42	.669	.743	.552	.410	.9004	88	.9993	.0349	.0012	.0000	28.630
43	.682	.731	.535	.391	.9325	89	.9998	.0175	.0003	.0000	57.280
44	.695	.719	.517	.372	.9656	90	1.0000	0.0000	.0000	.0000	Infinite
45	.707	.707	.500	.354	1.0000						

FIGURE 324

Polyphase motor: An ac motor that is designed for either three- or two-phase operation. The two types are alike in construction, but the internal connections of the coils are different. Three-phase motors vary from fractional horsepower size to several thousand horsepower. These motors have a fairly constant speed characteristic and are made in designs giving a variety of torque characteristics. Some have a high starting torque; others have a low starting torque. Some are designed to draw a normal starting current; others draw a high starting current. They are made for practically every standard voltage and frequency and are very often dual-voltage motors. Three-phase motors are used to drive machine tools, pumps, elevators, fans, cranes, hoists, blowers, and many other machines.

Two-phase motors are like three-phase motors in all respects, except for the number of groups and the connections of the groups. As in the three-phase motor, the number of groups is equal to the number of phases multiplied by the number of poles.

Potential transformer: A transformer that is used to insulate instruments and the operator from line voltage and act as a multiplier for the instruments. In order to obtain maximum safety for men and apparatus, one secondary or low-voltage lead must be grounded. The metal case, if any, must be grounded, and connections must not be changed with the voltage on. The primary or high-voltage winding of the transformer must be connected to the high-voltage circuit, and the secondary or low-voltage winding to the instruments. Finally, the line voltage should not exceed 125 percent of the rated voltage of the transformer.

In order to obtain accuracy, the frequency of the circuit should not be less than the lowest rated frequency by more than a few cycles. The circuit frequency should never be more than 125 cycles. The line voltage should range from 70 to 110 percent of the rated voltage, and the impedance of the secondary burden should not be less than the value indicated by the volt-ampere rating at rated secondary voltage. The volt-amperes taken by the burden at rated voltage should not exceed the volt-ampere rating on the nameplate.

The voltmeters generally used with potential transformers are rated at 130 or 150 V. Replacing such a voltmeter with a lower-rated one for taking readings when the voltage is considerably below the transformer rating may often be as unsatisfactory as estimating the reading at the lower end of the 130- or 150-V voltmeter. In such cases, a more suitable transformer should be obtained.

Power: The rate at which work is done. Power is equal to the amount of work done divided by the time required to do it. This unit does not show how much work has been done. It merely indicates how rapidly, or at what rate, the work is being done. The fundamental unit of electrical power is the watt. When the power in an electrical circuit is 1 W, this means that the work is being done in that circuit at the rate of 1 J/s, or 0.74 ft · lb/s. Note that the watt is not a quantity unit, but a rate unit.

Larger power units are the horsepower and the kilowatt. The horsepower represents a rate of doing work equal to 746 W, or 746 J/s, or 500 ft · lb/s. Note that time is always a factor in the measurement of power.

Power amplifier: An amplifier that delivers useful amounts of power to a load such as a speaker or antenna. The prime function of this device is to produce maximum output power. Due to transistor operation limitations (such as maximum current, voltage, and power dissipation ratings), the conditions for maximum power gain do not necessarily coincide with those for maximum power output. The maximum power dissipation rating of a transistor is very important in the operation of a power amplifier, for it is this rating that limits the power output obtainable from any specific transistor.

For all practical purposes, the schematic diagram of a power amplifier is similar to that of any low-power or medium-power power amplifier, with the major difference being the higher power rating, large physical construction, and mounting methods of the power transistor. One other difference is that the power amplifier, being designed for maximum output power rather than maximum gain, will usually have a much smaller value of load impedance than the preceding stages.

Power factor: The number by which the apparent power in the circuit (volts times amperes) must be multiplied in order to ascertain true power. When an ac circuit contains inductance, the current lags behind the voltage. When it contains capacitance, the current rises ahead of the voltage. In each case, the current and voltage reach their maximum values at different times and the product of the current and voltage at any instant is less than it would be if the two were in phase with each other.

If the voltage and current were measured separately, the voltmeter and ammeter would give the individual mean effective values. If they are measured by a wattmeter, the instrument indicates their combined effect synchronously, not the product of their effective values, which occur at different instants. Consequently, the wattmeter indication will be less than the product of the separate voltmeter and ammeter readings. The ratio of the power to this product is the power factor of the circuit. Expressed as a formula

$$\text{Power factor} = \frac{\text{Watts}}{\text{Amperes} \times \text{Volts}}$$

This gives rise to the two methods of rating electrical apparatus, one on the basis of watts or kilowatts, and the other on the basis of volt-amperes or kilowatt-amperes. The former represents actual power, usually in kilowatts, while the latter represents the apparent power, usually in kilovolt-amperes, generated, transmitted, or used by the apparatus. The latter rating is coming into more general use, since it more adequately represents the voltage and current conditions to which the apparatus is subjected.

Power gain: The ratio of power at some point in the radiation field of an antenna over the power at the same point of a single-dipole antenna located in the same position and fed in the same way as the antenna being measured.

An antenna with high directivity has a high power gain, and vice versa. The power gain of a single dipole with no reflector is one. An array of several dipoles in the same position as the single dipole and fed with the same line would have a power gain of more than one, with the exact figure depending on the directivity of the array.

Power gain, antenna: The amount by which power is increased by the action of an antenna. This may be expressed as the simple ratio of power output

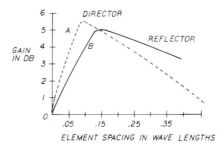

FIGURE 325

(ERP) to power input or in decibels. Figure 325 shows the effect of the element spacing on the power gain of an array as compared to the field strength of a half-wave antenna alone. Curve *A* shows the power gain for director spacings in wavelengths between a half-wave director and the driven half-wave radiator. Curve *B* shows the power gain for half-wave reflector spacing between the reflector and the driven half-wave radiator.

The graph shows that when a single parasitic element is used, there is little difference in the gain of the array if the parasitic element employed is used either as a reflector or as a director. When the parasitic element is tuned to work as a reflector, the spacing that gives maximum gain is about 0.15 wavelength, and this maximum gain peak is fairly broad. The director will give slightly more gain than the reflector, but the difference between the two is less than 0.5 dB.

Consequently, on the basis of gain, there is little choice between the two types of operation. The broader curve of the reflector does offer a distinct advantage, in that the spacing is less critical. Similarly, a smaller error in reflector spacing will reduce the gain of the array less than the same spacing error in a director type of array. For this reason, when a single element is used, the array is usually a reflector type of array.

Several parasitic elements can be used in conjunction with a driven antenna to increase further the directivity and power gain. The theoretical power gain of directional antenna arrays composed of an excited element and various numbers of parasitic elements is given in Fig. 326.

Power output: Identified by the symbol P_o, the power deliverable by an amplifier, generator, or other circuit into a specific load. Output power is usually measured in watts and is divided into the total power output to arrive at an overall power efficiency rating. For instance, an amplifier delivers 100 W of power to the load and the power consumed at the input is 150 W.

Power efficiency is equal to 100 divided by 150, or 66.6 percent.

Power rating: Measured in watts, an indication of the rate at which a device converts electrical energy into another form of energy, such as light, heat, or motion. An example of such a rating is noted when comparing a 150-W lamp to a 100-W lamp. The higher wattage rating of the 150-W lamp indicates that it is capable of converting more electrical energy into light energy than the lamp of the lower rating. Other common examples of devices rated in this manner are soldering irons and small electric motors.

Type	Driven Element Length	Reflector Length	1st Director Length	2nd Director Length	3rd Director Length	Spacing Between Elements	Approx. Gain DB	Approx. Radiation Resistance (Ω)
3-ELEMENT	$\dfrac{472}{F(MC)}$	$\dfrac{500}{F(MC)}$	$\dfrac{445}{F(MC)}$	—	—	.15-.15	7.0	20
3-ELEMENT	$\dfrac{472}{F(MC)}$	$\dfrac{495}{F(MC)}$	$\dfrac{450}{F(MC)}$	—	—	.25-.25	8.5	35
4-ELEMENT	$\dfrac{472}{F(MC)}$	$\dfrac{497}{F(MC)}$	$\dfrac{450}{F(MC)}$	$\dfrac{440}{F(MC)}$	—	.2-.2-.2	9.5	20
5-ELEMENT	$\dfrac{472}{F(MC)}$	$\dfrac{497}{F(MC)}$	$\dfrac{450}{F(MC)}$	$\dfrac{440}{F(MC)}$	$\dfrac{430}{F(MC)}$	.2-.2-.2-.2	10.0	15

FIGURE 326

In some electrical devices, the wattage rating indicates the maximum power the device is designed to dissipate, rather than the normal operating power. A 150-W lamp, for example, dissipates 150 W when operated at the rated voltage printed on the bulb. In contrast, a device such as a resistor is not normally given a voltage or current rating. A resistor is given a power rating in watts and can be operated at any combination of voltage and current, as long as the power rating is not exceeded. In most circuits, the actual power dissipated by a resistor will be considerably less than the resistor's power rating. In well-designed circuits, a safety factor of 100 percent or more is allowed between the actual dissipation of the resistor in the circuit and the power rating listed by the manufacturer. The wattage rating of the resistor is the maximum power the resistor can dissipate without damage from overheating.

Resistors of the same resistance value are available in different wattage values. Carbon resistors, for example, are commonly made in wattage ratings of .125, .25, .50, 1, and 2 W. The larger the physical size of a carbon resistor, the higher its wattage rating, since a larger amount of material will radiate heat more easily.

Power transformer: A device designed to change an ac voltage or a periodically varying dc voltage from one value to another without any change in frequency. A power transformer is shown in Fig. 327. It consists of a primary winding connected to the source of energy, a laminated iron core, and one or more secondary windings. Theoretically, any winding may be used as the primary, provided the proper voltage and frequency are applied to it. The laminated iron core serves as an efficient means of magnetically coupling together the primary and secondary windings.

A periodically varying voltage applied to the primary winding produces a varying current that, in turn, develops a varying flux in the iron core. This varying flux cuts all windings, inducing in each of them a voltage proportional to the number of turns.

The ratio of the primary voltage to any secondary voltage is practically equal to the ratio of the primary turns to the secondary turns, as indicated by the formula

$$\frac{E_p}{E_s} = \frac{N_p}{N_s}$$

The voltage induced in the primary winding by the growing and dying core flux is practically equal to the applied voltage. Moreover, this induced voltage directly opposes the applied voltage. Therefore, the current drawn from the supply is small.

When a secondary circuit is completed, current circulates around the iron core in the opposite direction to the primary current, reducing the core flux and the counter voltage of the primary. This action causes the current in the primary to vary in accordance with the secondary load. It is through this action that the transformer automatically adjusts itself to changes in secondary load.

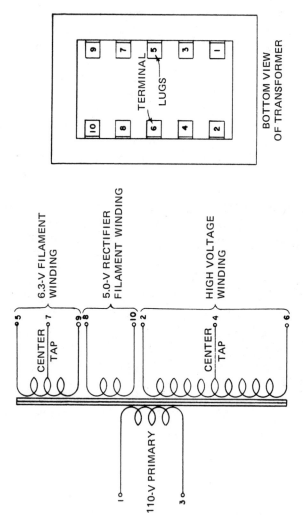

FIGURE 327 A power transformer.

424

Precision: **1.** The degree of exactness with which a quantity is stated.

2. The degree of discrimination or amount of detail; e.g., a 3-decimal digit quantity discriminates among 1000 possible quantities.

A result may have more precision than it has accuracy; e.g., the true value of pi to 6 significant digits is 3.14159; the value of 3.14162 is precise to 6 figures, given to 6 figures, but is accurate only to about 5.

Pressure regulator: A device that is widely used in industrial applications when there is a need for pressure sensing. A pressure regulator is made up of a control relay and a bourdon-type pressure gage. In operation, contacts on the gage energize the relay, which causes it to be opened or closed. In this way, the relay contacts can be used to control a large motor starter, thus avoiding any possible burning of the gage contacts. A regulator designed for standard operation opens the circuit at high pressure and closes it at low pressure. Special regulators are available and perform in the same manner, but in reverse. In other words, the circuit closes at high pressure and opens at low pressure.

Pressure regulators are manufactured to meet those needs of industry with regard to the control of hydraulic and pneumatic machines, such as machine tools, welding equipment, motor-driven pumps, air compressors, etc. Due to the fact that pressure regulators are used in such a wide variety of applications, there is a great deal of selection. Gage-type pressure regulators are used to provide accurate control of pressure or vacuum conditions on systems. These regulators are used as pilot control devices with magnetic starters. They are quite similar to pressure switches, in that they govern operation of pump or compressor motors in the same manner.

Printed circuit board: A flat dielectric surface upon which conductors and other circuits components have been printed. To be most effective, the printed circuitry should adhere permanently to the base plate. The printed wiring should have low resistance and be able to pass the desired current without overheating. The printed resistors, capacitors, and inductors should remain stable under their rated loads.

By using a mold of the desired shape, it is possible to produce uniform units, as far as physical dimensions are concerned. Since printed circuit boards are readily adapted to plug-in units, the elimination of terminal boards, fittings, and tie points results in a substantial reduction in the overall size of equipment.

Printed circuit boards come in various sizes, depending on the type of circuitry being used. Common sense dictates that a printed circuit board using transistors will be smaller than one used for the same function utilizing electron tubes. Regardless of the size of the printed circuit board, one can be sure that the same circuit(s) built in a conventional manner would occupy a much larger space. The functions of a printed circuit board are to replace as many of the passive circuit components as possible.

Printer, high-speed: A printer that operates at a speed more compatible with the speed of computation and data processing so that it may operate on-line. At the present time, a printer operating at a speed of 250 lines per minute, 100 characters per line, is considered to be high speed.

Product detector: A detector used for SSB reception whose output amplitude is proportional to the product of the amplitude of the reinserted carrier and the SSB signal. Figure 328 illustrates a transistor product detector. Transistors Q_1 and Q_2 form a balanced mixer circuit. The bias for these transistors is obtained from the voltage divider formed by R_1 and R_2 and is applied to the bases of Q_1 and Q_2 through emitter resistors R_6 and R_7.

The IF signal is applied to the base of each transistor 180° out of phase by transformer T_1. The carrier reinsertion oscillator signal is coupled in phase to the emitters of Q_1 and Q_2 through capacitors C_2 and C_3, respectively. Resistors R_4 and R_5 provide isolation between the emitters of Q_1 and Q_2. The IF signal and the reinserted carrier are heterodyned in the transistors. The resulting output consists of the sum and difference of the original input frequencies, as well as the two original frequencies. Capacitors C_4 and C_5 bypass the sum and original frequencies to ground. Since the circuit is balanced, the outputs from transistors Q_1 and Q_2 (developed across transformer T_2) are 180° out of phase with each other. This results in additional cancelling of the reinserted carrier, because it was applied in phase to the emitters of the transistors. Transformer

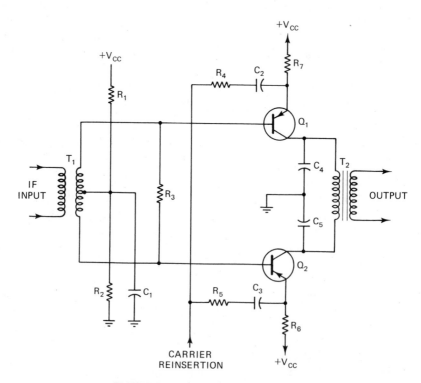

FIGURE 328 Transistor product detector.

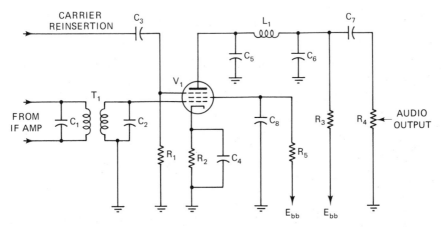

FIGURE 329 Electron tube product detector.

T_2 has an audio-frequency response that will attenuate any of the RF signals not previously cancelled. The difference between the two input signals is the desired intelligence, and this difference is developed across transformer T_2.

Figure 329 illustrates an electron-tube product detector. The sideband signal from the IF amplifier is applied to the control grid of V_1 through transformer T_1. The reinserted carrier is applied to the suppressor grid. The two input signals are heterodyned within the tube, and the output will contain frequencies equal to the sum and difference of the input signals, as well as the original input frequencies. All frequency components, with the exception of the difference frequency, will be attenuated by the low-pass filter made up of C_5, L_1, and C_6. The difference frequency, which is the desired audio signal, will be developed across R_4.

It should be noted that in both detectors, the reinserted carrier amplitude must be appreciably greater than the sideband IF amplitude. A carrier too small in relative amplitude would result in distortion of the output signal.

Propagation path: The line or lines between a transmitting and receiving point over which radio-frequency energy travels during a specific transmission. Propagation paths will vary with the frequency transmission and the condition of the earth's atmosphere. The latter often bends the electromagnetic waves back toward the earth, and several different paths may result from one transmission.

For example, a portion of the transmitted signal may travel over the earth's surface near the ground (ground wave), while another portion may travel far into the atmosphere before being bent back to a distant point around the curvature of the earth. Still another portion of the signal may leave the earth's atmosphere and travel into space.

Some multipath transmissions may be divided into two similar paths, with one traveling around the earth in an east to west direction and the other traveling west to east.

Pull-up resistor: A component that is used in an electronic circuit to raise the impedance value presented at the input or output. This resistor is often wired in series with the input or output device terminals and can be used to match impedances between circuits. A low-impedance input can be modified to present a high impedance by installing a pull-up resistor of high ohmic value in series with the input and the device or circuit to be connected at this point.

Pull-up resistors are often used in preamplifier circuits at audio frequencies to allow for the use of low-impedance microphones with high-impedance outputs, or vice versa.

Pulse-repetition frequency: That frequency that largely determines the maximum range and, to some degree, the accuracy of a radar set. The actual time elapsing between the beginning of one pulse and the beginning of the next, called the *pulse-repetition period,* is the reciprocal of the pulse-repetition frequency (PRF). Thus, for example, if the PRF is 400 Hz, the pulse-repetition period is 1/400 s, or 2500 μs. When the PRF is too high (the period between pulses is too short), the echo from the farthest target may return to the receiver after the transmitter has emitted another pulse, making it impossible to tell whether the observed pulse is the echo of the pulse just transmitted or the echo of the preceding pulse. Such a condition is referred to as *range ambiguity.*

Although the pulse-repetition rate must be kept low enough to attain the required maximum range, it must also be kept high enough to avoid some of the pitfalls a single pulse might encounter. If a single pulse were sent out by a transmitter, atmospheric conditions might attenuate it, the target might not reflect properly, or moving parts (such as a propeller) might throw it out of phase or change its shape. Thus, information derived from a single pulse would be highly unreliable. By sending many pulses one after another, however, many good pulses will return. The equipment will integrate or sum up the good points of all the pulses and present a clear, reliable picture.

Therefore, equipment is designed in such a manner that many pulses (10 or more) are received from a single object. In this way, effects of fading are somewhat reduced. At short ranges, the repetition frequency is increased in order to ensure accurate measurements. Many echoes are received from one target, and the integrating effect is increased.

Tactical employment of a radar set determines, to a large degree, the PRF to be used. Long-range search sets require a pulse rate slow enough to allow echoes from targets at the maximum range to return to the receiver before the transmitter is again pulsed. Higher pulse rates are used in aircraft interception sets where the maximum range is less.

The following equation will readily determine the PRF or the range, if one of them is known.

$$PRF = \frac{\text{Speed of light}}{2 \times \text{Range}}$$

Multiplying the range by 2 converts the range to radar range. Thus, a radar with a PRF of 800 Hz could operate to a maximum range of about 100 mi without range ambiguity. Most radar equipments use 200 yd as the standard radar mile. The difference between it and the nautical mile (6076 ft) is approximately 1 percent.

In theory, it is desirable to strike a target with as many pulses of energy as possible during a given scan. Thus, the higher the PRF, the better. A high PRF combined with a narrow pulse improves angular resolution and range rate accuracy by sampling the position of the target more often.

Punch: 1. To shear a hole by forcing a solid or hollow, sharp-edged tool through a material into a die.

2. The hole resulting from (1).

Punch card: A heavy, stiff paper of constant size and shape suitable for punching in a pattern that has meaning, and for being handled mechanically. The punched holes are sensed electrically by wire brushes, mechanically by metal fingers, or photoelectrically by photocells.

Push button: A small type of switch actuated by finger pressure. Typical examples include doorbells, elevator controls, and the like.

Push-pull amplifier: A circuit designed for increased power output and which utilizes two tubes or transistors operated 180° out of phase with each other in opposite halves of the symmetrical circle.

The class A push-pull amplifier consists essentially of two transistors connected back to back, with both transistors biased for class A operation. Figure 330 shows a simplified circuit for a class A push-pull power amplifier.

Through the use of a transformer phase splitter (T_1), two signals 180° out of phase are applied as inputs to the push-pull amplifier. R_1 limits the base bias current to establish the desired operating point of Q_1 and Q_2, while the other

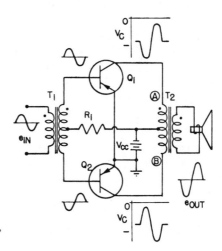

FIGURE 330   Class A, push-pull, power amplifier.

half represents the collector load for Q_2. T_2 also provides impedance matching between the relatively high-output impedances of the transistors and the low impedance of the speaker voice coil.

On the positive alternation of the input signal, the potential on the base of Q_1 will increase in a positive direction, while the potential on the base of Q_2 will increase in a negative direction. Since both transistors are of the PNP type, the potentials applied to their respective base elements will cause the conduction of Q_1 to decrease and the conduction of Q_2 to increase.

Current flows from the center tap of T_2 toward point B. Under quiescent operating conditions, these currents are equal and the magnitude and polarity of the voltage drops they produce are such that there is no difference of potential between points A and B. The center tap of T_2 is fixed at the maximum negative value in the circuit by virtue of being connected to V_{CC}. Thus, the potentials at points A and B are less negative by some amount than the center tap. As stated, the positive alternation of the input signal causes the conduction of Q_1 to decrease. In other words, point A becomes more negative because it is approaching the value of the potential at the center tap. The collector voltage waveform of Q_1, which is actually a graph of the potential at point A, is seen to be increasing in a negative direction at this time.

The positive alternation of the input signal also causes the conduction of Q_2 to increase. This causes the potential difference between point B and the center tap to increase. Thus, the potential at point B is increasing in a positive direction (as shown by the collector voltage waveform of Q_2). Since point A is becoming more negative and point B is effectively becoming more positive, there is a potential difference developed across the entire primary winding. By transformer action, the potential difference is coupled to the secondary of T_2 and appears as the waveform, e_{out}.

On the negative alternation of the input signal, the reverse of the above action occurs, as can be seen by the collector voltage waveforms of Q_1 and Q_2, and the polarity of the voltage developed across the primary of T_2 is reversed. The power output from this class A push-pull circuit is more than twice that obtainable from a single-ended class A power amplifier. An added advantage of this circuit is that due to the push-pull action of the output transformer, all even harmonics are eliminated in the output (if the transistor circuits are balanced). Since the distortion is caused mainly by second harmonics, elimination of these harmonics will result in a relatively distortion-free output signal.

The class A push-pull power amplifier finds its greatest application where minimum distortion is the primary consideration and high-output power and efficiency are deemed less important. Figure 331 shows a simplified circuit of a class B push-pull amplifier. The emitter-base junctions are zero biased. In this circuit, each transistor conducts on alternate half-cycles of the input signal. The output signal is combined in the secondary of the output transformer. Maximum efficiency is obtained even during idling (no input signal) periods, because neither transistor conducts during this period.

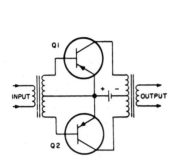

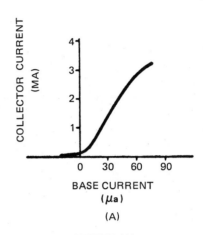

FIGURE 331   Class B, push-pull
amplifier with zero input bias.

FIGURE 332

An indication of the output current waveform for a given signal current input can be obtained by considering the dynamic transfer characteristic for the amplifier. It is assumed that the two transistors have identical dynamic transfer characteristics. This characteristic for one of the transistors is shown in Fig. 332. The variation in output (collector) current is plotted against input (base) current under load conditions. Since two transistors are used, the overall dynamic transfer characteristic for the push-pull amplifier is obtained by placing two of the curves (Fig. 332) back to back. The two curves are shown back to back and combined in Fig. 333.

Note that the zero line of each curve is lined up vertically to reflect the zero bias current. In Fig. 334, points on the input base current (a sine wave) are

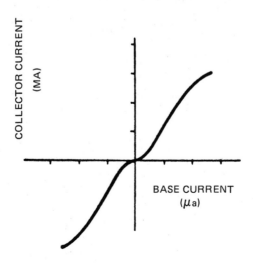

FIGURE 333

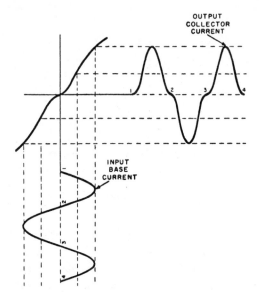

FIGURE 334 Dynamic transfer characteristic curves of class B, push-pull amplifier with zero bias, showing input and output current waveforms.

projected onto the dynamic transfer characteristic curve. The corresponding points are determined and projected as indicated to form the output collector current waveform. Note that severe distortion occurs at the crossover points; that is, at the points where the signal passes through zero value. This is called *crossover distortion*. This type of distortion becomes more severe with low signal input currents. Crossover distortion can be eliminated by using a small forward bias on both transistors of the push-pull amplifier.

    A class B push-pull amplifier with a small forward bias applied to the base-emitter junctions is shown in Fig. 335. A voltage divider is formed by resistors R_2 and R_1. The voltage developed across R_1 supplies the base-emitter bias for both transistors. This small forward bias eliminates crossover distortion.

    In the voltage divider (Fig. 335), electron current flow from the battery is in the direction of the arrow. This current establishes the indicated polarity across resistor R_1 to furnish the required small forward bias. Note that no bypass capacitor is used across resistor R_1. If a bypass capacitor were used (Fig.

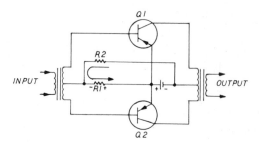

FIGURE 335

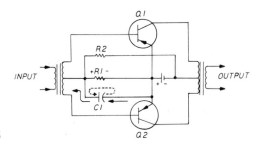

FIGURE 336

336), the capacitor would charge (solid line arrow) through the base-emitter junction of the conducting transistor (during the presence of a signal) and discharge (dashed line arrow) through resistor R_1. The discharge current through resistor R_1 would develop a dc voltage with the polarity indicated. This is a reverse bias polarity that could drive the amplifier into class C operation, with the resultant distortion even more severe than crossover distortion. The capacitor must not be used.

A study of the dynamic transfer characteristic curve of the amplifier demonstrates the limitation of crossover distortion. In Fig. 337a, the dynamic transfer characteristic curve of each transistor is placed back to back for zero bias current bias conditions. The two curves are back to back and not combined. The dashed lines indicate the base current values when forward bias is applied to obtain the overall dynamic characteristic curve of the amplifier. With forward bias applied, the separate curve of each transistor must be placed back to back and aligned at the base bias current line (dashed line). The zero base current lines (solid lines) are offset (Fig. 337).

In Fig. 338, points on the input base current (a sine wave) are projected onto the dynamic transfer characteristic curve. The corresponding points are

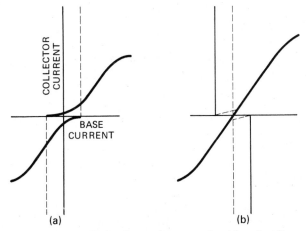

FIGURE 337 Dynamic transfer curves, low bias, class B.

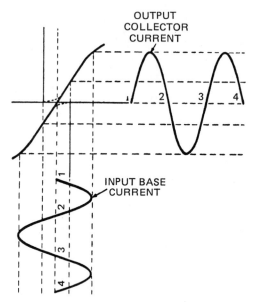

OUTPUT
COLLECTOR
CURRENT

INPUT BASE
CURRENT

FIGURE 338

determined and projected as indicated to form the output collector current waveform. Compare this output current waveform with that shown in Fig. 334. Note that crossover distortion does not occur when a small forward bias is applied.

Quarter-wave antenna: Any antenna that uses a main element for receiving or transmitting that is cut to an electrical quarter-wavelength at the operating frequency. To make these elements resonant, they are usually tuned against ground or a ground plane. The ground connection makes up the missing antenna element link required for resonance.

Raceway: An electrical wiring system in which two or more individual conductors are pulled into a conduit (pipe) or similar housing for the conductors after the system has been completely installed. The basic raceways are

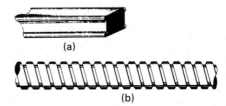

(a)

FIGURE 339 Flexible metallic conduit. (b)

rigid-steel conduit, electrical metallic tubing (EMT), and PVC (polyvinyl chloride) plastic. Other raceways include surface metal moldings (Fig. 339a) and flexible metallic conduit (Fig. 339b).

These raceways are available in standardized sizes and serve primarily to provide mechanical protection for the wires run inside and, in the case of metallic raceways, to provide a continuously grounded system. Metallic raceways, properly installed, provide the greatest degree of mechanical and grounding protection and provide maximum protection against fire hazards for the electrical system. However, they are more expensive to install.

Radar: An electronic system that may be used to detect the presence of objects such as airplanes or ships in darkness, fog, or during a storm. The word radar is an abbreviation for *radio detection and ranging*. Radar may also be used to determine bearing and distance. In special types of radar, elevation and speed may also be indicated.

Radar is one of the greatest scientific developments that emerged from World War II. Its development was mothered by necessity, that of detecting the enemy before being detected. The basic principles on which its functioning depend are relatively simple, and the seemingly complicated series of electrical events encountered can be resolved into a logical series of functions that, taken individually, may be identified and understood.

Naval scientists pioneered in finding practical uses for radar. Those uses were made chiefly for detecting and destroying an enemy and his armaments. This is still the most important use today. Civilian uses followed those for military purposes. For example, some familiar civilian uses are radar speed determination on highways for controlling traffic, radar weather prediction, commercial radar air navigation, and safeguarding air vehicles and merchant ships from the hazards of collisions.

Search radars can be classified as air search and surface search. These equipments are used for early warning networks and for general navigational purposes. Search radars produce detection at maximum ranges, while sacrificing some degree of accuracy and resolution. Fire-control radars, integral parts of certain gun and missile fire-control systems, are used after targets have been located by search radars. Special radars are used for specific purposes, which include recognition or identification.

The principle upon which radar operates is very similar to the principles of sound echoes or wave reflection. If one shouts in the direction of a cliff or some other sound-reflecting device, one hears the shout return from the direction of the cliff. What actually takes place is that the sound waves generated by the shout travel through the air until they strike the cliff. They are then reflected or bounced off. Some are returned to the originating spot, which makes it possible for the person to hear the echo. Some time elapses between the instant the sound originates and the time when the echo is heard, because sound waves travel through air at approximately 1100 ft/s. The farther the person is from the cliff, the longer this time interval will be. If a person is 2200 ft from the cliff and shouts, about 4 s elapse before the person hears the echo; that is, 2 s for the sound waves to reach the cliff and 2 s for them to return.

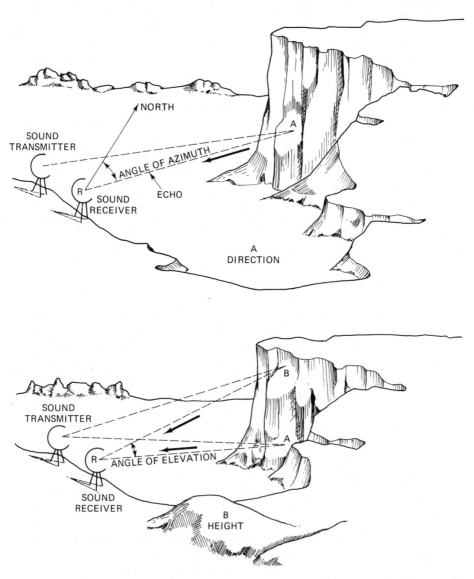

FIGURE 340 Determination of direction and height.

If a directional device is built to transmit and receive sound, the principles of echo, together with a knowledge of the velocity of sound, can be used to determine the direction, distance, and height of the cliff shown in Fig. 340. A sound transmitter that can generate pulses of sound energy can be so placed at the focus of the reflector that it radiates a beam of sound. The sound receiver can be a highly directional microphone located inside a reflector (at its focal point and facing the reflector) to increase the directional effect. The microphone is connected through an amplifier to a loudspeaker.

To determine the distance and the direction of the cliff, the transmitting and receiving apparatus are placed so that the line of travel of the transmitted sound beam and the received echo will very nearly coincide. They would coincide exactly if the same reflector could be used for both transmitting and receiving, as is done in radar systems. The apparatus (both the transmitter and receiver) is rotated until the maximum volume of echo is obtained. The horizontal distance to the cliff can then be computed by multiplying one-half of the elapsed time in seconds by the velocity of sound. This will be essentially the distance along the line RA (Fig. 340a). If the receiver has a circular scale that is marked off in degrees, and if it has been properly oriented with a compass, the direction or azimuth of the cliff can be found. Thus, if the angle indicated on the scale is 45°, the cliff is northeast from the receiver position.

To determine height (Fig. 340b), the transmitter and receiver antennas are tilted from the horizontal position (shown by dotted lines) while still pointing in the same direction. At first, the echo is still heard, but the elapsed time is increased slightly. As the angle of elevation is increased, an angle is found where the echo disappears. This is the angle at which the sound is passing over the top of the cliff and is therefore not reflected back to the receiver. The angle at which the echo just disappears is such that the apparatus is pointing along line RB. If the receiver is equipped with a scale that permits a determination of the angle of elevation, the height of the cliff (AB) can be calculated from this angle and either the distance RA or RB by the use of one of the basic trigonometric ratios.

All radar sets work on a principle very much like that described for sound waves. In radar sets, however, a radio wave of extremely high frequency is used instead of a sound wave. The energy sent out by a radar station (Fig. 341) is similar to that sent out by an ordinary radio transmitter.

The radar station has one outstanding characteristic different from a radio, in that it picks up its own signals. It transmits a short pulse and receives those echoes. This out-and-back cycle is repeated 60 to 4000 times/s, depending on the design of the set. If the outgoing wave is sent into clear space, no energy is reflected back to the receiver. The wave and the energy that it carries simply travel out into space and are lost for all practical purposes.

If, however, the wave strikes an object, such as an airplane (Fig. 341), a ship, a building, or a hill, some of the energy is sent back as a reflected wave. If the object is large compared to a quarter-length of the transmitted energy, a strong echo (but only a fraction of the transmitted energy) is returned to the

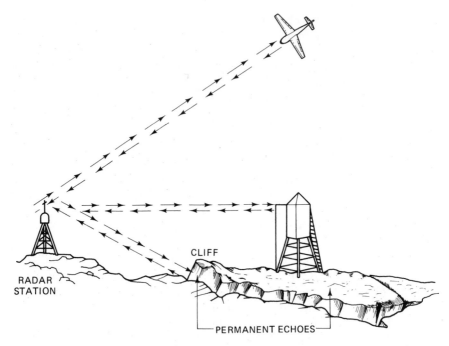

FIGURE 341   Transmission and reflection of radar pulses.

antenna. If the object is small, the reflected energy is small and the echo is weak.

Radio waves travel at the speed of light, approximately 186,000 land mi/s, or 162,000 nautical mi/s. A physical concept of this speed may be gained by considering the circumference of the earth as approximately 21,770 nautical mi. A radio wave could encircle the earth in approximately 134,400 μs, or in slightly less than one-seventh of a second.

Most of the radio waves of the UHF and SHF bands are only slightly affected by the earth's atmosphere and travel in straight lines. Accordingly, there will be an extremely short time interval between the sending of the pulse and the reception of its echo. It is possible, however, to measure the interval of elapsed time between the transmitted and received pulse with great accuracy, even to one ten-millionth of a second (1×10^{-7} s). The forming, timing, and presentation of these pulses are accomplished by a number of special circuits and devices.

The directional antennas employed by radar equipment transmit and receive the energy in a fairly sharply defined beam. Therefore, when a signal is picked up, the antenna can be rotated until the received signal is maximum. The direction of the target is then determined by the position of the antenna.

The echoes received by the radar receiver appear as marks on an oscilloscope. This scope may be marked with a scale of miles, degrees, or both.

Hence, from the position of a signal echo on the scope, an observer can tell the range and bearing of the corresponding target.

Radial distribution primary: A means of providing for the distribution of power in industrial plants. This system has poor voltage regulation because of the higher voltage drop in the longer secondary feeders. The flexibility of the radial system is in proportion to the number of secondary feeders employed.

Radio relay, tropospheric: Communications that are carried on by tropospheric scatter. This principle is shown in Fig. 342. Radio communications between two points that are separated by the curvature of the earth often take place by means of tropospheric scatter. The troposphere is the lower layer of the earth's atmosphere, extending to about 60,000 ft at the equator. This portion of the atmosphere has the ability to bend some high-frequency transmissions back toward the earth, touching down at a point beyond the curvature that would block line of sight communications between the transmit and receive stations.

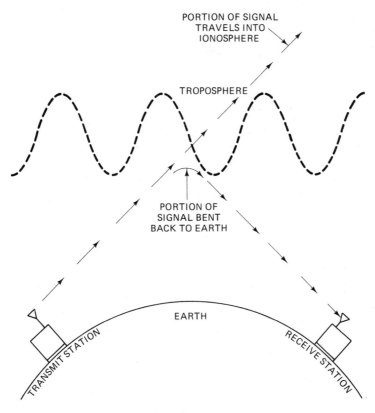

FIGURE 342

Radiosonde: A balloon-carried combination of radio transmitters and transducers that is used for sending signals to a ground monitoring station and that reveals such atmospheric conditions as temperature, humidity, and pressure. The name denotes a radio type of sonde, which is a device for gathering meteorological data at high altitudes.

Rate generator, servo-system: A small ac or dc generator that develops an output voltage proportional to the generator rpm whose phase or polarity is dependent upon the direction of rotation. Direct-current rate generators usually have permanent magnetic field excitation, whereas the ac units are excited by a constant ac supply. The most common type of ac rate generator is the drag-cup type, constructed similarly to that shown in Fig. 343. The generator has two stator windings 90° apart and an aluminum or copper-cup rotor. The rotor rotates around a stationary soft-iron magnetic core. One stator winding is energized by a reference ac source. The other stator winding is the generator output or secondary winding.

The voltage applied to the primary winding creates a magnetic field at right angles to the secondary winding when the rotor is stationary, as shown in Fig. 344. When the rotor is turned, it distorts the magnetic field so that it is no longer 90 electrical degrees from the secondary winding. Flux linkage is created with the secondary winding, and a voltage is induced. The amount of magnetic field that will be distorted is determined by the angular velocity of the rotor. Therefore, the magnitude of the voltage induced in the secondary winding is proportional to the rotor's velocity.

The direction of the magnetic field's distortion is determined by the direction of the rotor's motion. If the rotor is turned in one direction, the lines of flux will cut the secondary winding in one direction. If the motion of the rotor is reversed, the lines of flux will cut the secondary winding in the opposite direction. Therefore, the phase of the voltage induced in the secondary winding, measured with respect to the phase of the supply voltage, is determined by the direction of the rotor's motion. Other types of ac rate generators

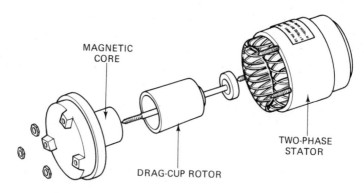

MAGNETIC
CORE

TWO-PHASE
STATOR

DRAG-CUP ROTOR

FIGURE 343   Drag-cup servomotor.

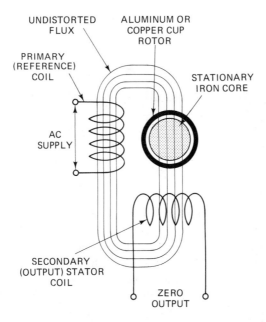

FIGURE 344

have a squirrel-cage rotor. Otherwise, their construction and principles of operation are identical to the drag-cup type.

The dc rate generator employs the same principles of magnetic coupling between the reference winding and the output winding as the ac generator. The dc rate generator, however, has a stationary primary magnetic field. This field is usually supplied by permanent magnets. The amount of voltage induced in the rotor winding is proportional to the magnetic flux lines the winding cuts. The polarity of the output voltage is determined by the direction in which the rotor cuts the lines of magnetic flux.

Rate generators are used in servo-systems to supply velocity or damping signals and are sometimes mounted on the same shaft with and enclosed within the same housing as the servo-motor.

Ratio detector: A commonly used FM detector that is relatively insensitive to amplitude variations of the input signal and does not usually have to be preceded by a limiter. Unlike the Foster-Seeley discriminator, the diodes in a ratio detector are connected in series with respect to the tuned input circuit. At the FM center frequency, both diodes have equal applied voltages, so they conduct equally. Conduction takes place on the half-cycles of the input signal when the top of the input transformer secondary (L_2) is negative. The conduction path is through diode CR_1, resistors R_1 and R_2, diode CR_2, and back to the tuned circuit, as shown by the solid arrows in Fig. 345. During conduction, capacitor C_4 charges, with a polarity as shown, to the input signal voltage. During the half-cycles of the input signal when the diodes are cut off,

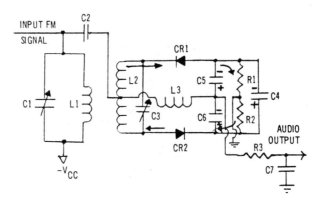

FIGURE 345 Ratio detector.

C_4 attempts to discharge through R_1 and R_2. However, the RC time constant of the combination is long in relation to both the IF and audio-modulation frequencies. Consequently, C_4 loses very little of its charge.

While the diodes are conducting, still at the center frequency of the input signal, capacitors C_5 and C_6 also charge. Since they are connected directly across capacitor C_4, the total charge on C_5 and C_6 is the same as that on C_4. Both C_5 and C_6 have equal values, so when both diodes conduct equally, the total charge divides equally between them. As can be seen, the audio-output voltage is taken from the junction of C_5 and C_6 and is effectively the center voltage that exists across both C_5 and C_6. This means that at the center frequency, the voltage at their junction is effectively zero, since the charges on C_5 and C_6 are equal.

When the input signal shifts above or below the center frequency, one of the diodes increases in conduction, while the other decreases. This changes the charge balance between C_5 and C_6, but has no effect on the C_4, R_1, and R_2 circuit, since the total current does not change. When the charge on one of the capacitors is greater than the other, the voltage at their junction changes above or below zero according to the charges.

Since the output is taken in relation to ground, C_5, C_6, R_1, and R_2 form a bridge circuit, with the audio voltage being produced between the two junctions. Essentially, the output at the junction of C_5 and C_6 depends on the ratio of the charges on C_5 and C_6. Resistor R_3 and capacitor C_7 filter the IF carrier. However, no matter how the relative charges on C_5 and C_6 vary, the total voltage across them is still the same as that across the C_4, R_1 and R_2 circuit. Capacitor C_4 keeps the total voltage effectively constant by charging when the total voltage tends to increase and discharging when the total voltage tends to drop. The total voltage only tends to change with amplitude variation. By maintaining that voltage steady, C_4 counteracts the effect of sudden changes in amplitude.

Actually, the overall operation of C_4 is a little more complex. For

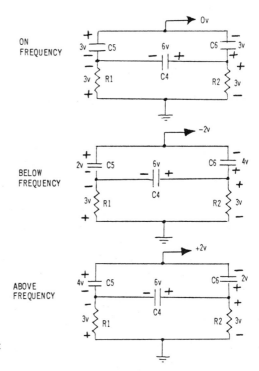

FIGURE 346 Ratio detector equivalent
circuits.

example, when the carrier amplitude rises and the output voltage tends to rise,
C_4 charges and causes a higher average current to flow through the entire
circuit, including the transformer. This loads the transformer down, lowering
its Q. As a result, the phase-shift sensitivity of the transformer is reduced. This
reduces the discriminator action of the circuit to lower the output and help
compensate for the amplitude increase. The opposite occurs when the amplitude of the carrier drops.

Since R_1 and R_2 are equal, the voltage drop across one will equal the
voltage drop across the other, and the amplitude of this drop will be determined by the charge on C_4. However, the charges on C_5 and C_6 vary according
to the FM signal. With this bridge circuit, the capacitor frequency of the
charges opposes the resistor voltage drops, and the result is the output. The
algebraic sum of each branch (capacitive or resistive) is always zero, since C_5
and C_6 vary oppositely. This is illustrated in Fig. 346.

RC coupling: The most common method of coupling two stages of
audio amplification. RC coupling in both transistor and electron-tube circuits is
shown in Fig. 347. The RC network (shown within the dashed line) used
between the stages consists of a load resistor (R_1) for the first stage, a dc
blocking capacitor (C_1), and a dc return resistor (R_2) for the input element of
the second stage. Because of the dissipation of dc power in the load resistors,

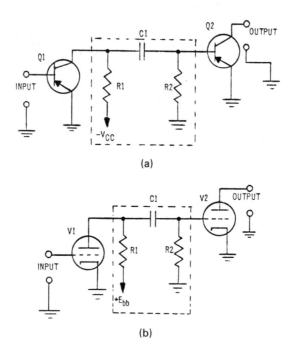

(a)

(b)

FIGURE 347   RC coupling.

the efficiency (ratio of ac power out to dc power delivered to the stage) of the RC-coupled amplifier is low.

The dc blocking capacitor prevents the dc voltage component of the output of the first stage from appearing on the input terminal of the second stage. To prevent a large single voltage drop across the dc blocking capacitor, the reactance of the capacitor must be small compared to the input resistance of the following stage, with which it is in series. Since the reactance must be low, the value of capacitance must be high. However, because of the low voltages used in transistor circuits, the physical size of the capacitor may be kept small. Physically larger capacitors are required in electron-tube circuits due to the higher voltages. The resistance of the dc return resistor is usually much larger than the input resistance of the second stage. The upper limit of the value of this resistor is dictated by the dc bias considerations in the case of the transistor circuit and by the shunting capacitance in the case of the electron-tube circuit.

The frequency response of the RC-coupled audio amplifier is limited by the same factors that limit frequency response in other RC-coupling circuits. In other words, the very low frequencies are attenuated by the coupling capacitor whose reactance increases with low frequencies. The high frequency response of the amplifier is limited by the shunting effect of the output capacitance of the first stage and the input capacitance of the second stage.

RC coupling is used extensively in audio amplifiers because of good frequency response, economy of circuit parts, and the small physical size that can be achieved with this method of coupling.

Reactance: The opposition offered to the flow of alternating current by a pure capacitance, pure inductance, or a combination of the two. When a voltage is applied to a pure inductance (L), current cannot flow immediately, because it is opposed by a voltage of opposite polarity (the CEMF generated by the moving magnetic field of the inductor). The current reaches its maximum value some time after the voltage has been applied. Voltage applied to an inductance, therefore, leads current by 90° in a pure inductance. If unavoidable resistance is present, the phase angle is proportionately less than 90°. The opposition thus offered by an inductance is termed *inductive reactance* (X_L). For a given value of inductance, the strength of the CEMF is proportional to the rate of change of the applied voltage. Therefore, the higher the frequency, the higher the CEMF and the higher the reactance.

When a voltage is applied to a pure capacitance (C), as to an ideal lossless capacitor, a current flows into the capacitor, decreasing in value until the capacitor becomes fully charged, whereupon the flow stops. The voltage across the capacitor is thus zero when the current is maximum, and vice versa. Current flowing into a capacitor is proportional to the rate of change of voltage. For an ac voltage, this rate of change is maximum when the cycle is passing through zero, and is zero when the cycle is maximum. Voltage across a pure capacitance, therefore, lags current. From the other point of view, current leads voltage. The current leads by 90°. If unavoidable resistance is present, the phase angle is proportionately less than 90°. The opposition thus offered by a capacitance is termed *capacitive reactance*.

Reactance-tube FM transmitter: An electron tube operated so that its reactance varies with the modulation signal and thereby varies the frequency of the oscillator stage. This is a practical method of obtaining FM transmission. The reactance-tube system of frequency modulation is shown in Fig. 348. In this circuit, the reactance tube is connected in parallel with the oscillator tank and functions like a capacitor whose capacitance is varied in accordance with the audio signal, as in the capacitor microphone system of frequency modulation.

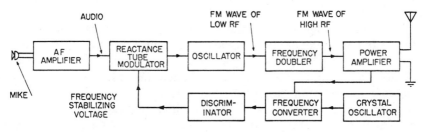

FIGURE 348 Block diagram of a reactance-tube FM transmitter.

The frequency of the AF signal determines the number of times per second that the oscillator tank frequency changes. On the other hand, the amplitude of the AF signal determines the extent of the oscillator frequency change or the amount of deviation. The frequency of the oscillator is thus changed, and the resulting FM signal is passed through a frequency doubler to increase the carrier frequency and the deviation frequency. A power amplifier feeds the final signal to the antenna. The transmitter is kept within its assigned frequency limits by comparing the output of the transmitter with that of a standard crystal-controlled oscillator and feeding back a suitable correcting voltage from a frequency converter and discriminator (frequency detector) stage.

Receiver alignment: The process of adjusting the tuned circuits of electronic equipment to produce resonance at frequencies required or specified for such adjustments. Alignment is necessary to keep a receiver operating at peak performance. Indications of a need for alignment include poor sensitivity and a difference in the frequency received from that of the dial setting. Alignment should also be performed whenever circuit components are replaced.

Before attempting alignment, it must be ensured that the receiver is otherwise trouble free and all available technical literatures pertaining to the equipment have been consulted. In addition, a thorough knowledge of the test equipment to be used is quite important.

Prior to actual alignment, all auxiliary functions provided in the receiver that may interfere with proper output indication or circuit resonance should be disconnected. This includes AGC, silencer or squelch action, noise and output limiters, etc. As this requirement may vary from receiver to receiver, the appropriate technical manuals should be consulted. Also, the antenna should be disconnected and replaced with a dummy antenna.

Alignment is accomplished by injecting a modulated signal from a signal generator into the receiver at appropriate points and adjusting the various tuned circuits for a maximum detector rectified output voltage. This voltage may be measured with the use of a high-impedance electronic dc voltmeter so as to minimize disturbance to the detector circuit. Alternately, the amplitude of the intelligence at the output of the receiver may be used as an indication of proper alignment. Figure 349 will be used as a representative receiver in block diagram form.

In general, alignment is best begun in the circuit farthest from the antenna. In the case of a superheterodyne receiver, this is the last IF transformer preceding the detector. In some receivers, it is best to disable the local oscillator stage prior to attempting IF alignment (S_1 in Fig. 349). Again, this requirement varies from receiver to receiver.

A modulated signal, usually at the nominal band-center frequency of the particular IF system, is injected at the input electrode of the last IF amplifier stage, and the last or output IF transformer is adjusted for maximum output

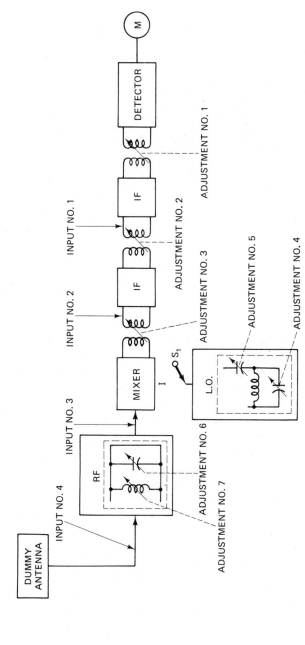

FIGURE 349 Receiver block diagram showing signal input and adjustment points for alignment.

indication (input No. 1 and adjustment No. 1 in Fig. 349). The amplitude of the input from the signal generator should be adjusted to provide a signal output level that is well above the noise level at the output indicator and also well below the level that would cause overdriving or saturation. If either of these conditions are not met, it will be difficult to get an accurate indication of proper alignment.

After the last or output IF transformer is aligned, the point of signal injection is moved to the input electrode of the preceding IF stage (input No. 2), and the next-to-last IF transformer is adjusted (adjustment No. 2). If there are more than two IF stages, this procedure is continued until the entire IF strip is aligned. Naturally, the amplitude of the injected signal must be decreased as the point of injection moves closer to the antenna and more circuits come into alignment.

Alignment of the first or input IF transformer is achieved with the injected signal placed on the input electrode of the mixer (input and adjustment No. 3). It should be noted that it is necessary to recheck the overall alignment of the IF strip after the individual stages have been aligned.

The injection frequency used for IF alignment is usually equal to the nominal band center of the particular IF system. In some cases, stagger tuning for example, the injected signal to alternate IF transformers must be adjusted above or below the nominal band center to achieve maximum response of the IF transformers. The end result of such alignment is a wide IF bandpass.

Following the IF alignment, the local oscillator is adjusted. This is done by injecting a modulated signal of the upper alignment frequency specified for the particular receiver onto the input electrode of the mixer (input No. 3) and adjusting the shunt trimmer capacitor of the oscillator tank circuit for optimum output indication (adjustment No. 4). It is now necessary to check oscillator alignment near the low frequency end of the tuning band. The signal generator is reset to the lower alignment frequency specified for the receiver, still inserted at input No. 3, and the series padder capacitor of the oscillator tank is then adjusted to produce maximum output (adjustment No. 5). When this process is completed, the shunt trimmer capacitance should be rechecked for maximum output at the high-frequency alignment point. Finally, the low-end padder adjustment should again be checked.

The RF stage(s) or preselector is aligned last. To accomplish this, a dummy antenna is connected to the receiver, and a signal of the upper alignment frequency is injected into the antenna input terminal (input No. 4). The RF stage is then adjusted for maximum output indication (adjustment No. 6). The signal generator is then set at the lower alignment frequency, and the RF-tuned circuits are adjusted for maximum output (adjustment No. 7). The above steps are repeated until no further improvement is obtained.

Special problems are encountered in aligning modular constructed receivers. The impedance of each module must be matched to that of the next. If an impedance mismatch exists, the tuned circuits cannot be properly aligned.

In addition, a greater amount of signal amplitude is required to get an output indication from improperly matched modules. The connection of test equipment to a module may cause a mismatch to occur. Therefore, care must be taken to ensure that the test equipment used forms a proper impedance match with the modules being aligned.

Coarse alignment may be performed on each individual module of a modular constructed receiver. A fine overall alignment must be performed with a signal applied to the input of the receiver and all the modules connected. This takes into account the effect of the interconnections of each module. The alignment procedure for a specific receiver must be obtained from the manufacturer's technical manual for that receiver.

Receiver, crystal: A simple type of receiver whose detector is a diode made from a Galena crystal. Figure 350 shows a block diagram of this simple receiver. The antenna performs the function of reception. Selection of the desired signal is accomplished by tuning the frequency selection circuit. The detector separates the audio component from the RF component. The audio component is then passed on to the earphones, which perform the function of reproduction by converting the electrical signal into sound waves.

The basic crystal set has several disadvantages that limit its usefulness. Among these are poor sensitivity and selectivity. Also, the earphones can be used by only one person at a time. This last problem led to another function being added to the receiver—audio amplification. The audio amplifier produced enough power to operate a speaker so that everyone could hear the reproduced sound waves at the same time. Figure 351 shows the block diagram of the receiver with a stage of audio amplification added.

By adding stages of audio amplification, weaker stations that could not be heard when the headphones were used alone could now be heard. This was due to the amplifier responding to a smaller audio-output component than the detector, but this did not improve receiver sensitivity. Of course, any noise present would be amplified along with the signal, and if the noise was larger in amplitude than the signal, it would drown out the signal. These facts limited the number of stages of audio amplification.

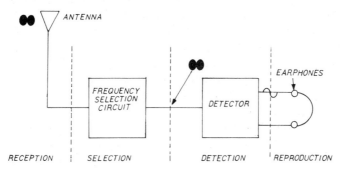

FIGURE 350

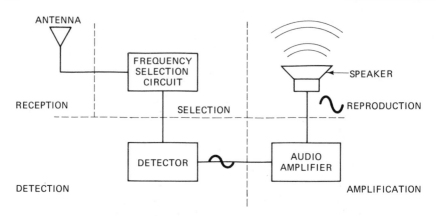

FIGURE 351 Audio amplified receiver.

Receiver, facsimile: The complete device or system that selects, amplifies, and demodulates the picture signal picked up through radio transmission, wires, or cable. The receiver circuitry demodulates the information elements of the signal to reproduce a picture.

Receiver, manual gain control: A circuit incorporated in the RF section of a receiver to provide maximum sensitivity. While high sensitivity is one of the parameters of a good receiver, in some cases, high sensitivity may be a liability. For example, the signal received from a nearby station can be strong enough to overload the RF sections of the receiver. This may cause the audio output to become distorted to the point of complete loss of intelligibility. To overcome this problem, manual gain control of the RF section is utilized. By using a manual gain control, maximum sensitivity is realized and weak input signals are provided with maximum amplification. When a strong input signal is received, the RF gain may be reduced to prevent overloading. Typical manual gain control circuits for a receiver are illustrated in Fig. 352.

C_1 is an emitter-cathode bypass capacitor. R_1 and R_2 develop emitter-cathode bias for the amplifier. C_2 provides dc isolation between the tank and the base of Q_1 in the transistor version. Gain control is nothing more than a manual bias adjustment. When the wiper arm of R_2 is set at point B, minimum forward bias is applied to the transistor and maximum bias is developed in the tube circuit. This causes both amplifiers to operate closer to cutoff and thereby reduces their gain. When the control is moved toward point A, the opposite effects occur. R_1 limits the maximum conduction of the devices when R_2 is short-circuited. In transistor circuits, an alternate biasing method may be encountered where the transistor is operated near saturation. In this case, a large change in gain is again a function of bias.

Receiver, radio frequency: A device or system operated at the end of a communications link and which accepts a signal and processes or converts it for local use. A receiver must perform certain basic functions in order to be useful. These functions, in order of their performance, are reception, selec-

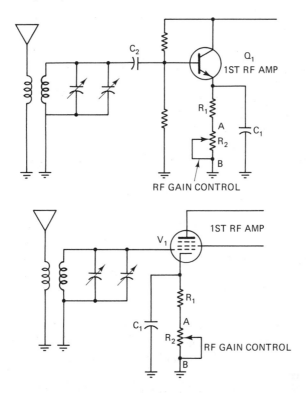

FIGURE 352 Typical RF gain controls.

tion, detection, AF amplification, and reproduction. Reception involves having the transmitted electromagnetic wave pass through the receiver antenna in such a manner as to induce a voltage in the antenna. Selection involves being able to select a particular station's frequency from all the transmitted signals that happen to be induced in the receiver's antenna at a given time. Detection is the action of separating the low-frequency intelligence from the high-frequency carrier. AF amplification involves amplifying the low-frequency intelligence (audio in the case of a radio) to the level required for operation of the reproducer. Reproduction is the action of converting the electrical signals to sound waves, which can then be interpreted by the ear as speech, music, etc. The ability of a receiver to reproduce the signal of a very weak station is a function of the receiver's sensitivity. In other words, the weaker a signal that can be applied to a receiver and still achieve the same value of signal output, the better that receiver's sensitivity rating. The ability of a receiver to select and reproduce a desired signal from among several closely spaced stations or from among interfering frequencies is determined by the receiver's sensitivity. In other words, the better a receiver is at differentiating between desired and undesired signals, the better the receiver's selectivity rating.

Figure 353 shows the block diagram of a simple receiver, which will

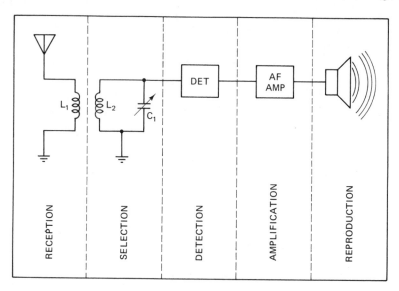

FIGURE 353 Simple receiver block diagram.

perform all the functions required of a receiver. It also illustrates the functions performed by the various sections of the receiver. The input to the receiver is the electromagnetic wave propagated from the antenna of the transmitter. This wave will pass through the antenna of the receiver and induce a small ac voltage. The section of the receiver formed by the antenna and L_1 perform the function of reception. L_1 is the primary of the input transformer, and the voltage induced in L_1 is coupled to the secondary L_2. L_2 and C_1 form a tuned circuit, with C_1 being variable to permit tuning across the broadcast band. Thus, the tuned input circuit performs the function of selecting a specific frequency from among those present in the antenna circuit. The output of the tuned circuit is a modulated RF signal.

This modulated RF signal is then fed to the detector circuit, where the function of detection (rectification and filtering) is performed. The output of the detector circuit is a weak audio signal. The audio signal from the detector is too weak to satisfactorily operate a speaker. Therefore, it is fed to an audio-frequency amplifier to increase its amplitude. The output of the AF amplifier is fed to the speaker, which performs the function of reproduction, or converting the electrical signals back to the form of the original input to the transmitter (in this case, sound waves).

Reciprocating pump: A pump in which the flow, or discharge, varies directly with the speed, regardless of the head. Reciprocating pumps in small sizes are generally single cylinder (simplex), whereas in larger sizes, they may be two cylinder (duplex) or three cylinder (triplex). Any of these may be built as either a single- or double-acting pump; that is, with either one or two impulses per cylinder per revolution.

Where a uniform flow is important, the triplex double-acting pump, which gives six impulses per revolution, is used. For high heads, the efficiency of a reciprocating pump is higher than that of any other type of pump if the pump is in good condition and its efficiency varies very little with its size. Triplex pumps at a head of 200 ft average about 75 percent efficiency. At heads from 300 to 1000 ft, their efficiency is 80 percent or over.

Since reciprocating pumps operate at low speed, some gearing is generally included in the design to provide a higher-speed drive shaft. In order to use a reasonable motor speed, an additional speed reduction is generally needed in gearing, chains, or belts. Since reciprocating pumps start against the maximum head, the motor must deliver more than full-load torque at starting to overcome the head and pump friction. Squirrel-cage motors with high starting torque, NEMA design C, are generally used in the medium sizes, while for large pumps in isolated positions, wound-rotor motors are frequently specified to give high starting torque with a minimum of starting current. If a pump is provided with a bypass connection so that it may be unloaded when starting, motors with standard starting torque characteristics may be used. When the pump is up to speed, the bypass closes and the pump then discharges against its full head.

Recorder: Any device that has the capability of storing electronic information for playback at a later time. The medium of recording may be paper, magnetic tape, magnetic core, special semiconductor circuits, or wire.

Recording: In electronics terminology, a hard copy printout of electronic values or parameters. It is also the act of putting audio and video information on magnetic tape. Magnetic tape is a prime medium for electronic recording and provides a convenient method of storage for audio and video information, which may be played back at any time. Magnetic tape may also be used for the storage of computer information. This is also a form of recording whereby the magnetic fields on the tape are arranged in an order that corresponds with the input information.

Recording head: In audio- and videotape recording, an electromagnet that is driven by the input information in the form of electric currents. Changes in current alter the magnetizing properties of the head, which rests against the magnetic tape. As the tape travels across the head, magnetic fields are arranged in a manner that corresponds with the input information.

The reciprocal of the recording head is the playback head, which is a transducer that changes the low-level magnetic fields from the recorded tape to electrical impulses. The tape travels across the playback head at varying speeds (depending on the recorder), and all information is converted back to audio or video outputs.

Recording instrument: Any of a number of measuring devices designed to read electronic values. The information is then put into hard copy form by a chart recorder. Usually, the chart recorder and metering instrumentation are housed in one device package. The electronic information that is input to the recording instrument causes one or more styli to move up and down on a piece of graph paper in relationship to the changing electronic values. The paper is contained on a small reel, which is attached to a motor drive. Each time the recording intrument is activated, the motor drive moves the paper at a steady rate. In this manner, the imprints of the styli provide accurate metering information with relationship to time. The chart traces may be stored for future reference.

Recording paper: Graph paper with specially calibrated scales used with recording instruments. It is usually seen in the form of a tightly wound paper reel and is pulled through the instrument by a motor drive.

Rectification: The process of changing alternating current to direct current. The transmission of electrical energy over great distances was made economical through the use of alternating current. Alternating current could be transmitted over great distances with a minimum of power dissipation within the transmission line. Most electron tubes and many other electrical devices require a steady source of dc voltage. This voltage may be provided by a dc generator or by changing alternating current to direct current. The devices used to accomplish rectification are called *rectifiers*.

Rectifier: A device that will conduct current of a specific polarity, either positive or negative, in one direction but not in the other. Solid-state rectifiers are made from germanium or silicon crystal materials, with silicon being used almost universally for power supply applications. These rectifiers are rated in many ways, but there are three values of interest in dc power supply applications.

The peak reverse voltage (PRV) is the voltage that the diode must be able to withstand when it is not conducting current. In application, this current will vary with the load and can be almost three times the ac voltage delivered by the transformer secondary, depending on the type of rectifier circuitry used.

The second important rating in rectifier selection is the peak current rating of the rectifier. This is the maximum rating of current that can be passed through the device. The third rating is surge current, sometimes listed as the I_s factor, which may be several hundred times the peak current rating (I_{rep}). I_s ratings set the maximum surge that the device can safely handle. This high current occurs for a small fraction of a second when voltage is initially applied to a dc power supply circuit. An average current is also applied to solid-state rectifiers and sets the proper continuous operating limits of the device.

Solid-state rectifiers are very advantageous when used in the construction of dc power supplies. They require no external power, are very efficient, and are very compact when compared to the vacuum-tube rectifier of a few years ago, which required filament voltage, larger amounts of mounting space, and sometimes, external cooling devices such as a fan or blower to remove heat.

Silicon rectifiers are available in a wide range of current and voltage ratings, with devices of less than 600 PRV rating capable of handling up to 400 A and more. Higher voltage rectifiers of around 1000 PRV rating may have current ratings on the order of 1 to 2 A. Owing to the small physical size of solid-state rectifiers, it is possible to stack the units to provide current ratings and PRV ratings in multiples of the discrete component ratings. Commercially stacked units are available with ratings on the scale of 15,000 peak reverse voltage (PRV) with current handling capabilities of 1 A.

Rectifier circuit: An integral assembly of one or two rectifying devices. Rectifiers are used when it is necessary to provide direct current for a load and the available source produces alternating current. The rectification should cause minimum losses of available power, and it should produce as steady a direct current as possible. The rectified output current is always pulsating. It is the purpose of the proper connection of rectifiers to produce such a combination of pulses that they will overlap and result in a direct current with minimum ripple, or variations.

Various types of rectifiers may be used in rectifier circuits. Each type is represented by a standard symbol, as shown in Fig. 354. Symbols (a) to (e) indicate types of rectifier tubes or tanks. The dot in the circle is an indication that the tube is gas filled. The gas may be mercury vapor. The symbol at (f) represents a semiconductor rectifier; the direction of the arrowhead indicates the forward direction of the current through the rectifier.

Standard connections are used to combine rectifiers into rectifier circuits.

Reflectometry, frequency domain: A fast, simple, and reliable technique developed to locate defects in microwave cables and waveguide systems connecting receivers, transmitters, and antennas. Although the name sounds complicated, the technique is easy to use and permits direct readout of cable distance (in feet) to the impedance discontinuity (fault).

This system has developed an impressive record of reliability, greatly

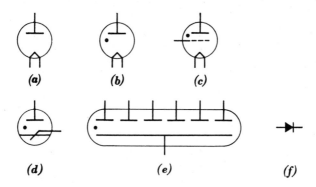

(a) High-vacuum hot-cathode tube
(b) Gas-filled hot-cathode tube without grid
(c) Gas-filled hot-cathode tube with grid
(d) Mercury-pool tube with ignitor
(e) Multianode mercury-pool tube
(f) Semiconductor rectifier

FIGURE 354

reduced service time, and improved service standards. Because the system checks cables at their actual operating frequencies, discontinuities outside those frequencies will not affect the test. When measurements indicate a fault, its location (in terms of distance in feet from the point of test) can be precisely determined. Repairs, therefore, can be made quickly and effectively.

Before this technique was developed, cables were tested primarily by means of time-domain reflectometry, a system that has several severe limitations. For example, TDR makes measurements covering a spectrum determined by its pulse characteristics and identifies all discontinuities or parameter changes, including those outside the operating frequency range that do not affect system operation. With FDR, however, the analysis is made at the actual operating frequencies.

While FDR works in waveguide and band-limited systems including transmission networks that contain filters, TDR cannot work in such systems because it requires a transmission line that passes the whole spectrum from the fundamental frequency (2 to 5 mHz) to the highest harmonic (15 GHz). Waveguides that act as high-pass filters cannot transmit TDR pulses. Similarly, TDR cannot see through low-pass or bandpass filters, because these eliminate the low-frequency harmonics.

FDR identifies defective systems by using insertion-loss (attenuation in the line) and return-loss (voltage standing wave ratio or VSWR) measurements to classify the system as good or in need of repair. These measurements are made with the test setup shown in Fig. 355. Such a test configuration provides simultaneous measurement of both return loss and insertion loss (or gain) of passive (or active) devices.

If the input and output connectors are accessible, an insertion-loss check verifies input-to-output performance across the band. For insertion-loss measurement, the network analyzer (using its B and REF channels) indicates the ratio of output to input directly in decibels. For tests of long cables whose ends are accessible, the system allows measurements from a connector end as far as 2000 ft from the test setup.

In some systems, however, either the input or the output connector, such as a cable to a distant superstructure point in a plane or missile, may be inaccessible. For such systems, a return-loss measurement made on the accessible connector will provide a total system check. In high-gain systems in which an active device isolates input and output, a return-loss check can be made from each end. Another useful application of return-loss measurement occurs when the cable feeds the output of a high-power microwave tube, which can be destroyed by excessive reflected energy. For return-loss measurements, the network analyzer (using the A and REF channels) indicates the ratio of reflector power to incident power directly in decibels. As shown in Fig. 355, the signals in each case are sampled by directional couplers.

Comparison of each measured signal with the incident power supplies automatic compensation for any swept-source power variation across dB

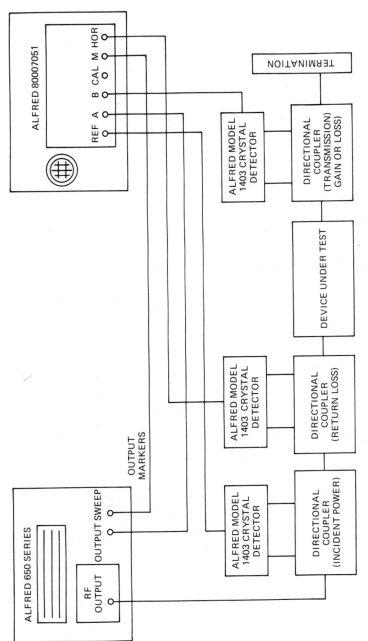

FIGURE 355 Test setup for VSWR and insertion performance.

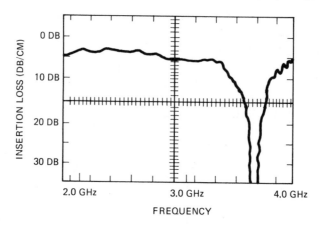

FIGURE 356   Insertion-loss measurement.

versus frequency on the network analyzer cathode-ray tube (CRT). To provide measurement capability for long cable systems having high losses, the instrument furnishes direct dB readout over a 60-dB dynamic range.

An example of insertion-loss measurement is given in Fig. 356. Here, a loss of less than 10 dB is acceptable. The cable, however, needs repair, because a fault that produces an insertion loss greater than 35 dB at a frequency of 3.56 GHz is present. An example of a return-loss measurement for the same cable is shown in Fig. 357. Here, a loss of 11 dB, which corresponds to a VSWR of less than 1.8, is acceptable. At 3.56 GHz, then, the return loss is equal to 5 dB, which corresponds to a VSWR of 3.6.

The dual-channel network analyzer permits display of both measurements simultaneously when the complete test setup shown in Fig. 355 is used. For cable-length and fault-location measurements, a waveguide or coaxial tee is incorporated in the test setup (Fig. 358). The system is calibrated with the

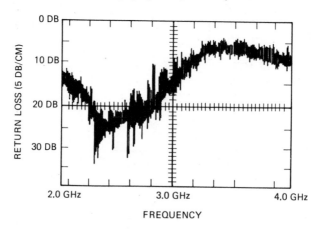

FIGURE 357   Return-loss measurement.

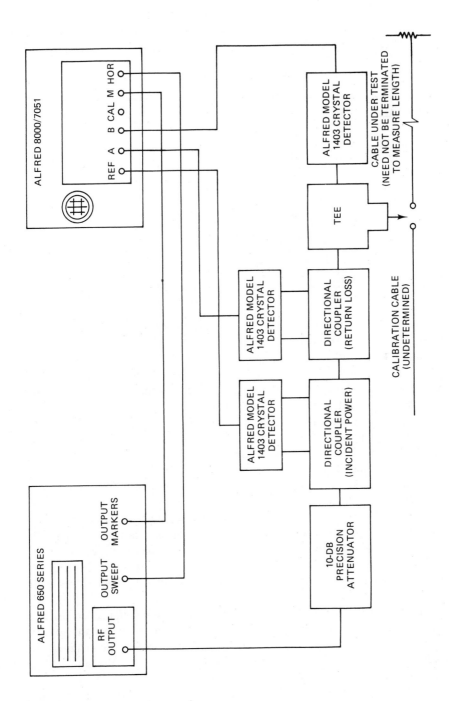

FIGURE 358 Test setup for fault-location measurement.

459

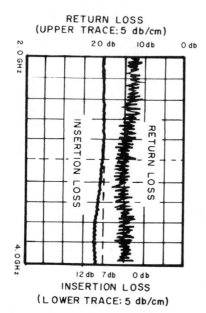

RETURN LOSS
(UPPER TRACE: 5 db/cm)

2.0 GHz

20 db   10 db   0 db

INSERTION LOSS

RETURN LOSS

4.0 GHz

12 db 7 db   0 db
INSERTION LOSS
(LOWER TRACE: 5 db/cm)

FIGURE 359 Dual-channel display of a repaired cable.

calibration cable, and the cable under test is then connected to the tee. The resultant CRT display on the network analyzer consists of a stationary pattern containing a series of half-dome ripples. A count of the total number of these ripples indicates the number of feet from the cable end to the fault.

The FDR display is from the cable system that produced the needs repair indications in Figs. 356 and 357. Multiplying the 5-2/3 ripples that appear on the oscilloscope by the display calibration factor of 2 ft per ripple will quickly identify the location of the fault (11-1/3 feet from the cable-end connector). Figure 359 illustrates a dual-channel display after faults (Figs. 356 and 357) have been corrected. The insertion loss is less than 10 dB, and the return loss is greater than 11 dB, indicating that performance is now satisfactory.

Reflectometry, time domain: A measurement concept used in the analysis of wide-band systems. The art of determining the characteristics of electrical lines by observation of reflected waveforms is not new. For many years, power transmission engineers have located discontinuities in power transmission systems by sending out a pulse and monitoring the reflections. Discontinuity is any abnormal resistance or impedance that interferes with normal signal flow.

TDR is particularly useful in analyzing transmission systems because the amplitude of the reflected signal corresponds directly to the impedance of the discontinuity, and the distance to the discontinuity can be determined by measuring the time required for the pulse to travel down the line to the reflecting impedance and back to the monitoring oscilloscope.

The TDR analysis consists of the insertion of a step or pulse of energy into a system and the subsequent observation at the point of insertion of the energy

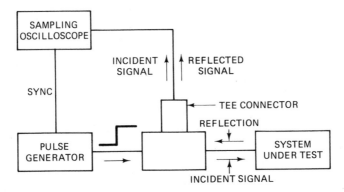

FIGURE 360   Typical time domain reflectometer.

reflected by the system. Several arrangements are possible, but the following procedure is used with the newer, specialized reflectometers. See Fig. 360.

A fast (or incident) step is developed in the pulse generator. This step then passes through a tee connector and is sent into the system under test. The sampling oscilloscope is attached to the tee connector and the incident step, along with the reflected waveform, is displayed on the CRT. Analysis of the magnitude, duration, and shape of the reflected waveform will determine the type of impedance variation in the system under test.

TDR discontinuities are clearly separated in time on the CRT. It is easy to see the mismatch caused by a connector even if a bad discontinuity is present elsewhere in the system. By using the aforementioned analysis, it is possible to establish which connector is troublesome and in what way. Once it is determined that a discontinuity appears in a waveform, it is simple to locate it in the system. A timesaving way of doing this is to calibrate the system so that 1 cm on the horizontal axis is the equivalent of a certain number of feet for the transmission system under test. The limiting factor is the system rise time, and any closely spaced discontinuities will appear as a single discontinuity.

The finite rise time also limits the size of the distinguishable reactive impedance response. For example, a small shunt capacity in a 50-Ω system will cause the waveform to depart from the ideal response (Fig. 361).

The maximum observable line length is a function of the repetition rate chosen. This rate determines the duration of the pulse after its rise. For example, a 200-kHz repetition rate permits the use of TDR devices with up to 1000 ft of air-dielectric cable, or 670 ft of polyethylene-dielectric coaxial cable. The speed at which a wave travels through a transmission system is determined by the system's velocity constant. Thus, a wave travels faster through air than through polyethylene. This explains the difference in maximum lengths of coaxial cable that can be checked using a particular repetition rate on the TDR. The longer the cable, the lower the repetition rate must be.

For applications where the output of a test system is of interest, a dual-

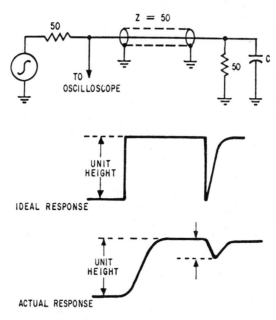

FIGURE 361 Small shunt capacity in system degrades ideal response.

channel sampling oscilloscope can be used in a TDR configuration to display both the input and transmission characteristics of the system (Fig. 362). The step generator is connected to the sampled channel, and an accurate 50-Ω termination is connected to the output for the initial calibration. This termination can be a 1 percent film-type load resistor with short lead lengths. Prior to removal of the 50-Ω termination and connection of the test system, the vertical gain of the oscilloscope is adjusted to give a 10-cm step.

The magnitude and location of reflections can be viewed more closely by

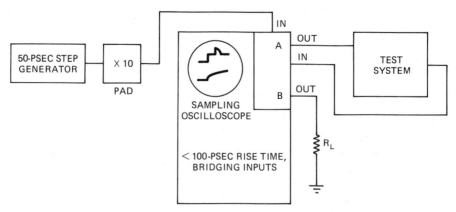

FIGURE 362 Dual-channel oscilloscope analysis of reflection and transmission characteristics.

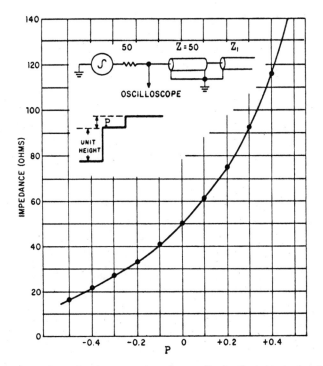

FIGURE 363   Curve for the conversion of normalized reflected pulse height into system impedance.

selection of an appropriate time scale and vertical magnification. Calibration of the horizontal reflection allows rapid physical location of points of interest. Calibration of the vertical axis in terms of the reflection coefficient (the change in amplitude of the flat portion of the incident waveform) is caused by the reflected signal. For example, p of Fig. 363 allows direct interpretation of cable impedance at the discontinuity. With the sampling oscilloscope calibrated, cable impedance can be determined from the height of the reflected step. The graph shown in Fig. 363 may be used to translate step heights into impedance.

Reflex klystron: A klystron tube that utilizes the same grids for both bunching and catching. When the basic klystron is used as an oscillator, it is critical to adjust. For that reason, the reflex klystron was developed. In the reflex klystron, the collector plate is replaced by a repeller plate. The potential applied to the repeller plate is negative. Electrons moving toward this plate will be repelled back in the direction of their origin. The repeller plate is the most negative element in the tube. A diagram showing the reflex klystron is shown in Fig. 364.

The reflex klystron is similar in operation to the basic klystron. Electrons accelerated by the accelerator grid will be velocity modulated as they pass through the cavity grids. The cavity grids in the reflex klystron perform this

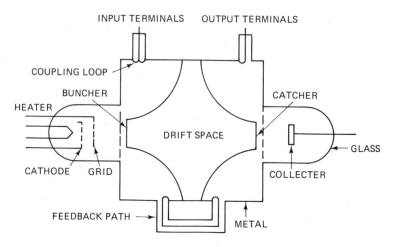

FIGURE 364   Klystron tube with resonator.

function in the same manner that it is performed in the basic klystron. The electrons, after passing through the cavity grids, will move at different velocities. Since the repeller plate is made highly negative, the electrons progressing toward it will stop and reverse their direction. The high-velocity electrons will come physically closer to the repeller plate than either the medium- or low-velocity electrons.

After repulsion, the electrons will be directed back toward the cavity grids. In the reflex klystron, bunching occurs immediately before the electrons come under the influence of the RF field about the cavity grids. The distance that the electrons move before they are repelled by the negative repeller plate is a function of the voltage values of the accelerating grid, the dc value of the voltage applied to the cavity grids, the dc voltage applied to the repeller plate, and the magnitude of the RF voltage coupled to the cavity grids by the cavity resonator. The voltages applied and the physical construction of the klystron should be of such values that the electrons will return to the cavity grids in bunches.

The potential of the cavity grids when the repelled electrons return is important. The bunched electrons should be returned when the potential applied to the cavity grids is such that the energy of the returning bunches will be absorbed. The maximum absorption of energy will occur when the bunched electrons reach the midpoint between the cavity grids in coincidence with the maximum positive peak of RF voltage between these grids. As the electron bunch reaches the midpoint, the grid nearest the repeller plate must be positive in relation to the other buncher grid for correct alignment of the electrostatic field. The electron bunch will be decelerated in this field, thus expending some of its energy in sustaining RF oscillations within the grid cavity.

Under these conditions, electrons leaving the cathode will receive maximum acceleration from the cavity field, while returning electron bunches will receive maximum deceleration. If the grids are separated by approximately one half-wavelength, the electron bunch would pass through the first grid (the one nearest the repeller plate) as its RF potential is zero and changing from negative to positive. The electron bunch would pass through the second grid when its potential is zero and is changing from negative to positive. After the returning electron bunches have given their energy to the cavity, they are absorbed by the cavity grid nearest the cathode and are returned to the power supply.

The cavity grids perform a dual function of velocity modulation and that of a catcher grid. The output from the tube is taken by use of the coupling loop shown in Fig. 365.

By proper adjustment of the negative voltage applied to the repeller plate, the electrons that have passed through the bunching field may be made to pass through the resonator again at the proper time to deliver energy to this circuit. Thus, the feedback needed to produce oscillation is obtained and the tube construction is greatly simplified. Spent electrons are removed from the tube by the positive accelerator grid or by the grids of the resonator. The operating frequency of the tube can be varied over a small range by changing the voltage on the repeller plate. This potential determines the transit time of the electrons between their first and second passages through the resonator. However, the output power of the oscillator is affected considerably more than the frequency by changes in the magnitude of the repeller voltage. This is because the output power depends on the fact that the electrons are bunched at exactly the decelerating half-cycle of oscillating grid voltage. The volume of the resonant cavity is changed to change the oscillator frequency. The repeller voltage may be varied over a narrow range to provide minor adjustments in frequency.

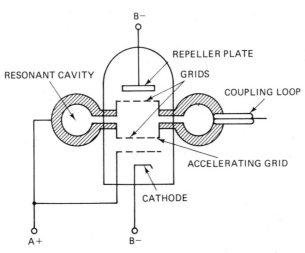

FIGURE 365 Reflex klystron.

Refraction: The bending of a light ray. When a light ray passes at an angle from a less-dense to a more-dense medium, it is bent toward the normal. In other words, the angle of refraction is less than the angle of incidence. Likewise, when a light ray passes from a more-dense to a less-dense medium, it is bent away from the normal. In other words, the angle of refraction is greater than the angle of incidence. Both of these conditions are shown in Fig. 366a.

This *index of refraction* is a term used to describe how much bending will take place as electromagnetic waves pass through a given substance. The higher the index of refraction, the more the bending. The index of refraction is the ratio of the velocity of light waves in a vacuum to the velocity of light in the substance being considered.

In the case of radio waves being refracted in the ionosphere, the path is gradually curved because there is no sharp point of transition between layers of

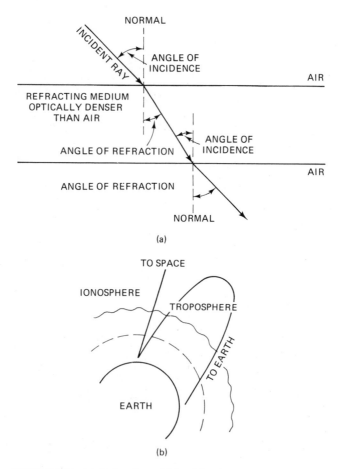

(a)

(b)

FIGURE 366 (a) Refraction of light; (b) refraction in ionosphere.

different densities. Those rays that make a large angle with respect to the horizontal along the earth may be refracted a small amount and pass on through the ionosphere to outer space. Those that make a smaller angle will travel a greater distance in the ionosphere and may be bent to such an extent that they will return to earth. A simplified illustration of ionospheric refraction is given in Fig. 366b. A certain amount of refraction also takes place in the troposphere (below the ionosphere) because of the proximity of warm and cold air masses.

Regeneration: 1. The process of returning a part of the output signal of an amplifier to its input circuit in such a manner that it reinforces the excitation and thereby increases the total amplification.

2. Periodic restoration of stored information.

Regenerative braking: A system that uses the principle of regenerating back into the power supply, and like dynamic braking and plugging, is applicable to both ac and dc motors. Regenerative braking requires no changes in the control circuits, but simply an overhauling load, which tends to run the motor above its rated speed or to act as a generator. Naturally, regenerative braking cannot be used below the synchronous speed of a motor.

If the synchronous or no-load speed of an induction motor or a dc motor is 1200 rpm and the full-load speed is 1150 rpm, at approximately 1250 rpm, it will develop a generating torque equal to approximately its full-load rating. A synchronous motor will not change in speed, but will change from a motor to a generator, depending on the torque requirements. Consequently, regenerative braking is useful only as a means of lowering loads, such as by large mine hoists, or by quarry hoists on inclines. When it is desired to stop the load (and hence, the motor), dynamic braking, mechanical braking, or both types of braking are required.

Figure 367 shows the speed-torque curves of a wound-rotor induction motor. The indicated speeds are obtained by resistance control during motoring and also during regenerative braking. As can be seen, the synchronous speed of the motor is 1200 rpm. At all speeds below this value, the motor operates in the same manner as a wound-rotor induction motor. At all speeds above 1200 rpm, regenerative braking action occurs.

For example, curve 1 in Fig. 367 represents the relation between speed and torque at a starting resistance that allows the lowest speed when the machine is motoring, but allows the highest speed when regenerative braking is taking place and the motor is lowering a load. The highest speed when motoring and the lowest speed when braking are obtained with all resistance cut out. These conditions are represented by curve 4 in Fig. 367. Curves 2 and 3 are simply intermediate curves between those indicated at curves 1 and 4.

Register: A device that provides temporary storage of a binary word during computation. Registers are composed of flip-flops and are generally found in the control and arithmetic units of digital computers. There may be several registers in a large computer. However, a simple arithmetic unit may

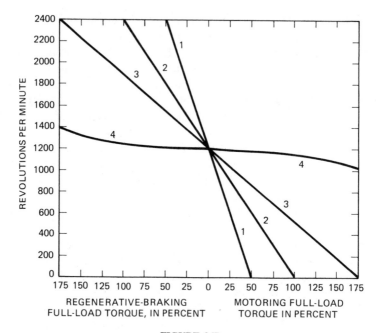

FIGURE 367

contain only one or two. The number depends on the amount and variety of other storage facilities that are included. It is also somewhat dependent on the speed at which the computer is designed to operate.

Register, control: A register that holds the identification of the instruction word to be executed next in time sequence following the current operation. The register is often a counter that is incremented to the address of the next sequential storage location, unless a transfer or other special instruction is specified by the program.

Relay: A magnetically operated switch that can be used to control circuits distant from the operating point; control a relatively high-voltage or high-wattage circuit by means of a low-power, low-voltage circuit; or obtain a variety of control operations not possible with ordinary switches. Whether the circuits controlled will be closed or opened when the relay coil is energized will depend on the arrangement and connection of the relay contacts.

When current flows through the relay coil, it magnetizes the iron core with a polarity that depends on the connection of the coil to the source. This pole induces a pole of opposite sign in the iron section of the movable assembly, and the attraction between these operates the relay switch. If the current through the coil is reversed, both poles are reversed and attraction always occurs. From this, it can be seen that relays may be designed to operate from either direct or alternating current.

While relays may vary widely in mechanical construction, they all operate on the same principle. Figure 368 shows some of the differences in design.

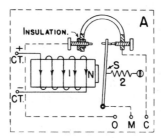

WESTERN UNION RELAY

C = NORMALLY CLOSED CONTACT.
O =   "   OPEN   "
M = MOVING CONTACT.
I = MAGNET COIL WITH TERMINALS CT.
2 = SPRING.

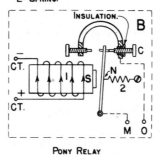

PONY RELAY

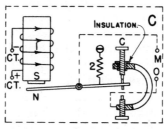

DIXIE RELAY

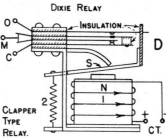

CLAPPER TYPE RELAY.

FIGURE 368 Relay designs.

Reluctance: The opposition to magnetic flux offered by a magnetic material. The equation for reluctance is

$$R = \frac{cm}{\mu A}$$

where R equals reluctance, cm equals length in centimeters, μ equals permeability, and A equals area in square centimeters. Reluctance is equivalent to resistance in an electric circuit.

Repulsion-start motor: A single-phase motor ranging in size from approximately 0.10 to 20 hp and which has high starting torque and a constant-speed characteristic. It is used in commercial refrigerators, compressors, pumps, and other applications requiring high starting torque.

Repulsion-start, induction-run motors are of two different designs. In one, known as the brush-lifting type, the brushes are automatically moved away from the commutator when the motor reaches approximately 75 percent of full speed. This type generally has the radial form of commutator.

In the other, called the *brush-riding type,* the brushes ride on the commutator at all times. This type has the axial form of commutator. The brush-riding arrangement is used almost exclusively on smaller motors, whereas the brush-lifting arrangement is used for both small and large motors. In their other operating principles, these motor types are identical.

Resistance: The ability of a substance to oppose the flow of electrical current. When voltage is applied to a circuit, the electromotive force starts electrons in motion for current flow. An electronic circuit is normally composed of various components that affect the flow of current. However, one component, the conductor, is designed to pass current unimpeded to the various other components in the circuit. This conductor usually takes the form of a piece of wire. The term *conductor* is correct when describing this wire, because it does conduct the flow of electricity more than it impedes its flow. Any practical conductor, however, also exhibits a resistance to the flow of current. This is known as the resistive properties of the wire, and any other components will also have a certain resistance to the flow of current.

There are three types of material that make up electronic components. These are conductors, nonconductors, and semiconductors. A conductor tends to conduct the flow of electricity, while offering a minimal amount of resistance. A nonconductor restricts the flow of current more than it conducts. A semiconductor material is one that lies between conductors and nonconductors. It tends to pass current as well as it resists its flow. A resistor is a device that offers a lumped and specific value of resistance to the flow. This can be thought of as being a nonconductor of a rated value.

The unit of resistance is the ohm. The value of 1 Ω is the amount of resistance that, if connected between the two terminals of a 1-V battery, will produce a current flow of 1 A. It can be seen, then, that all of the terms used to describe values in an electronic circuit are interrelated and dependent upon one another for proper understanding.

Resistor: A lumped resistance that is contained in one discrete package. Resistance is measured in ohms, which is an indication of a material's resistivity to the conduction of electrical current. Even conductors have a certain degree of resistance to this flow.

The resistivity is measured in ohms and is calculated from the ohms value of a cube of the material being rated measuring 1 cm per side. Figure 369 shows a chart of various metals with their resistivity factors. The lower this figure, the higher the conductivity and the lower the resistance.

Conductance and resistance are often discussed together, because conductance is the reciprocal of resistance. The unit of conductance is the mho, which is the word *ohm* spelled backward.

When electrical current encounters a resistance, heating occurs and the voltage or potential force of this current is dropped by a degree that corresponds to the degree of resistance encountered by the flow. Resistors are manufactured in many different physical sizes, which are rated at different power or wattage values for different applications. Because of this power factor, a 1-Ω resistor may be 50 times the size of a 1,000,000-Ω resistor, or vice versa. The physical size of a resistor is an indication of its power-handling capability, not its value of resistance in ohms.

Resistors may be combined in electrical circuits for increased wattage and an increase or decrease in resistance, depending on whether the resistors are connected in series or parallel, respectively.

When heating effects become more than the resistor is rated to get rid of safely, the entire device may burn up. More often, however, the device will operate at a temperature that is much higher than its intended operation calls

RELATIVE RESISTIVITY OF METALS

MATERIAL	RESISTIVITY COMPARED TO COPPER
ALUMINUM (PURE)	1.6
BRASS	3.7–4.9
CADMIUM	4.4
CHROMIUM	1.8
COPPER (HARD-DRAWN)	1.03
COPPER (ANNEALED)	1.00
GOLD	1.4
IRON (PURE)	5.68
LEAD	12.8
NICKEL	5.1
PHOSPHOR BRONZE	2.8–5.4
SILVER	0.94
STEEL	7.6–12.7
TIN	6.7
ZINC	3.4

FIGURE 369

for, and physical and chemical changes will take place in the material that forms the resistive element. When resistors heat up to above-normal temperatures, the resistance value changes, sometimes to a high degree. This change can cause component damage in other parts of a major electronic circuit and will often cause erratic operation of critical pieces of equipment.

On the practical side, resistors come in several wattage ratings and in many mounting styles and configurations. Common wattage ratings for the small carbon resistors include 0.25, 0.5, 1, 2, and, sometimes, 4 W. Wire-wound resistors, which use resistance wound on an insulated form to package their lumped resistance, are available in wattage ratings from about 5 to 200 W and more.

Many carbon resistors identify their resistance value in ohms by a color-coded finish. These colored bands identify the first number, the second number, and the zeros that follow, also called the *multipliers*. Figure 370 shows the color-code chart that will apply to all resistors with banded markings. Wire-wound resistors do not usually have coded markings. Rather, they carry their ohmic value and wattage rating in printed form on the resistor case.

The resistance of most resistors has a tolerance value that may range from 5 to 10 and sometimes to 20 percent. This means that the value of resistance may differ from the stated value by the indicated percentage. For noncritical applications, 10 percent resistors are used; while 5 percent components are chosen for the portions of electronic circuits where closer tolerance must be maintained. For meter-multiplier purposes where extreme accuracy must be maintained, 1 percent resistors are available in specific values. The closer tolerance components are usually more expensive. The tolerance of the component is either color-coded on the case (see Fig. 370), or it may be stamped onto the case of a larger resistor along with the resistance value.

Resonance: The state in which the natural response frequency of a circuit coincides with the frequency of an applied signal, or vice versa, yielding

	RESISTOR		COLOR CODE
COLOR	SIGNIFICANT FIGURE	DECIMAL MULTIPLIER	TOLERANCE (%)
BLACK	0	1	
BROWN	1	10	
RED	2	100	
ORANGE	3	1,000	
YELLOW	4	10,000	
GREEN	5	100,000	
BLUE	6	1,000,000	
VIOLET	7	10,000,000	
GRAY	8	100,000,000	
WHITE	9	1,000,000,000	
GOLD	–	0.1	5
SILVER	–	0.01	10
NO COLOR	–		20

FIGURE 370

intensified response. Resonance may also refer to the state in which the natural vibration frequency of a body coincides with an applied vibrational force, or vice versa, yielding reinforced vibration of the body.

Resonant line: A transmission line that has standing waves of current and voltage. The line is of finite length and is not terminated in its characteristic impedance. Therefore, reflections are present. A resonant line, like a tuned circuit, is resonant at some particular frequency. The resonant line will present to its source of energy a high- or low-resistive impedance at multiples of a quarter-wavelength. Whether the impedance is high or low at these points depends on whether the line is short- or open-circuited at the output end. At points that are not exact multiples of a quarter-wavelength, the line acts as a capacitor or an inductor.

A resonant transmission line may assume many of the characteristics of a resonant circuit that is composed of lumped inductance and capacitance. The more important circuit effects that resonant transmission lines have in common with resonant circuits having lumped inductance and capacitance are:

1. Series resonance: Resonant rise of voltage across the reactive circuit elements and low impedance across the resonant circuit.
2. Parallel resonance: Resonant rise of current in the reactive circuit elements and high impedance across the resonant circuit.

Retransmit: A procedure involving the reception of a previously transmitted radio-frequency signal, demodulating it, and then feeding the output information into another RF transmitter. This is the equivalent of connecting the audio output of a receiver to the input of a radio transmitter when the first is detecting some other broadcast signal.

Figure 371 shows a basic retransmit operation. Repeater stations depend on retransmit capability to perform their communications function. The transmission from a mobile station is detected by the repeater receiver. It is demodulated at this point, and the resulting audio information is fed to the input of the repeater transmitter, where the actual retransmission occurs. Often, the retransmitted signal is at a frequency different from that of the original transmission.

RF amplifier: An amplifier used for both receiving and transmitting. There are two general classes of RF amplifiers—the untuned amplifier and the tuned amplifier.

In the tuned amplifier, response is desired over a broad frequency range, and its sole function is amplification. In the tuned RF amplifier, very high amplification is desired over a small range of frequencies or at a single frequency. Thus, in addition to amplification, selectivity is also desired to separate the unwanted from the wanted signals. The use of the tuned RF amplifier is generally universal, while that of the untuned amplifier is relegated to a few special cases. Consequently, when RF amplifiers are mentioned, they are ordinarily assumed to be tuned unless specified as otherwise.

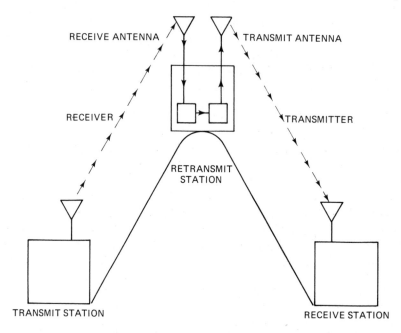

FIGURE 371   Basic retransmit operation.

In receiving equipment, the RF amplifier circuit serves to both amplify the signal and select the proper frequency. In addition, it serves to fix the signal-to-noise ratio. A poor RF amplifier will make the equipment able to respond only to large input signals, whereas a good RF amplifier will bring in the weak signals above the minimum noise level (determined by the noise generated in the receiver itself only) and thus permit reception that would otherwise be impossible.

RF amplifiers and IF amplifiers are almost identical. Both are actually RF amplifiers, but the IF amplifier operates at a fixed frequency that is usually lower than the frequency of the RF amplifier. The RF amplifier normally consists of only one stage, whereas the IF amplifier uses a number of cascaded stages to obtain high gain with the desired selectivity at the fixed IF. In some cases, more than one stage of RF amplification is used to obtain greater sensitivity.

As might be expected, since the RF amplifier is employed over the entire RF spectrum, careful attention to design parameters is necessary to obtain proper operation. In the medium-, low-, and high-frequency ranges, conventional transistors, tubes, and parts are used. In the VHF, UHF, SHF, and microwave ranges, specially designed transistors, tubes, and parts are required to obtain optimum results.

Figure 372 shows the position of the RF amplifier in relation to the other stages of the RF superheterodyne receiver. The RF amplifier stage receives its

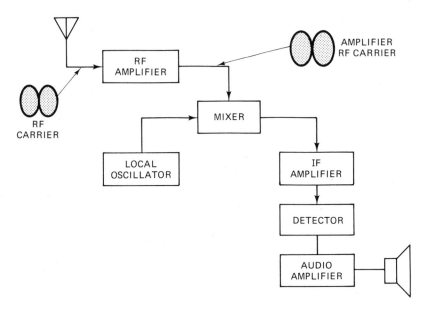

FIGURE 372

input from the antenna and sends an amplified output to the mixer stage for conversion to the receiver's intermediate frequency (IF).

The RF stage or preselector is optional in a superheterodyne receiver, but practically all receivers have at least one RF stage to amplify the RF signal and improve their signal-to-noise ratio. Noise can originate externally and/or be generated within the receiver. Although most of the receiver's gain is in the IF section, the RF gain is of primary importance in determining the ratio of received signal voltage to noise voltage generated in the receiver. The amplitude of receiver noise produced by the RF amplifier is on the order of microvolts. The mixer stage generates the majority of receiver noise voltage. Enough RF amplification is needed, therefore, with a low-noise level in the RF amplifier to supply an adequate signal with a high signal-to-noise ratio for the input to the mixer stage. As a result, the signal level at the mixer input is the limiting factor in the ability of the receiver to reproduce an acceptable output with a weak signal input from the antenna.

Most of the receiver's selectivity (ability to reject adjacent channel frequencies) is in the IF section, but RF amplification assists in rejecting interfering RF signals that can produce beat frequencies within the IF bandpass of the receiver. This is especially important in rejecting image frequencies. Filters and wavetraps are sometimes used in the RF input circuit to reject unwanted image signals. The advantage of additional RF stages lies in the improved selectivity, which is possible against image and other undesired frequencies.

In the transistor RF amplifier, the effect of the base-collector capacitance

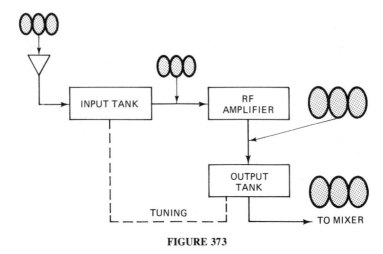

FIGURE 373

and the development of negative resistance through a change in internal parameters can cause oscillation. Neutralization circuits are used to prevent this oscillation and provide maximum gain.

Figure 373 shows a block diagram of an RF amplifier and its associated tank circuits. Amplification and isolation are provided by the action of the transistor or tube used in the preamplification section. On the other hand, selectivity and image rejection (and to a certain extent, signal-to-noise ratio) are mainly a function of the bandwidth of the stage, which is determined primarily by the tuned circuits and coupling networks of the stage.

In Fig. 373, the input tank selects the desired signal from among the many signals received by the antenna. From the output tank, the desired signal is applied to the RF amplifier. The RF amplifier increases the amplitude of the desired signal but does not change its frequency. From the amplifier, the signal is coupled by the output tank to the mixer stage. The output tank improves the selectivity and the signal-to-noise ratio.

Ringing oscillator: A circuit that produces periodic oscillations. This circuit (Fig. 374) produces a short series of oscillations each time an input gate is applied. The oscillations are normally used as distance marks in radar indicators.

The ringing oscillator utilizes a parallel-resonant LC tank circuit to produce an output. An active device is used as a switch to gate the oscillations. The frequency of oscillations in the tank is determined by the values of inductance and capacitance in the tank circuit. The pulse-repetition frequency is determined by the rate at which the switch is opened and closed. The pulse width is determined by the time lapse between opening and closing the switch.

In the quiescent condition, the device is biased by R_1 (Fig. 374), causing A_1 to conduct heavily (near saturation). Current flow is from ground through L_1 and A_1 to source voltage, building up a magnetic field around the inductor.

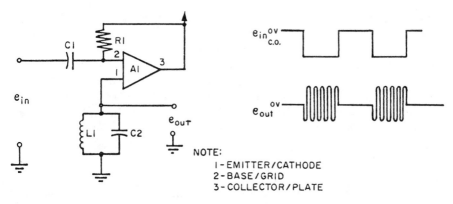

NOTE:
1 - EMITTER/CATHODE
2 - BASE/GRID
3 - COLLECTOR/PLATE

FIGURE 374

The circuit will remain in the quiescent state with no output until the negative gate is applied.

The negative input gate, applied through C_1 to the base grid of A_1, instantaneously drives the device into cutoff. A_1 will remain cutoff for the duration of the input gate. Current flow through A_1 ceases, causing the magnetic field around L_1 to collapse. The collapsing field induces a voltage in the inductor of a polarity that keeps current flowing in the same direction through the coil. Since A_1 is cutoff, this continuing current charges C_2 negative with respect to ground. When the magnetic field has completely collapsed, no further voltage is induced in the inductor, and current flow ceases. Consequently, there is no longer an induced voltage to maintain the charge on C_2, and it discharges through L_1. The discharge current builds up a magnetic field around L_1, which is of opposite polarity to the original field. When C_2 is completely discharged, current flow attempts to cease, causing the field around L_1 to collapse, charging C_2 in the opposite direction with respect to ground. This flywheel action continues for the duration of the negative gate.

The tank circuit is designed to have a very high Q (low loss), so that damping of the waveform will be negligible. If the negative gate were sufficiently long, damping of the waveform would occur due to the dc losses inherent in the tank. The maximum possible pulse width obtainable from a given oscillator would be determined by the Q of the tank. During normal operation, the tank will only oscillate for the duration of the negative gate. The frequency of the RF oscillations in the tank is determined by the values of L and C.

When the negative input gate ceases, A_1 will again conduct heavily, causing a steady current through L_1, and preventing tank oscillations. The only output at this time will be a slight, positive voltage with respect to ground due to the dc resistance of the inductor.

Ripple: The ac component of the dc output from a power supply, which is produced as a product of rectification. The output voltage of a rectifier contains a ripple component whose frequency depends on the number of phases of the rectifier. The magnitude of the ripple depends on the number of rectifier phases, the amount of phase control, and the loading of the rectifier. This ripple is of no consequence in most electrochemical applications and in motor applications where the rectifier has six or more phases.

In electric railway applications where there are exposed communication lines near the trolley, it is usually necessary to filter the rectifier output and reduce the ripple to prevent interference with the communications system. In applications employing small motors with single-phase rectifiers, there may be overheating of the motors due to the ripple. In such applications, oversized motors are used.

Rise time: In digital integrated circuit ratings, the transition time measured with the waveform under consideration changing from the defined low level (logic 0) to the defined high level (logic 1). The elapsed time is measured between 10 and 90 percent points of the waveform.

Rotor coils: Coils usually made of hard-copper strap bent on edge, although some designs call for aluminum or aluminum alloy. The conductor is bare, and turns are insulated from each other by strips of mica or other class B material placed between them. Since the slots are radial in most designs, each turn of each coil must be formed to slightly different dimensions. These dimensions are obtained on a machine specially designed for this purpose.

Because of their tapered shape, the coils must be placed in the slots one turn at a time. The coil is suspended in an overhead rack so that one turn at a time may be dropped down to the slot. The turn insulation extends beyond the coil ends, which are insulated from the rotor body by sheets of mica or asbestos wrapped around the rotor.

Routine: A set of coded instructions arranged in proper sequence to direct the computer to perform a desired operation or sequence of operations. A subdivision of a program consisting of two or more instructions that are functionally related; therefore, a program.

Satellite communication system: A communication system that uses earth-orbiting vehicles to relay radio transmissions. Communication satellites are of two types—passive and active. A passive satellite merely reflects radio signals back to earth, whereas an active satellite amplifies received signals and retransmits them back to earth. This increases the signal strength at the receiving terminal compared to that available from a passive satellite. For this reason, ground transmitters need less power and smaller antennas.

The basic design of a satellite communication system depends, to a large degree, on the orbit of the satellite. An orbit is identified by its shape and the inclination of its orbital plane in relation to the earth's equatorial plane. In general terms, an orbit is either elliptical or circular, and its inclination is classified as inclined, polar, or equatorial. A special type of orbit used in satellite communications is the *synchronous orbit.*

A satellite in a circular orbit at a height of approximately 19,300 nautical mi above the earth is in synchronous orbit. At this altitude, the satellite's period of rotation is 24 h (the same as the earth's). Thus, the satellite orbits in synchronism with the earth's rotation. Satellites in this type of orbit appear to hover motionlessly in the sky.

A typical satellite with a communication transmitter and receiver installed is shown in Fig. 375. One earth, shipboard, or aircraft terminal transmits to the satellite on a frequency referred to as the up-link frequency. The satellite receives, amplifies, and translates the signal to the down-link frequency and transmits it back to earth, where it is received by the receiving terminal.

Satellite communications relieve the crowded high-frequency spectrum for long-range communications. Frequencies in the UHF band or above are considered to be line-of-sight frequencies. However, if a satellite is substituted for the ionosphere to either reflect signals back to earth or amplify and transmit them, frequencies higher than those in the HF band may be utilized. Longer ranges are available, depending on the height of the satellite, and greater reliability is attained, as satellite communication frequencies are not depen-

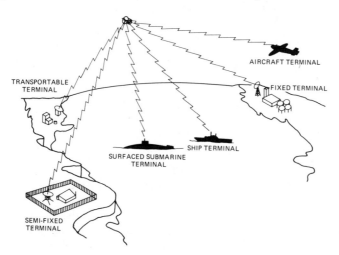

FIGURE 375 Satellite communications.

dent upon the ionosphere and are affected only slightly by atmospheric conditions.

Limitations of a satellite communication system are determined by the satellite's technical characteristics and orbital parameters. Active communication satellites are limited by satellite transmitter power and, to a lesser extent, receiver sensitivity.

The availability of a satellite to act as a relay station between two communication terminals depends on the locations of the terminals and the orbital parameters of the satellite. All satellites except those in synchronous orbit will be in view of any given pair of terminals only part of the time.

Saturable reactor: An inductor whose magnetic flux can reach saturation level easily. Adjustment of impedance in ac circuits is obtained by the use of saturable reactors. They may be used either in the ac circuit leading to the rectifier transformer or between the transformer and rectifier to control the dc voltage.

When a saturable reactor is partially saturated, its circuit action is very similar to that of a grid when used to delay the firing of a mercury-arc rectifier. When the reactor is fully unsaturated, the rectifier output voltage will drop to a few percent of its maximum value, provided the reactors are designed for the full voltage of the rectifier transformer.

When the reactor is completely saturated, it causes no delay in firing. There is, however, a reduction in voltage from the value that would be obtained if the reactor were not in the circuit because of the considerable value of the reactance of the reactor even when saturated. For this reason, the rectifier transformer must be designed for a voltage at least 20 to 30 percent higher than if some other means of voltage control were used.

Saturable-reactor voltage control may be applied to all types of rectifiers except mechanical rectifiers. If the commutating reactors of mechanical rectifiers are suitably excited, a limited amount of voltage control may be obtained. For semiconductor rectifiers, the saturable reactor is the only practical means of getting rapid-response voltage control.

A modification of voltage control by saturable reactors consists of reactors in connection with a transformer winding that supplies the rectifiers with a small part of the ac voltage, such as 10 to 15 percent. When this is done, the size of the reactor and its effect on power factor are greatly reduced.

Saturation curve: A curve that indicates variations of the generated voltage of an alternator corresponding to the variations of the field current. When analyzing an alternator that is being operated with a constant load, it is helpful to use saturation or characteristic curves of the alternator. The magnitude of the voltage induced in a coil is proportional to the number of turns in the coil, the amount of magnetic flux linking the coil, and the rate at which the flux linkages change with respect to time.

For a particular alternator, the number of turns in the coils and the rate of change of the flux linkages with the coils are fixed if the machine is operating at its rated speed. But the amount of flux linking the coils can be changed by varying the strength of the magnetic field. This is done by changing the amount of current flowing through the field coils.

Although the voltage induced in the armature winding is directly proportional to the amount of flux, the flux is not directly proportional to the exciting current over the full operational range of the alternator. This is due to the fact that the iron portions of the magnetic circuit present greater reluctance to flux as the concentration of flux, or flux density, becomes higher. At high flux densities, the iron becomes saturated by the magnetic field. Because of the saturation of the iron, greater increases in the magnetizing force or exciting current are required to achieve a given increase in flux.

Scatter radiation: The propagation of the electromagnetic energy from a radio-frequency transmitter to many different points on the earth by atmospheric bending. A portion of the signal will be deflected back to earth by the troposphere, which is the lowest layer in the earth's atmosphere. Tropospheric scatter typically brings the signal back to earth a few hundred miles from the original point of transmission. The portion of the transmission that is not bent by the troposphere travels in a straight line toward the upper atmosphere, or ionosphere, where it is bent back to the earth, touching down 1000 mi or more away from the point of transmission. Scatter radiation is often referred to as *skip*.

Schmitt trigger: A two-stage pulse-shaping circuit often used to form NAND gates. This circuit is amplitude sensitive and is designed to produce an output only when its input signal exceeds a prescribed reference level. The output of a Schmitt trigger circuit is a rectangular waveform of constant amplitude whose pulse width is equal to that period of time during which the input signal exceeds a preset reference level. Examples of inputs to and corresponding outputs from a Schmitt trigger are illustrated in Fig. 376.

Figure 377 illustrates a Schmitt trigger circuit. An examination of this circuit reveals that it is basically an emitter-cathode-coupled bistable trigger circuit. Coupling from the second stage (A_2) to the first stage (A_1) is obtained by

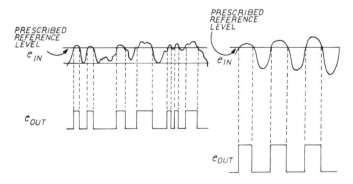

FIGURE 376

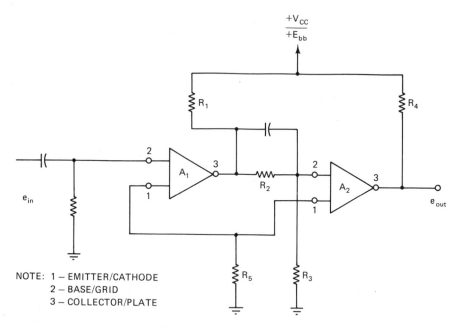

FIGURE 377   Schmitt trigger circuit.

the use of an unbypassed common emitter-cathode resistor rather than direct resistive coupling. Also, there is no voltage divider to bias *A*.

Scott connection: A commonly used method of phase transformation. Devices intended for operation on two-phase systems may be operated from three-phase circuits by means of transformers connected to effect a phase transformation. Similarly, devices intended for three-phase circuits may be operated from a two-phase system by means of transformers.

Two specially designed single-phase transformers are used for Scott connection. One is called the *main transformer*, 1; the other is called the *teaser transformer*, 2. On the side that is connected to the three-phase system, *A*, *B*, *C*, the main transformer has a center tap or 50 percent tap, *T*; and the teaser transformer has a tap *A* at 86.6 percent of the winding. Tap *A* is connected to line *A* of the three-phase side. For the sake of simplicity, it is assumed that both transformers have a 1 to 1 turn ratio.

The Scott connection is obtained by connecting one end of the tapped winding of the teaser transformer to the center tap *T* of the main transformer (Fig. 378). The 86.6 percent tap of the teaser transformer and the two ends of the tapped winding of the main transformer represent the leads *A*, *B*, and *C* connected to the three lines of a three-phase system. The ends of the untapped windings of both transformers represent the leads of A_1, A_2, B_1, and B_2 connected to a four-wire two-phase system. If A_2 and B_1 are connected together, the connection provides a three-wire two-phase system.

If a two-phase voltage source is applied to the two-phase side of the Scott connection (leads A_1, A_2, B_1, and B_2), the untapped windings will each carry

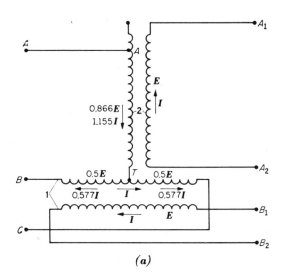

FIGURE 378 Transformer operation.

100 percent of their rated voltage E; these voltages are displaced 90° from each other. Across the terminals *A-B, B-C,* and *C-A* of the three-phase side, there will be available three phase voltages, each equal to the voltage *E* applied to each of the untapped windings. These voltages are displaced 120° from each other. Thus, the two-phase system has been transformed to a three-phase system by the two transformers of the Scott connection. The phase voltages do not change in value.

Similarly, if a three-phase voltage source is applied to terminals *A, B,* and *C,* two phase voltages will be available across the leads A_1-A_2 and B_1-B_2. Each phase voltage *E* will be of the same value as each phase voltage *E* on the three-phase side, but the voltages will be displaced 90°, instead of 120°.

Secondary cell: A cell in which the electrodes and the electrolyte are altered by the chemical action that takes place when the cell delivers current. These cells may be restored to their original condition by forcing an electric current through them in the opposite direction to that of discharge. The automobile storage battery is a common example of the secondary cell.

Secondary emission: Emission of electrons from a material caused by the impact of particles striking its surface. If a stream of electrons flowing at a high velocity strikes a material, the force may be great enough to dislodge other electrons from the surface. The dislodged electrons are called *secondary electrons* to distinguish them from the *primary electrons* that caused the secondary emission. Although secondary emission occurs to some extent in most tubes, it is used as a source of electrons in only a few specialized electron tubes.

Selectivity: The ability of a circuit to pass signals of one frequency while rejecting all others. This term is especially applicable to radio-frequency receivers and describes their ability to discriminate against frequencies differing from that of the desired signal. Overall selectivity in receivers will depend on the individual selectivity of each tuned circuit and the number incorporated in the overall unit. The receiver's ability to reject adjacent channel signals is called *skirt selectivity* and is determined by the receiver bandwidth at high attenuation.

Selenium diode: A rectifier device that is created by joining an N-type region with a P-type region to form a single wafer of semiconductor material. In this case, the semiconductor is specially processed selenium. This is a non-metallic element whose symbol is *Sd*. The finished device is capable of passing current in one direction while blocking its flow in the other.

Selenium diodes were used in early solid-state power supplies and were often incorporated in units that used large, finned heat sinks for cooling purposes. They are nearly obsolete today, having been replaced by modern junction diodes made from silicon.

Self-induction: The action that occurs in a coil to induce a voltage. A coil having an electric current flowing through it has a magnetic field around it. As long as the current remains steady, no perceptible change takes place in the magnetic field. Any variation in the strength of the current, however, will produce a change in the positions of the magnetic lines of force surrounding the coil. The action that causes this voltage is known as *self-induction*.

Semiconductor: A material that has a conductivity lower than that of a conductor, but higher than that of an insulator. Semiconductors include compounds such as copper oxide, zinc oxide, indium antimonide, and gallium arsenide, and elements such as germanium and silicon when doped with certain impurities.

Series motor: A motor in which the field coils are connected in series with the armature. With low-flux density in the field iron, the series field strength is proportional to the armature current. A series motor is shown in Fig. 379. If the supply voltage is constant, the armature current and the field flux will be constant only if the load is constant. If there were no load on the motor, the armature would speed up to such an extent that the windings might be thrown from the slots and the commutator destroyed by the excessive centrifugal forces. For this reason, series motors are never belt-connected to their loads. The belt might break, and the motor would then overspeed and

SERIES   **FIGURE 379**

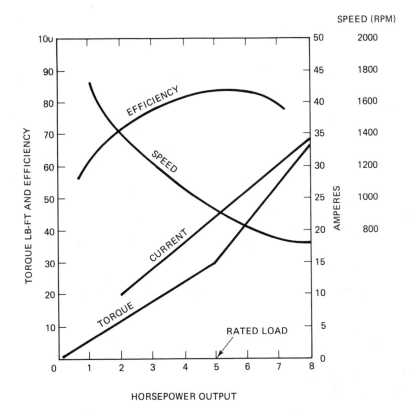

FIGURE 380 Characteristic curves of a series motor.

destroy itself. Series motors are always connected to their loads directly or through gears.

Figure 380 shows that as the load increases, the armature speed slows down and the counter-electromotive force (CEMF) is reduced. The current through the armature is increased and the field strength is likewise increased. This reduces the speed to a very low value. The armature current, however, is not excessive because the torque developed depends on both the field flux and the armature current.

Figure 381 shows that when a heavy load is suddenly thrown on a shunt motor, it attempts to take on the load at only slightly reduced speed and CEMF. The flux remains essentially constant; therefore, the increased torque is proportional to the increase in armature current. With heavy overload, the armature current becomes excessive and the temperature increases to a very high value. The shunt motor cannot slow down appreciably on heavy load as can the series motor. Hence, the shunt motor is more susceptible to overload.

The series motor is used when there is a wide variation in both torque and speed, such as in traction equipment, blower equipment, hoists, cranes, etc.

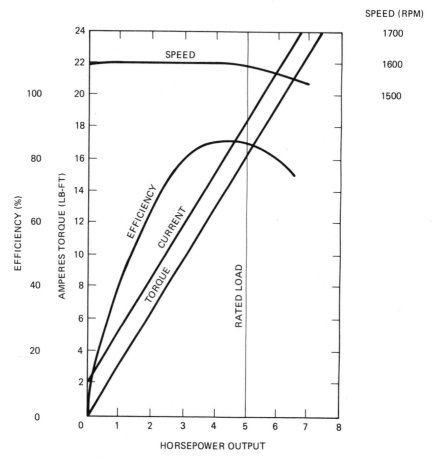

FIGURE 381   Characteristic curves of a shunt motor.

Series regulator: A circuit in which regulation is achieved by division of voltage between the regulator and the load impedance, depending on the needs of the load. Figure 382 shows a series regulator circuit. Efficiency of series regulators is high under light-load conditions and low under full-load conditions. The series regulator has no inherent overload protection. A short circuit in the load would cause heavy current through the regulator circuit, overloading most types.

A schematic diagram of a solid-state series voltage regulator is shown in Fig. 383. Q_1 functions as a variable resistance between the source and load impedance. CR_1, a zener diode, in conjunction with R_1, maintains a constant voltage at the base of Q_1. The unregulated input voltage is applied across the series network of Q_1 and RL. The fixed base voltage is of sufficient value to forward bias Q_1. Under normal operating conditions, the base bias is fixed at a value that will produce the desired voltage across the load impedance. The load voltage is equal to the unregulated input voltage minus the drop across Q_1.

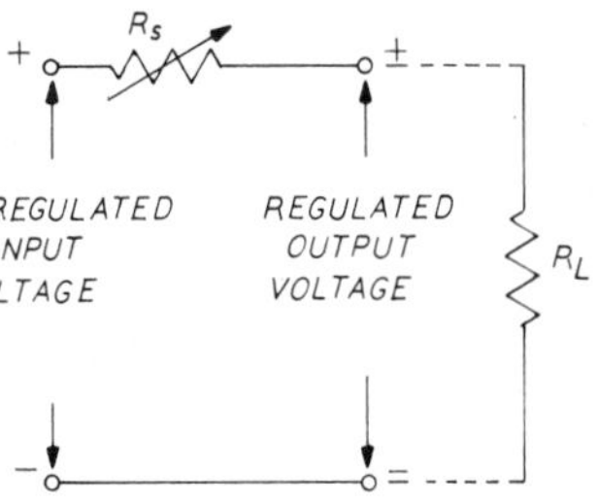

UNREGULATED
INPUT
VOLTAGE

REGULATED
OUTPUT
VOLTAGE

R_L

FIGURE 382   *SERIES   REGULATOR*

Assume that the line voltage increases. This will cause a momentary increase in load voltage, which causes the emitter of Q_1 to appear more negative with respect to ground, decreasing the forward bias of Q_1. This decrease in forward bias causes the internal resistance of the transistor to increase, producing an increased drop across Q_1, thereby returning the voltage drop across RL to its normal level. The action of the regulator is reversed, with a decrease in applied line voltage.

Assume an increase in load (more current). Voltage across the load impedance will decrease and the emitter of Q_1 will become less negative, increasing the forward bias. This reduces the transistor's internal resistance, effectively reducing the voltage drop across Q_1. This allows more current flow through RL, thus returning the output voltage to its normal level. A decrease in load will cause a reverse reaction of the regulator.

An electron-tube version of a series regulator is shown in Fig. 384. The operation is similar to a solid-state series regulator. V_1 acts as a variable resistance between the source and the load impedance to compensate for changes in line and load voltage. V_2 maintains the grid of V_1 at a fixed reference voltage. If the output voltage increases, the cathode potential goes more positive with respect to ground, thereby increasing the bias. This increase in bias causes V_1 to decrease conduction, decreasing the drop across RL to the desired level. The opposite action will occur for a decrease in output voltage.

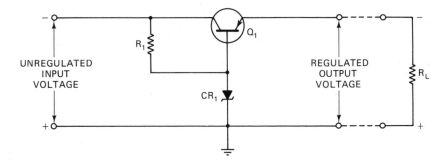

FIGURE 383   Solid state series voltage regulator.

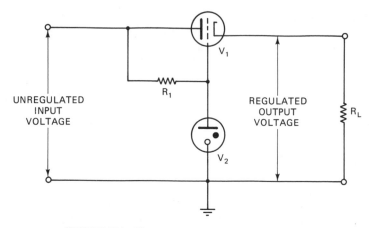

FIGURE 384 Electron tube series voltage regulator.

A disadvantage of simple series regulators is that they do not rapidly respond to small changes in voltage. The effectiveness of a series regulator is improved by the addition of circuitry that detects and amplifies small changes, thus allowing the regulator to respond more rapidly. The shunt-detected series regulator is such a circuit.

Series resonance: That resonance occurring in a circuit composed of an inductor, ac generator, and capacitor wired in a series. When the reactances in a series circuit cancel and, as far as the source is concerned, the circuit appears to contain only resistance, the circuit is said to be in a *condition of resonance.* When resonance is established in a series circuit, certain conditions will prevail:

1. The inductive reactance will be equal to the capacitive reactance.
2. The circuit impedance will be minimum.
3. The circuit current will be maximum.

When resistive, inductive, and capacitive elements are connected in series, their individual characteristics are unchanged. That is, the current through and the voltage drop across the resistor are in phase, while the voltage drops across the reactive components (assuming pure reactances) and the current through them are 90° out of phase. However, this is not true of their combined characteristics, and a new relation must be recognized with the introduction of the three-element circuit. This pertains to the effect on total line voltage and current when connecting reactive elements in series whose individual characteristics are opposite in nature, such as inductance and capacitance. Such a circuit is shown in Fig. 385a.

Note that current is the common reference for all three-element voltages because there is only one current in a series circuit and it is common to all elements. The common series current is represented by the dashed line in Fig.

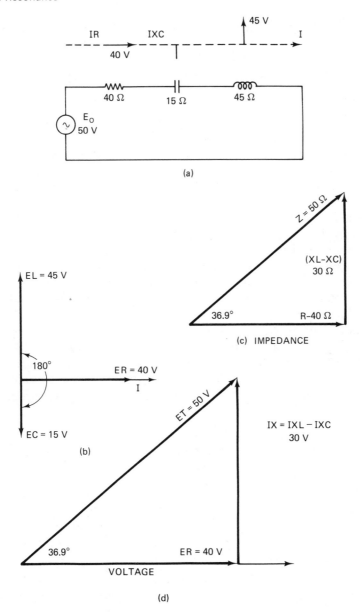

FIGURE 385 Resistance, inductance, and capacitance in series.

385a. The voltage vector for each element, showing its individual relation to the common current, is drawn above each respective element. The total voltage E_t (Fig. 385d) is the vector sum of the individual voltages of IR, IX_L, and IX_C.

The three element voltages are arranged for summation in Fig. 385b. Since IX_L and IX_C are each 90° away from I, they are 180° from each other.

Vectors in direct opposition (180° out of phase) may be subtracted directly. The total reactive voltage E_X is the difference of IX_L and IX_C.

The final relationship of line voltage and current, as seen from the source, is shown in Fig. 385d. Had X_C been larger than X_L, the voltage would lag rather than lead. When X_C and X_L are of equal value, line voltage and current will be in phase.

Service: The conductors and equipment for delivering energy from the electrical supply system to the wiring system of the premises served.

Service entrance: The conductors and equipment for delivering energy from an electrical supply system, such as the power company's lines, to the wiring system of a building or premise. All of the parts of a typical overhead service entrance are shown in Fig. 386. An underground system is shown in Fig. 387. Certain NEC requirements are also noted on these drawings for quick reference by the electrical inspector.

Service equipment: The necessary equipment, usually consisting of a circuit breaker or switch and fuses, and their accessories, located near the point of entrance of supply conductors to a building or other structure, or an otherwise defined area, and intended to constitute the main control and means of cutoff for the supply.

Shaded-pole motor: A single-phase induction motor in which the stator windings are not distributed around the core, but are wound in coils around the salient, or projection, pole cores as in dc motors. The rotor in a shaded-pole motor is of the squirrel-cage type, and the motor does not have a starting switch. Starting torque is obtained in some single-phase motors by

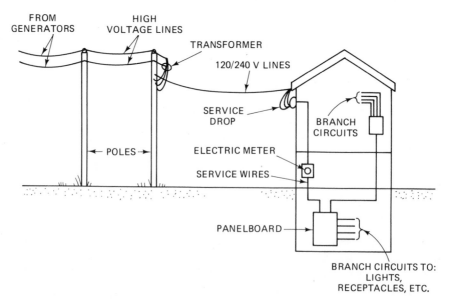

FIGURE 386 Various sections of a typical overhead service entrance.

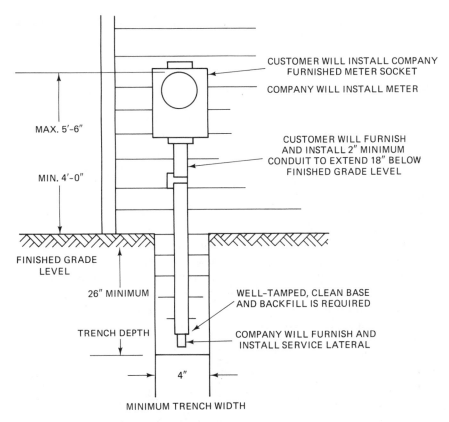

MAX. 5'-6"

MIN. 4'-0"

CUSTOMER WILL INSTALL COMPANY
FURNISHED METER SOCKET

COMPANY WILL INSTALL METER

CUSTOMER WILL FURNISH
AND INSTALL 2" MINIMUM
CONDUIT TO EXTEND 18" BELOW
FINISHED GRADE LEVEL

FINISHED GRADE
LEVEL

26" MINIMUM

TRENCH DEPTH

4"

WELL-TAMPED, CLEAN BASE
AND BACKFILL IS REQUIRED

COMPANY WILL FURNISH AND
INSTALL SERVICE LATERAL

MINIMUM TRENCH WIDTH

FIGURE 387 Various sections of a typical underground service entrance.

applying the principle of shaded poles, where the magnetic flux in the two sections of a pole are displaced slightly, both in time and space.

This motor is called *shaded* because one portion of the pole is shaded; that is, it is designed with a short-circuited strap, or shading coil, which produces a flux out of phase with the flux in the main portion of the pole. The short-circuited strap around the shaded portion of the pole has a current induced in it when the flux changes in that portion of the pole. The induced current flows in the direction necessary to oppose the change in flux that induces it. That is, the current in the shading coil delays the change in flux, whether increasing or decreasing, in the shaded section of the pole.

On the other hand, the flux through the main part of the pole changes in phase with the applied current. The time displacement between the fluxes in the two parts of the pole results in a crude form of rotating field. The effect of the time and space displacement between the fluxes is to produce a flux through the air gap between the stator and rotor that shifts toward the shading coil. The direction of rotation of a shaded-pole motor is from the unshaded portion of the pole toward the shaded portion.

The shading coil may be made of aluminum, copper, or brass in the shape of a hairpin. It is put in place from one side of the core, and the open end is then formed over and welded in order to make a closed circuit. The size of the metal loop is an important part of the design and greatly affects the locked-rotor torque.

Thus, if for any reason a shading coil has to be replaced, it is not satisfactory to replace, for example, a shading coil made of 0.040×0.162-in.^{2} material with one made of 0.063×0.250-in.^{2} material. Also, if the original shading coil is made of brass, it should not be replaced by one of copper without changing the area to obtain the same resistance as that of the original coil. Instead of the metal loop, a number of turns of wire with the two ends joined together may be used as a shading coil. If the motor is repaired, no change should be made in the wire size of the number of turns in the shading coil.

Shockley diode: A four-layer semiconductor device that blocks forward current flow until a specific value of potential is applied between its anode and cathode. This potential, positive on the anode with respect to the cathode, is called the *breakover voltage*. Once the breakover voltage is reached, the diode will continue to conduct until its anode voltage is decreased below a specific minimum voltage, which is called the *diode turn-off voltage*. Once the anode voltage has been decreased below the diode turn-off voltage, the diode will not conduct until the anode voltage is again raised to the breakover voltage.

A pictorial diagram, schematic symbol, and equivalent transistor circuit of a Shockley diode are shown in Fig. 388. P_1, N_1, and P_2 (Fig. 388a) form the equivalent of one transistor, while N_1, P_2, and N_2 form a second equivalent transistor. The mechanism of operation of the Shockley diode is reverse breakdown of junction 2 (J_2 in Fig. 388a).

Shot effect: A type of noise that occurs in UHF amplifiers due to variations in the emission of electrons from the cathode. It can be minimized by operating the tube well below emission saturation.

Shunt capacitor: A capacitor that is located between two circuit elements running parallel to each other. The word *shunt* means *to parallel*, but a shunt capacitor actually connects two parallel elements. Figure 389 shows a

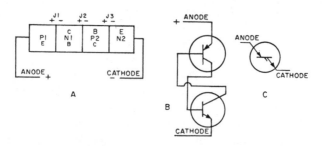

FIGURE 388   Shockley diode diagrams.

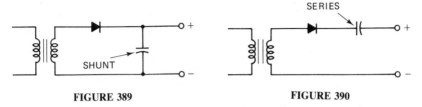

FIGURE 389   FIGURE 390

simple electronic circuit to which a shunt capacitor has been added between the positive and negative power supply leads. Figure 390 shows the same circuit with the capacitor connected in series. The same capacitor is now known as a *series component*. The actual connection into the electronic circuit determines whether or not the capacitor is shunt-mounted.

Shunt capacitors are often used in radio-frequency circuits in bypassing applications. The shunt is often between the active side of the circuit and ground. Capacitor value is chosen so that the component will pass an undesirable frequency while appearing as a high resistance to frequencies that are to be conducted to other circuit portions. The frequencies that will pass through the capacitor are shunted to ground, thus removing them from the active circuit.

Shunt motor: A motor that maintains nearly constant speed from no load to full load. The shunt field winding consists of many turns of small wire and is connected in parallel with the armature winding or across the line. The diagrams in Fig. 391 show the proper connection for the armature and field.

The characteristic curves in Fig. 392 show that the torque developed by a shunt motor varies with the armature current. This is true because the torque is proportional to the armature and field flux. The field maintains constant strength because it is connected across the line, and the armature flux will vary with the armature current. The torque of a shunt motor is considered to be fair in comparison to other dc motors. It will start about 50 percent overload before being damaged by excessive current.

The shunt motor maintains nearly constant speed from no load to full

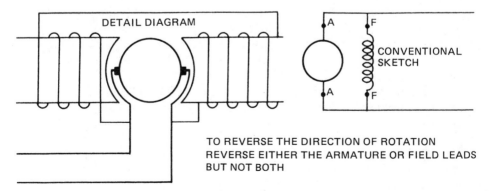

FIGURE 391

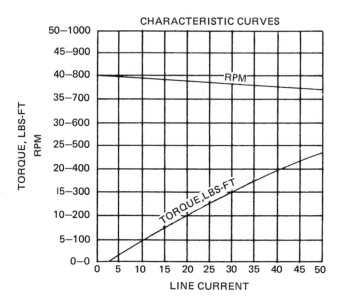

FIGURE 392

load because the shunt field strength is constant. The characteristic curve shows that the speed varies about 10 percent from no load to full load, which gives this motor very good speed regulation.

The shunt motor is widely used where it is desired to control the speed above and below normal speed. A shunt field rheostat connected in series with the shunt field will cause the motor to increase in speed. A resistor connected in series with the armature will cause the motor to decrease in speed.

Shunt motors sometimes have a few turns of heavy wire wound on each field pole and connected in series with the armature. This winding produces the same polarity as the shunt field winding and produces a more stable operation when the motor is carrying a fluctuating load.

Shunt regulator: A circuit that is used when fairly good electronic regulation and simplicity are requirements in power supply design. Shunt regulators are not as efficient regarding voltage regulation as their series regulator counterparts, but they perform an adequate job while offering the advantages of simple design and relatively inexpensive components.

Sideband: With respect to a carrier, one of the additional frequencies generated by the modulation process. In simple amplitude modulation, the two sidebands are $fc + fm$ and $fc - fm$, where fc is the carrier frequency and fm is the modulation frequency.

Side lobe: That amount of radio-frequency energy that is transmitted from the sides of a directional antenna. Figure 393 shows a typical radiation pattern of a three-element yagi beam. The principle lobe (maximum area of radiation) occurs off the front of the antenna, while minor lobes appear at the back and sides.

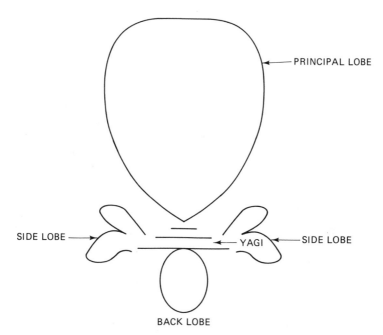

PRINCIPAL LOBE

SIDE LOBE

YAGI

SIDE LOBE

BACK LOBE

FIGURE 393   Typical radiation pattern of a three-element yagi beam.

Most modern beam antennas are designed so that side and back lobe radiation is kept to a minimum. As more and more power appears here, the front lobe is decreased accordingly. A term known as *side lobe rejection* is often used to describe *beam antenna performance*. A rejection figure of -20 dB indicates that the signal strength from the sides of the antenna is 20 dB below the signal strength from the front.

Signal diode: A diode used for light duty, such as for modulation, demodulation, or detection. Signal diodes fall into various categories, such as general-purpose, high-speed switch, and parametric amplifiers. These devices are used as mixers, detectors, and switches, as well as in many other applications.

Signal-to-interference ratio: In the reception of radio-frequency transmissions, a comparison of the strength of the desired signal to the strength of other signals on the same frequency. For example, if two transmitters are producing an output on the same frequency, the signal-to-interference ratio would be a comparison of the strength (at the receiver) of transmit station A to transmit station B, assuming that the first station is the one from which reception is desired. Any undesired signal is classified as interference and may emanate from many sources unrelated to productive radio-frequency transmissions. This can include atmospheric static, electrical noise, etc.

Signal-to-interference ratio is normally rated in decibels. The expression 15-dB S/I indicates that the signal strength of the desired transmission at the receiver is 15 dB greater than the highest interference signal strength.

Signal-to-noise ratio: The ratio of the desired signal amplitude to the noise voltage amplitude. Noise voltages having the same frequencies as the desired signal will receive a proportionate amount of amplification. Thus, the amplitude of the voltage induced in an antenna must be sufficiently large in relation to the amplitude of the noise voltages in order to overshadow their effects.

The effect of noise voltages whose frequencies do not lie within the bandwidth of the desired signal can be minimized by decreasing the bandwidth (increase in selectivity) of the input circuits as much as possible without decreasing the overall bandwidth below the required amount. Therefore, decreasing the bandwidth results in an increased signal-to-noise ratio.

Silicon-controlled rectifier: A three-terminal solid-state device that can conduct current in one direction only. The silicon-controlled rectifier or SCR is classified as a reverse blocking triode thyristor and is most often used to control or switch direct current. The SCR is able to handle far greater power than transistors or vacuum tubes. However, although the plate current of the tube can be controlled by varying the grid voltage and the collector current of the transistor by its base bias, the gate of the SCR can turn the anode current full on but not off. In order to stop the anode current, it is necessary to interrupt it at least momentarily. Since it cannot be used to control the load current directly, the SCR must be used indirectly for this purpose.

Sine wave: A curve that can be seen on an oscilloscope that shows the behavior of a current or voltage. An ac functions in a circuit by starting at zero. It then builds up in one direction until it reaches a maximum, declines to zero, builds up in the reverse direction, and again declines to zero. These fluctuations are repeated over and over and constitute a sine wave.

While the curve is being traced, the passing of time is shown by the movement of the spot of light from left to right, and the amount of voltage or current at any instant is shown by the distance above or below the horizontal line. The range above the line represents pressure or flow in one direction. The range below the line shows pressure or flow in the other direction. The values above the line arbitrarily are called *plus,* and the values below the line are called *minus,* so they can be used mathematically. If a sine wave of voltage is applied to a resistance, the resulting current will also be a sine wave. This follows Ohm's law, which states that current is directly proportional to the applied voltage.

Figure 394 shows a sine wave of voltage and the resulting sine wave of current superimposed on the same time axis. Notice that as the voltage increases in a positive direction, the current increases along with it. When the voltage reverses direction, the current reverses direction. At all times, the voltage and current pass through the same relative parts of their respective cycles at the same time. When two waves, such as those shown in Fig. 394, are precisely in step with one another, they are said to be in phase. To be in phase, the two waves must go through their maximum and minimum points at the

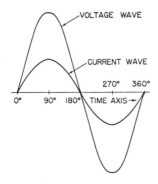

FIGURE 394 Voltage and current waves in phase.

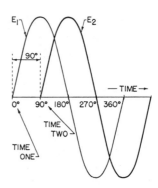

FIGURE 395 Voltage waves 90° out of phase.

same time and in the same direction. In some circuits, several sine waves can be in phase with each other. Thus, it is possible to have two or more voltage drops in phase with each other and also in phase with the circuit current.

Figure 395 shows a voltage wave E_1 considered to start at 0° (time 0). As voltage wave E_1 reaches its positive peak, a second voltage wave E_2 starts its rise (time one). Since these waves do not go through their maximum and minimum points at the same instant of time, a phase difference exists between the two waves, and they are said to be out of phase. For the two waves in Fig. 395, this phase difference is 90°.

To describe the phase relationship between two waves further, the terms *lead* and *lag* are used. The amount by which one wave leads or lags another is measured in degrees. Referring again to Fig. 395, wave E_2 is seen to start 90° later in time than wave E_1. Thus, wave E_2 lags wave E_1 by 90°. This relationship could also be described by stating that wave E_1 leads wave E_2 by 90°.

It is possible for one wave to lead or lag another by any number of degrees except 0 or 360°, in which condition the two waves are in phase. Thus, two waves may differ in phase by 45°, but two waves differing by 360° would be considered in phase.

A phase relationship that is quite common is the one shown in Fig. 396.

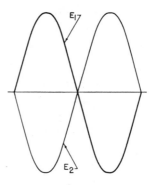

FIGURE 396 Two waves 180° out of phase.

The two waves illustrated have a phase difference of 180°. Notice that although the waves pass through their maximum and minimum values at the same time, their instantaneous voltages are always of opposite polarity. If two such waves existed across the same component, they would have a cancelling effect on each other. If the two waves are equal in amplitude, the resultant wave would be zero. However, if they have different amplitudes, the resultant wave would have the polarity of the larger and be the difference of the two.

To determine the phase difference between two sine waves, locate the points on the time axis where the two waves cross the time axis traveling in the same direction. The number of degrees between the crossing points is the phase difference. The wave that crosses the axis at the later time (to the right on the time axis) lags the other.

Single-sideband receiver: A receiver in which a special type of detector and a carrier reinsertion oscillator are used. The carrier reinsertion oscillator must insert a carrier in the detector circuit at a frequency that corresponds almost exactly with the relative position of the carrier in the original spectrum.

The filters used in SSB receivers serve special purposes. Many SSB signals may exist in a small portion of the frequency spectrum. Filters supply the selectivity necessary to receive adequately only one of the many signals which may be present. They may also select USB or LSB operation, as well as reject noise and other interference.

The oscillators in a SSB receiver must be extremely stable. In some types of SSB data transmission, a frequency stability of ±2 Hz is required. For simple voice communication, a deviation of ±50 Hz may be tolerable. SSB receivers may employ additional circuits to enhance frequency stability, improve image rejection, or provide automatic gain control (AGC). However, the circuits contained in the basic receiver shown in Fig. 397 will be found in all single sideband receivers.

The need for extreme frequency stability may be understood if one considers the fact that a small deviation in local oscillator frequency from the correct value will cause the IF produced by the mixer to be displaced from its correct value. In AM reception, this is not too damaging, since the carrier and sidebands are all present and will all be displaced an equal amount. Therefore,

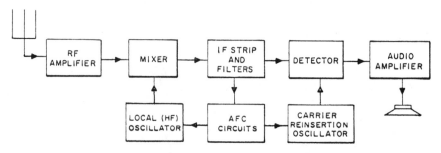

FIGURE 397   Basic SSB receiver.

the relative positions of carrier and sidebands will be retained. However, in SSB reception, there is no carrier present in the incoming signal.

The carrier reinsertion oscillator frequency of the carrier is determined by the local oscillator frequency. For example, assume that a transmitter with a suppressed carrier frequency of 3 mHz is emanating a USB signal. Also assume that the intelligence consists of a 1-kHz tone. The transmitted frequency will be 3001 kHz. If the receiver has a 500-kHz IF, the correct local oscillator frequency should be 3500 kHz. The IF output of the mixer would be 499 kHz. If a carrier were present, it would correspond to an IF frequency of 500 kHz. Therefore, the carrier reinsertion oscillator frequency should be 500 kHz in order to preserve the frequency relationship of the carrier and sideband at 1 kHz.

If the local oscillator frequency drifts to 3500.5 kHz, then the IF output of the mixer will become 499.5 kHz. However, the carrier reinsertion oscillator would still be operating at 500 kHz. This will result in an incorrect audio output of 0.5 kHz. If the intelligence transmitted was a complex signal such as speech, it would be unintelligible due to the displacement of the side frequencies caused by the local oscillator deviation.

Even with the correct local oscillator frequency, distortion may occur due to a shift in carrier reinsertion oscillator frequency. This would also cause a displacement in the relative positions of the carrier and sideband.

The carrier reinsertion oscillator may be any type of stable oscillator. It is tuned to the correct frequency of the IF bandpass. Usually, a crystal oscillator will be used, as this type of oscillator provides good stability.

Although a conventional diode detector may be used for SSB reception, the type of detector most commonly employed is the *product detector*. It is so called because under ideal operating conditions, its output amplitude is proportional to the product of the amplitude of the reinserted carrier and the SSB signal. Figure 398 illustrates a transistor product detector.

Transistors Q_1 and Q_2 form a balanced mixer circuit. The bias for these transistors is obtained from the voltage divider formed by R_1 and R_2 and is applied to the bases of Q_1 and Q_2 through the secondary of transformer T_1. The emitter operating voltage is applied to Q_1 and Q_2 through emitter resistors R_6 and R_7.

The IF signal is applied to the base of each transistor 180° out of phase by transformer T_1. The carrier reinsertion oscillator signal is coupled in phase to the emitters of Q_1 and Q_2, through capacitors C_2 and C_3, respectively. Resistors R_4 and R_5 provide isolation between the emitters of Q_1 and Q_2.

The IF signal and the reinserted carrier are heterodyned in the transistors. The resulting output consists of the sum and difference of the original input frequencies, as well as the two original frequencies. Capacitors C_4 and C_5 bypass the sum and original frequencies to ground. Since the circuit is balanced, the outputs from transistors Q_1 and Q_2 (developed across transformer T_2), are 180° out of phase with each other.

This results in additional cancelling of the reinserted carrier, because it

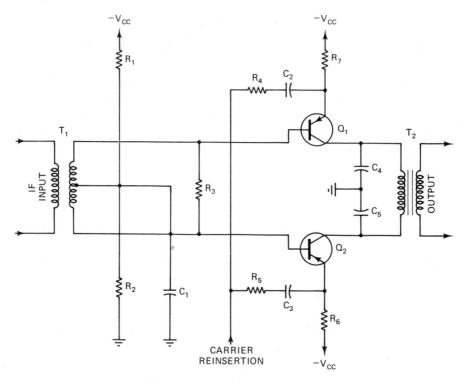

FIGURE 398 Transistor product detector.

was applied in phase to the emitters of the transistors. Transformer T_2 has an audio-frequency response that will attenuate any of the RF signals not previously cancelled. The difference between the two input signals is the desired intelligence, and this difference is developed across transformer T_2.

Figure 399 illustrates an electron-tube product detector. The sideband signal from the IF amplifier is applied to the control grid of V_1 through transformer T_1. The reinserted carrier is applied to the suppressor grid. The two input signals are heterodyned within the tube, and the output will contain frequencies equal to the sum and difference of the input signals, as well as the original input frequencies. All frequency components, with the exception of the difference frequency, will be attenuated by the low-pass filter made up of C_5, L_1, and C_6. The difference frequency, which is the desired audio signal, will be developed across R_4.

It should be noted that in both detectors, the reinserted carrier amplitude must be appreciably greater than the sideband IF amplitude. A carrier too small in relative amplitude would result in distortion of the output signal.

Skin effect: The effect that represents the tendency of air conductors to carry the circuit current on the surface, or skin, of the conductors rather than uniformly throughout their cross section. As a result of this tendency, many

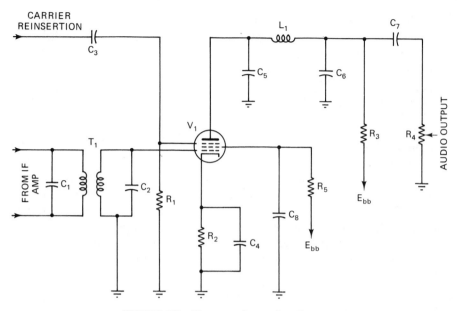

FIGURE 399   Electron tube product detector.

electrical conductors are made of hollow tubing in order to save the added weight and expense of the unused central portion of the solid conductor. The ac resistance of a conductor is approximately proportional to the frequency and length of the conductor and inversely proportional to its diameter.

The effective ac resistance of electrical conductors is frequently higher than their dc resistance, especially when they are being used in high-frequency circuits, as in radio transmitters and receivers. Direct current is distributed uniformly through the cross-sectional area of a homogeneous conductor. For example, if a conductor having a cross-sectional area of 1000 circular mil is carrying 1 A of direct current, 1/1000th of an A (1 mA) is flowing in each circular mil of cross-sectional area.

However, when the current in the conductor varies in amplitude, this uniform distribution throughout the conductor's cross section is no longer obtained. The accompanying magnetic field is strongest near the center of the conductor and weaker at the circumference. The varying field induces a voltage in the conductor that opposes the change in current. The voltage induced in that portion of the conductor near the center is greater than the voltage induced in the outer surface of the conductor. The total opposition to the current flow includes the effect of this induced EMF and is greater near the center of the conductor than at the surface. Therefore, the current is divided inversely with the opposition (more of the current flowing near the circumference and less near the center of the conductor).

The overall result of this action is a decrease in the available area of cross section to conduct the current and an increase in conductor resistance. The decrease in area and increase in resistance become pronounced at high frequencies, at high current densities, and at high magnetic flux densities.

Sky wave: That portion of a radio wave that moves upward and outward and is not in contact with the ground. This is shown in Fig. 400. A sky wave behaves similarly to a ground wave. Some of the energy of the sky wave is refracted (bent) by the ionosphere so that it comes back toward the earth. Some energy is lost in dissipation to particles of the atmospheric layers. A receiver located in the vicinity of the returning sky wave will receive strong signals even though it may be several hundred miles beyond the range of the ground wave.

One of the things that has a great effect on the sky wave is the ionosphere, that portion of the atmosphere located approximately 40 to 50 mi from the earth. The ionosphere has many characteristics. Some waves penetrate and pass entirely through it into space and never return. Other waves penetrate, but bend. Generally, the ionosphere acts as a conductor and absorbs energy in varying amounts from the electronic wave. The ionosphere also acts as an electronic mirror and refracts (bends) the sky wave back to the earth, as illustrated in Fig. 401. Here, the ionosphere does by refraction what water does to a beam of light.

The ability of the ionosphere to return an electronic wave to the earth depends on the angle at which the sky wave strikes the ionosphere, the frequency of the transmission, and ion density. When the wave from an antenna strikes the ionosphere at an angle, the wave begins to bend. If the frequency and angle are correct and the ionosphere is sufficiently dense, the wave will eventually emerge from the ionosphere and return to the earth. If the

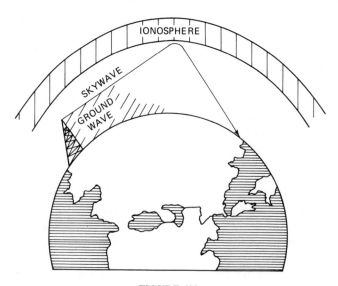

FIGURE 400

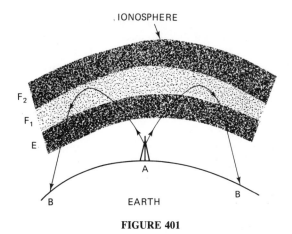

FIGURE 401

receiver is located at either of the points B, the transmission from point A will be received.

The sky wave in Fig. 402 is assumed to be composed of rays that emanate from the antenna in three distinct groups that are identified according to the angle of elevation. The angle at which the group 1 rays strike the ionosphere is too nearly vertical for the rays to be returned to earth. The rays are bent out of line, but pass completely through the ionosphere and are lost.

Group 3 rays strike the ionosphere at the smallest angle that will be refracted and still return to the earth. At any smaller angle, the rays will be refracted but will not return to the earth.

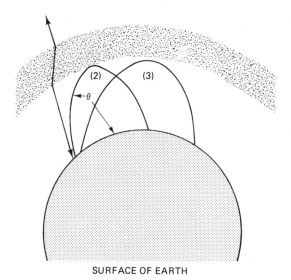

FIGURE 402

As the frequency increases, the initial angle decreases. Low-frequency fields can be projected straight upward and will be returned to the earth. The highest frequency that can be sent directly upward and still be refracted back to the earth is called the *critical frequency*. At sufficiently high frequencies, regardless of the angle at which the rays strike the ionosphere, the wave will not be returned to the earth. The critical frequency is not constant, but varies from one locality to another with the time of day, the season of the year, and the sunspot cycle.

Because of this variation in the critical frequency, nomograms and frequency tables are issued to predict the maximum usable frequency (MUF) for every hour of the day for every locality in which transmissions are made. Nomograms and frequency tables are prepared from data obtained experimentally from stations scattered all over the world. All the information is pooled, and the results are tabulated in the form of long-range predictions that remove some of the guesswork from transmissions.

Sleeve antenna: A broadband antenna that is similar in appearance to the whip antenna, but has a large-diameter sleeve at its base. Critical dimensions are determined by the lowest frequency to be used by the antenna.

Figure 403 shows a sleeve antenna. The large-diameter sleeve may be welded in place, and the sleeve diameter, in feet, should be equal to 30 divided by the frequency in megahertz. The sleeve section length is found by dividing 75 by the frequency in megahertz. The same method is used to find the diameter and the length of the whip section.

As Fig. 403 illustrates, the impedance at the base of the antenna is matched to the 50-Ω transmission line impedance by a quarter-wave transformer section. The impedance of this section is found by multiplying the antenna base impedance by the transmission line impedance, and then taking the square root of the result. The impedance of the whip section is 238 Ω, and the transmission line impedance is 50 Ω. Multiplying these values and taking the square root of the product gives an impedance of 109 Ω for the quarter-wave section.

When used for the higher frequencies, sleeve antennas have several desirable characteristics. They are superior to single-wire or regular whip antennas in both vertical pattern and impedance characteristics when fed against ground. The impedance characteristic of a sleeve antenna is such that satisfactory standing wave ratios (3 to 1 or less) can be obtained over a wide frequency range, and radiation at low angles is much greater than a regular whip.

Slip: In a synchronous motor, the difference between the rotor and stator speeds. At no load, an induction motor runs at practically synchronous speed. With a load, the motor is below synchronous speed by a percentage known as the *slip*. That is, if the synchronous speed is 1800 rpm and the full-load speed is 1700 rpm, the slip at full load is 100/1800, or 5.5 percent.

The slip of any induction motor depends on the voltage drop in the

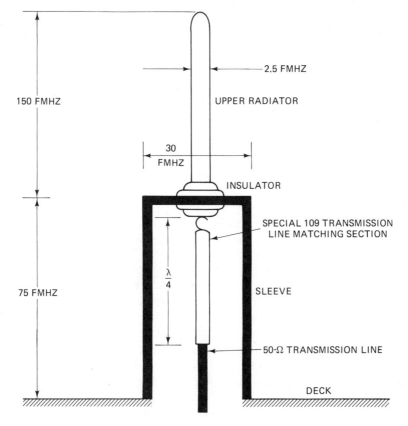

FIGURE 403 Sleeve antenna.

secondary circuit; i.e., on the secondary resistance times the current squared. The greater the secondary resistance, the higher the starting torque with a given current. The greater the slip, the greater the losses, and the lower the efficiency.

Slot antenna: A radiator formed by cutting a narrow slot in a large metal surface. Such an antenna is shown in Fig. 404. The slot length is a half-wave, while the width is a small fraction of a wavelength. Such an antenna is frequently compared to a conventional half-wave dipole consisting of two flat metal strips whose sizes are such that they would just fit into the slot cut out of the large metal sheet. This type is called the *complementary dipole*. This comparison is made because the radiation pattern produced by the slot antenna, which cut into an infinitely large metal sheet, and that of the complementary dipole antenna are the same.

However, several important differences exist between the slot antenna and its complementary antenna. First, the electric and magnetic fields are interchanged. In the case of the dipole antenna shown at the bottom of Fig.

LARGE METAL SHEET

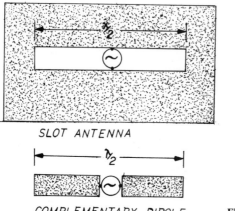

SLOT ANTENNA

COMPLEMENTARY DIPOLE   **FIGURE 404**

404, the electric lines are horizontal while the magnetic lines form loops in the vertical plane. With the slot antenna, the magnetic lines are horizontal and the electric lines are vertical. The electric lines are built up across the narrow dimension of the slot. As a result, the polarization of the radiation produced by a horizontal slot is vertical. If a vertical slot is used, the polarization is horizontal.

A second difference is that the direction of the lines of electric and magnetic force reverse abruptly from one side of the metal sheet to the other. In the case of the dipole, the electric lines have the same general direction, while the magnetic lines form continuous, closed loops.

When energy is applied to the slot antenna, currents flow in the metal sheet. Radiation then takes place from both sides of the sheet. In the case of the complementary dipole, however, the currents are more confined, so a much greater magnitude of current is required to produce a given power output. From this, it can be seen that the current distribution of the dipole resembles the voltage distribution of the slot. The edges on the slot have a high voltage concentration and relatively low current distribution. The complementary dipole has a quite high current concentration and relatively low voltage.

A coaxial line is frequently used to feed the slot antenna. The outer conductor of the feed line is bounded to the metal sheet, while the inner conductor is connected to the opposite side of the slot, as shown in Fig. 405. Note that the coaxial line is not connected at the center of the slot length, since a severe impedance mismatch would occur.

The impedance at the center of a slot whose electric length is a half-wave is 530 Ω. This impedance is reduced to a lower value as the input connection is moved toward the end of the slot. At a point about 1/20 of a wavelength from either end, the input impedance falls to about 50 Ω. A coaxial line may thus be connected here without serious mismatch. The slot length may be increased to

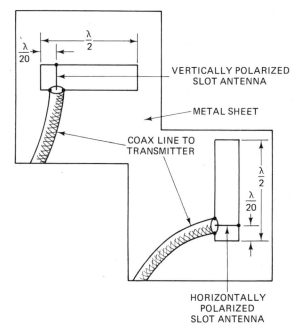

FIGURE 405 Feeding slot antenna with coaxial line.

a full wave if it is desired to connect the coaxial line to the center of the slot. With a full-wave slot, the input impedance at the center of the slot is 50 Ω.

Frequently, it is desired to produce radiation from only one side of the metal sheet in which the slot is cut. This is conveniently arranged by boxing in the slot with a section of waveguide. If the depth of the waveguide is a quarter-wave (Fig. 406), the waveguide will introduce no reactance. Under these conditions, the input impedance at the center of the slot is about 1000 Ω, and the impedance at a point about 1/20 of a wavelength from either end is about 100 Ω.

Another convenient method of making a slot antenna radiate from only one side of the metal sheet in which it is cut is to energize the slot by means of a waveguide. This does away with the coaxial feeder altogether.

Solar cell: A photovoltaic power transducer that converts light to electricity. It is called a *cell* because it is a self-contained source of dc voltage. Like batteries, solar or photovoltaic cells may be connected in series or parallel to provide practical power levels.

Figure 407 shows a circular photovoltaic cell. It is made by combining two ultrathin layers of silicon crystal that have been treated with certain impurities. One material is negative, while the other is positive. When these two materials are sandwiched, one atop the other, a PN junction is formed. This is where the photoelectric effect takes place. This phenomenon is the absorption of photons to create equal numbers of positive and negative charges.

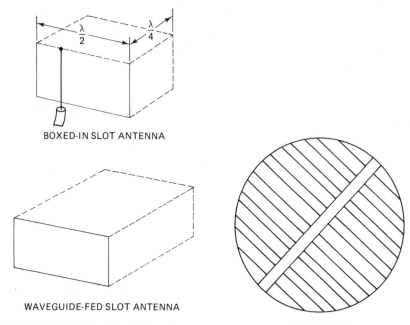

BOXED-IN SLOT ANTENNA

WAVEGUIDE-FED SLOT ANTENNA

FIGURE 406   Unidirectional slot antennas.   **FIGURE 407**

Solder: A metallic alloy with a low melting point and which is used to join metallic surfaces. Solder used in electrical work comes in two forms—solid solder bars and the rosin-core type, resembling wire wound on spools. Figure 408 shows the two types. The rosin-core solder is the type most commonly used. The rosin core of the solder, when melted, flows on the surfaces of the splice or joint, preventing their contact with air and thus preventing oxidation. Without oxidation, the solder and surfaces will cohere strongly.

Solenoid valve: A mechanical device that is used to control the flow of fluids, such as water, oil, etc. It may also be used to control the amount of gas or

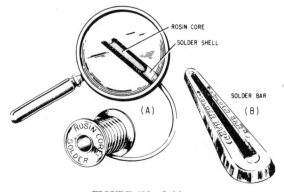

FIGURE 408   Solder.

air. Valves were at one time manually operated, but the current trend is toward those devices that are operated electrically and that are placed quite close to those devices that they control in order to keep piping to a minimum. In this manner, remote control can also be used, since it will only be necessary to run a pair of control wires between the valve itself and the device used for control.

A solenoid valve consists of an electromagnet or solenoid with its plunger or core and a valve with an opening into which a disc or plug can be placed to regulate the flow of water, oil, gas, etc. The plunger, which is magnetic in nature, controls the movement of the valve as it is drawn into the solenoid when the coil is energized. The valve, which is automatic, will operate only when current is applied to the solenoid and will automatically shut down or return to its original position when current is cut off. These devices most often operate a single pole switch, contact, or solenoid coil.

Solenoid valves can be either two- or four-way valves, depending on their usage. Two-way solenoid valves are commonly used to control the flow of methyl chloride, Freon, and sulfur dioxide in both air-conditioning and refrigeration applications. They are magnetically operated, in contrast to four-way valves, which are electrically operated. A four-way valve may be used to control a double-acting cylinder. In operation, when the coil is deenergized, one side of the piston is at atmospheric pressure, and the line pressure acts upon the other side. When electrical energization occurs at the valve magnet coil, the valve operates simultaneously to exhaust the high-pressure side of the piston to atmospheric pressure. This effect results in the piston and its load reciprocating in response to the movement of the valve. Four-way valves are used in the operation of pneumatic cylinders on press clutches, spot welders, machine and assembly jig clamps, and tools and lifts, which are commonly used in the large factories and plants throughout the country.

Sort: To arrange items of information according to rules dependent upon a key or field contained in the items or records. For example, to digital sort is to first sort the keys on the least significant digit, and to resort to each higher-order digit until the items are sorted on the most significant digit.

Sound: Acoustical energy that normally lies within the range of the human hearing response. Sound is always produced by vibration of a body or element and is usually conducted by the air. These vibrations at the sound transmitter create a sympathetic set of corresponding vibrations at the sound receiver, which, in the case of the human body, is the eardrum.

Basically, sound waves do not differ greatly from radio-frequency waves or even light waves, at least when represented graphically. Sound waves, however, are at a much lower frequency than waves that fall into the other spectrums. Human response to sound waves approaches 20 Hz at the low end to 15 kHz or more at the high end.

Space-charge effect: In an electron tube, the charge of electrons in the space between the cathode and plate. If the cathode in Fig. 409 were cold, there would be no electron emission. Consequently, the voltage gradient, or

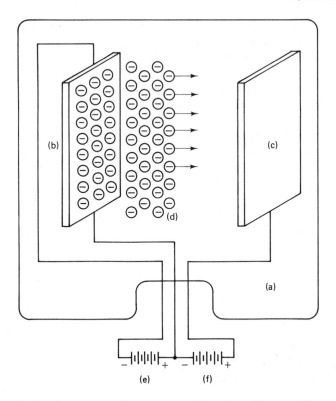

FIGURE 409   Space charge: (1) vacuum bulb; (2) cathode; (3) anode; (4) electrons; (5,6) storage batteries.

difference of potential per unit length of space, between the anode and cathode would be uniform. Thus, the potential of a point halfway between the anode and the cathode would be half the potential between the anode and the cathode.

When electrons are emitted from the hot cathode, this voltage accelerates them toward the anode, and their speed increases as they get farther from the cathode. As a result, electrons will be closer together near the cathode than at any other point. Since the electrons all have negative charges, they repel each other and the effect is an increase in the voltage gradient near the cathode.

Because of the space-charge effect, the voltage gradient at the surface of the cathode may be so low that not all the electrons emitted at the cathode will be attracted to the anode. The space-charge effect may be overpowered by increasing the anode voltage until substantially all the electrons emitted by the cathode will be attracted to the anode, and no further increase in anode current will result from a further increase in the anode voltage.

If the emission of the cathode is increased by increasing its temperature, the current will again be limited by the space charge at a higher level than before, and it will be necessary to increase the anode voltage further to attract all the electrons emitted by the cathode. Thus, at low anode voltages, the anode current is limited by the space charge; and at high anode voltages, it is limited by the cathode emission.

Specific gravity: The ratio of the weight of a certain volume of liquid to the weight of the same volume of water. The specific gravity of pure water is 1.000. Sulfuric acid has a specific gravity of 1.830; this sulfuric acid is 1.830 times as heavy as water. The specific gravity of a mixture of sulfuric acid and water varies with the strength of the solution from 1.000 to 1.830. As a storage battery discharges, the sulfuric acid is depleted, and the electrolyte is gradually converted into water. This action provides a guide in determining the state of discharge of the lead-acid cell. The electrolyte that is usually placed in a lead-acid battery has a specific gravity of 1.350 or less.

Spectrophotometer: A photoelectric instrument used for chemical analysis. The instrument operates by allowing light to pass through a material that is under analysis. The emerging light rays are broken up into a spectrum that is examined by the photoelectric circuit. This, in turn, provides drive to a chart recorder, which plots a spectrogram.

Speed regulation: The characteristic of an electric motor to maintain its speed when a load is applied. This characteristic will remain the same as long as the applied voltage does not vary. The speed regulation of a motor is a comparison of its no-load speed to its full-load speed and is expressed as a percentage of full-load speed. The lower the speed regulation percentage figure of a motor, the more constant the speed will be under varying load conditions, and the better will be the speed regulation. The higher the speed regulation figure, the poorer is the speed regulation.

Squirrel-cage rotor: A rotor in an alternating current motor that is composed of a laminated iron core, into which a number of straight copper bars have been implanted. Squirrel-cage rotors are more rugged and require less maintenance than wound rotors, but may also give trouble due to open circuits or high-resistance points in the rotor circuit. Symptoms of such conditions are slowing down under load and reduced starting torque. Such conditions can usually be detected by looking for evidence of heating at the end ring connections, particularly noticeable when shutting down after operating under load.

Standard deviation: Used in statistical analysis, the square root of the mean of squares of deviation from the mean proper.

Standing wave: A stationary distribution of current or voltage along a line that is due to interactions between a wave that has been transmitted down the line and a wave that has been reflected back. It is characterized by maximum-amplitude points (loops) and minimum-amplitude or zero points (nodes).

Standing wave ratio: The ratio of the effective voltage at a loop to the effective voltage at a node, or the effective current at a loop to the effective current at a node. It is also equal to the ratio of the characteristic impedance of the line to the impedance of the load, or vice versa.

When the line is terminated in a perfect match, all of the energy sent down the line is absorbed by the load and none is reflected. Under these conditions, no standing waves are present. The maximum and minimum values are the same. Thus, the standing wave ratio (SWR) is equal to 1.0.

Two mismatched lines are shown in Fig. 410. In each of these lines, the characteristic impedance, Z_O, of the line is 300 Ω, and the load impedance is 60 Ω. The ratio of the effective current at (a) to the effective current at (b) is equal to y/x or 5/1, which is also equal to 300/60.

In Fig. 410, the line impedance is less than the impedance of the load. The ratio of the effective current at (d) to the effective current at (e) is equal to 1500/300 or 5/1. In the first example, the SWR is equal to the ratio of the Z_0 of the line to the Z of the load. In the second example, it is equal to the ratio of the Z of the load to the Z_O of the line. In both examples, the SWR is equal to the ratio of the effective current at a loop to the effective current at a node.

In general, the higher the SWR, the greater the mismatch between the line and the load. A knowledge of the position of the current and voltage loops

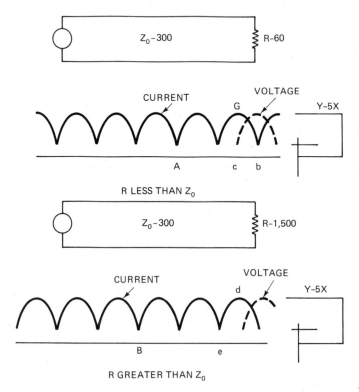

FIGURE 410   Mismatched lines showing standing-wave ratio.

and nodes along the line will indicate whether the load resistance is less than or greater than the characteristic impedance. For example, in Fig. 410, there are a voltage node and a current loop at the load. This occurs because the load resistance is less (approaching a shorted condition) than the characteristic impedance of the line. Thus, it is a simple matter (by the use of an RF measuring device) to determine whether the load resistance is greater or smaller than Z_O. If the load resistance is greater than Z_O, the output end of the line will appear more like an open circuit and RF measuring devices will indicate maximum effective voltage and minimum effective current at that point.

Stator: The nonmoving or stationary coil in a motor into which the rotor is inserted. Stators of all types of alternators are relatively similar. The horizontal-shaft salient-pole type will be used for discussion. The stationary part of the alternator consists essentially of a laminated iron core and its supporting structure. The stator winding is placed in slots provided on the inner cylindrical surface of the laminated core. End shields or guards that extend over the stator-coil ends and protect them from mechanical injury are attached to the supporting structure.

Steel conductor: Copper-covered steel wire of conductivity 40 percent of that of copper, but of much higher strength, which is used for long-span distribution lines. A combination of this material with strands of copper in the same conductor makes a satisfactory conductor if less strength and more conductivity than that of the copper-covered steel alone are wanted.

Steel tower construction: A method of construction used mainly for high-voltage transmission lines over long distances and over mountainous terrain. Such systems are used primarily for the support of conductors, not the mounting of transformers, switches, and other equipment. This equipment is normally located at substations or switchyards. Steel tower construction is also used for the supporting structures of overhead trolley systems for electrified railroads, and modified tower structures are sometimes used for recreational floodlighting at football and baseball stadiums.

The basic components of a steel tower consist of a base or footing, the main tower structure, bridge or cross-arm members, insulators and vibration dampeners, and tower grounding. The size and type of footing depend on the type and height of the tower and the nature of the ground and terrain. Footings are usually completely detailed in the contract drawings, and specifications and details such as the ones in Fig. 411 will be included on the working drawings.

Step-voltage regulator: A form of voltage regulator much like a variable autotransformer, having a movable-brush tap that can be adjusted to give a constant output voltage of from 0 to 117 percent of input voltage. A schematic diagram of connections is shown in Fig. 412. By changing the position of the movable-brush tap 1 on the autotransformer winding 2 from the lowest point 3 to the highest point 4 of the winding, the output voltage between leads 5 can be adjusted from 0 to 140 V if the supply voltage is 120 V.

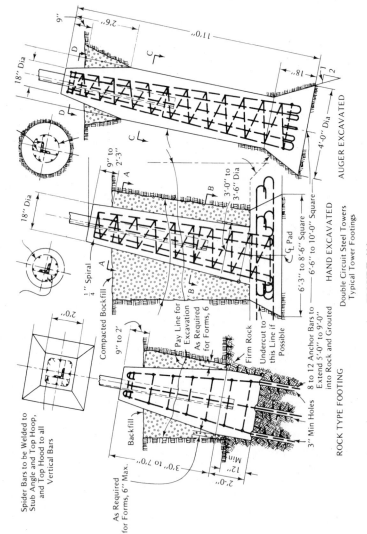

Spider Bars to be Welded to
Stub Angle and Top Hoop,
and Top Hood to all
Vertical Bars

Compacted Bockfill

$\frac{1}{4}$" Spiral

Pay Line for
Excavation
As Required
for Forms, 6

Backfill

18" Dia

9" to
2'-3"

3'-0" to
3'-6" Dia

18" Dia

9" to 2'

2'-0"

9"

2'-6"

11'-0"

18"

4'-0" Dia

AUGER EXCAVATED

6'-3" to 8'-6" Square
6'-6" to 10'-0" Square

℄ Pad

Firm Rock

Undercut to
this Line if
Possible

HAND EXCAVATED

8 to 12 Anchor Bars to
Extend 5'-0" to 9'-0"
into Rock and Grouted

3'-0" to 7'-0"

3" Min Holes

12"
Min

2'-0"

As Required
for Forms, 6" Max.

ROCK TYPE FOOTING

Double Circuit Steel Towers
Typical Tower Footings

FIGURE 411

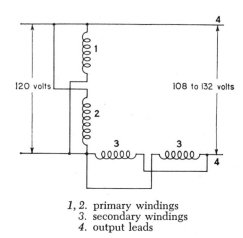

1, 2. primary windings
3. secondary windings
4. output leads

FIGURE 412   Connections for Regulator

Storage:

1. The term preferred to *memory*.

2. Pertaining to a device in which data can be stored and from which it can be obtained at a later time.

3. A device consisting of electronic, electrostatic, electrical, hardware, or other elements into which data may be entered, and from which data may be obtained as desired.

4. The erasable storage in any given computer.

Storage oscilloscope: An oscilloscope that retains a CRT pattern until erased by manual activation. This device utilizes a cathode-ray storage tube that is specially constructed to retain information in the form of images on a special electrode until erased by an electronically generated input signal.

Storage oscilloscopes are useful in scientific evaluation and diagnosis, especially when applied to signals that are instantaneous and nonrepetitive in nature. Such a signal may be captured in the storage tube and retained for long-term observation. When all information has been obtained from a single trace, the storage tube is electronically erased and is ready to receive another input.

Strain gage: A device used for the determination of displacements, distortions, or deformations in machines and structures. The transducers for this type of measurement may be in the form of resistors, capacitors, or inductive devices so arranged that their characteristic resistance or reactance is changed proportionally to the strain or deformation in the machine part being tested. These transducers, particularly those of the resistance type, may be very inexpensively constructed and are frequently permanently attached to or built into the structures to be investigated.

S unit: A measurement term that reflects the strength of incoming signals to a radio receiver. These measurements range from S-1 to S-9, with the higher numbers representing stronger signals. Past the S-9 point, measurements continue in decibels over S-9. A very strong signal might be read as 20 dB over S-9. A value of S-9 is typically equal to a received signal strength of 50 microvolts. With the descension of each S unit, the signal strength drops by 6 dB. Therefore, S-8 is 6 dB over S-9, S-7 is 12 dB below S-9, and so on.

S units are measured on an S meter. These devices are often seen on receivers and transceivers designed for amateur radio applications. In common practice, many S meters are not accurately calibrated to read S-9 with a signal input of 50 μV, although these adjustments can be made using a calibrated signal generator. Most often, S meters are used as relative indicators of signal strength.

For example, one receive signal causes the meter indicator to swing to S-8, while another signal garners a reading of S-7. Regardless of the actual signal strength in microvolts, which causes the meter to read S-8, it can be determined that the second signal of S-7 is 6 dB below the strength of the first one. Here, the first reading forms an indication to which all others are compared. All readings, then, are relative.

Sun lamp: A special kind of mercury lamp that, by selective absorption of the glass bulb, limits the generated rays to those found in natural sunlight, causing suntan but eliminating the far ultraviolet rays. The radiated energy causing suntan is measured in E-Viton units. The E-Viton is a quantitative unit of ultraviolet watts radiated, which is weighted with respect to the effectiveness of various wavelengths in producing skin reddening. One E-Viton corresponds to the quantity of radiant energy that produces as much reddening of the skin as 10 μW or 0.00001 W of energy at a wavelength of 2967 A. Sun lamps are rated in E-Viton units.

The self-contained RS reflector sun lamp with built-in starting switch and filament ballast is quite popular. It will operate satisfactorily only on alternating current circuits on either 50- or 60-cycle current within the range of 110 to 125 V. It takes about 3 min to reach full ultraviolet output on starting and approximately 5 min for restarting. Part of the restarting time is required for the bimetallic switch to cool down and close, and no light is produced in this interval.

The RS sun lamp delivers 35 E-Vitons/in.^{2} at a distance of 30 in. from the axis of the lamp. This amount of energy is sufficient to produce a mild sunburn on untanned skin in 5 to 10 min, the equivalent of 15 to 18 min of exposure to midsummer sun.

The life of the RS lamp on normal ac circuits in household use has been estimated at approximately 600 applications, or 1000 h when operated 5 h per start.

Since the sun's rays reaching the earth contain no rays shorter than 2800 A, the bulbs of sun lamps are made of special glass, which, like the atmosphere, absorbs the shorter rays.

Superheterodyne receiver: A radio-frequency receiver in which the received signal from the first detector is mixed with the signal of a local oscillator in order to extract the intermediate frequency. The superheterodyne receiver was developed to overcome many of the disadvantages of the TRF receiver. The essential difference between the two is that in the TRF receiver, the RF amplifiers preceding the detector are tunable over a band of frequencies, whereas in the superheterodyne receiver, the corresponding amplifiers are tuned to one fixed frequency called the *intermediate frequency* (IF).

The principle of frequency conversion by heterodyne action is employed to convert any desired station frequency within the receiver range to the intermediate frequency. Thus, an incoming signal is converted to the fixed intermediate frequency before detecting the audio-signal component, and the IF amplifier operates under uniformly optimum conditions throughout the receiver range. The IF circuits may thus be made uniformly selective, high in voltage gain, and of satisfactory bandwidth to contain all of the desired sideband components associated with the AM carrier.

A block diagram of a superheterodyne receiver is shown in Fig. 413. Shown below corresponding sections of the receiver are the waveforms of the signal at that point. The RF signal from the antenna first passes through an RF amplifier (preselector), where the amplitude of the signal is increased. A locally generated (local oscillator), unmodulated RF signal of constant amplitude is then mixed with the carrier frequency in the mixer stage. The mixing or heterodyning of these two frequencies produces an intermediate frequency signal that contains all of the modulation characteristics of the original signal. The intermediate frequency is then amplified and fed to a conventional detector for recovery of the audio signal. The detected signal is amplified in the AF section and then fed to a headset or loudspeaker. The RF amplifier, detector, AM amplifier, and reproducer are basically the same as those in a TRF set.

The RF stage amplifies the small ac voltages induced in the antenna by the

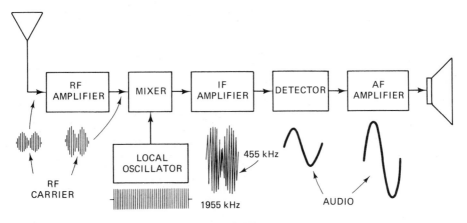

FIGURE 413

electromagnetic wave from the station transmitter. Utilizing a tuned circuit between the antenna and the input of the RF amplifier permits selection of the desired station frequency from among the many present in the antenna. In addition to amplifying the RF signal, the RF amplifier isolates the local oscillator from the antenna. If the antenna were connected directly to the mixer stage, a part of the local oscillator signal might be radiated into space. If the mixer stage were connected directly to the antenna, unwanted signals called *images* might be received, because the mixer stage produces the intermediate frequency by heterodyning two signals whose frequency difference equals the intermediate frequency.

The image frequency always differs from the desired station frequency by twice the intermediate frequency. The image frequency is higher than the station frequency if the local oscillator frequency tracks above the station frequency, as generally used for lower frequencies. The image frequency is lower than the station frequency if the local oscillator tracks below the station frequency, as generally used for the highest frequencies. This is shown in Fig. 414.

For example, if a superheterodyne receiver has an intermediate frequency of 1500 kHz (Fig. 414) and the local oscillator has a frequency of 1055 kHz, the output of the IF amplifier may contain two signals, one from the 1500-kHz station and the other from an image station of 590 kHz. The same receiver near the low end to a 90-kHz station has a local oscillator frequency of 1355 kHz. The output of the IF amplifier contains the station signal and may contain an image signal [1810-kHz image signal = 900-kHz station signal − (2 × 455-kHz IF signal)]. Thus, the 1810-kHz signal is an image that may be heard simultaneously with the 900-kHz station signal.

It may also be possible for any two signals having sufficient strength and separated by the intermediate frequency to produce unwanted signals in the reproducer. The selectivity of the preselector (RF stage preceding the mixer stage) tends to reduce the strength of these unwanted signals. However, there is a practical limit to the degree of selectivity obtainable in the preselector due to the fact that the RF stage must have a wider bandwidth than the bandwidth

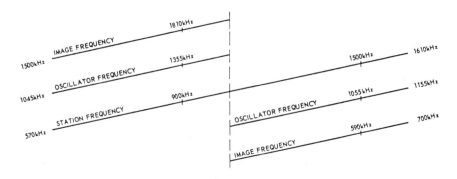

FIGURE 414

of the desired signals. The ratio of the amplitude of the desired station signal to that of the image is called the *image rejection ratio* and is an important characteristic of a superheterodyne receiver.

The function of the local oscillator stage is to produce a constant amplitude sine wave of a frequency that differs from the desired station frequency by an amount equal to the intermediate frequency of the receiver. Although the oscillator may be operated either above or below the station frequency, in most broadcast band receivers, the oscillator is operated above the station frequency. In order to allow selection of any frequency within the range of the receiver, the tuned circuits of the RF stage and the local oscillator are variable. By using a common shaft or ganged tuning for the variable component of the tuned circuits, both may be tuned in such a manner as to maintain the difference between the local oscillator frequency and the incoming station frequency equal to the receiver IF.

The function of the mixer stage is frequency conversion by heterodyne action. The input to the mixer consists of two signals—the modulated RF signal and the unmodulated local oscillator signal. The mixer then combines or mixes these two signals. As a result of this mixing action, the output of the mixer will contain four major frequencies, plus many very minor frequencies. The four major frequencies are the original signal frequency, the local oscillator frequency, the sum of the signal and oscillator frequencies, and the difference of the signal and oscillator frequencies. The additional frequencies present are produced by combinations of the fundamentals and harmonics of the signal and oscillator frequencies. Of the frequencies present in the output of the mixer, only the difference frequency is used in amplitude-modulated receivers. The output circuit of the mixer stage contains a tuned circuit that is resonated at the difference frequency. The output of the mixer is fed to the IF amplifier and consists of a modulated intermediate frequency signal.

Superheterodyne receivers employ one or more IF amplifiers, depending on design and quality of the receiver. Transformers are usually used for interstage coupling in the IF section. The IF circuits are permanently tuned to the difference frequency between the incoming RF signal and the local oscillator. As previously stated, all incoming signals are converted to the same frequency by the mixer stage, and the IF amplifier operates at only one frequency. The tuned circuits, therefore, are permanently adjusted for maximum signal gain consistent with the desired bandpass and frequency response. These stages operate as class A voltage amplifiers, and practically all of the selectivity (adjacent channel and interference frequencies, not image frequency) of the superheterodyne receiver is determined by them.

The values of intermediate frequencies for broadcast receivers range from 130 to 485 kHz, with the most popular value being around 455 kHz. The output of the IF amplifier is fed to the detector. The receiver functions of detection, AF amplification, and reproduction are performed by the detector, AF amplifier, and speaker.

Since the IF stages operate at a single frequency, the superheterodyne receiver may be designed to have better selectivity and sensitivity across the entire broadcast band and better gain per stage than the TRF receiver. Although a superheterodyne receiver contains more tuned stages than a TRF receiver, the majority of the stages are fixed tuned at one frequency. This reduces the tracking problem and makes alignment of the tuned stages much easier. The major disadvantage of this receiver is the reception of image frequencies.

Superregenerative receiver: A regenerative detector circuit in which regeneration is periodically increased almost to the point of oscillation and then decreased. This quenching action takes place at a supersonic rate (typically at 50 or 100 kHz), so that the quenching is inaudible. The result is that much more regeneration is afforded, without the detector going into oscillation, than is possible by simply increasing the regeneration manually. An extremely sensitive detector is the result.

Suppressor: A general term describing any device, component, or circuit that limits an electronic function. Some suppressors are actually RF filters that suppress radio interference. Noise suppressors (also called *noise eliminators*) are often incorporated in radio receivers to cut down on unwanted interference from electrical sources. More commonly, a suppressor is the third grid in a pentode vacuum tube.

Suppressor grid: A grid in a pentode vacuum tube. Three grid elements are positioned between the plate and cathode electrodes. In addition to the control grid and the screen grid that are found in tetrode vacuum tubes, a third grid (the suppressor grid) is incorporated. The suppressor grid is often connected to the cathode within the tube, but may also be brought out to a connecting pin on the tube base. Either way, the suppressor grid is negative with respect to the minimum plate voltage. The secondary electrons that would travel to the screen grid if there were no suppressor are diverted back to the plate due to the negative charge. The plate current is not reduced and the amplification possibilities are increased using pentode tubes.

Figure 415 shows a schematic representation of a typical pentode vacuum tube with each element identified.

Surface metal raceway: One of the exposed wiring systems that is quite extensively used in existing buildings when new wiring or extensions to the old system are installed. Although it does not afford as rugged and safe a protection to the conductors as rigid conduit or EMT, it is a very economical and quite dependable system when used under the conditions for which it was designed. The main advantage of surface metal raceway is its neat appearance where wiring must be run on room surfaces in finished areas.

Surface metal raceways may be installed in dry locations, except where subject to severe physical damage or where the voltage is 300 V or more between conductors (unless the thickness of the metal is not less than 0.040 in.). Surface metal raceways should not be used in areas that are subjected to corrosive vapors, in hoistways, or in any hazardous location. In most cases, this system should not be used for concealed wiring.

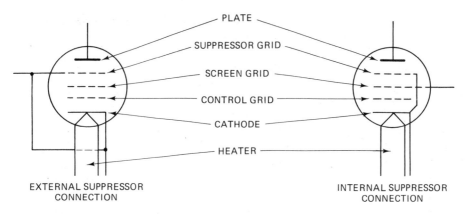

PLATE
SUPPRESSOR GRID
SCREEN GRID
CONTROL GRID
CATHODE
HEATER

EXTERNAL SUPPRESSOR
CONNECTION

INTERNAL SUPPRESSOR
CONNECTION

FIGURE 415

Many of the rules for other wiring systems also apply to surface metal raceways. The system must be continuous from outlet to outlet, it must be grounded, and all wires of the circuit must be contained in one raceway.

In planning the installation of a surface metal raceway system, the electrical contractor should make certain that all materials are provided before work is begun. One missing fitting could delay an entire project. Proper tools should be available as well. For example, a surface metal raceway bending tool enables electricians to bend certain sizes of the raceway like rigid conduit or EMT.

Switch: A device that is used for opening and closing circuits. Switches are constructed in such a way that the person operating them cannot come into electrical contact with the circuit being energized or deenergized. This type of construction is used to prevent operating personnel from receiving an electrical shock.

Switches are manufactured in hundreds of different types and current ratings. Some switches open a single conductor or circuit, while others are designed to open or close many circuits simultaneously. The schematic symbol for a switch used to open or close a single conductor is shown in Fig. 416.

Switch, bolted pressure: A load-break disconnect. These types of switches are suitable for service entrance equipment and are usually furnished in a general purpose dead-front enclosure. They have a fuse access door with a

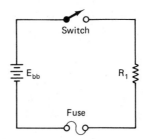

Switch

E_{bb}

R_1

Fuse

FIGURE 416   Circuit showing the use of
a fuse and switch.

lock. Incoming line terminals are located at the top and load terminals are at the bottom. All neutrals are full capacity.

Switchgear: Any devices used for electrical switching, interrupting, local and remote control, metering, protection, and regulation. The term also covers all of the various assemblies of these devices with their associated interconnections, supporting structures, and accessories.

Switchgear has a part in virtually all phases of the generation, transmission, distribution, and utilization of electric power. Large circuit breakers provide the chief safety factor of high-voltage distribution systems. These quickly isolate system faults to permit the remainder of the system to operate normally. Through load and frequency controls, switchgear devices stabilize the overall system operation and give maximum protection to equipment and personnel.

Symbols, electrical and electronic: In the preparation of working drawings for the electrical and electronic industries, symbols that are used to simplify the work of preparing such drawings. Most engineers, designers, and draftsmen use symbols adopted by the United States of America Standards Institute (USASI) for use on electrical and electronic drawings. However, many designers and draftsmen frequently modify these symbols to suit their own particular requirements for the type of work they normally encounter. For this reason, most drawings have a symbol list or legend drawn and lettered on each set of working drawings. The symbols are made with straight edges, templates, and French curves. An explanatory note is then lettered adjacent to the symbol.

The list in Fig. 417 represents a good set of electrical symbols for building construction. The list in Fig. 418 should suffice for the majority of the draftsman's needs on electronic drawing. If additional data are occasionally needed on a given project, notes may be used on the drawings, or a new symbol can be devised to cover the situation at hand. Note that many of the symbols have the same basic form, but their meanings are different because of some slight difference in the symbol.

Synchro: A device that is used as a means of transmitting the angular position information of a rotary device (handwheel) to one or more remote indicators, such as radar antenna to radar repeater, handwheel to receiver indicator dials, etc. In most cases, synchros are used where it would be impractical or quite involved to use the mechanical equivalent systems, as shown in Fig. 419. Synchro transmitters, receivers, control transformers, and differentials used in various systems are pictured in schematic symbols using their external connections and relative positions of their windings. Figure 420a shows the symbols used to indicate wiring connections. Figures 420c, d, e show the symbols that are generally used for purposes of explaining the theory of operation of the synchro mechanisms.

Synchros are electromagnetic devices that resemble ac electric motors. They are composed of a rotor and stators. The name rotor is given to the

A

ASSEMBLY, SUBASSEMBLY

It is an assembly of items that is mounted and prewired as a unit, which can not be identified in a specific group or which may contain items made up of other parts.

AR

AMPLIFIER

B

MOTOR

1. General (fan, blower)

2. Series Field

3. Application: Engine Starting Motor

SER FLD

BT

BATTERY

The long line is always positive, polarity must be identified in addition (shown as multicell).

$$\overset{BT}{\underset{+}{\dashv | | \vdash}}$$

C

CAPACITOR

If it is necessary to identify the capacitor electrodes, the curved element shall represent the outside electrode in fixed paper-dielectric and ceramic dielectric capacitors, and the low potential element in feed through capacitor.

1. General

$$\overset{C}{\dashv\vdash}$$

2. Polarized, Electrolytic Capacitor

$$\overset{C}{\underset{+}{\dashv\vdash}}$$

3. Feed-through Capacitor (with terminals shown on feed-through element for clarity.)

Commonly used for bypassing high frequency current to chassis.

CB

CIRCUIT BREAKER

1. General

old drawings)

2. Circuit Breaker with thermal overload device.

3. Circuit Breaker with magnetic overload device.

4. Circuit Breaker with thermal magnetic overload device.

5. Application: 3-pole circuit breaker with thermal magnetic overload device in each pole and trip coil (shown with boundary lines)

FIGURE 417

523

CR

RECTIFIER, DIODE

Triangle points in direction in which rectifier conducts current easily.

1. Diode, Metallic Rectifier, Electrolytic Rectifier, Asymmetrical Varisitor.

2. Application: Full-Wave Bridge Type Rectifier.

3. Controlled Rectifier (SCR)

4. Bidirectional Diode (Suppressor)

5. Zener Diode

6. Tunnel Diode

CT

CURRENT TRANSFORMER
1. General

2. Current Transformer with polarity marking.
Instantaneous direction of current into one polarity mark corresponds to current out of the other polarity mark.

DS

SIGNALING DEVICE except meter or thermometer.
1. Audible Signaling device.
1.1 bell

1.2. buzzer

1.3. howler

2. Visual Signaling Device (indicating, pilot, signal or illuminating lamp)
2.1. incandescent lamp

2.2. neon lamp
2.2.1 alternating-current type

2.2.2. direct-current type
NOTE: Polarity mark is not part of the symbol.

E

ELECTRICAL SHIELDING, PERMANENT MAGNET, SPARK PLUG, MISCELLANEOUS ELECTRICAL PARTS.
1. Electrical shield (short dashes) normally used for eletric or magnetic shielding. When used for other shielding, a note should so indicate.

2. Permanent magnet

$$\frac{E}{\boxed{PM}}$$

3. Spark plug

FIGURE 417 **Continued**

4. Miscellaneous Electrical part
4.1. engine choke
4.1.1. thermal

4.1.2. magnetic

4.1.3. thermal magnetic

4.2. fuel pump

4.3 sending units (oil, water, etc.)

FUSE

F

G

GENERATORS
1. General

2. Field, Generator
2.1. compensating or commutating

3 winding symbol curve

2.2 series

coil symbol

2.3. shunt or separately excited.

coil symbol

3. Winding Symbols
3.1. 1-Phase   3.2. 3-Phase wye

(old drawings

          Indicates slip rings
                            or collector rings.)

3.3 3-Phase delta

4. Application: charger generator and cranker.

5. Application: revolving armature generator (shown as single phase, 3 wire)

6. Application: revolving field generator (shown as 3-phase wye, 4 wire

7. Application: Magneto

H

HARDWARE (bolts, nuts, screws, etc.) if applicable.

HR

HEATER, manifold, glow plug, general.

J

RECEPTACLE-fixed or stationary connector.
The connector symbol is not an arrow head. It is larger and the lines are drawn at a 90° angle.
1. Female Contact

female connector symbol

2. Male contact

FIGURE 417   Continued

3. Application: 4-conductor connector with 3-male contacts and 1-female contact with individual contact designations.

or

if no confusion results from its use by disregarding the type of contacts in the receptacle, it may be shown as

4. Receptacles of the type commonly used for power-supply purposes (convenience outlets)
4.1. female contact

4.2. male contact

5. Application: 3-conductor polarized connector with female contacts.

6. Application: 3-conductor polarized connector with male contacts.

RELAY, CONTACTOR, SOLENOID (electrically or thermally operated)
1. Coil
1.1. basic operating

1.2. time delay

2. Contacts
2.1. basic contact assemblies
2.1.1. closed contact (break)

2.1.2. open contact (make)

2.1.3 transfer

2.2. contacts with time delay feature.
2.2.1. closed contact, time delay opening

2.2.2. open contact, time delay closing

2.2.3 closed contact, time delay closing

Note: contacts at left are for wiring diagrams. Contacts at right for schematic diagrams & wiring diagrams of contactors.

2.2.4 open contact, time delay opening

3. Application: Relay with transfer contacts

INDUCTOR, REACTOR
1. Air core

2. Iron Core (if desired to distinguish magnetic-core inductors)

FIGURE 417 Continued

526

3. Saturating core

4. Saturable-core inductor (reactor)
NOTE: explanatory words & arrow are not part of the symbol shown

DC WINDING

M

METERS, GAUGES, CLOCKS with calibrated dials
1. clock, electric timer
1.1. motor

1.2. transfer contacts

2. Indicating meters, gauges, etc.

*replace the asterisk by one of the following letter combinations, depending on the function of the meter.

A	Ammeter
AH	Ampere-hour
F	Frequency meter
MA	Milliammeter
OP	Oil Pressure
PF	Power Factor
T	Temperature
TT	Total Time (Running Time)
V	Voltmeter
W	Wattmeter
WH	Watthour meter

MP

MECHANICAL PARTS including nameplates - if applicable

P

PLUG- affixed to a cable, cord or wire
The connector symbol is not an arrowhead. It is larger and the lines are drawn at a 90° angle.
1. Female contact

2. Male contact

3. Application: 4-conductor connector with 3-male contacts and 1-female contact with individual contact designations.

or

if no confusion results from its use by disregarding the type of contacts in the plug, it may be shown as

4. Plugs of the type commonly used for power-supply purposes (mating connectors)
4.1. female contact

4.2. male contact

5. Application: 3-conductor polarized connector with female contacts.

6. Application: 3-conductor polarized connector with male contacts

Q

TRANSISTOR
1. General
1.1. NPN

1.2. PNP

2. Unijunction
2.1. N-type base

FIGURE 417 Continued

2.2. P-type base

3. Field-effect
3.1. N-type base

3.2. P-type base

R

RESISTOR

do not use both styles of symbols on the same diagram
1. General (fixed)

R
—⌁⌁— OR —[R]—

2. Tapped

R
—⌁⌁— OR —[R]—

3. Adjustable contact

R
—⌁⌁— OR —[R]—

4. Rotary type adjustable

The preferred method of terminal indentification is to
designate with the letters "CW" the terminal adjacent to

the movable contact when it is in an extreme clockwise
position as viewed from the knob end.

Rheostat

R
—⌁⌁— OR —[R]—
CW   CW

5. Non linear

R
—⌁⌁— OR —[R⁄]—

RT

THERMISTOR; THERMAL RESISTOR

"T" indicates that the primary characteristic of the element within the circle is a function of temperature

RT
—(⌁T)—

VARISTOR, SYMMETRICAL

resistor, voltage sensitive (silicon carbide, etc)

 OR —[RV]— OR

S

SWITCH

1. Thermal cutout, thermal flasher

S
—⌐_—

2. Switch
2.1. momentary-fixed contact on momentary switch

▼———

2.1.1. open contact (make) (ignition points)

2.1.2. closed contact (break)

S
o—◣—

2.1.3. 2-open contacts (make)

2.1.4. push button, open contact (make)

S |
▲ ▲

2.1.5. push button, closed contact (break)

S |
▼ ▼

2.2. locking or maintained-fixed contact for maintained switch.

o

2.2.1 open contact (make)

S
o— o

2.2.2. closed contact (break)

S
o—o

2.2.3. 2-open contact (make)

S o
o— o OFF

FIGURE 417   Continued

T
TRANSFORMER - ignition coil
1. Iron core

2. Air core

TB
TERMINAL BLOCK-MARKER STRIP

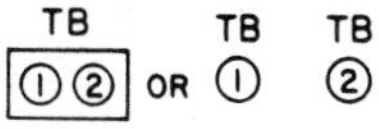

TB   TB   TB
① ② or ①   ②

TC
THERMOCOUPLE

VR
VOLTAGE REGULATOR, CHARGE, CURRENT

W
CONDUCTORS, CABLE, WIRING,
BUSBAR, ETC.

1. Conductive path or conductor; wire

2. Two conductors or conductive paths

3. Three conductors or conductive paths

4. Crossing of paths or conductors not connected. The crossing is not necessarily at a 90° angle.

5. Splice

6. Junction of connected paths, conductors, or wires (other than a terminal)

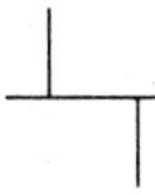

7. Terminal
may be added to each point of attachment to the connecting lines to any one of the graphic symbols.

O

8. Shielded single conductor

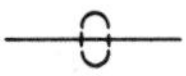

9. Shielded 2-conductor cable with shield grounded

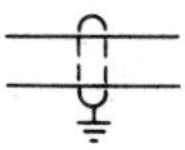

10. 2-conductor cable

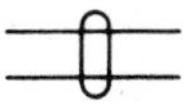

11. Grouping of leads
Normally, bend of line indicates direction of conductor joining cable

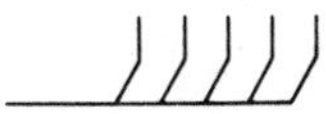

OR

12. Associated or future (short dashes)

X
FUSEHOLDER, SOCKET, LAMPHOLDER

Z
NETWORK, General
Where specific letters do not fit, when considered a part.

FIGURE 417   Continued

529

2.3. application: 3-position, 1-pole, circuit closing (make), off. momentary circuit closing (make).

2.4. application: 2-position, 1-pole, momentary circuit closing (make), circuit closing.

2.5. selector switch
2.5.1. 4-position with non-shorting contacts

2.5.2. 4-position with shorting contacts

2.6 master or control switch

A table of contact operation must be shown on the diagram. A typical table is shown below.

DETACHED CONTACTS SHOWN ELSEWHERE ON DIAGRAM

CONTACT	POSITION		
	A	B	C
1 – 2			X
3 – 4	X		
5 – 6			X
7 – 8	X		

X indicates contact closed

FOR WIRING DIAGRAM

2.7. flow actuated switch
2.7.1. closing on increase in flow

2.7.2. opening on increase in flow

2.8. liquid level actuated switch
2.8.1 closing on rising level

2.8.2 open on rising level

2.9. temperature actuated switch (thermostat)
2.9.1. closing on rising temperature

2.9.2 opening on rising temperature

2.10. pressure or vacuum actuated switch
2.10.1. closing on rising pressure

2.10.2 opening on rising pressure

2.11. centrifugal actuated switch (overspeed)
2.11.1. closing on speed

2.11.2. opening on speed

FIGURE 417   Continued

530

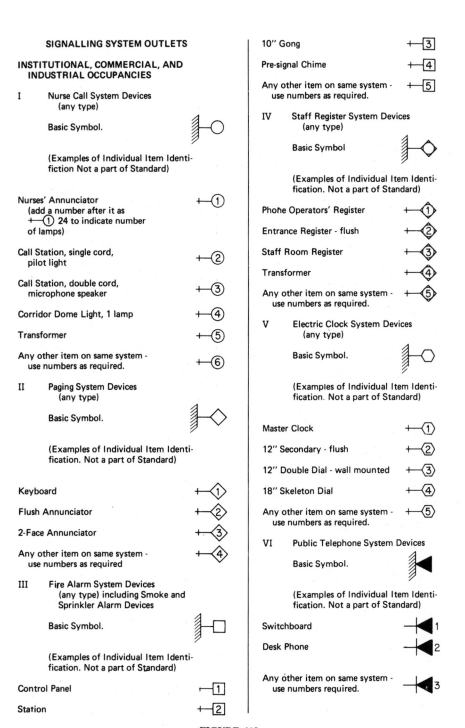

SIGNALLING SYSTEM OUTLETS

INSTITUTIONAL, COMMERCIAL, AND INDUSTRIAL OCCUPANCIES

I   Nurse Call System Devices
(any type)

      Basic Symbol.

      (Examples of Individual Item Identifiction Not a part of Standard)

Nurses' Annunciator
(add a number after it as
┼─① 24 to indicate number
of lamps)

Call Station, single cord,
pilot light

Call Station, double cord,
microphone speaker

Corridor Dome Light, 1 lamp

Transformer

Any other item on same system -
use numbers as required.

II   Paging System Devices
(any type)

      Basic Symbol.

      (Examples of Individual Item Identification. Not a part of Standard)

Keyboard

Flush Annunciator

2-Face Annunciator

Any other item on same system -
use numbers as required

III   Fire Alarm System Devices
(any type) including Smoke and
Sprinkler Alarm Devices

      Basic Symbol.

      (Examples of Individual Item Identification. Not a part of Standard)

Control Panel

Station

10" Gong

Pre-signal Chime

Any other item on same system -
use numbers as required.

IV   Staff Register System Devices
(any type)

      Basic Symbol

      (Examples of Individual Item Identification. Not a part of Standard)

Phone Operators' Register

Entrance Register - flush

Staff Room Register

Transformer

Any other item on same system -
use numbers as required.

V   Electric Clock System Devices
(any type)

      Basic Symbol.

      (Examples of Individual Item Identification. Not a part of Standard)

Master Clock

12" Secondary - flush

12" Double Dial - wall mounted

18" Skeleton Dial

Any other item on same system -
use numbers as required.

VI   Public Telephone System Devices

      Basic Symbol.

      (Examples of Individual Item Identification. Not a part of Standard)

Switchboard

Desk Phone

Any other item on same system -
use numbers required.

FIGURE 418

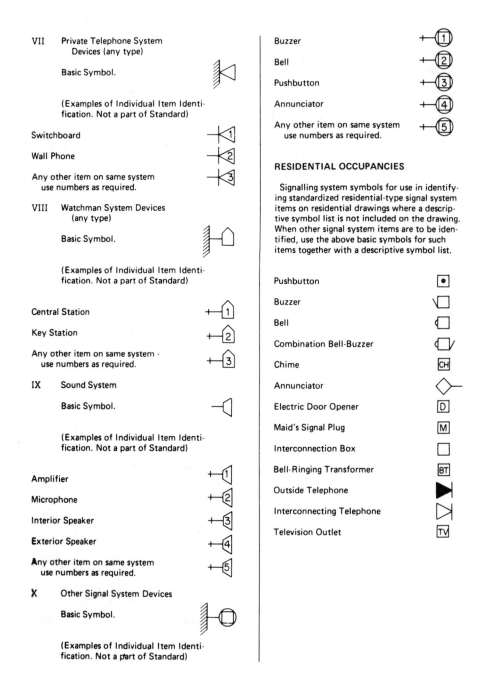

VII Private Telephone System
     Devices (any type)

     Basic Symbol.

     (Examples of Individual Item Identi-
     fication. Not a part of Standard)

Switchboard

Wall Phone

Any other item on same system
     use numbers as required.

VIII Watchman System Devices
      (any type)

      Basic Symbol.

      (Examples of Individual Item Identi-
      fication. Not a part of Standard)

Central Station

Key Station

Any other item on same system -
     use numbers as required.

IX Sound System

    Basic Symbol.

    (Examples of Individual Item Identi-
    fication. Not a part of Standard)

Amplifier

Microphone

Interior Speaker

Exterior Speaker

Any other item on same system
     use numbers as required.

X Other Signal System Devices

   Basic Symbol.

   (Examples of Individual Item Identification. Not a part of Standard)

Buzzer

Bell

Pushbutton

Annunciator

Any other item on same system
     use numbers as required.

RESIDENTIAL OCCUPANCIES

Signalling system symbols for use in identifying standardized residential-type signal system
items on residential drawings where a descriptive symbol list is not included on the drawing.
When other signal system items are to be identified, use the above basic symbols for such
items together with a descriptive symbol list.

Pushbutton

Buzzer

Bell

Combination Bell-Buzzer

Chime

Annunciator

Electric Door Opener

Maid's Signal Plug

Interconnection Box

Bell-Ringing Transformer

Outside Telephone

Interconnecting Telephone

Television Outlet

FIGURE 418 Continued

ELECTRICAL SYMBOLS

SWITCH OUTLETS

Single-Pole Switch	S
Double-Pole Switch	S_2
Three-Way Switch	S_3
Four-Way Switch	S_4
Key-Operated Switch	S_K
Switch and Fusestat Holder	$S_F H$
Switch and Pilot Lamp	S_P
Fan Switch	S_F
Switch for Low-Voltage Switching System	S_L
Master Switch for Low-Voltage Switching System	S_{LM}
Switch and Single Receptacle	⊖S
Switch and Duplex Receptacle	⊜S
Door Switch	S_D
Time Switch	S_T
Momentary Contact Switch	S_{MC}
Ceiling Pull Switch	Ⓢ
"Hand-Off-Auto" Control Switch	HOA
Multi-Speed Control Switch	M
Push Button	•

RECEPTACLE OUTLETS

Where weather proof, explosion proof, or other specific types of devices are to be required, use the upper-case subscript letters. For example, weather proof single or duplex receptacles would have the uppercase WP subscript letters noted alongside of the symbol. All outlets should be grounded.

Single Receptacle Outlet	⊖
Duplex Receptacle Outlet	⊜
Triplex Receptacle Outlet	⊕
Quadruplex Receptacle Outlet	⊕
Duplex Receptacle Outlet- Split Wired	⊖
Triplex Receptacle Outlet- Split Wired	⊖
250 Volt Receptable Single Phase Use Subscript Letter to Indicate Function (DW-Dishwasher; RA-Range, CD - Clothes Dryer) or numeral (with explanation in symbol schedule)	⊖
250 Volt Receptacle Three Phase	⊜
Clock Receptacle	Ⓒ
Fan Receptacle	Ⓕ
Floor Single Receptacle Outlet	⊟
Floor Duplex Receptacle Outlet	⊟
Floor Special-Purpose Outlet	◿ *
Floor Telephone Outlet - Public	◀
Floor Telephone Outlet - Private	◁

Example of the use of several floor outlet symbols to identify a 2, 3, or more gang floor outlet:

⊟◀◁.

Underfloor Duct and Junction Box for Triple, Double or Single Duct System as indicated by the number of parallel lines.

Example of use of various symbols to identify location of different types of outlets or connections for underfloor duct or cellular floor systems:

Cellular Floor Header Duct

---

*Use numeral keyed to explanation in drawing list of symbols to indicate usage.

FIGURE 418 Continued

CIRCUITING

Wiring Exposed (not in conduit) —— E ——

Wiring Concealed In Ceiling or Wall ————————

Wiring Concealed in Floor — — — —

Wiring Existing* - - - - - - - -

Wiring Turned Up ————————o

Wiring Turned Down ————————•

Branch Circuit Home Run to Panel Board. 2 1

    Number of arrows indicates number of circuits. (A number at each arrow may be used to identify circuit number.)**

BUS DUCTS AND WIREWAYS

Trolley Duct*** | T | | T |

Busway (Service, Feeder, or (Plug-in)*** | B | | B |

Cable Trough Ladder or Channel*** | C | | C |

Wireway*** | W | | W |

PANELBOARDS, SWITCHBOARDS AND RELATED EQUIPMENT

Flush Mounted Panelboard and Cabinet***

Surface Mounted Panelboard and Cabinet***

Switchboard, Power Control Center, Unit Substations (Should be drawn to scale)***

Flush Mounted Terminal Cabinet (In small scale drawings the TC may be indicated alongside the symbol)***

Surface Mounted Terminal Cabinet (In small scale drawings the TC may be indicated alongside the symbol)***

Pull Box (Identify in relation to Wiring System Section and Size)

Motor or Other Power Controller (May be a starter or contactor)***

Externally Operated Disconnection Switch***

Combination Controller and Disconnection Means***

POWER EQUIPMENT

Electric Motor (HP as indicated)

Power Transformer

Pothead (Cable Termination)

Circuit Element, e.g., Circuit Breaker

Circuit Breaker

Fusible Element

Single-Throw Knife Switch

Double-Throw Knife Switch

Ground

Battery

Contactor | C |

Photoelectric Cell | PE |

Voltage Cycles, Phase   Ex: 480/60/3

Relay | R |

Equipment (Connection (as noted)

---

*Note: Use heavy-weight line to identify service and feeders. Indicate empty conduit by notation CO (conduit only).
**Note: Any circuit without further identification indicates two-wire circuit. For a greater number of wires, indicate with cross lines, e.g.:

—|—|—|— 3 wires;   —|—|—|—|— 4 wires, etc.

Neutral wire may be shown longer. Unless indicated otherwise, the wire size of the circuit is the minimum size required by the specification. Identify different functions, of wiring system. e.g., signalling system by notation or other means.
***Identify by Notation or Schedule

FIGURE 418 Continued

534

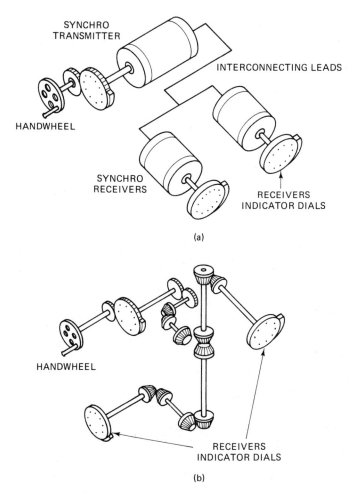

SYNCHRO
TRANSMITTER

INTERCONNECTING LEADS

HANDWHEEL

SYNCHRO
RECEIVERS

RECEIVERS
INDICATOR DIALS

(a)

HANDWHEEL

RECEIVERS
INDICATOR DIALS

(b)

FIGURE 419 Data transfer with synchros and data transfer with gears.

movable element of the device. It is similar to the armature in a motor. Stators
are stationary windings mounted about, but not in contact with, the rotor
winding at fixed intervals.

The laminations of the rotor core are stacked together and rigidly
mounted on a shaft. Slip rings are mounted on, but insulated from, the shaft.
The ends of the coil are connected to these slip rings. Brushes riding on the slip
rings provide continuous electrical contact during rotation, and low-friction
ball bearings permit the shaft to turn easily.

There are two common types of synchro motors now in use—the salient-
pole and the drum or wound rotor. The salient-pole rotor, which is frequently
called a *dumb-bell* or *bobbin* because of the shape of its core laminations, is

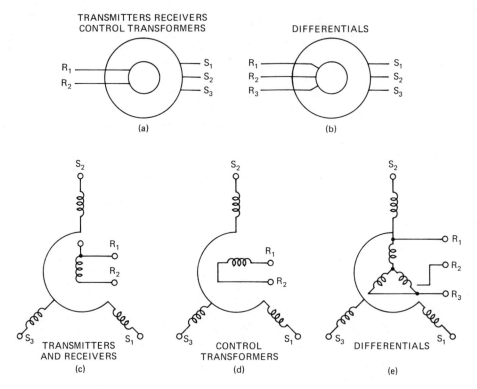

FIGURE 420 External connections and relative positions of windings for synchros.

shown in Fig. 421. During one complete excitation cycle, the magnetic polarity of the rotor changes from zero to maximum in one direction, to zero, reverses to maximum in the opposite direction, and returns to zero.

The drum or wound rotor is shown in Fig. 422. This type of rotor is used in most synchro control transformers. The control transformer is the output of a synchro control system. The winding of the wound rotor may consist of three

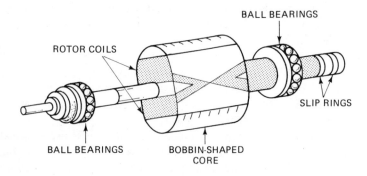

FIGURE 421 Salient-pole rotor.

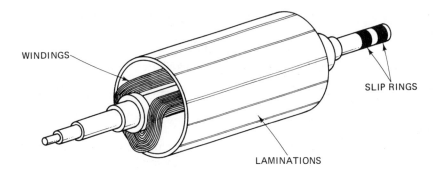

FIGURE 422 Drum or wound rotor.

coils so wound that their axes are displaced from each other by 120°. One end of each coil terminates at one of three slip rings on the shaft, while the other ends are connected together.

The stator of a synchro is a cylindrical structure of slotted laminations on which three Y-connected coils are wound with their axes 120° apart. Figure 423 shows a typical stator assembly. Stator windings are not connected directly to an ac source. Their excitation is supplied by the ac magnetic field of the rotor. The rotor is mounted so that it may turn within the stator. A cylindrical frame houses the assembled synchro. Standard synchros have an insulated terminal block secured to one end of the housing at which the internal connections to the rotor and stator terminate and to which external connections are made.

Synchronism: The simultaneous occurrence of any two events. Thus, two alternating currents or pressures are said to be in synchronism when they have the same frequency and are in phase.

Synchronous converter: A rotating machine similar to a motor or generator consisting of a field and an armature. One side of the armature is connected to an ac system through slip rings; the other side has a commutator

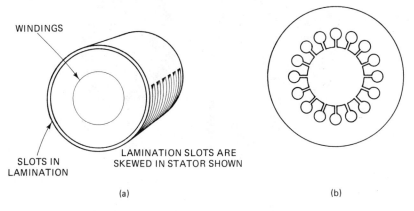

FIGURE 423 Typical stator and stator lamination.

connected to a dc system. The alternating current is converted to direct current by the commutator.

Converters are used less frequently than stationary rectifiers because their efficiency is much lower than that of a rectifier. Maintenance expenses and installation costs of the rectifier are low, making it more desirable than costly converters. For these reasons, together with the fact that they operate without noise, rectifiers can be used in locations that otherwise would not be considered.

A typical application of a converter is in emergency power and lighting systems where a dc motor fed from battery banks drives an ac generator. Another arrangement might have an ac motor driving a dc generator that is used to charge batteries.

The synchronous converter, also called a *rotary converter,* is a single machine that can convert large amounts of ac to dc power, or vice versa. It is considered more efficient and economical than a motor-generator converter.

Synchronous motor: A motor that requires a separate source of dc for the field and special starting components, including a salient-pole field with starting grid winding. The rotor of the conventional synchronous motor is essentially the same as that of the salient-pole ac generator. The stator windings of induction and synchronous motors are essentially the same. The stator of a synchronous motor is shown in Fig. 424a.

If supplied with proper voltage, a dc generator operates satisfactorily as a dc motor and there is practically no difference in the construction and rating of the two. Similarly, an ac generator becomes a synchronous motor if electric power is supplied to its terminals from an external source. Synchronous motor rotating fields are generally of the salient-pole type, as shown in Fig. 424b.

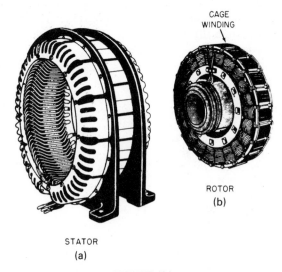

CAGE
WINDING

ROTOR
(b)

STATOR
(a)

FIGURE 424

Assume that two ac generators of the salient-pole type are operating in parallel and feeding the same bus. If the prime mover is disconnected from one of the generators, it will become a synchronous motor and continue to run at the same speed, drawing its power from the other ac generator.

With the exception of certain modifications to make its operation more efficient and also to make it self-starting, the synchronous motor is very similar to the salient-pole rotating-field ac generator. The rotor fields of both are separately excited from a dc source, and both run at synchronous speeds under varying load conditions. An eight-pole generator rotor revolving at 900 rpm generates 60 cycles/s. Likewise, an eight-pole synchronous motor supplied with 60-cycle current will rotate at 900 rpm.

A polyphase current is supplied to the stator winding of a synchronous motor and produces a rotating magnetic field the same as an induction motor. A direct current is supplied to the rotor winding, thus producing a fixed polarity at each pole. If it could be assumed that the rotor had no inertia and that no load of any kind were applied, the rotor would revolve in step with the revolving field as soon as power was applied to both of the windings. This, however, is not the case. The rotor has inertia; and in addition, there is a load.

The reason a synchronous motor has to be brought up to synchronous speed by special means may be understood from a consideration of Fig. 425. If the stator and rotor windings are energized, as the poles of the rotating magnetic field approach rotor poles of opposite polarity (Fig. 425a), the attracting force tends to turn the rotor in the direction opposite to that of the rotating field. As the rotor starts in this direction, the rotating field poles are leaving the rotor poles (Fig. 425b), and this tends to pull the rotor poles in the same direction as the rotating field. Thus, the rotating field tends to pull the rotor poles first in one direction and then in the other, with the result that the starting torque is zero.

Some type of starter must be used with the synchronous motor to bring the rotor up to synchronous speed. Although a small induction motor may be used, this is not generally done. Sometimes, if direct current is available, a dc motor coupled to the rotor shaft may be used. After synchronous speed is

TENDENCY OF ROTOR   TENDENCY OF ROTOR
TO TURN COUNTER-   TO TURN CLOCKWISE
CLOCKWISE

FIGURE 425 Synchronous motor.   (a)   (b)

attained, the dc motor is converted to operate as a generator to supply the necessary direct current to the rotor of the synchronous motor.

In general, however, another method is used to start the synchronous motor. A cage-rotor winding is placed on the rotor of the motor to make the machine self-starting, as in an induction motor. At start, the dc rotor field is deenergized and a reduced polyphase voltage is applied to the stator windings. Thus, the motor starts as an induction motor and comes up to a speed that is slightly less than synchronous speed. The rotor is then excited from the dc supply (generally, a dc generator mounted on the shaft), and the field rheostat is adjusted for minimum line current.

If the armature has the correct polarity at the instant synchronization is reached, the stator current will decrease when the excitation voltage is applied. If the armature has the incorrect polarity, the stator current will increase when the excitation voltage is applied. This is a transient condition, and if the excitation voltage is increased further, the motor will slip a pole and then come in step with the revolving field of the stator.

If the rotor dc field winding of the synchronous machine is open when the stator is energized, a high ac voltage will be induced in it because the rotating field sweeps through the large number of turns at synchronous speed. Therefore, it is necessary to connect a resistor of low resistance across the rotor dc field during the starting period. During this period, the dc field is disconnected from the source and the resistor is connected across the field terminals. This permits alternating current to flow in the dc field winding. Because the impedance of this winding is high compared with the inserted external resistance, the internal voltage drop limits the terminal voltage to a safe value.

Both the alternating currents induced in the rotor field winding and the cage-rotor winding during starting are effective in producing the starting torque. The torques produced by the rotor dc field winding and the cage-rotor winding at different speeds are shown by the curves T_r and T_s, respectively, in Fig. 426. Curve T is the sum of T_r and T_s and indicates the total torque at

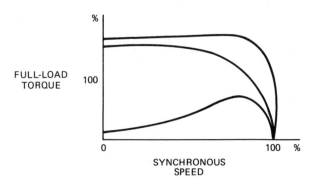

FIGURE 426

different periods during the starting period. Note that T_r is very effective in producing torque as the rotor approaches synchronous speed, but that both windings contribute no torque at synchronous speed because the induced voltage is zero and no dc excitation is applied to the dc winding.

The power factor of an induction motor depends on the load and varies with it. The power factor of a synchronous motor carrying a definite load may be unity or less than unity, either lagging or leading depending on the dc field strength.

T

Tabulator: A machine that reads information from one medium, e.g., cards, paper tape, or magnetic tape, and produces lists, tables, and totals on separate forms or continuous paper.

Tachometer: A speed indicator used to indicate or record the speed of a machine. The scales may be calibrated to read directly in revolutions per minute or feet per minute of any driven mechanism, such as a conveyor used in a fabric or paper mill drive. Tachometer scales may also be calibrated to indicate miles per hour.

An electrical tachometer consists of a small generator belted or geared to the unit whose speed is to be measured. The voltage produced in the generator varies directly as the speed of the rotating part of the generator. Since this speed is directly proportional to the speed of the machine under test, the amount of the generated voltage is a measure of the speed. The generator is electrically connected to an indicating or recording instrument that is calibrated to indicate units of speed. An electrical tachometer is particularly well suited to the remote indication of the speed of a unit.

Tank circuit: A parallel LC circuit that operates by a periodic transfer of energy from one component to the other during the first half-cycle and reverses this procedure during the second cycle of operation. In other words, a complete cycle of operation has taken place when the energy of one component has been transferred to the other component and then back to the original component. The current that flows within the tank circuit is called *circulating current*. The current that flows in the external circuit between the source and the tank circuit is called *line current*.

Since the circulating current is common to both components, the inductor and capacitor can be said to act as if they were connected in series as far as tank circuit action is concerned. In order to understand the circulating current and the action of energy transfer between the components in a tank circuit fully, a step-by-step analysis is made of an ideal tank circuit in Fig. 427.

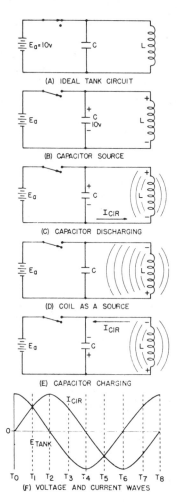

(A) IDEAL TANK CIRCUIT

(B) CAPACITOR SOURCE

(C) CAPACITOR DISCHARGING

(D) COIL AS A SOURCE

(E) CAPACITOR CHARGING

(F) VOLTAGE AND CURRENT WAVES

FIGURE 427   Tank circuit action.

When a voltage is applied to a coil, during the first instant, a maximum CEMF will be produced and virtually no current will flow through the coil. In other words, the coil represents an open circuit or very high impedance. A capacitor, on the other hand, represents a very low-impedance circuit during the first instant because maximum current flows. It is also known that if a capacitive circuit contains no series resistance, the charging time of the capacitor will be almost instantaneous.

It can be seen that if the switch in Fig. 427a were closed, the following sequence of events would occur. The instant the switch is closed, the coil will develop a CEMF almost equal to the applied voltage, and virtually no current will flow through the inductive branch. Since the coil will appear as an open component at the first instant, the source will only see a capacitor connected across its terminals. The capacitor, having no resistance in series with it, will charge to the source potential almost instantaneously.

At the exact instant the capacitor has completed its charge, the switch is opened and all circuit action is stopped (merely for the purpose of explanation). The action so far has taken only an instant. In fact, the capacitor has assumed its charge so rapidly that no significant current has started to flow through the inductive branch. Therefore, no field has been established around the coil.

The conditions now existing in the circuit are illustrated in Fig. 427b. With E_a disconnected, the capacitor will now act as the source. The voltage and current conditions are illustrated in Fig. 427f. At time zero, voltage across the tank (E_{tank}) is maximum, and since the capacitor has not started to discharge, the circulating current (I_{cir}) is zero.

The capacitor will now commence to discharge through the coil. The coil will produce a CEMF of polarity opposite to the capacitor's polarity. This CEMF, the capacitor voltage, and hence, the circuit voltage (E_{tank}) will decrease toward zero as the capacitor discharges. Since CEMF across the coil (the opposition of the coil to current flow) is decreasing, the current flow (I_{cir}) will increase. Figure 427c shows the action of the circuit after the capacitor has been discharging for a short time. The capacitor's charge has decreased, while the increasing current flow has begun to establish a field around the coil. The voltage and current conditions correspond to time one (t_1) in Fig. 427f.

The above action will continue until the capacitor has discharged completely. Since the circuit is assumed to have no losses, all the energy that was contained in the capacitor has now been transferred to the inductor. Originally, this energy was stored in the electrostatic field of the capacitor, and it is now stored in the electromagnetic field of the inductor. Since E_{tank} leads the circulating current by 90°, I_{cir} will be maximum when E is minimum, as at t_2 in Fig. 427f.

At the instant the capacitor is discharged completely, the circulating current will attempt to cease flowing. Since inductance opposes any change in current, the field of the coil will begin collapsing in order to maintain current flow in the same direction. The inductor is now acting as the source for the circuit, and in accordance with the inductive theory, the collapsing field will induce a voltage in the coil with the polarity shown in Fig. 427d.

It can be seen from the current waveform in Fig. 427f that the rate of change of current at t_2 is zero. However, an instant after t_2, the rate of change of current will begin increasing along with induced voltage (E_{tank}). As the inductor field continues to collapse, keeping I_{cir} flowing, the capacitor is assuming a charge with a polarity opposite to its original polarity. This is illustrated in Fig. 427e and the waveform of Fig. 427f at t_3, where the current is still flowing in the positive direction (even though decreasing), while the voltage has changed polarity and is increasing in a negative direction.

The above action will continue until t_4, when the inductor field has completely collapsed and the capacitor is fully charged. At this time, the circuit has completed a half-cycle of operation, the voltage is again maximum (negative), and the current is minimum. The conditions of the circuit are similar to those of Fig. 427b, with the exception that all polarities are reversed.

An instant after t_4, the capacitor will begin to discharge again, but this time in the direction opposite to the original discharge. The arrow in Fig. 427c would be reversed. The action of the circuit between t_4 and t_6 in Fig. 427f will be similar to the action described between t_0 and t_2. In other words, the capacitor will discharge completely and the increasing current will establish a field around the coil. At t_6, the circuit will again appear as in Fig. 427d, with the polarity of the coil reversed. The inductive field will again collapse in order to maintain current flow in the same direction (clockwise). At t_7, the action of the circuit is similar to Fig. 427e, where the inductor field is partially collapsed and the capacitor is partially charged (polarities opposite to those shown). This action continues until t_8, where the circuit is again equal to Fig. 427b, with the capacitor fully charged to the original potential and the inductive field fully collapsed.

The capacitor has assumed its original potential because this circuit, being ideal, is assumed to have no losses. At t_8, the conditions of the circuit are identical to those of t_0. In other words, the circulating current is zero and the tank voltage is maximum.

The tank circuit has gone through a complete cycle of operation. The waveforms in Fig. 427f are seen to be in the form of a sine wave.

Tap changer: A tap-changing mechanism that regulates the induced voltage in a transformer. The tap changer operates either when the transformer is deenergized (no-load tap changer or tap changer for deenergized operation), or when the transformer is energized (load tap changer). The changing of the tap position on the transformer winding changes the number of turns active in this winding and thus regulates the voltage in the winding.

The standard location of the tap-changer operating handle of power transformers is on the side of the case, so that its position can be observed from the ground. Some transformers have the tap-changer handle located on the cover or accessible through a handhole in the cover.

Telemetering: The transmission of a measurement over long distances, usually by electromagnetic means. Telemetering is an operation not only of increasingly complex electrical transmission networks, but of pipelines for oil and gas, of waterworks systems, and of widely extended industrial plants requiring that accurate information regarding the magnitude of measured quantities and the positions of mechanisms be available at remote locations, such as a central control room or a dispatcher's office. Telemetering is actually a displaying of measurements performed at a remote location and transmitted electrically to the place of display.

The prefix *tele* means distance, so telemetering is a measuring at a distance. While telemetering covers situations in which the indicating devices are located as much as several hundred miles from the circuit being measured, the distance factor alone does not determine whether or not a measuring system is a telemetering system. For example, a measuring system in which the measuring instruments are a few hundred feet from their auxiliary devices is not a telemetering system if the instruments are connected directly to their shunts, multipliers, or transformers. Another measuring system in which the elements are located within a distance of only a few feet is considered a telemetering system if there is no interconnecting physical link.

Any telemetering system is made up of a measuring element such as a voltmeter or ammeter, a transmitting element or transmitter that develops an electrical signal representative of the measured magnitude, a receiving element or receiver that translates the transmitted signal into a suitable indication or record, and a telemetering circuit or link.

The function of the elements in a telemetering system is illustrated in Fig. 428. The quantity to be measured may be an electrical quantity such as current, voltage, or power; or it may be pressure, liquid flow, water level, temperature, or another nonelectrical quantity. The quantity is first measured and then converted by the transmitter (1) into an electrical signal proportional to the measured quantity. The signal should be of the form most suitable for transmission over the telemetering link. The distance over which the signal is to be transmitted may vary from a few feet to hundreds of miles.

The connecting link (2) is also called the *telemetering channel.* This term is used to identify one or more conductors that carry the telemetering signal. Power lines or telephone lines may be used as channels for carrier currents of high frequency, but care must be taken so that the telemetering signal does not interfere with the signal already carried by the lines, and that it is not affected by that signal.

In recent years, there has been a sharp increase in the application of telemetering systems to measurements in the fields of aircraft, meteorology, and guided missiles. In such applications, there can be no interconnecting physical link, so microwave transmission that does not require conductors between stations is used. Remote measurement using microwaves or radio as the connecting link is generally known as *mobile telemetering.* Except for the connecting link, mobile telemetering is similar to telemetering that sends the signal over wires.

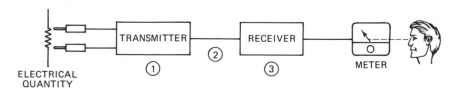

FIGURE 428   Telemetering.

The last element in a telemetering system is the receiver (3). The receiver may be of the recording or indicating type. The indications of variations in the transmitted signal are interpreted as variations of the quantity being measured. The receiver is placed in an office or central headquarters, where the information about the telemetered machinery is needed.

Telemetering systems are classified according to the means used to represent the measured quantities into five distinct groups: current systems, impulse systems, position systems, frequency systems, and voltage systems.

A current system of telemetering is one in which the measured quantities at the transmitting end are translated into representative values of current and sent over the telemeter lines to a point at which the receiving apparatus is located.

In the impulse system of telemetering, a current (usually direct) is employed in a closed-circuit arrangement. This current is interrupted so that either the rate corresponds to the quantity being measured or the spacing between current impulses is proportional to the quantity being measured, or the duration of the current impulses is varied according to the quantity being measured.

The position system generally requires a link composed of at least three conductors and operates by varying the ratio of two or more voltages in the three-wire circuit. The receiving instrument is responsive to this ratio and tends to duplicate the position of the transmitting member that controls the relative value of the voltages.

In the frequency system of telemetering, the frequency of an ac source is varied according to the quantity being measured. The receiving instrument generally takes the form of a frequency meter calibrated in terms of the measured quantity. Because of the ease with which the frequency of an electronically controlled circuit can be adjusted and maintained, this system enjoys a growing popularity in many branches of telemetering.

In the voltage system, voltages generally on the order of millivolts and proportional to the quantity being measured are indicated at a remote position by a millivoltmeter, or more commonly by a self-balancing potentiometer.

Telephone system: A communications system in which conductors carry information between two telephone instruments, each capable of sending and receiving.

The telephone instruments do not actually carry the sound themselves, but rather produce it by means of electric current impulses. The basic telephone circuit uses the principle that a change in resistance in any series circuit will cause a change in current in that circuit.

The device that is used to change sound into changing resistance is called a *transmitter*. This is contained in the part of a telephone that one talks into. Thus, a typical telephone circuit will consist of a series circuit in which the transmitter is used to cause changing current whenever it is spoken into.

To have sound transmitted to another location, a device that produces sound whenever a changing current is passed through it must be used. This device is called a *receiver*. It is an electric magnetic with a moving piece of metal called a *diaphragm*. The voice transmission is accomplished by means of transmission lines (telephone lines). They are sent or transmitted in the form of electrical energy through a telephone line, from the transmitter of one telephone to the receiver of another telephone.

Sound is transmitted by means of waves in the air. These air waves may be set up by one's voice, the clapping of hands, or anything that causes a disturbance of the air. Different sounds have waves of different volume and frequency. For example, a loud sound, such as the report from a gun, has waves of greater energy than a low or feeble sound of, say, a whisper. A high-pitched sound has waves of high frequency, and a low sound has waves of a lower frequency. These waves, upon striking the eardrum, cause the drum to vibrate and transmit impressions of various sounds to the body's nerves and brain, thus enabling the person to hear.

In order to be heard by the ordinary human ear, sound waves must be between 16 and 15,000 waves/s; that is, 16 to 15,000 Hz. Sounds within these frequencies are audible to the average human ear. Sound waves travel at about 1100 ft/s through the air and about 4700 ft/s under water. Ordinary sounds can be heard at distances of only a few hundred feet to only a few miles for the very loudest sounds.

Electricity travels at the rate of 186,000 mi/s and can be transmitted over hundreds of miles without much loss. Therefore, if sound wave energy is changed into electrical impulses and these impulses are used to reproduce the sounds at a distance, both the speed and the distance at which sound can be transmitted can be greatly increased. This is exactly what a telephone system does.

In modern equipment, when the telephone handset is lifted to make a call, switch contacts in the telephone set close to cause a circuit to be established between the telephone set and the central office equipment. The central office equipment, in turn, connects the telephone set through its switching equipment to a dial tone generator, which sounds a dial tone, letting the calling party know that the circuit has been established so that dialing can commence.

A telephone ringer provides an audible ringing bell signal to a called party on an incoming telephone call. The ringer in a telephone operates when ringing current is placed on the telephone line from an ac ringing generator, which is located in the telephone central office. The ac ringing voltage is between 85 and 105 V at a low frequency of between 20 and 66 Hz. The ringer consists of a coil of wire wound on a magnetic core, an armature, a permanent magnet, and a bell.

Central offices or telephone exchanges serve to connect telephones of one line to those of other lines. There are thousands of these central exchanges throughout this country and abroad to handle the many millions of telephones in use. The exchange in one city or town handles the calls of the subscribers in that city and surrounding territory. The exchange is connected by transmission lines to exchanges of other cities and can complete a circuit for one of its own subscribers through the exchange of another town, to a telephone several thousand miles away, or in fact, to practically any area in the world.

The vast network requires many types of elaborate exchange circuits. Almost all telephone exchanges in use today are of the automatic type. There may be a few rural communities that still use manual exchanges, but these are rapidly being replaced with the automatic types. The general function of either type is to receive a signal from the calling party, get a connection, and ring the party on any other line as quickly as possible. With the automatic exchange, these operations are performed by electrical and mechanical equipment.

On automatic equipment, when the calling party dials, a number of impulses are sent to magnets and relays at the central office, causing them to move a selector element, which picks out the desired line. Other parts of the mechanism then test the line to determine whether it is busy or not. If it is clear, an automatic switch starts ringing the called party.

Most telephone circuits in use today utilize one insulated wire. The other side of the circuit is completed through the earth by carefully made ground connections. Some lines which utilize a two-wire metallic talking circuit use a ground circuit for ringing.

When telephone lines run parallel to power lines, they often pick up, by magnetic induction, an interfacing hum. To avoid this, the pairs of wires should occasionally be crossed into opposite positions on the poles so that one wire will not be closest to the transmission line throughout its entire length.

This crossing of wires to prevent induced interference is known as *transposition*. Sometimes it is done to avoid induction from other telephone wires. Transposing the wires frequently and evenly will balance out most induction. Telephone wires should never be left close enough to high-voltage power lines so that there would be danger of their coming in contact with each other.

Satisfactory telephone operation depends, to quite an extent, on proper line construction. Therefore, all telephone lines should be made with the proper materials, the wires properly spliced with low-resistance joints, ground connections kept in good condition, and all proper precautions taken.

Television transmission: Television synchronizing pulses, generated by a crystal-controlled oscillator operating at a frequency of 31.5 kHz, which are locked to the 60-Hz local power line frequency with an automatic frequency control (AFC) circuit. The equalizing pulses are formed by the 31.5-kHz signal, permitting a pulse repetition rate of 31,500 pulses/s. Utilizing a 2 to 1 multi, the original frequency is divided by 2 to provide the 15,750-Hz pulse for horizontal scanning. Through additional frequency-dividing multis, the 60-Hz pulse for comparison to the power line frequency is obtained. The AFC circuit assures correction of the 31.5-kHz oscillator when necessary. The entire system is phase-locked to the line frequency, which prevents ripple in the receiver from becoming noticeable to the viewer.

By the use of several pulse-shaping circuits, the sync pulses are reformed with the steep sides, and flattops are required in order to meet FCC standards for good operation. The blanking pulses are also formed in the sync generator, along with numerous other timing pulses used by cameras and studio amplifying equipment. Only 75 percent of the carrier envelope is used for the camera signal and blanking, the remainder being reserved for the sync pulses. Since these are located in the blacker-than-black region, they are never visible on the television screen.

A vertical sync pulse is transmitted at the termination of each field to guarantee the start of vertical sweep at the camera and receiver. It is necessary for the receiver to be able to segregate vertical and horizontal sync pulses for routing to their respective scanning generators. The vertical pulse is given a duration time that is 19 times as long as the horizontal pulse. Thus, the two can be readily separated with integrating and differentiating circuitry. The rise time and decay time are important factors in evaluating the usefulness of the pulse in a television receiver.

Although vertical and horizontal sync pulses are used for two widely different purposes, they differ only in duration. Both are removed from the square-wave portion of the video signal by the differentiator and integrator. Since the sides of the square wave are vertical, time is not involved. Only during the on time of the pulse are the derived pulses generated with the polarity reversed during the instantaneous vertical portions. The application of a square-wave voltage to the series RC circuit causes an inrush of current, forming differentiated pulses. The effect of this current on the voltage developed across the capacitor forms the integrated pulses. The differentiated pulse, therefore, is the instantaneous IR drop across a resistor due to the capacitor charging current. The integrated pulse is the voltage developed across the capacitor over a period of time due to the cumulative current flowing in said capacitor.

Blanking pulses (needed to remove retrace lines) are transmitted at the end of each scanned line and at the end of each field. These pulses swing the grid of the CRT negative to the black level, cutting the beam off until retrace is completed. Sixteen percent of the horizontal line scanning time is used by the horizontal blanking pulse, and 8 percent of the frame scanning period is used by the vertical blanking pulse. Thus 42 horizontal lines are blanked out, leaving 483 active.

Terminal: A connection point in a circuit that may be a specialized mechanical component designed specifically for the connection of a conductor, or any general point of a circuit to which a connection is made. A terminal is often a solder lug, metal post, or other type of conductor, to which a solder connection may be made or an input device connected. Terminals are often associated with input and output stages of a circuit and especially with the connection of a power source.

Terminal block: A block used as a connecting and distributing point for electrical circuits. Their construction is simple and in most cases, they consist of a strip of insulation with connections already mounted. Little preventive maintenance is required on terminal blocks.

Terminal board: An insulated board usually enclosed in some type of metal housing carrying several lugs, tabs, or screws as terminals. A terminal board is shown in Fig. 429. These devices are used quite extensively in communications systems, security/fire alarm systems, and in control wiring of all types.

Tesla: A unit of magnetic flux density whose symbol is T. One tesla equals 1 Wb/m^2, or 10^4 G. This unit is named for Nikola Tesla, who was instrumental in developing the tesla coil.

Thermal noise: A type of noise that occurs in UHF amplifiers due to thermal agitation of the electrons within the conductor. The ultra high frequency requires the use of good noise characteristics and good gain.

Thermal protector: A protective device for assembly as an integral part of a motor or motor compressor and which, when properly applied, protects the motor against dangerous overheating due to overload and failure to start.

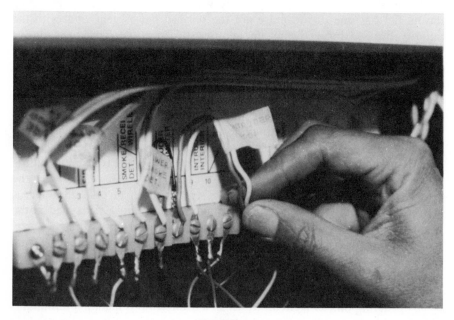

FIGURE 429

Thermal runaway: A heating effect especially applicable to transistors and which is caused by the combination of collector current drain and the temperature buildup that results. Temperature rise is a problem in transistors, because as the temperature increases, it tends to increase the flow of electrons in both N- and P-type semiconductor materials. This causes more current to flow through the transistor for the same applied voltage. As more current flows, more heat develops. This process pyramids until the transistor is destroyed.

Thermal runaway can be created by a defective transistor or by improperly operating circuits connected to the transistor. If the former is the cause, device replacement is necessary. When the circuit is at fault, correction of the defect(s) should bring about normal operation. As the operating temperature of the transistor decreases, the electron flow returns to normal.

Thermal transducer: A device used to convert a temperature change into an electrical current that corresponds to the fluctuation. Monitoring devices that measure the potential set up by the transducer are used as indication devices. Thermal transducers measure temperature by the contact potentials generated when two dissimilar metals are joined. Thermal transducers are used as remote temperature sensors in electronic devices and especially in biomedical applications. Here, the transducer is usually very small and may be inserted into the human physiological system.

Thermionic emission: The emission of electrons by a heated conductor. When a piece of metal is heated, the electrons become agitated and travel about at greatly increased speeds. Those electrons that are near the surface try occasionally to leave the metal. When they encounter the mass of the relatively enormous air molecules, they are repelled and return to the metal. If, however, the hot metal is placed in a vacuum, the free electrons find it much easier to leave the metal and get out into the surrounding space. They soon lose their energy and return to the metal, since there is nothing to attract them elsewhere. This activity is what is referred to as *thermionic emission.*

Thermistor: A special semiconductor device that functions as a thermally sensitive resistor whose resistance varies inversely with temperature. Thermistors have large negative temperature coefficients; that is, as the temperature rises, resistance of the thermistor decreases; and as the temperature drops, their resistance increases. The resistance of a thermistor is varied not only by ambient temperature changes, but also by heat generated internally by the passage of current.

Since the thermistor is basically a variable resistor, it is usually constructed from semiconductor material of greater resistivity than is used in transistors or semiconductor diodes. Therefore, the thermistor's response to ambient temperature variations does not track equally with that of the transistor semiconductor, so compensation is achieved only at a few points of correspondence. As a result, the thermistor's greatest usage is in the field of temperature controls and measurements and power-measuring equipment based on heating effect, such as RF-measuring microwave equipment.

Although the thermistor is a semiconductor, it does not have intrinsic amplification capability, as does the transistor. It is used in thermal compensating circuits for transistor stabilization.

Thermocouple: A device that produces an electrical potential purely by thermal means. A thermocouple circuit provides a means for the replenishment of electrons on a continuous basis from the material that gains the abundance of electrons to the material that exhibits an electron deficiency. The direction of electron movement is dependent on the potential difference between the metals used in its construction, which is dependent on the difference of thermal effect or temperature at their junctions.

Thermocouple-type meter: A type of instrument whose junction is heated electrically by the flow of current through a heater element. It does not matter whether the current is alternating or direct, because the heating effect is independent of current direction. The maximum current that may be measured depends on the current rating of the heater, the heat that the thermocouple can stand without being damaged, and the current rating of the meter used with the thermocouple. Voltage may also be measured if a suitable resistor is placed in series with the heater.

A simplified schematic diagram of one type of thermocouple is shown in Fig. 430. The input current flows through the heater strip via the terminal blocks. The function of the heater strip is to heat the thermocouple, which is composed of a junction of two dissimilar wires welded to the heater strip. The

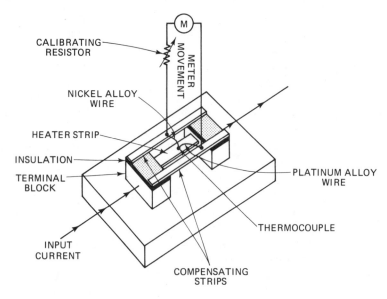

FIGURE 430

open ends of these wires are connected to the center of two copper compensating strips. The function of these strips is to radiate heat so that the open ends of the wires will be much cooler than the junction end of the wires, thus permitting a higher voltage to be developed across the open ends of the thermocouple. The compensating strips are thermally and electrically insulated from the terminal blocks.

The heat produced by the flow of line current through the heater strip is proportional to the square of the heating current. Because the voltage appearing across the two open terminals is proportional to the temperature, the movement of the meter element connected across these terminals is proportional to the square of the current flowing through the heater element. The scale of the meter is crowded near the zero end and is progressively less crowded near the maximum end of the scale. Because the lower portion of the scale is crowded, the reading is necessarily less accurate. For the sake of accuracy in making a given measurement, it is desirable to choose a meter in which the deflection will extend at least to the more open portion of the scale.

The meter used with the thermocouple should have low resistance to match the low resistance of the thermocouple, and it must deflect full scale when rated current flows through the heater. Because the resistance must be low and the sensitivity high, the moving element must be light.

A more nearly uniform meter scale may be obtained if the permanent magnet of the meter is constructed so that as the coil rotates (needle moves up scale), it moves into a magnetic field of less and less density. The torque then increases approximately as the first power of the current instead of as the square of the current, and a more linear scale is achieved.

If the thermocouple is burned out by excessive current through the heater strip, it may be replaced and the meter recalibrated by means of the calibrating variable resistor.

Thevenin's theorem: A theorem that states that any linear network of impedance and sources, if viewed from any two points in the network, can be replaced by an equivalent impedance Z_{th} in series with an equivalent voltage source E_{th}. According to this theorem, any linear dc circuit, regardless of its complexity, can be replaced by a Thevenin's equivalent, which is shown in Fig. 431. The process whereby a Thevenin's equivalent circuit is developed for a given network is best illustrated by an example.

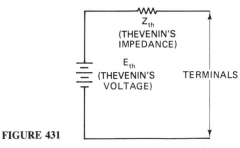

FIGURE 431

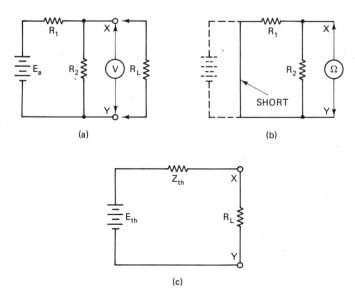

FIGURE 432

Assume that the circuit in Fig. 432a is to be used to develop a Thevenin's equivalent circuit. Basically, the problem consists of finding values for E_{th} and Z_{th}. These quantities can be found using the following procedure.

1. Disconnect the section of the circuit considered as the load (R_L in Fig. 432a).
2. By measurement or calculation, determine the voltage that would appear between the load terminals with the load disconnected (terminals X and Y). This open circuit voltage is called Thevenin's voltage (E_{th}).
3. Replace each source within the circuit by its internal impedance. (A constant voltage source such as a battery is replaced with a short, while a constant current source is replaced with an open.) See Fig. 432b.
4. By measurement or calculation, determine the impedance (resistance) the load would see, looking back into the network from the load terminals. See Fig. 432b. This is Thevenin's impedance (Z_{th}).
5. Draw the equivalent circuit consisting of R_L and Z_{th} in series connected across source E_{th} (Fig. 432c). Solve for the load current and voltage.

Thyratron: A gas triode or gas tetrode tube that is employed primarily for electronic switching and control purposes. The thyratron differs from the vacuum tube with regard to the actions each performs. Basically, the thyratron control grid is the main controlling element. Anode current starts to flow abruptly when grid voltage reaches a specified value. At this point, the grid provides no further control. Anode current will continue to flow until anode voltage is interrupted or reversed. This means that grid control is possible for starting the flow of anode current, but cannot cease its flow once it has begun.

Thyratrons were often used as rectifiers in variable voltage power supplies several decades ago. These were capable of handling output voltages in the kilovolt range while passing a moderate amount of current. Today, the thyratron has been all but eliminated by solid-state rectifiers including diodes, silicon-controlled rectifiers (SCRs), and integrated circuits.

Thyristor: Any semiconductor switch whose bistable action depends upon PNPN regenerative feedback. Thyristors can be two-, three-, or four-terminal devices, and both the unidirectional and bidirectional devices are available.

The silicon-controlled rectifier (SCR) is by far the best known member of the thyristor family. Because it is a three-terminal device (anode, cathode, and control gate) that can conduct current in one direction only (anode to cathode), the SCR is classified as a reverse-blocking triode thyristor.

Some other reverse-blocking triode thyristors are the silicon unilateral switch (SUS), the light-activated silicon-controlled rectifier (LASCR), the complementary SCR (CSCR), the gate turn-off switch (GTO), and the programmable unijunction transistor (PUT).

Time, acceleration: The time between the interpretation of instructions to read or write on tape and the transfer of information to or from the tape into storage, or from storage onto tape, as the case may be.

Time-division multiplexing: The simultaneous transmission of several intelligible signals using only a single transmitting carrier signal. To a great extent, the maximum permissible number of intelligible transmissions taking place in the radio spectrum per unit of time is being increased through the use of multiplexing. Multiplexing involves the transmission of several intelligible signals on the same frequency during the same period of time normally required for the transmission of a single signal. Either of two methods of multiplexing may be used. These are time-division and frequency-division multiplexing.

Toggle switch: A switch that possesses a mechanism that snaps into the on or off position at the opposite extremes to which the controlling lever is moved. The basic toggle switch contains a lever that is connected to a spring-loaded toggle linkage. This produces a rapid contact transfer.

Shown in Fig. 433, the toggle switch is often used for such noncritical purposes as on/off control of primary power to an electrical or electronic device and other general make-break purposes. There are several different types that offer the standard on/off function. Others may have a center position that may be switched to the left to engage one circuit or to the right to engage another.

FIGURE 433 Toggle switch.

Toggle switches are normally rated as to voltage and current-handling capabilities and for particular switching arrangements. For example, a SPST switch has a single pole-single throw action. A DPST type is actually two SPST devices in parallel. The DPST switch (double pole-single throw) contains two poles or sets of contacts. Most toggle switches must be manually turned to the off position after being activated, but some are designed with momentary actions, which are engaged only when held in the on position by the hand. When released, these switches automatically return to the normally off position by means of spring tension.

Toggle switches are commercially available in a variety of physical sizes. The larger devices are usually designed with contacts and internal mechanisms large enough to handle moderate to heavy current. Miniature switches may be rated at 1 A or less and are intended for use in small solid-state circuits.

Torque: A twisting force that is generally associated with electric motors. The action of a basic motor can be better understood by referring to Fig. 434. A loop of wire is shown here suspended in a magnetic field. This loop of wire is known as an *armature*. The loop is connected to a commutator-brush assembly. The purpose of this assembly is to provide a contact area between the movable loop and the stationary dc source. Notice that one portion of the loop is connected to the commutator segment designated X, and the other portion of the loop is connected to the commutator segment designated segment Y. The commutator segments are insulated from one another. Segment Y is connected to brush A. Brush A, in turn, is connected to the negative terminal of the

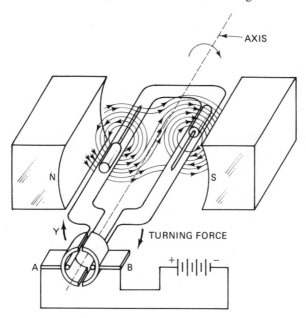

FIGURE 434 Basic DC motor action.

battery. As the commutator segments turn, they will each be in contact with one of the brushes.

Assume the starting position illustrated. Commutator segment Y is in contact with brush A, and the other segment, X, is in contact with brush B. When current is permitted to flow in the direction indicated, fields are established about both sides of the loop of wire. The direction of current flow causes a magnetic field to exist in a direction which will cause the loop to start to rotate in a clockwise direction. In the left-hand loop, the fields are aiding and repelling at the bottom and opposing and attracting at the top. Therefore, the left-hand loop will move in an upward and clockwise direction.

The current flowing through the right-hand portion of the loop causes an opposing and attracting field at the bottom, resulting in a downward and clockwise motion. Therefore, the action of the left-hand portion is aided by the action of the right-hand loop. A shaft is mounted at the axis of the loop, allowing it to rotate freely. Initially, when the loop starts to rotate, it does so with a twisting force, which is known as *torque*.

The torque will cause the conductors (loop) to rotate in the field. However, when the conductors reach a certain point in their travel (90° after the indicating starting position), they will be parallel to the lines of force established by the magnet. It would seem that at this position, all motion would stop. This statement would be true if the conductors did not possess momentum. The actual condition is that the conductors possess sufficient momentum to ride past this point. Notice the position of both loops and the commutators at this time, as shown in Fig. 435. The loop is perpendicular to the lines of flux and the commutator bars are not making contact with the brushes. However, the momentum of the loop will be sufficient to cause the loop to pass through this point and reestablish contact with the commutator bars. Notice also that the commutator bars will now be connected to the opposite brushes. However, current flow will still be into brush A and out of brush B.

The armature conductors for a motor are assembled in coils and connected to the commutator assembly. Current flows in one direction in the conductors under the north pole and in the opposite direction in the conductors under the south pole. To develop a continuous motor torque, the current in a coil must reverse when the coil passes the dead center position (top and bottom). The function of the commutator is to reverse the current flow at the

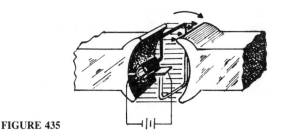

FIGURE 435

proper time to maintain current flow in all conductors under a given pole. The total torque is the arithmetic sum of the individual torques contributed by all the armature conductors.

When the speed of a motor is constant, the generated torque due to the armature current is just equal to the retarding torque caused by the combined effect of the friction losses in the motor and the mechanical load.

Transducer: A device that converts energy from one form to another; e.g., a quartz crystal embedded in mercury changing electrical energy to sound energy, as is done in sonic delay lines in computer storage systems.

Transformer: In its simplest form, two coils or windings on a single magnetic core. Transformers are used on alternating currents to either increase or decrease the voltage in one of the windings in relationship to the other winding. This ability of the transformer makes it possible to use generators that produce moderately high voltages and to change over to a very high voltage and proportionately small current in the transmission lines. These high voltages can then be reduced at the point of usage for safe levels of voltages.

The elementary principle of transformer operation is shown in Fig. 436. Here, there are two windings on opposite sides of a ring-like core made of iron. The source of alternating current and voltage is connected to the primary winding of the transformer, and the secondary winding is connected to the circuit in which there is to be a higher voltage and smaller current, or else a larger current and smaller voltage than in the primary. If there are more turns on the primary than on the secondary winding, the primary voltage will be higher than the secondary voltage. The secondary current will also be propor-tionately larger than the primary current.

Alternating current is continually changing, continually increasing and decreasing in value. Every change of alternating current in the primary winding of the transformer produces a similar change of flux in the core. Every change of flux in the core and every corresponding movement of magnetic field around the core produces a similarly changing movement of magnetic field around the core, produces a similarly changing electromotive force in the secondary winding, and causes an alternating current to flow in the circuit that is connected to the secondary. The ratio of the voltage in the secondary as compared to that in the primary coil depends on the amount of magnetic flux, the frequency of the alternating current, and mainly on the number of turns in the coils.

The only current that flows in the primary coil or windings is the magnetizing current necessary to set up the flux in the closed magnetic circuit.

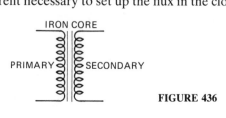

FIGURE 436

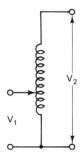

FIGURE 437   Autotransformer.

This current is usually a very small percentage of the full-load primary current of the transformer.

In a well-designed transformer, there is very little magnetic leakage. Such a leakage will cause a decrease of secondary voltage when the transformer is loaded. When a current flows through the secondary in phase with the secondary voltage, a corresponding current flows through the primary in addition to the magnetizing current. The magnetizing effects of the two currents are equal and opposite.

Autotransformers (Fig. 437) have only one coil, any portion of which may be used as the primary and any portion of which may be used as the secondary. The ratio of transformation depends on the portions used. That is, if the whole winding is used as the primary winding and one-third as the secondary and the losses are neglected, the voltage of the primary will equal three times the voltage of the secondary, and the current of the secondary equals three times the current of the primary.

Since the primary and secondary coils are in direct electrical contact in an autotransformer, either coil is exposed to the pressure that may exist in the other. Therefore, it would not be safe to use an autotransformer to reduce high voltage for any purpose where the high voltage would be dangerous. Autotransformers for a given capacity are considerably smaller and less expensive to build than conventional transformers. They are used where there is little difference between the primary and secondary voltages or where there is no danger incurred by the use of either.

In a series, or current, transformer (Fig. 438), the primary coil is

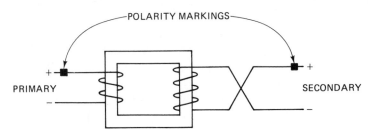

FIGURE 438   Series transformer.

connected in series with the circuit, and the secondary is connected in series with the device to which the current is supplied. As the current through the primary increases, the magnetization in the core increases, thus increasing the voltage in the secondary. If the resistance of the secondary circuit is fixed, the secondary current will increase in proportion to the secondary voltage and also in proportion to the current in the main circuit. Current transformers are used frequently on ac metering devices for operating wattmeters where the current is 200 A or higher.

Electric service voltages of 480/277 V are quite common on large commercial and industrial buildings. Therefore, some provision must be made to furnish normal 120-V power for operating convenience outlets and other equipment rated at 208 or 120 V. The most convenient way of accomplishing this transformation is to use dry transformers to reduce the 480/277-V system to 208/120 V. These transformers normally feed lighting and power panelboards for the branch circuit distribution.

A power transformer is a device that is normally used in conjunction with rectifier and filter circuits to ultimately provide a dc output voltage. The term *power transformer* is a general one and may include those devices that accept ac power from the mains and output many different voltages at the secondary. Some use multiple secondary windings to produce a moderate to high voltage and several lower voltages. Some may contain only a single secondary winding, which may output values in the low, medium, and high range. Again, the term is most often applied to transformers used to drive rectifier circuits that output direct current.

When transformer ratios do not give quite the right voltage or when voltage drop is excessive, a booster transformer may be used in the primary circuit to raise or alter the voltage the required amount. Assume that an electric service gives 110/220 V and is a three-wire, single-phase connection. A 10 percent booster transformer connected in each phase will raise the normal voltage to approximately 120/240, the currently accepted rating. When the secondary is connected in reverse order, the transformer becomes a choke or bucking transformer, reducing the line voltage instead of raising it.

The low-voltage source of an electric bell system is usually a transformer. If the ac source is not reliable and if it is important that the signal be completely dependable, a storage battery should be used. The battery can be kept charged with a rectifier.

When alternating current is available in a building, the primary winding of a step-down transformer is connected to the distribution source. At the secondary terminals, the low voltage of the required value can be obtained for application in signal circuits. This transformer may be called a low-voltage transformer because it reduces the lighting circuit voltage, usually 110 to 120 V, to a very low voltage such as 8 or 10 V. Another name for such a transformer is a *bell-ringing transformer,* because it is used with bell systems.

The terminals are plainly marked on the transformer in order that mistakes in connecting them may be avoided. The transformer secondary voltage for signal circuits in an ordinary building is 8 V; for larger installations, the voltage may be 12, 16, or 24 V. Step-down transformers are made in various ratings, but the ordinary bell-ringing transformer is rated at 100 V-ac and 50 W, since it operates at a 50 percent power factor. A transformer need not be separately fused on the 110-V side, provided the transformer is of a design that inherently limits to a low value the current that can flow through the transformer, even when short-circuited.

Cascade transformers are used primarily for very high ac voltage applications that require little power, such as high potential tests, for the measurement of high ac potentials, and for the acceleration of electrons and positive ions to high energy. Each transformer unit, except the last, is provided with a tertiary winding, which supplies power to the unit following it. Transformers beyond the first are insulated from ground for the voltage of the preceding units. The units are mounted in a single stack on porcelain cylinders, progressive insulation being supplied by the transformer tanks themselves.

The rating selected for a transformer installation should handle the normal full load of the system that it feeds, allowing for demand factors. Most transformers will handle as much as 200 percent of their rated load over long periods of time, but this practice will shorten the life of the transformer, as well as increase transformer losses.

Although transformers require less care than almost any other type of electrical apparatus, neglect of certain fundamental requirements may cause problems. Transformers must be installed so that they receive proper cooling, the oil must be kept very clean and dry, electrical connections and mechanical parts must be kept tight, excessive loads must not be carried for any length of time, protection from excess impulse voltages must be provided, and external parts must be protected by paint or another approved method.

Transformer coupling: An efficient means of transferring energy from one stage to the next in an amplifier. Figure 439 shows a transformer-coupled amplifier with a tuned primary. Two coils labeled L_1 and L_2 constitute an air core transformer. The plate tank is tuned to the operating frequency. When a positive voltage of sufficient amplitude is applied to the grid of the tube, plate current will flow. The output of the amplifier will be a sine wave due to the flywheel effect in the plate tank circuit.

A characteristic of a parallel resonant circuit is that it offers maximum impedance to the resonant frequency, causing plate current to be minimum. Therefore, line current is minimum but circulating current within the tank is maximum. The maximum circulating current passing through inductor L_1 will cause a magnetic field to fluctuate about it. Since inductor L_2 is in close proximity, it will also be under the influence of the same field. Therefore, a voltage will be induced into inductor L_2 through basic transformer action. The quantity of coupling between L_1 and L_2 may be varied by physically moving the secondary of the transformer closer to or farther away from the primary,

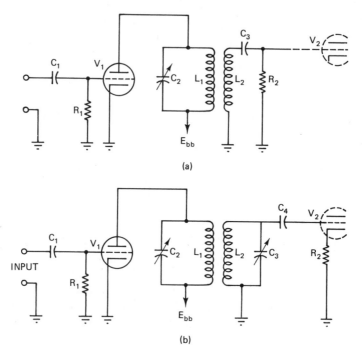

FIGURE 439 Transformer-coupled amplifier.

depending on the amount of drive required by the next stage. If capacitors are placed across the primary and secondary windings of the transformer in a transformer-coupled network, a double-tuned transformer coupling system is obtained (Fig. 439b).

Coil L_1 is the primary and coil L_2 is the secondary of the transformer. C_2 tunes L_1 to resonance at the signal frequency. A large signal voltage is produced across the high impedance of the parallel resonant circuit formed by L_1 and C_2. The large circulating tank current in the primary of the transformer creates a magnetic field which induces a voltage in the secondary winding, L_2. The voltage across the secondary winding is applied to the grid of V_2.

Transistor: An active semiconductor device composed of a minimum of three sandwiched layers of specially treated silicon or germanium material. Transistors are made of the same types of materials as solid-state rectifiers. Germanium and silicon material are basically stable, having very few free electrons that govern conductivity. In preparing these materials for solid-state devices, certain impurities are introduced, which cause a deficiency or abundance of electrons, thus increasing conductivity. Material that conducts due to a lack of electrons is called *P-type*. Material with an abundance of electrons is called *N-type*. These two types of materials are combined in transistors and all solid-state devices to arrive at desired operating characteristics.

A schematic representation of a bipolar transistor is shown in Fig. 440. At

FIGURE 440   (a)   (b)

A, an NPN transistor is shown, with B representing the PNP variety. These designations indicate the polarity of the device as well as the crystal-type layering. A PNP transistor is composed of an N-type crystal sandwiched between two layers of P-type material. The reverse is true of an NPN transistor, where the P-type material is positioned at the center.

Figure 440 also shows the designations of the leads or wires that connect the transistor to the rest of the electronic circuit. The base is usually the control leg or input to the device, while the collector (or sometimes the emitter) serves as the output leg. In the correct circuit, a small voltage applied at the base may be amplified into a larger voltage at the transistor output. In a control circuit, the proper voltage at the base will cause the transistor to conduct and pass an applied current from the emitter onto the collector. This type of circuit may be used for electronic switching and solid-state relay purposes.

Some other types of transistors, such as FETs, do not have a base. Instead, the controlling lead is called a *gate*. Figure 441 shows two types of FETs schematically with a P- and N-channel device at *A* and *B,* respectively.

Transistor amplifier: An amplifier whose active elements are composed of transistors as opposed to vacuum tubes. The basic transistor amplifier is described by three terms—amplification, amplifier, and classes of operation. The amplifier is the device used to increase the current, voltage, or power of an input signal without appreciably altering its essential quality. Amplification is the result of controlling a relatively large quantity of current (output circuit) with a small quantity of current (input circuit).

The transistor amplifier may be connected in any one of three basic configurations. These circuits are the common emitter, common base, and common collector. The term *common,* as used here, means that the element named is part of, or common to, both the input and output circuits.

Transistor amplifiers may be operated class A, class B, class AB, and class C. A class A amplifier is one in which the emitter-base junction bias and the alternating voltages applied to the emitter-base junction are such that the collector current (I_C) in specific transistors flows at all times. It is operated so

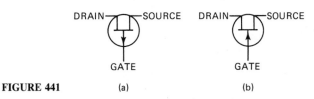

FIGURE 441   (a)   (b)

that the waveshape of the output is the same as that of the signal applied at the input.

A class B amplifier is one in which the emitter-base bias is approximately equal to the cutoff value, so that I_C is approximately zero when no excitation voltage is applied to the emitter-base junction. I_C flows in a specific transistor for approximately one-half of each cycle when an alternating signal voltage is applied to the emitter-base junction.

A class AB amplifier is one in which the emitter-base bias and the alternating voltages applied to the emitter-base junction are such that I_C in a specific transistor flows for appreciably more than half but less than the entire electrical cycle.

A class C amplifier is one in which the emitter-base is appreciably greater than the cutoff value. I_C in the transistor is zero when no alternating voltage is applied to the emitter-base junction and flows in a specific transistor for appreciably less than one-half of each cycle when an alternating signal voltage is applied to the emitter-base junction.

Transistor oscillator: Any oscillator circuit that uses transistors exclusively as the active component. Any oscillator that can be made using vacuum tubes can also be duplicated with transistors. These solid-state devices are often preferred in oscillator circuits, which are used to determine the final output frequencies of RF transmitters because they are made stable more easily than vacuum tube types.

Vacuum tube oscillators normally operate at higher temperatures than do transistor types, and heat fluctuations can cause them to stray from center frequency. Most oscillator circuits of the type used for frequency control must be highly stable. This indicates firm mechanical mounting of components and temperature stabilization. Transistors are physically small devices and can be firmly mounted on small circuit boards. They are not as susceptible to frequency drift due to mechanical shocks because they contain no movable parts. Technically, vacuum tubes have no moving parts either, but moderate to severe vibrations can set up a corresponding vibration in the tube electrodes, which can cause frequency shifts.

Transmission line: A line that conducts or guides electrical energy from the input or transmitter end of the line to the output end of the line. If this function is to be performed with minimum loss, such elements as impedance matching and line losses must be considered.

Transmission lines may be classified as resonant or nonresonant lines, each of which may have advantages over the other under a given set of circumstances. There are various types of transmission lines, such as the parallel two-wire line, the twisted pair, the coaxial line, and waveguides. The use of a particular type is dependent on the frequency, the voltage, the amount of power, the efficiency required, or the kind of installation to be used.

Transmit: To reproduce information in a new location, replacing whatever was previously stored.

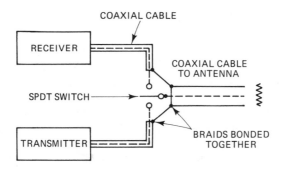

FIGURE 442   Transmit/receive switch.

Transmit/receive switch: Any device that allows a single antenna to be used with a radio transmitter and receiver by switching the feedline input between these two devices. During the transmit operation, the antenna feedline is switched to the transmitter output terminal. During the receive function, the feedline is switched to the receiver antenna input contacts.

Figure 442 shows a simple, manually operated transmit/receive switch for use with a coaxial cable feedline. This circuit incorporates a single-pole double-throw switch, which feeds the transmitter in one position, while the receiver is connected in the opposite position.

Modern transmit/receive switches often use electromechanical relays to perform the same function. Quite frequently, an automatic changeover is provided by triggering the relay by means of sampling a small amount of the RF output of the transmitter. When the transmitter is keyed, the sampled RF is rectified and triggers a small relay. When the contacts of this device close, a circuit is completed to the antenna changeover relay, which closes its contacts and connects the antenna feedline to the transmitter output. When the transmitter returns to the off state, both relays open again and the feedline is reconnected to the receiver.

Electronic transmit/receive switches are also available which contain no moving parts. The feedline is normally connected to both units; but when the transmitter is activated, an electronic blocking circuit prevents RF from entering the receiver.

Many modern transmitters contain internal electronic transmit/receive switches that handle the antenna changeover function. All single unit transceivers use similar circuitry as well.

Transmit/receive switches, whether mechanical, electromechanical, or electronic, must be built from components that are rated to pass and block RF energy efficiently. Additionally, each must be rated to withstand the maximum amount of RF voltage and current that is produced by the transmitter. Most transmit/receive switches are rated to handle a specific amount of RF power measured in watts. High-powered switches are usually larger and more expensive than those that are designed to switch lower power levels.

Transmitter: An electronic device used for generating and amplifying a radio-frequency carrier for transmission through space from an antenna. The radio transmitter and its antenna form one link of a communications system. All that is needed to complete the system is a radio receiver and an antenna. A transmitter may be a simple, low-power unit for sending voice messages a short distance, or it may be a highly sophisticated unit utilizing thousands of watts of power for sending many channels of data (e.g., voice, teletype, television, telemetry, etc.) simultaneously over long distances.

Transmitters operate from the very low frequency (VLF) band to the ultra high frequency (UHF) band. In the VLF band (10 to 30 kHz), signals can be transmitted over long distances and even through magnetic storms that blank out higher radio-frequency channels. This band is used for radio-telegraph transmissions, time standard transmissions, and for radio navigation.

The low frequency (LF) band (30 to 300 kHz) is used for long-range direction finding, medium- and long-range communications, and aeronautical radio navigation.

In the medium frequency (MF) band (300 to 3000 kHz), relatively long distances can be covered. The international distress frequency, 500 kHz, is in this band, as well as commercial broadcasting, amateur radio, loran, maritime, and aircraft communications.

The high frequency (HF) band includes frequencies from 3 to 30 mHz. This band is used for aeronautical, mobile communications, amateur radio, police communications, international broadcasting, and maritime communications.

The very high frequency (VHF) band extends from 30 to 300 mHz. It is used for aeronautical radio, navigation and communications, early warning radar, television, amateur radio, and FM broadcasting stations.

The ultra high frequency (UHF) band extends from 300 to 3000 mHz and is used for short-range communications such as between ships, amateur radio, television, and radar, as well as other services. This is only a partial list of the uses of each of the frequency bands.

One of the simplest types of radio transmitters is the amplitude-modulated transmitter. This transmitter is designed to send audio intelligence. An AM transmitter has four essential components:

1. A generator of RF oscillations.
2. A means of amplifying and, if necessary, multiplying the frequencies of these oscillations.
3. A method of modulating the RF carrier with the audio intelligence.
4. A power supply to provide the operating potential to the various transistors and/or electron tubes.

A block diagram of a basic transmitter used for amplitude-modulated transmission is illustrated in Fig. 443. The signal paths are represented by light

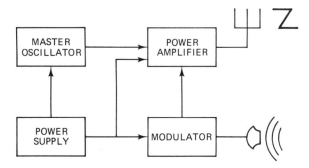

FIGURE 443 Basic AM transmitter.

lines, and the power supply paths to individual functional units are shown by heavy lines. The oscillator is the basic frequency-determining element of the transmitter. It is here that the RF signal is generated. If the oscillator fails to function, no RF signals will be produced.

Frequently, the transmitter output frequency is so high that it is difficult to maintain a stable oscillator. To overcome this difficulty, the oscillator is operated on a submultiple of the transmitter's output frequency. A process called *frequency multiplication* is used to increase the transmitter frequency.

Present-day transmitters may contain several oscillators to perform various functions. In general, only one of these is used to generate the basic transmitter radio frequency. This oscillator is usually called the *master oscillator* (MO) to distinguish it from any other oscillator in the transmitter.

The power amplifier (PA) is operated in such a manner that it greatly increases the magnitude of the RF current and voltage. The output of the modulator is coupled to the power amplifier to produce the desired AM signal. The output from the power amplifier is fed to the antenna via RF transformers and transmission lines. The modulator receives the intelligence to be transmitted. This normally weak signal is amplified to a level sufficient to modulate the RF in the power amplifier.

Transmitters require dc voltages ranging from hundreds of voltage negative to thousands of volts positive. Additionally, they need ac voltages at smaller values than those available from a normal power source. It is the function of the power supply to furnish these voltages at the necessary current ratings. In frequency modulation, the modulating signal combines with the carrier in such a way as to cause the frequency of the resultant wave to vary in accordance with instantaneous amplitude of the modulating signal.

Figure 444 is the block diagram of a narrow band frequency modulation (NBFM) transmitter. The modulating signal is applied to a reactance tube, causing the reactance to vary. The reactance tube is connected across the tank circuit of the oscillator. With no modulation, the oscillator generates a steady center frequency. With modulation applied, the reactance tube causes the frequency of the oscillator to vary around the center frequency in accordance

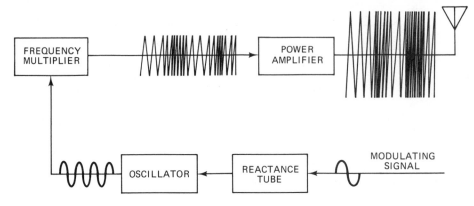

FIGURE 444   Narrow band FM transmitter.

with the modulating signal. The output of the oscillator is then fed to a frequency multiplier to increase the frequency and then to a power amplifier to increase the amplitude to the desired level for transmission.

Transmitter, facsimile: A device or system that generates signals depicting graphic material. The facsimile transmitter sends signals to a receiving point by means of wire lines or radio transmission. Here, the electrical signals are demodulated for information content, and a reproduction of the scanned picture is printed on paper.

Traveling wave tube: A vacuum tube designed to operate at microwave frequencies and which contains input and output couplers, a collector, transmission line, and electron gun. The traveling wave tube (TWT) is particularly suitable for amplifying microwave frequencies with a minimum of circuitry. The tube makes use of the distributed interaction between an electron beam and a traveling wave signal induced into a helix, as shown in Fig. 445. An electron gun emits a narrow beam of electrons having an acceleration proportional to an anode voltage on the order of kilovolts. This beam is directed through a long, loosely wound helix and is collected by an electrode (collector). A strong magnetic axial focusing field is supplied by a solenoid (or permanent magnet) to prevent the electron beam from spreading in a conical pattern and to guide the beam through the center of the helix. The RF signal (2) to be amplified is coupled to the input end of the helix adjacent to the electron gun. Under appropriate operating conditions, an amplified signal (11) appears at the output of the traveling wave tube.

The induced RF signal travels around the turns of the helix and thus, has its lineal velocity retarded by an amount equal to the ratio of the axial length of the helix to the length of the wire forming the helix. The potential difference between the cathode and anode is adjusted so that the electron beam travels a little faster than the RF signal along the axis of the tube. As the RF signal travels around the helix, the alternating characteristics of the signal force the electrons in the beam to slow down and speed up.

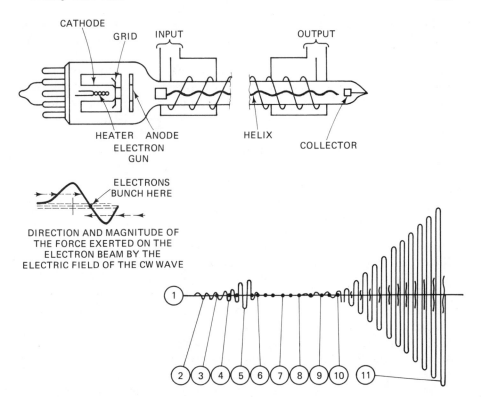

FIGURE 445   Traveling wave tube and amplification pattern.

The electrons thus travel in accelerated bunches (3). Amplification of the signal on the helix begins as the field formed by the electron bunches interacts with the electric field of the RF signal (4). A newly formed electron bunch adds a small amount of energy to the RF signal on the helix. This slightly amplified RF signal continues to increase the density of electron bunching, and the accelerated bunches, in turn, add still more energy to the RF signal. Amplification increases as acceleration of the electron beam causes the electron bunches to travel more nearly in phase with the electric field of the RF signal. The additive effect of the two fields exactly in phase produces the greatest resultant amplification (5) and (11). Thus, as the signal wave proceeds toward the output end of the helix, its amplitude increases exponentially.

The total interchange of energy between the electron beam and the helix wave is such that large amounts of power amplification can be achieved, from 20 to 40 dB in a single tube. Spurious modulation and hum generated within TWT amplifiers are usually 35 dB below the output signal level, and the noise figure is less than 30 dB. A broadband TWT can amplify any type of RF signal.

To prevent spurious oscillation in a TWT, it is necessary to dampen internal feedback caused by reflections arising from slight impedance mismatches at the input and output terminals. Energy reflected from the output terminal travels back to the gun end of the tube. Upon reflection at the gun end, the new wave constitutes a spurious oscillation of feedback energy that is further amplified along with the desired signal. This condition is controlled by an attenuator (a conductive coating of aquadag on the tube wall) near the input end of the helix to absorb any backware (6) propagated along the helix. The attenuator also absorbs some of the forward or desired growing wave present on the helix. However, the bunching of the primary wave electrons is, to a first approximation, unaffected by the presence of the attenuator (7).

Hence, as the electron bunches emerge from the attenuator region, they induce a new forward traveling wave (8) on the helix at the input side of the attenuator. This wave then travels toward the helix output nearly synchronously with the electron bunches.

Triac: A thyristor that contains three terminals, and when in the conducting state, will pass current in either direction. The triac is the electrical equivalent of two silicon-controlled rectifiers connected in reverse parallel. Shown schematically in Fig. 446, the device is most often used to switch and control alternating current circuits.

Triaxial speaker: A speaker consisting of three independent dynamic loudspeakers housed in one assembly. Separate speakers are provided for low, middle, and high frequency ranges. Triaxial speakers are commonly seen in the better quality stereo systems. The low-frequency speaker is often referred to as the *woofer* and is designed to respond best to all frequencies below a certain predetermined point. The high-frequency speaker is called the *tweeter* and responds to all frequencies above a certain point. The midrange speaker covers the frequency gap between the previous two.

Triaxial speaker systems use electronic changeover networks, which automatically duct that portion of the complex signal that falls within a specific frequency range to the appropriate speaker. Such a system provides a more natural sounding output than do those systems that incorporate only a single speaker.

Trimmer capacitor: A type of variable capacitor that allows the value of capacitance to be varied over a prescribed range. The trimmer capacitor consists of two plates separated by a dielectric other than air. The capacitance is

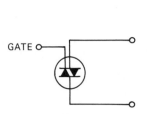

FIGURE 446 Triac.

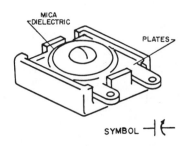

FIGURE 447 Trimmer capacitor.

varied by changing the distance between the plates. This is ordinarily accomplished by means of a screw that forces the plates closer together. Figure 447 shows a trimmer capacitor.

Triode: A vacuum tube containing a cathode, plate, and a single grid element. In 1907, Lee DeForest opened the door to what proved to be the birth of a new age in the history of man's technical advancement. By placing a small wire mesh or grid into a diode, he discovered that electron flow from the cathode to the plate could be controlled by varying the grid potential. Since three active elements or electrodes were involved, the device was called a *triode*.

The operation of the triode in controlling current flow was considered analogous to the action of a valve in controlling water flow. Due to this concept, the term *valve* was used originally in place of *vacuum tube*. Even today, some countries such as England retain the word valve.

The triode electron tube is similar in construction to the diode, the primary difference being the addition of a grid-like electrode place in the area between cathode and plate. This electrode is called the *control grid* and enables the tube to amplify by controlling the flow of plate current. The construction features of a typical triode are shown in Fig. 448. Electrical connections to the cathode, grid, and plate are made through the base pins and support wires. The cathode sleeve is insulated from the heater and is connected by means of a short lead to one of the base pins. Note that the grid is located much closer to the cathode than to the plate.

Although in certain triodes gas is inserted in the tube, those which are used for voltage amplification are of the high vacuum tube. In such tubes, the air pressure within the glass or metal envelope is reduced to approximately one one-hundred millionth that of atmospheric pressure. Even with this high degree of evacuation, a large quantity of air molecules still remain, which could interfere with tube operation if means were not provided to eliminate them. As with diode vacuum tubes, a getter is used to combine with the residual gas or

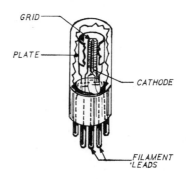

GRID
PLATE
CATHODE
FILAMENT LEADS

FIGURE 448   *CUT-AWAY VIEW OF A TRIODE*

LADDER TYPE

ELLIPTICAL
HELIX

FIGURE 449   Typical grid structures.

any gas that may be liberated from the heated elements after the tube is placed into operation.

The control grid of a triode is usually made of fine wire coiled into a helix and mounted on two supporting wires. Typical grid structures are shown in Fig. 449. Low-power triode grids are usually made of manganese nickel, while high-power triode grids are made of molybdenum compounds to reduce grid emission. Such emission is undesirable and can occur due to the influence of the intense electrostatic field between plate and grid when plate voltage is very high (as in high-power tubes). The plate (anode) of triodes used for low-power applications is made of nickel or iron. The plate is pressed out of sheet material in the form of a cylinder or other shape and completely surrounds the grid. For higher power uses, the plate may be constructed of graphite or copper. In many cases, the plate is provided with a rough surface or is blackened to increase its ability to radiate heat caused by electron bombardment.

The physical structure and electrical symbol of an indirectly heated triode are shown in Fig. 450. It can be seen that the oxide-coated cathode is located in the center of the tube structure. Surrounding the cathode is the helical-shaped grid structure. The plate, in the form of a cylinder, completely encircles the grid. Regardless of the shape of the grid or plate, the grid must be placed so that the electrons emitted from the cathode will pass between the wires of the grid on their way to the plate. In the triode symbol, the heater is shown beneath the cathode. It is common practice to omit the heater from this symbol to simplify schematic drawings. Where complete equipment schematics are encountered, the heaters of all the tubes will usually be shown near their source of power.

Tropospheric propagation: The effect the elements of the tropo-sphere have on the transmission of radio waves. This specifically applies to the bending of a transmitted signal that originates on the earth's surface and travels into the tropospheric region, where it is often directed back toward another point on the earth. The troposphere is the portion of the earth's atmosphere that lies between the earth's surface and the stratosphere. The troposphere extends from ground level to approximately 7 mi up.

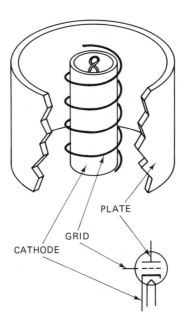

PLATE

GRID

CATHODE

FIGURE 450 Physical structure and schematic symbol of a triode vacuum tube.

Propagation of radio waves by signal bending in the troposphere often results in the transmission and reception of these signals over a greater distance than would be experienced through ground wave communications alone. Radio waves that are transmitted at very low angles with respect to the earth's surface are often bent downward slightly and in a curved path due to the tropospheric effect. Due to the impact of the troposphere on the original transmission, radio waves are often bent around the curvature of the earth. Figure 451 shows this process.

The actual effect of the troposphere on RF signals will depend upon the physical condition of this portion of the atmosphere and the frequency of transmission. Tropospheric bending is most likely to occur in weather conditions which bring about clear, cool days. When the temperature rises, tropospheric effect begins to decrease.

Each time the signal bending process takes place, there is a degree of attenuation of the radio signal. Again, the physical condition of the troposphere will determine the amount of attenuation. Cool weather conditions tend to bring about more significant bending with minimal attenuation.

Truth table: A chart used in connection with logic circuits to illustrate the states of the inputs and outputs of a given stage under all possible signal conditions. It provides a ready reference for use in analyzing the operating theory of the circuit and is useful in developing the overall signal flow diagram.

Tuned-plate tuned-grid oscillator: An oscillator that utilizes a tuned circuit in both the plate and the grid circuits, and may be employed in a wide range of frequencies, from the low to the ultra high. Because of reduced

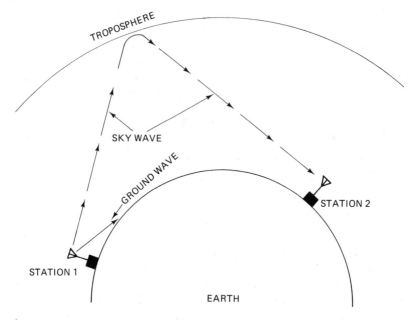

FIGURE 451

feedback between plate and grid at low frequencies, the TPTG oscillator is not as satisfactory at low frequencies as some other types of circuits. Figure 452 shows a TPTG oscillator circuit. The feedback necessary to sustain oscillations is coupled from the plate circuit to the grid circuit by means of the interelectrode capacitance between the plate and the grid (C_{gp}).

In the equivalent ac circuit (Fig. 453), both parallel tanks are below resonance a small amount, so they appear as highly inductive coils L_1' and L_2'.

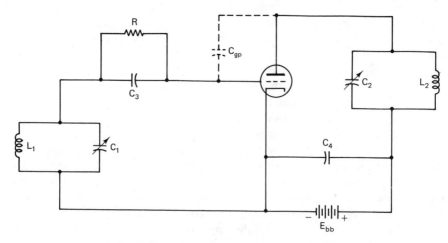

FIGURE 452   Tuned-plate tuned grid oscillator.

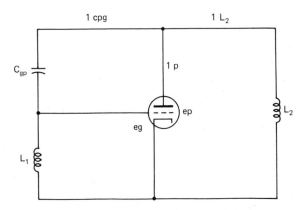

FIGURE 453   Simplified equivalent circuit.

The plate-grid capacitance C_{gp} is small and X_{cgp} is greater than $X_{L_1'}$. The left-hand branch is capacitive. The right-hand branch is inductive, and at the operating frequency, $X_C - X_{L_1'} = X_{L_2'}$.

Plate voltage e_p (Fig. 454) is the reference vector; e_p appears across both parallel branches. The current i_{L2} lags e_p by 90° because it flows in the inductive circuit L_2'. The current i_{cgp} leads e_p by 90° because it flows in the left-hand branch, which contains more capacitive reactance than inductive reactance. The grid voltage e_g leads i_{cgp} by 90° because it appears across the inductive portion of L_1' of the left-hand branch.

Notice that i_p is in phase with e_g and 180° out of phase with e_p, the relation necessary to sustain oscillations. The tank having the highest Q determines the oscillator frequency. If the plate tank is more heavily loaded than the grid tank, the grid tank will have the higher Q and will determine the oscillator frequency.

Tuned radio-frequency receiver: Also known as a *straight-through receiver*, a circuit consisting of a tuned radio-frequency amplifier connected to

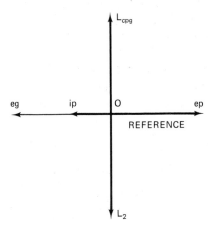

FIGURE 454   Vectors.

an RF detector and an audio-frequency amplifier. The sensitivity of the basic receiver is a function of the detector. The detector requires a minimum level to perform its function. If signal level is below this minimum, detection will not take place.

It was discovered that the addition of tuned RF amplifier stages increased not only the sensitivity of the set, but also its selectivity. Figure 455 shows a block diagram of a tuned radio-frequency (TRF) receiver.

The amplitude of the AM signal at the input of the receiver is relatively small. The RF does not change its basic shape. The detector separates the audio component from the RF component and passes the audio component to the audio-frequency amplifier. The AF amplifier increases the amplitude of the signal to a value sufficient to operate the loudspeaker.

The sensitivity and selectivity of a TRF receiver are improved by increasing the number of RF stages. Figure 456 shows the block diagram of a TRF receiver having two stages of RF amplification. The tuned tanks of each RF stage must be tuned to the same frequency. Therefore, when changing frequency, it is necessary to change all of the tanks to the new frequency. When two or more stages are used, the tuning capacitors are ganged (gang tuning) on the same shaft. This is indicated by a dashed line connecting the tanks in Fig. 456.

The improved selectivity of the TRF receiver is due to the presence of several tuned stages. Figure 457 shows the overall response curve, stage by stage, for Fig. 456. Curve 1 represents the response curve of the first tuned circuit. Curve 2 is the overall response curve of the first RF amplifier stage (including the first and second tuned circuits). Curve 3 is the overall response curve of the two RF amplifiers in cascade (including the three tuned circuits). Notice that slope of the sides becomes steeper for each succeeding stage. This results in better rejection of unwanted signals.

Notice also that the passband of each succeeding stage is becoming narrower. This will cause some of the sidebands to be rejected and lower the fidelity of the receiver. To overcome this problem, the bandwidth of each tuned stage is made wider than the previous stage (i.e., the bandwidth of the second tuned circuit, etc.). This is done by lowering the Q of the tank circuit, which, in turn, reduces the gain of the stage. This puts a practical limit on the number of stages that may be added to the TRF receiver.

The TRF receiver offers another advantage, that of improved signal-to-noise ratio. This is due to the better selectivity. Noise signals outside the passband of the receiver will be greatly attenuated or rejected and cannot interfere with the desired signal.

Tungsten: One of the first materials to be widely used for vacuum tube cathodes. Tungsten has the advantage of mechanical ruggedness but must be heated to a very high temperature (2227°C) to obtain a sufficient amount of emission. Because tungsten has a high work function (4.53 eV), its efficiency is poor. Emitter efficiency is measured in milliamperes of emission per watt of

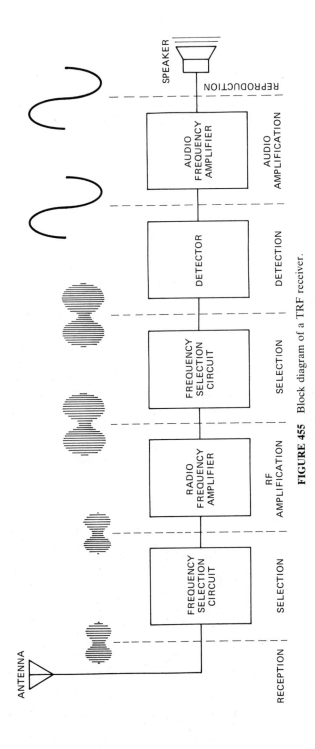

FIGURE 455 Block diagram of a TRF receiver.

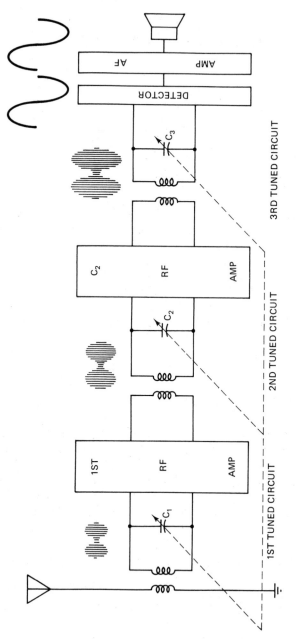

FIGURE 456 TRF receiver with two stages of RF amplifications.

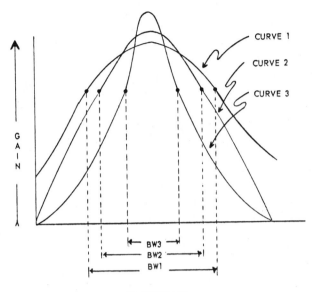

GAIN

CURVE 1
CURVE 2
CURVE 3

BW3
BW2
BW1

FIGURE 457

heating power. Tungsten emitters have an efficiency of approximately 7 mA/W. Due to their low efficiency, tungsten cathodes are seldom used in modern electron tubes.

Tunnel diode: A PN junction diode that characteristically displays negative resistance when in the forward conduction mode. The tunnel diode, also called the *Esaki diode,* is one of the most significant solid-state devices to emerge from the research laboratory since the transistor. It is smaller and faster in operation than either the transistor or the electron tube. Also, it offers a host of additional features.

The high switching speed of the tunnel diode, coupled with its simplicity, and stability, makes it particularly suitable for high-speed operation. Tunnel diodes operate effectively as amplifiers, oscillators, and converters at microwave frequencies. In addition, they have extremely low power consumption and are relatively unaffected by radiation, surface effects, or temperature variations.

The two-terminal nature of the tunnel diode should be considered. On one hand, this feature allows the construction of circuits that are very simple and consequently provide a savings in size and weight. This also provides a significant improvement in reliability. On the other hand, the lack of isolation between the input and the output can be a serious problem in some applications. Below the microwave frequency region, transistors are usually more practical and economical. However, at microwave frequencies, tunnel diodes have several advantages and are highly competitive with other high-frequency devices.

A tunnel diode is a small, two-lead device having a single PN junction,

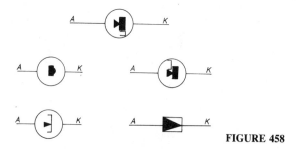

FIGURE 458

which is formed from very heavily doped semiconductor materials. It differs from other PN junction diodes in that the doping levels are from one hundred to several thousand times higher in the tunnel diode. The high impurity level in both the N- and P-type materials results in an extremely thin barrier region at the junction. The effects that occur at this junction produce the unusual current-voltage characteristics and high-frequency capabilities of the tunnel diode.

There are various schematic symbols used to represent the tunnel diode, as shown in Fig. 458. The MIL-SRD symbol is shown at the top of the figure. Fig. 459 compares the current-voltage characteristic of the tunnel diode with that of a conventional PN diode. The broken line shows the curve for the conventional diode and the solid line indicates the characteristics of the tunnel diode.

In the standard PN diode, the forward current does not begin to flow freely until the forward voltage reaches a value on the order of half a volt. This corresponds to point *Y* in Fig. 459. This forward voltage is sometimes referred to as the *offset voltage*. In the reverse direction, the conventional diode has a

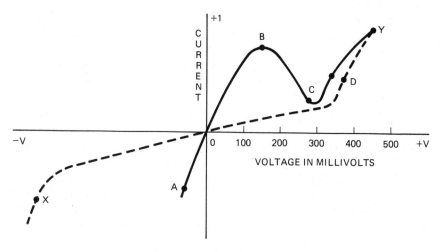

FIGURE 459

high resistance to current flow until the breakdown region is reached. The breakdown region is shown at point *X* in Fig. 459.

The tunnel diode is much more conductive near zero voltage. Appreciable current flows when a small bias is applied in either the forward or the reverse direction. Because the active region of the tunnel diode is at a much lower voltage than the standard semiconductor devices, it is an extremely low power device.

As the forward bias on the tunnel diode is increased, the current reaches a sharp maximum, as shown at point *B* in Fig. 459. This point is referred to as *peak voltage* and *peak current point*. The curve then drops to a deep minimum at point *C*. These are called *valley point voltage* and *valley point current*, respectively. The curve now increases exponentially with the applied voltage and finally coincides with the characteristic curve of a conventional rectifier. This is shown at point *Y* on the characteristic curve. The drop in current with an increasingly positive voltage (area *B* to *C*) gives the tunnel diode the property of negative resistance in this region. This negative resistance enables the tunnel diode to convert dc power supply current into ac circuit current and thus permits its use as an oscillator.

Tunnel diodes use much higher doping levels than conventional diodes. Typically, the transistor region in tunnel diodes (potential barrier) is 100 times less than that of a standard PN junction diode. This difference in width and doping level accounts for the major difference in the tunnel diode and other types of junctions. Because of the narrowness of the potential barrier, charge differences may easily make themselves felt across the junction when the junction voltages are applied.

Figure 460 shows a graph that indicates the rate of electron orbital distribution that occurs in P- and N-type materials. The graphs have been greatly expanded and distorted for the sake of clarity. Notice that initially, the N-material electrons undergo the greatest orbital distortion. This is the area on the graph from the origin to point *A*. During this portion of the graph, the electrons of the N material have a greater orbital distortion than those in the P material. Consequently, energy is transferred across the relatively narrow junction. This corresponds to the area from 0 to *B* in Fig. 459. In this area,

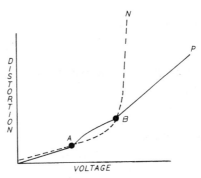

FIGURE 460

relatively large currents are produced with the application of very small voltages.

At point *A* in Fig. 460, the rate of orbital distortion of the electrons in the P material has increased and caught up with the rate of distortion in the N material. The electron orbits in the P material are now being distorted faster than the N-material electron orbits can distort and intrude across the junction. The total transfer of energy must therefore decrease. The area from *A* to *B* in Fig. 460 corresponds to the area *B* to *C* in Fig. 459. This is the negative resistance region of the tunnel diode.

At point *B* of Fig. 460, the orbital distortion of the N-material electrons has once again caught up with the distortion rate in the P material. Beyond point *B,* the N-material electron orbits are distorting faster than those in the P material. There is therefore an increase in the effective intrusion of the N-material electron orbits and a corresponding increase in current. From this point on, the operation of the tunnel diode is the same as that of a standard PN junction.

Twin lead: A transmission line that consists of two flexible conductors molded in a flexible, flat ribbon of plastic. It is often called a *flat ribbon line* and is most often used for connecting television antennas to their receivers. Television twin lead usually has a nominal impedance of 300 Ω and is nonshielded. It is especially applicable to home television installation because of its flexibility and relatively low cost. Since this line is not shielded, it is often necessary to use offset insulators when the cable must travel over large metallic objects.

In recent years, twin-lead cable has been replaced to some degree by 72-Ω coaxial cable for television reception applications. The latter is also flexible and does not require offset insulators due to its continuous shield. However, coaxial cable is more expensive than twin lead and is not as efficient a conductor of VHF and UHF signals. For this reason, twin lead is still popular for many receiving applications.

Unidirectional microphone: A microphone that offers maximum response to sound waves that approach its element from only one direction. Typically, maximum response is had by directing sound waves to the front of the microphone. Minimum response or maximum rejection is usually obtained from the back and sides.

Unidirectional microphones are often used in the professional recording of music where a vocalist must stand in close proximity to the musicians. The maximum area of response is directed away from the band and allows the singer's voice to be better regulated during the taping process. The microphone element offers maximum sensitivity in the direction from which the vocalist's sound waves arrive while suppressing the pickup from the band.

Unidirectional microphones are often called noise-cancelling microphones and are used in radio-communications work, especially in mobile and other high-noise applications. The same principle applies, as maximum response to the human voice is obtained while rejection of incoming noise from other directions results.

Unijunction transistor: A three-terminal semiconductor device, sometimes called a double-based diode, which can be triggered by, or an output can be taken from, each of the three terminals. Once the unit is triggered, the emitter current increases regeneratively until it is limited by the power supply. Thus, the action of the unijunction transistor is similar to that of the gas thyratron tube. It can be employed in a variety of circuits, but finds its greatest usefulness in the switching and pulsing fields.

Universal series motor: A motor that operates on the magnetic interaction between the armature and field poles. This motor runs in the same direction whether the current flows in on line A or line B in Fig. 461, since reversing the flow of current in the line wires changes the polarity of both armature and field poles at the same instant as shown at C and D. Therefore, if such a motor is supplied with ac, the torque developed will always be in the same direction. It is called a *universal motor* because it operates on both ac and dc.

To operate satisfactorily on ac, all parts of the magnetic circuit must be laminated to prevent undue heating from eddy currents, and element windings are usually desirable on the armature to ensure acceptable commutation. On

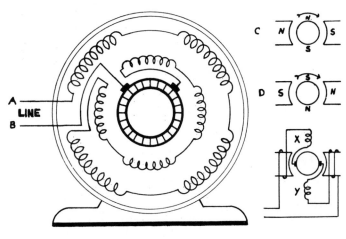

FIGURE 461

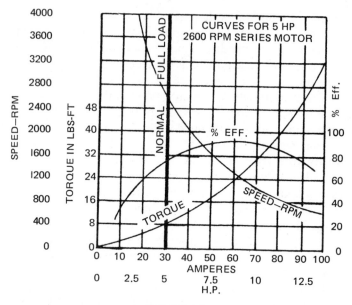

FIGURE 462

the larger motors, compensating windings are employed to improve operation and reduce sparking.

This motor will produce about four times normal full-load torque with two times normal full-load current. The torque produced increases very rapidly with an increase in current, as the curves in Fig. 462 indicate. The variation in speed from no load to full load is so great that complete removal of load is dangerous in motors of this type, except those having fractional horsepower ratings.

This motor is widely used in fractional horsepower sizes for fans, vacuum cleaners, kitchen mixers, milk shakers, and portable equipment of all types, such as electric drills, hammers, sanders, saws, etc. Higher ratings are employed in traction work and for cranes, hoists, etc. In general, they are suitable for applications where high starting torque or universal operation is desired.

Vacuum tube: An electronic device consisting of two or more electrode elements housed in an envelope from which virtually all air and gases are removed. Vacuum tubes are classified in accordance with the number of elements they contain. A diode contains two elements, while triodes contain three. Vacuum tubes have the ability to pass electrons and regulate their flow by external means. They act as an electron valve, with the actual electron flow controlled by a grid that lies between the two main elements, the plate and the cathode. For this reason, the electron tube is still called a *valve* in some parts of the world.

Van de Graaff generator: An electrostatic generator that produces large amounts of electric charge at very high voltages and is used mainly for lightning studies and atom smashing. This type of generator actually produces potentials of several million volts by transporting electrostatic charges from a continuously moving belt to a large hollow sphere, as shown in Fig. 463. Negative electric charge is sprayed onto an endless fabric belt between a comb of needle points and a rounded surface. The negative charge is mechanically transported by the pulley-driven belt to another comb-shaped collector, which transfers it to the outside of the metal dome.

Since the process is continuous and the belt can be run at high speed, enormous charges can be built up on the dome. In fact, the maximum voltage is limited only by direct electrical discharge from the metal shell and the quality of the insulation.

Variable capacitor: A capacitor whose value may be easily changed by altering the spacing of its electrodes. A variable or air-variable capacitor is sometimes known as a *tuning capacitor* because of its uses in the frequency-controlling stages of transmitters, receivers, and other devices.

Shown in Fig. 464, it consists of a number of movable plates meshing into a number of stationary plates. The movable plate assembly is called the *rotor,*

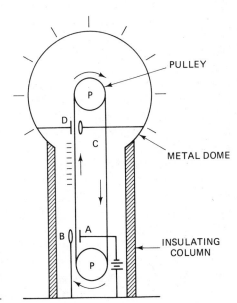

FIGURE 463 Van de Graaff generator.

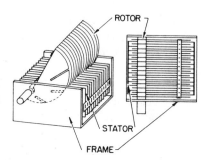

ROTOR

STATOR

FRAME

FIGURE 464

while the stationary plates form the stator. The rotor is mounted on a longitudinal shaft and can be rotated to bring different amounts of surface area of the rotor plates in parallel with those of the stator. As the meshing becomes more complete, the capacitance value increases. As the plates are drawn away from each other, a decrease in capacitance occurs. Air-variable capacitors may be ganged together for increased capacitance values.

The capacitance value of the air-variable capacitor is determined by the surface area of the rotor and stator plates and by their separation from each other. The separation, however, determines the voltage rating of the device, so high voltage variables must be spaced accordingly. To make up for the decreased capacitance this spacing brings about, the surface areas of the rotor and stator plates must be increased.

Variable capacitors are used in many different applications, many of them involving frequency control. They will be found in the variable-frequency oscillator sections of many types of transmitters and also in the driver and final amplifier portions of these same units. Like other capacitors, they may be wired in series for increased voltage ratings at the expense of decreased capacitance. Parallel connections of air variables provide increased overall capacitance ratings while maintaining the same voltage rating of a single unit. Single-section air-variable capacitors are usually rated to have a maximum capacitance of 10 to 500 pF. Minimum capacitance ratings are also important, especially in frequency control circuits, because even when the rotor and stator plates are completely separated, there will still be a small capacitive effect. This can be a few picofarads in small units to 50, 60, or more picofarads in high-capacitance units.

Variable resistor: A lumped resistance component whose value may be continuously varied. Variable resistors are available in many different component configurations. These units are used as volume controls in audio-frequency devices and have the electrical ability to control the flow of current. They are occasionally called *varistors*. Potentiometers and rheostats are also variable resistors, but in a more specialized form.

Veitch diagram: An approach to equation simplification in logic circuits. These diagrams provide a very quick and easy means of finding the simplest logic equation needed to express a given function. Veitch diagrams for

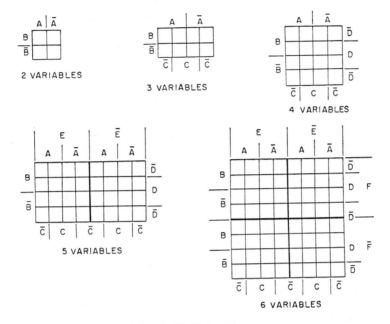

FIGURE 465 Veitch diagrams.

two, three, or four variables are readily constructed. Any number of variables may be plotted on a Veitch diagram, although the diagrams are difficult to construct and use when more than four variables are involved. Figure 465 shows examples of Veitch diagrams.

Velocity, angular: Circular velocity such as that of the armature of an alternator, which is symbolized by the lowercase Greek letter omega. Angular velocity is measured by determining the size of the angle the moving object generates per unit of time. Thus, angular velocity is measured in degrees per second, or radians per second, the latter being more common.

Very high frequency: That portion of the radio-frequency spectrum that lies in the range of 30 to 300 MHz. Radio waves at these frequencies respond differently to atmospheric bending (skip) depending on which end of the spectrum they are located. Near the lower limit of 30 MHz, VHF signals may be bent in the ionosphere and travel back to earth thousands of miles away from the transmit station location. As frequency increases, however, signals tend to pass through the atmosphere unaffected.

For this reason, the middle and upper portions of the VHF spectrum may be used for satellite and space communications. While microwave frequencies are often chosen for commercial communications between space and earth (including satellite television broadcasts), amateur radio operators sometimes use the 2-m band (140–148 MHz) and the 1.25-m band (220–225 MHz) as downlink frequencies between an orbiting amateur satellite and earth stations.

Another form of amateur radio activity called *moon bounce* or *eme* (earth-moon-earth) involves transmitting a VHF signal on the earth, bouncing it off the moon's surface, and retrieving the signal again at another earth station. This is not possible in the high frequency range (below 30 MHz), because the great majority of these signals are almost entirely reflected back to the earth's surface, never leaving the atmosphere.

VHF communications are used by many amateur and commercial earth-based stations. The FM broadcast band and television transmissions for channels 2 through 13 are contained within the VHR spectrum. Many utilities, public service departments, and business organizations utilize the VHF frequency spectrum for communications purposes. These include fire and police departments, commercial two-way users, and a myriad of other service-oriented entities.

VHF transmissions are especially applicable to mobile and portable stations, as the physical length of antennas is greatly decreased over what would be required in the high frequency band. A typical quarter-wave antenna for use at 150 MHz would be less than 20 in. in length. Small antenna size allows for the practical construction and use of multielement arrays, which offer high transmitting and receiving gain in a physically small package.

Most VHF communications are considered to be line of sight. This means that the sending and receiving stations must be in a direct line with each other and are not blocked by mountains or the curvature of the earth. As was previously mentioned, some bending may take place at the lower VHF frequencies (and even at the upper range of this spectrum under unusual atmospheric conditions), but trip skip above 100 MHz is quite rare.

Most commercial VHF transmitters use frequency modulation, although some AM equipment still exists. The latter was more popular several decades ago and is rarely seen in modern times.

Vibrator: An electromechanical device used to obtain a high ac or dc voltage from a comparatively low dc source. This device has certain advantages over the dynamotor type of power supply. For example, the vibrator is lighter and less expensive than the dynamotor. It is also more efficient. However, the vibrator can be used only when a limited amount of high voltage current is needed. Also, its life is relatively short, and it tends to produce radio interference (hash). It is extensively used in the power packs of lightweight mobile equipment.

Video: The transmission or reception of pictures or visual information by electronic means. The term especially applies to that portion of a television signal that carries the information needed to reconstruct a picture at the receiver.

Videotape recorder: A wide-band magnetic tape device used for producing videotape recordings with a camera, or for making a recording of television programs for playback at a later time. Strictly speaking, videotape recorders use reel-to-reel magnetic tape as a recording medium, although the term applies equally to video cassette recorders (VCRs), which house the reels in a tape cartridge.

Videotape recorders were formerly rather bulky and expensive devices, but in recent years, market demand has brought about decreased prices and new technologies. Today, the videotape recorder is a compact device designed to be placed near the television receiver. Additionally, electronic timers are used to activate the device at times when the owner is away in order to record television programs to be played back at a later time. Some models offer freeze-frame capabilities, which will allow a particular portion of a taped program to be halted for a closer visual inspection.

Through the use of videotape recorders, the instant replay has become an important part of sports broadcasts. This was impossible using the older motion picture film equipment of a few decades ago. Today, major sporting broadcasts are videotaped simultaneously with live coverage. When an instant replay is desired, the VTR tape is rewound and then rebroadcast.

Videotape recording requires no processing and operates in the same manner as audiotape recorders, in that information can be recorded and played back a few seconds later.

Video terminal: A computer installation that is designed to present programming results on a cathode-ray tube.

Vidicon: A specialized image or camera tube whereby a charge density pattern is stored and scanned by an electron beam. Typical vidicons use a scan beam composed of low-velocity electrons. The image is formed by photoconduction on a sensitized surface, where it is later scanned by the beam. This entire process takes place in milliseconds. The electrical output of the vidicon tube is then channeled to amplifying and processing circuits and is often converted to a radio-frequency output.

Virtual memory: In a computer, a memory technique that transfers information one page at a time between the primary and secondary memory. It adds only the page-swapping time to the operating time of the system. This technique allows the programmer to address storage without regard as to whether primary or secondary storage is being addressed.

The virtual memory technique has been utilized in large system development programs where an executive system allows the programmer to write programs as if memory capacity were unlimited. The executive keeps the programs on disks and out of use until they are required. Each disk is then loaded into the system when it is called for by the program.

Voice transmission: Sending the human voice by radio waves from a transmitting site. The radio wave is modulated to correspond with the frequency fluctuations of the human voice. The major energy content of the average human voice lies within a frequency range of 300 to 3000 Hz. This is the range needed to establish effective voice communications. At the receiving end, the radio wave is demodulated and the original voice transmission is recovered.

Voltage-controlled oscillator: Any oscillator that is capable of accomplishing an output frequency change by varying the amplitude of the control voltage input. Voltage-controlled oscillators (VCOs) may depend on a crystal or inductor to establish a basic frequency, but when the control voltage is applied, a phase shift is introduced and the frequency changes.

Voltage-controlled oscillators are usually very stable devices, as they do not depend on mechanical variations in order to change frequency. A normal adjustable oscillator will use a variable capacitor or variable inductor to swing the output frequency, but these components are mechanical in nature and their values may be changed by inadvertent shocks. The VCO contains fixed LC circuits. Therefore, it is far more mechanically stable and can withstand vibration and impacts that would tend to cause standard types of oscillators to change frequency in an undesirable manner.

Voltage-controlled oscillators are often used as frequency determining devices in VHF transmitters and receivers. In the Heathkit HW-2036A VHF transceiver, the VCO is the heart of the entire circuit and provides the proper frequencies for the transmitter and receiver injection. This circuit is part of a phase-locked loop, which feeds control voltage to the VCO and holds it as stable and accurate as a crystal-controlled oscillator. These latter types of oscillators are more stable than LC varieties but cannot be easily varied in frequency. By substituting a voltage-controlled oscillator driven by a phase-locked loop, frequency stability on the order of a crystal-controlled oscillator is obtained, along with the full ability to easily change from one frequency to another.

This is only one example of how a VCO is used in modern electronic devices. These circuits are most often seen in applications where frequency stability is quite critical and oscillator tuning is required.

Voltage quadrupler: Two half-wave voltage doubler circuits connected back to back and sharing a common ac input. In order to show how the voltage quadrupler works, Fig. 466 has been shown as two voltage doublers. The counterparts of one circuit are shown as a prime (') in the second circuit. (C_1 in the first circuit is the same as $C_{1'}$ in the other circuit; CR_1 and $CR_{1'}$ conduct at the same time, etc.)

When the circuit is first turned on, it will be assumed that the top of the

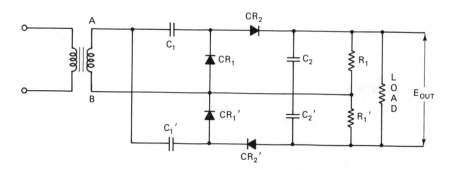

FIGURE 466   Voltage quadrupler circuit.

secondary winding, point A, is negative with respect to B. At this instant, CR_1 is forward biased and conducts, allowing C_1 to charge. On the next alternation, point B is negative with respect to point A. At this time, two things are going to occur. First, C_1, which was charged to approximately the peak voltage across the secondary winding, will aid the source. Since CR_2 is now conducting, C_2 will be charged to approximately twice the incoming voltage. Second, $CR_{1'}$ will conduct, charging $C_{1'}$ to approximately the peak voltage of the input. During the following alternation, $C_{1'}$ adds to the input, which allows $CR_{2'}$ to conduct, charging $C_{2'}$ to twice the input voltage.

When CR_2 conducts, C_1 will aid the input voltage in charging C_2 to approximately twice the peak voltage of the secondary and $CR_{2'}$, conducting at the same time, charges $C_{1'}$. On the next alternation, $CR_{1'}$ conducts, and since C_1 is in series with the input, it aids in charging $C_{2'}$ to twice the peak of the secondary. The voltages across C_2 and $C_{2'}$ add to provide four times the peak secondary voltage in the output.

Voltage regulator: A device used on electrical circuits to maintain a constant voltage output or to vary the voltage from a fixed input, which can be applied to regulate a total circuit or a specific load on any given circuit. Most voltage regulators are similar in construction to a wound-rotor induction motor. The primary or shunt windings are wound on a laminated steel rotor, and the secondary or regulating windings are wound on a laminated steel stator. By means of gearings and stops, the rotor is limited to 180° rotation electrically and mechanically. The primary winding is connected in shunt across the line. The secondary winding is connected in series with the load. Turning the rotor with respect to the stator changes the flux linkages and raises or lowers the output voltage, depending on the polarity determined by the position of the rotor with respect to the stator. Therefore, voltage regulators are capable of raising or lowering the voltage in equal amounts.

Standard voltage regulators are either self-cooled or fan-cooled (forced-air), depending on the size and application. Most voltage regulators require only normal maintenance that would be required on any electric motor. There are no brushes to replace nor hundreds of components to be concerned with.

Voltage tripler: A rectifier circuit that delivers a dc output voltage that is three times the peak value of the input voltage from the line or transformer secondary.

Figure 467 depicts a typical voltage tripler circuit with waveforms and circuit operation. Part A shows the complete circuit. In part B, C_1 is shown charging as CR_2 is conducting. In part C, C_3 is illustrated charging as CR_3 is conducting. Part D shows the charge path for C_2 while CR_1 is conducting. In part E, a comparison is made of the input signal and its effects on the voltages felt across C_1, C_2, C_3, and the load. The following explanation uses Fig. 467 as the operating device.

Close inspection of Fig. 467a should reveal that removal of CR_3, C_3, R_2, and the load resistor results in a voltage doubler circuit. The connection of

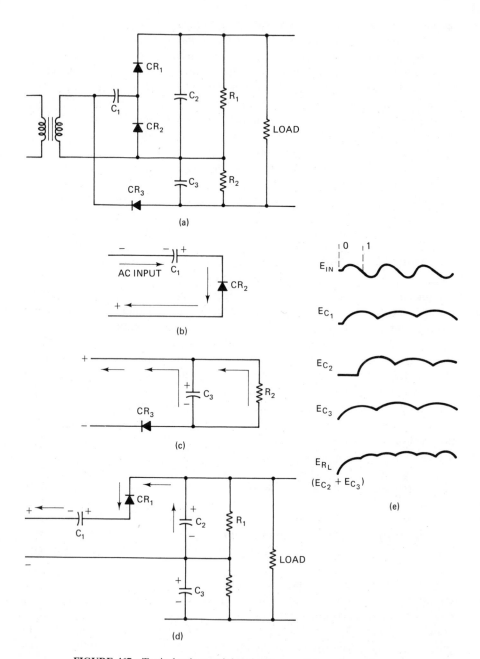

FIGURE 467 Typical voltage tripler circuit operation and waveforms.

circuit elements CR_3 and the parallel network of C_3 and R_2 to the basic doubler circuit is arranged so that they are in series across the load. The combination provides approximately three times more voltage in the output than is felt across the input. Fundamentally, then, this circuit is a combination of a half-wave voltage doubler and a half-wave rectifier circuit arranged so that the output voltage of one circuit is in series with the output voltage of the other.

Figure 467b shows how C_1 is initially charged. Assume that the input is such that CR_2 is conducting. A path for charging current is from the right-hand plate of C_1 through CR_2 and the secondary winding of the transformer to the left-hand plate of C_1. The direction of current is indicated by the arrows.

At the same time that the above action is taking place, CR_3 is also forward biased and is conducting, and C_3 is charged with the polarities indicated in Fig. 467c. The arrows indicate the direction of current. There are now two energized capacitors, each charged to approximately the peak value of the input voltage.

On the next half-cycle of the input, the polarities change so that CR_1 is now the conducting diode. Figure 467d indicates how capacitor C_1, now in series with the applied voltage, adds its potential to the applied voltage. Capacitor C_2 charges to approximately twice the peak value of the incoming voltage. As can be seen, C_2 and C_3 are in series and the load resistor is in parallel with this combination. The output voltage, then, will be the total voltage felt across C_2 and C_3, or approximately three times the peak voltage of the input.

Figure 467e indicates the action taking place using time and the incoming voltage. At time zero, the ac input is starting on its positive excursion. At this time, the voltage on C_1 and C_3 is increasing. When the input starts to go into its negative excursion, both C_1 and C_3 start to discharge and the voltage across C_2 is increasing. The discharge of C_1 adds to the source voltage when charging C_2 so that the value of E_{C_2} is approximately twice the value of the peak value of the input. Since E_{C_2} and E_{C_3} are in series across the load resistor, the output is their sum.

The charge paths for capacitors C_1, C_2, and C_3 are of comparatively low impedance when compared to their discharge paths. Therefore, even though there is some ripple voltage variation in the output voltage, the output voltage will be approximately three times the value of the input voltage. The ripple frequency of the output, since capacitors C_2 and C_3 charge on alternate half-cycles of the input, is twice that of the input ripple frequency.

Voltage waveshape: A graph of the voltage wave shown as a function of time or distance.

Voltmeter, electronic: A voltage-measuring device that utilizes one or more electron tubes and their associated circuitry to obtain a high input impedance. The device depends on the amplifying and rectifying properties of vacuum tubes to effect very precise measurements.

Volume: A description of the magnitude or power level of an audio-frequency wave which is expressed in volume units (VU). A VU meter is often used in critical audio applications to measure the level of audio-frequency energy that is passed into a system or between systems.

Volume compandor: A process of compressing audio energy at one point in a circuit in order to reduce the intensity range of signals. After this is accomplished, the signal passes through an expander circuit which improves the signal-to-interference ratio. Volume companding is used for noise reduction purposes. The compression stage passes the signal at a point ahead of the noise signal and expansion occurs at the compandor output.

Volume compression: The automation reduction of the gain of an audio-frequency amplifier. This process usually involves sensing circuitry, which is designed to decrease the drive of an audio-frequency amplifier when its output reaches or exceeds a certain preset level. This allows the volume to be reduced automatically in amplitude but preserves the initial waveform which was applied at the input. Here, only the intensity of the signal is decreased, while the quality and linearity are maintained.

VU meter: A fast-acting, alternating current meter employed to monitor the volume level in an audio channel in which the signal level is fluctuating. The scale of a VU meter is graduated in volume units, hence the name. Figure 468 shows a VU meter. VU meters, when used in commercial radio stations, monitor the output levels from the control board to the audio-processing equipment and to the modulator section of the transmitter. A fast-acting meter is required because input fluctuations to the audio control board are often complex waveforms with many near-instantaneous peaks. A slow-acting meter would tend to indicate only the average value of the audio input, and transmitter overdrive could result.

VU meters are often seen on many different types of audio equipment, including mixers, preamplifiers, and speech compressors. Even though a fast-

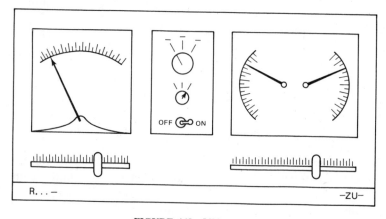

FIGURE 468 VU meter.

acting meter is used, instantaneous voice peaks cannot be seen simply because the mechanical indicator cannot move rapidly enough. To record high magnitude voice or audio peaks that are of extremely short duration, an oscilloscope is required.

Ward Leonard control system: A system using variable voltage control and providing smooth control of motor speed at all loads. It uses three rotating machines, as shown by the block diagram in Fig. 469. The dc motor to be controlled has its fields energized by the exciter, which is a small dc generator. The armature of the controlled motor is connected directly to the armature of the controlling dc generator. The exciter energizes the field of the controlling generator, and both the controlling generator and the exciter are mechanically driven by the ac motor supplied by a three-phase ac source. A rheostat in the field of the controlling generator changes the voltage generated in its armature. This variable voltage, when applied to the dc motor armature, causes changes in the motor speed.

The disadvantages of Ward Leonard speed control are the high cost of the three controlling machines and low efficiency. The main advantage is that it provides a step-less speed variation over a wide range of speeds. Ward Leonard speed control is applied in passenger elevators in high buildings, in hoists, in excavators, and in paper mills.

Watt-hour meter: A meter that is used by utility companies to register the amount of electricity consumed by its customers. A typical meter is shown in Fig. 470. Such meters are installed by electric utility companies on the customer's premises for the purpose of billing. A meter is connected so that all electricity used passes through it, and the meter records the exact amount of power consumed.

As shown in Fig. 470, there is a disc near the center of the meter that revolves when electricity passes through the meter coils. The coils are constructed on the same principle as a split-phase induction motor, in that the stationary current coil and the voltage coil are placed so that they produce a rotating magnetic field. The disc near the center of the meter is exposed to the rotating magnetic field. The torque applied to the disc is proportional to the power in the circuit, and the braking action of the eddy currents in the disc makes the speed of rotation proportional to the rate at which the power is consumed. The disc, through a train of gears, moves the pointers on the register dials to record the amount of power used directly in kilowatt hours (kWh).

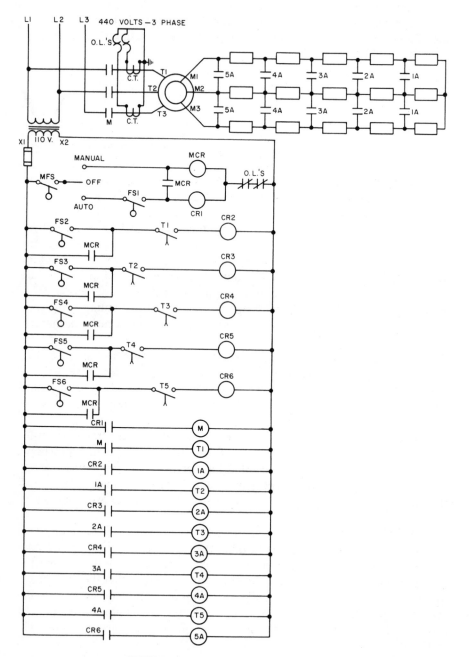

FIGURE 469 Automatic speed regulation.

FIGURE 470   A typical watt-hour meter.

The dial farthest to the right on the meter counts the kilowatt hours singly. For example, if 9 kW are consumed constantly in the circuit to which the meter is connected, the right-hand dial will move three digits in 20 min or nine digits in 1 h, showing that in 1 h, 9 kWh of electric power have been consumed. The second dial from the right counts by tens, the third dial by hundreds, and the left-hand dial by thousands.

Figure 471 gives an example of how to read a kilowatt-hour meter. Notice that some of the pointers move in a clockwise direction, while others move counterclockwise. The pointer or dial at the far right rotates in a clockwise direction; the second from the right moves in a counterclockwise direction; the third from the right moves clockwise; and the dial on the left moves counterclockwise. Therefore, the readings on the meter in this figure are 1, 1, 3, 8, which indicates 1138 kWh.

Some meters (Fig. 472) have more than four dials, the most common configuration being five. A five-dial meter is read in the same manner as a four-dial meter, in that the number which the pointer has just passed is read. The pointers in this figure show 2, 2, 1, 7, 9, or 22,179 kWh.

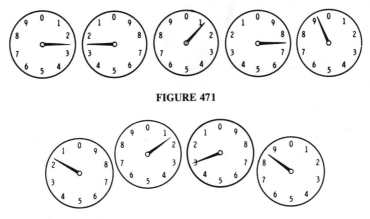

FIGURE 471

FIGURE 472

In order to record the reading of a watt-hour meter, a task required by meter readers employed by electric utility companies, it is only necessary to record the numbers indicated by the pointers or dial hands from left to right in order. Reading the dials from right to left gives a check for the first reading.

To determine the amount of electricity used in a given time, the previous reading of the meter should be subtracted from the final reading. The result is the amount of kilowatts used during the period. This value multiplied by the cost per kilowatt hour will give the amount of the consumer's bill.

Most utility companies use a sliding scale in calculating the customer's electric bill. There will be one rate for the first 50 or so kilowatt hours and a lower rate for the second 50 to 100 kW. A third step takes the rate still lower. Other reduced rates may be applied for electric water heating or electric space heating.

Wattmeter: An electrodynamometer type of meter used to measure electric power. The wattmeter consists of a pair of fixed coils known as current coils and a moving coil known as the *voltage (potential) coil.* See Fig. 473. The fixed coils are made up of a few turns of comparatively large conductor. The potential coil consists of many turns of fine wire. It is mounted on a shaft carried in jeweled bearings, so that it may turn inside the stationary coils. The movable coil carries a needle that moves over a suitably graduated scale. Flat coil springs hold the needle to a zero position.

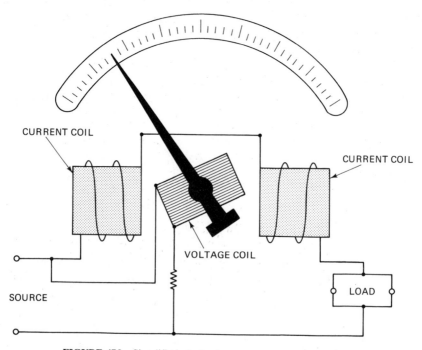

FIGURE 473   Simplified electrodynamometer wattmeter circuit.

The current coil (stationary coil) of the wattmeter is connected in series with the circuit (load), and the potential coil (movable coil) is connected across the line. When line current flows through the current coil of a wattmeter, a field is set up around the coil. The strength of this field is proportional to the line current and is in phase with it. The potential coil of the wattmeter generally has a high-resistance resistor connected in series with it. This is for the purpose of making the potential coil circuit of the meter as purely resistive as possible. As a result, current in the potential circuit is practically in phase with line voltage. Therefore, when voltage is impressed on the potential circuit, current is proportional to and in phase with the line voltage.

The actuating force of a wattmeter is derived from the interaction of the field of its current coil and the field of its potential coil. The force acting on the movable coil at any instant (tending to turn it) is proportional to the product of the instantaneous values of line current and voltage.

The wattmeter consists of two circuits, either of which will be damaged if too much current is passed through them. This fact is to be especially emphasized, because the reading of the instrument does not serve to tell the user that the coils are being overheated. If an ammeter or voltmeter is overloaded, the pointer will be indicating beyond the upper limit of its scale. In the wattmeter, both the current or potential circuit may be carrying such an overload that their insulation is burning and yet the pointer may be only part-way up the scale. This is because the position of the pointer depends on the power factor of the circuit as well as the voltage and current. Thus, a low-power factor circuit will give a very low reading on the wattmeter even when the current and potential circuits are loaded to the maximum safe limit. This safe rating is generally given on the face of the instrument.

A wattmeter is always distinctly rated, not in watts, but in volts and amperes. Figure 474 shows the proper way to connect a wattmeter into a circuit.

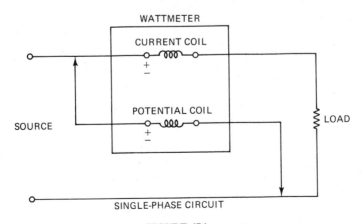

FIGURE 474

Wavelength: A measurement unit that describes the distance one complete alternating current wave will travel at the speed of light. Each unit is normally expressed in meters and the speed of light is approximately 300,000,000 m/s. Therefore, the wavelength of a 1-Hz signal will be approximately 300,000,000 m. A 100-Hz signal will have a wavelength of 300,000,000/100 m, or 3,000,000 m.

Wheatstone bridge: A device used for measuring the resistance of electrical circuits or components by comparison with a standard resistance of known value. The Wheatstone bridge consists of a box of resistance coils of various resistance in and out of the balancing circuits.

A schematic diagram of a Wheatstone bridge is shown in Fig. 475. The coil or line for which the resistance is to be measured is connected as R_x in the bridge circuit. The resistance circuits R_a, R_b, and R_c are known resistances and are called *bridge arms*. A and B are called *ratio arms* or *balance arms,* and C is called the *rheostat arm.*

Arms A and B usually have the same number of resistor units of similar value in ohms, while arm C has a number of resistors of different values. When the unknown resistance, R_x, is connected across the proper terminals and the bridge arms are balanced so that the galvanometer shows no reading when the switch or button closes the circuit, the value of R_x in ohms can be determined by the following formula

$$x = \frac{A}{B} \times xC$$

where x is the resistance in ohms of the device under test
  A is the known resistance in ratio arm A in ohms
  B is the known resistance in ratio arm B in ohms
  C is the known resistance in rheostat arm C.

The Wheatstone bridge has been used for many years to measure the resistance of coils or windings in electrical equipment; of lines, cables, and circuits; and the insulation around various wires and devices. However, it is not

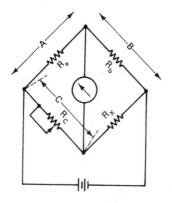

FIGURE 475 Wattmeter connection.

commonly used to measure a resistance smaller than 1 Ω or greater than 100,000 Ω.

Whip antenna: A vertical transmitting or receiving antenna usually equivalent to an electrical quarter-wavelength that is tuned against ground or an artificial ground system. The vertical whip antenna is most often used in the frequency range between 1.8 and 30 mHz. Long whips introduce mechanical problems, but the average whip length used with communications equipment is approximately 35 ft long. A whip antenna represents a resonant quarter-wave antenna at a frequency near 6 mHz. When the whip length is such that it is one quarter-wavelength at the operating frequency, the antenna will appear as a pure resistance at the point where energy is fed to the base of the antenna.

Any antenna displays a resistive component to an input signal. That portion of an antenna's total input impedance that causes the radiator (antenna) to consume power is frequently called the *radiation resistance.*

When the input impedance of an antenna is primarily comprised of radiation resistance, the efficiency of the antenna is high. However, at frequencies above or below antenna resonance, inductive or capacitance reactances dominate the input impedance, and antenna efficiency is low. When such reactances are unavoidable, as frequently occurs in whip antennas, opposite reactances are used to cancel out each other and to improve antenna efficiency.

White noise: Noise that has equal energy at all frequencies. When detected on a receiver, white noise will occur from the low end of the frequency range to the high end. This type of interference can be generated by many different types of electrical faults and is often heard as a frying or sizzling sound.

Wireless: An early name for radio and specifically, radio communications. This term is still used in some countries to describe any radio-frequency transmitter. As the name implies, these types of transmissions are made without the use of hard wiring between the transmit station and the receiver. Wireless communications are most often associated with telegraphy.

Wire recorder: A magnetic recorder that uses a round, stainless-steel wire as the recording medium. The diameter of the wire is approximately 0.0035 in. and takes the place of magnetic tape in more modern devices. Wire recorders do not enjoy great popularity today, but they were a major source of audio recording several decades ago. Some of the first speaking dolls and other toys used this method to reproduce audio sound. Some may still be manufactured today. However, magnetic tape technology has all but replaced this early method.

Work: The force applied to an object multiplied by the distance through which the object is moved. If the force applied to a given object is insufficient to move it, no work is done. The units for measuring work are the joule or the foot-pound, the same as those employed for energy.

Y

Yagi antenna: A directional antenna consisting of several parallel rods or wires, as shown in Fig. 476. One rod serves as a driven half-wave dipole radiator, while the other serves as a parasitic reflector, and one or more as parasitic directors.

Yield point: In many materials used in electrical/electronic applications, a point reached on the stress-strain diagram (Fig. 477) at which there is a marked increase in strain or elongation without an increase in stress or load. It is usually quite noticeable in ductile materials, but may be scarcely perceptible or possibly not present at all in certain hard-drawn materials such as hard-drawn copper.

Yoke: The ferromagnetic ring or cylinder that holds the pole pieces of a dynamo-type generator and acts as a part of the magnetic circuit. The system of coils used for magnetic deflection of the electron beam in cathode-ray tubes, such as television picture tubes, is known as the *yoke*. To understand how the electron beam can be deflected by a magnetic field, first consider that the stream of electrons traveling in one direction constitutes an electric dc current, regardless of whether they flow in a conductor or in free space. One basic law in electricity states that a magnetic field surrounds every electric current. When a current-carrying wire is placed in a uniform magnetic field, its own field interacts with the external magnetic field and the wire experiences a force that depends on the relative direction of the fields and the strength of the current.

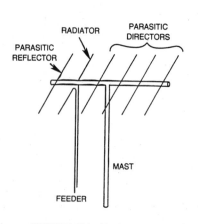

FIGURE 476   Yagi antenna.

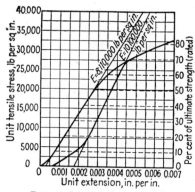

Repeated stress-strain curve, 795,000 cmils ACSR; 54 × 0.1214 in aluminum/7 × 0.1214 in steel.

FIGURE 477

The same is true for a beam of electrons, which is an electric current even though it is not carried in a wire or conductor.

Young's modulus: The ratio of internal stress to the corresponding strain or deformation. It is a characteristic of each material, form, and type of stressing. For deformations involving changes both in volume and shape, special coefficients are used. For conductors under axial tension, the ratio of stress to strain is called *Young's modulus*.

Z

Zener diode: A special type of solid-state diode that is used in the regulation circuitry of dc power supplies and other electronic applications. This silicon junction diode presents an almost constant voltage drop at applied voltages that are greater than its breakdown point, which is the point between conductance and nonconductance. This point is sometimes referred to as the *zener knee*. All voltages past this point are effectively shorted to ground. This causes an increase in current drain, which is thrown off as heat from a series resistor within the voltage line at a point preceding the zener diode.

Voltage ratings of zener diodes range from a fraction of a volt to several hundred volts, with power ratings typically to about 50 W. Physical mounting configurations may be in the form of a small, cylindrical case with wire leads, much like a resistor in appearance, or the case-mounted or stud variety, with provisions for transfer of heat to a metal mounting base for the high-wattage units.

Almost all forms of electronic voltage regulation use a zener diode or the zener diode principle to perform their jobs of stabilizing dc voltage output under varying load conditions. Like rectifier diodes, zeners may be placed in the regulator circuit in series and parallel configurations for increases in power-handling ability. However, the breakdown point will be altered and the components will exhibit different values when combined and operated as one unit. Two zener diodes in a series configuration will act as a unit with a 6-V breakdown point of each has a separate breakdown point of 3 V. A parallel configuration of the same two zener diodes would result in a breakdown point of approximately 1.5 V.

Zener diodes may be used in conjunction with resistors of low ohmic values to perform their regulating functions or as a reference source for more highly sophisticated regulation circuits that incorporate transistors, thyristors, and other solid-state devices.

Zinc: A metallic element with a symbol of Zn, the atomic number 30, and the atomic weight 65.38. Zinc is familiar as the negative electrode material in dry cells and as a protective coating for some sheet metals used in electronics. At ordinary temperature, it is brittle; but in the range of about 100 to 150°C, it becomes malleable and can be rolled into sheets and drawn into wire.

Zinc-alkaline battery: A relatively new type of storage battery introduced some years ago. Due to its high cost, it has gained very little commercial appeal. However, space projects are utilizing this type of battery due to its greater power output compared to its size and weight. Compared with the conventional lead-acid cell, the zinc cell has a watt-hour output per pound of cell over five times the output of the lead cell, while its watt-hour output per cubic inch of cell is three to four times as much.

Zinc anode: A wound-anode that has its anode composed of a corrugated zinc strip. A paper absorbent is wound in an offset manner so that it protrudes at one end and the zinc protrudes at the other. The zinc is amalgamated with mercury (10 percent), and the paper is impregnated with the electrolyte, which causes it to swell and produce a positive contact pressure.

Zinc-carbon battery: A primary cell in which the negative electrode is zinc and the positive electrode is carbon, and which may be either wet or dry.

Zinc coating: A coating widely used for the protection of metallic objects to prevent or postpone corrosion from various causes. There are four principle methods of applying zinc coating—hot dipping, cementation, spraying, and electroplating. The hot-dipped method is the type most often used today. If a very heavy coat is required, the object is usually electroplated.

Zinc oxide: A substance used as a phosphor coating on the screen of a very short persistence cathode-ray tube. The formula is ZnO. The fluorescence and phosphorescence are blue-green.

Zonal-cavity method: A method of lighting calculations that is used to determine the average maintained illumination level on the work plane in a given lighting installation and also to determine the number of lighting fixtures of a particular type required in a given area to provide the desired or recommended illumination level.